The Past and Present of Leprosy

Archaeological, historical, palaeopathological and clinical approaches

Proceedings of the International Congress on the Evolution and Palaeoepidemiology of the Infectious Diseases 3 (ICEPID), University of Bradford, 26th-31st July 1999 (under the Presidency of Professor Yves Coppens)

Edited by

Charlotte A. Roberts
Mary E. Lewis
K. Manchester

BAR International Series 1054
2002

Published in 2016 by
BAR Publishing, Oxford

BAR International Series 1054

The Past and Present of Leprosy

Proceedings of the International Congress on the Evolution and Palaeoepidemiology of the Infectious Diseases 3 (ICEPID) University of Bradford 26th–31st July 1999 (under the Presidency of Professor Yves Coppens)

ISBN 978 1 8417 1434 9

BAR Publishing is the trading name of British Archaeological Reports (Oxford) Ltd. British Archaeological Reports was first incorporated in 1974 to publish the BAR Series, International and British. In 1992 Hadrian Books Ltd became part of the BAR group. This volume was originally published by Archaeopress in conjunction with British Archaeological Reports (Oxford) Ltd / Hadrian Books Ltd, the Series principal publisher, in 2002. This present volume is published by BAR Publishing, 2016.

Printed in England

BAR
PUBLISHING

BAR titles are available from:

BAR Publishing
122 Banbury Rd, Oxford, OX2 7BP, UK
EMAIL info@barpublishing.com
PHONE +44 (0)1865 310431
FAX +44 (0)1865 316916
www.barpublishing.com

Dedicated to Johs Andersen,
Jal Mehta and
Vilhelm Møller-Christensen who
contributed so much to the study of
leprosy both past and present

President of the ICEPID

Professor Yves Coppens, France

President of the Leprosy Congress

Dr Keith Manchester, England

Presidents of the Scientific Committee

Dr John Cule, Wales
Professor Mirko Grmek, France
Professor Michel Lechat, Belgium
Dr Jal Mehta, India
Professor Don Ortner, United States
Professor John Stanford, England

Scientific Committee

Dr Pia Bennike, Denmark
Dr Joël Blondiaux, France
Professor Jane Buikstra, United States
Dr Luigi Capasso, Italy
Mrs Eve Cockburn, United States
Dr Helen Donoghue, England
Dr Olivier Dutour, France
Dr Rimantas Jankauskas, Lithuania
Dr Antónia Marcsik, Hungary
Dr Colin McDougall, England
Professor Tivadar Miko, Japan
Dr El Molto, Canada
Dr Anne Marie Moulin, France
Dr Gyorgy Pálfi, Hungary
Dr Michel Panuel, France
Dr Charlotte Roberts, England
Dr Juliet Rogers, England
Professor Dr Michael Schultz, Germany
Dr Mark Spigelman, England
Dr Ann Stirland, Engalnd
Professor Eugen Strouhal, Czech Republic
Dr Diane Trembly, United States
Dr Véronique Vincent, France
Professor Yo Yuasa, Japan
Dr Joe Zias, Israel

Organised by Charlotte Roberts, Olivier Dutour and Gyorgy Pálfi

Administered by John McIlwaine, Jason Maher and Kim Paley and assisted by Anthea Boylston, Lynda Isaac, Mary Lewis, Tanya Smith and Francis Thornton

Generously sponsored by The Bioanthropology Foundation, The British Academy, The Wellcome Trust and the University of Bradford

I.C.E.P.I.D. 3: Conference Delegates

Conference Background and Context

'Few diseases have a richer cultural heritage, and none so rich a mythology' (Fine, 1982:161, referring to leprosy)

Leprosy is a most fascinating disease medically, socially and culturally, a disease that is still very much with us today in some parts of the world. As a biological anthropologist who has a particular interest in palaeopathology and, more specifically, in the infectious diseases, it gave me great pleasure to be the host for the third Congress on the Evolution and Palaeoepidemiology of the Infectious Diseases. It was hoped that for those who attended the first and second Congresses on syphilis (Toulon, France) and tuberculosis (Szeged, Hungary), the delegates would gain as much out of this Congress as I did from the others.

When the Congress on the Past and Present of Leprosy was first discussed with my colleagues Dr Pálfi and Professor Dutour, the organisers of the previous two meetings, it was agreed that Bradford would be the most appropriate venue for such a Congress. Under the guidance (and much enthusiasm) of the President of this Congress, Dr Keith Manchester, research into the antiquity of leprosy in the British Isles began in the early 1980s at the University of Bradford. With grants from the Science and Engineering Research Council and the Wellcome Trust, Keith set the stage for prolific research activity on leprosy which has continued to this day. Later in the 1980s, the Calvin Wells Laboratory in the Department of Archaeological Sciences at Bradford also acquired a very important collection of around 350 skeletons from a Late Medieval cemetery from Chichester Sussex which, in its early years, was founded as a leprosy hospital. Many of these skeletons have bone changes of leprosy and feature in some papers in this volume. This Collection has, since its acquisition, provided teaching and research material within the Department, and for visiting international researchers. Very few leprosy hospital cemeteries have been excavated anywhere in the world which makes the Chichester site very valuable in documenting the many facets of this disease in the British Isles. In addition to this, the Calvin Wells Laboratory acquired radiographs of leprous patients through Keith Manchester's friend and colleague, the late Johs Anderson from Denmark; these radiographs are also used for teaching and research and, again, provide a very valuable resource. Keith also developed a strong working friendship with the late Dr Jal Mehta from Pune, India. Jal attended the Congress but unfortunately passed away recently. Calvin Wells, the General Practitioner from Norfolk (after whom the Bradford Laboratory is named) also published some of the first cases of skeletal evidence of leprosy in this country.

Should we therefore have needed to justify holding the Congress in Bradford, the work of Keith Manchester, the presence of the Chichester skeletons, and "modern" leprosy radiographs, holds some weight! However, as many people know, historical records also tell us that leprosy was a particular problem in this country in the later Medieval period (as well as in many parts of the rest of Europe), and therefore the history of leprosy in British contexts is important in the world view of leprosy in antiquity. As probably most people working in leprosy today and in the past know, Professor Møller-Christensen, following the excavation of the leprosy hospital cemetery at Naestved in Denmark, pioneered the establishment of diagnostic criteria for skeletal changes of leprosy; without his work our research in the palaeopathology of leprosy would be much the weaker. Keith Manchester and myself were very fortunate to meet Vilhelm in the early 1980s in Copenhagen and benefited from his great expertise in the bone changes of leprosy.

The Congress itself attracted a great deal of interest around the world and there were 23 countries represented. I thank all the authors for supporting the Congress and providing a fascinating collection of papers and posters, and for all the hard work the referees put into the papers that appear here. Additionally, I thank the authors for their great patience and forbearance with the production of this volume; the delay is the result of several factors. Any mistakes in the editorial process (apart from obvious omissions by the authors!) are the fault of the editors.

A wide range of disciplines were apparent at the Congress (as seen in the volume), ranging from people working with leprous patients today to those studying the medical history of this infection. The study of the evolution and palaeoepidemiology of leprosy, by necessity, requires a multidisciplinary approach; the Congress and its papers bear witness to this fact.

Finally, I thank Olivier Dutour and Gyorgy Pálfi for all their help in organising the Congress and supporting the production of this volume, the Presidents of the Congress, the Scientific Committee, our sponsors, and the many people at the University of Bradford who helped to make the Congress such a success.

Last but by no means least, I would personally like to thank Archaeopress (and especially Rajka) for their quick turnaround with this volume once I submitted it; their forbearance is much appreciated. Any errors are mine.

A final note, and one of great importance that was discussed regularly throughout the conference, are the terms used to describe and refer to people suffering from leprosy. Clinical leprologists emphasise the need to avoid use of the term "leper", preferring the terms "person suffering from leprosy, or Hansen's Disease" or "a leprous sufferer". Unfortunately, the English language uses the term "leper" as a derogatory term (due to historical associations of leprosy with punishment for sin etc.), even today when leprosy is a little better understood; this is something that is not helpful for those suffering from leprosy today. The editors of this volume have tried to omit reference to "leper" in any paper and alternative terms have been substituted. If "leper" does appear it is often in a direct quotation from a publication, or it is included only in inverted commas. We hope that researchers in the future will heed the advice of clinical leprologists and avoid the use of the term "leper".

Charlotte Roberts
(Co-Academic Coordinator of the Congress, and Host)
December 2001

Contents

History and palaeopathology of leprosy worldwide

Miscellanea

The International Leprosy Association (ILA) global project on the history of leprosy

Jo Robertson
45-47 Banbury Road
Oxford
OX2 6PE
United Kingdom

Email: jo.robertson@wuhmo.ox.ac.uk

Leprosy has a very special place in human history. More than almost any other disease, it has brought out both the best and the worst in humankind: personal courage, loving care, and scientific endeavour on the one side; discrimination against fellow human beings and organisational apathy on the other.

The starting point for this project, therefore, is a belief that the story of leprosy and how humanity has responded to it, offers important insights for many areas of human care and medical effort. The experience both of those affected by the disease and of those who have tried to respond to their needs and to the medical challenge, should not be lost or too easily forgotten. Furthermore, humanity has not yet reached the end of the leprosy story. The past remains an important source of learning for the work that remains to be done over the next several decades.

There has been great success in work against leprosy over the last fifty years. The level of activity seen in the past two decades especially has been unprecedented, but is likely to decline as most countries achieve a situation in which leprosy is relatively rare. It is important, therefore, to preserve information and material today while there are still many people, projects and institutions active in anti-leprosy work.

The aims of the project, therefore, are to:

- Create pathways so future researchers will be able to find their way to a wealth of material on all aspects of human activity against leprosy throughout the world.
- Fill gaps, especially in the contemporary history of anti-leprosy activity. For example, recording the experiences of individuals affected by leprosy, and assisting with the preservation of archives of major institutions active over the past half century.
- Be a source of material for all who will be working in leprosy in the next few years.

Scope and Resources of the Project

The Project is interested in all aspects of the impact of leprosy. In particular, it is intended to be global, interested in the experience of leprosy in all parts of the world. While also interested in all periods, its focus will be on the modern history from the publication in 1847 of *Om Spedalskhed* by Danielssen and Boeck, the first systematic description of the disease. The project will gather information on resources relating to leprosy in earlier periods but will not attempt to fill gaps in knowledge of those periods.

The Project benefits from a grant from the Nippon Foundation, provided through the World Health Organisation, to be used over a three-year period (1999-2002), is governed by a Steering Group chaired by Dr. Yo Yuasa, President of the ILA, and run by a Project Coordinator based at Oxford.

What the project will not do

The project will NOT produce a history book. Many historical records relating to leprosy already exist all around the world in many different places: Museums, leprosaria, libraries, earlier historical studies, and the records of institutions including the World Health Organisation, the International Leprosy Association and the Congresses, associations, national programmes, and NGOs in endemic countries. What is important is that these resources be known about, well-kept and accessible. Neither will the Project attempt to create an all-embracing archive. The aim is to encourage proper maintenance of records and materials in the places to which they relate.

Activities

Following consultation with a wide range of people, especially during the last International Leprosy Association Congress, the Steering Group decided that the aims of the Project would be achieved through four main areas of activity:

1. **Identifying existing historical resources and creating links between them**

- Finding out about existing libraries, museums and collections of material in whatever form they may be: written or visual, artefacts, or special items such as commemorative stamps.
- Devising a directory of such places and their contents.
- Using available technology to link these facilities: Web, CD-ROM, digitapes etc.

2. **Providing advice on the selection, cataloguing, and storage of historical materials**

- Encourage agencies to select and maintain historical records of their activities. It is important that all leprosy organisations keep records, from international institutions and donors to field programmes and leprosaria.
- Develop and promote advice on what to select and how to catalogue and store it.

3. **Compiling information on discriminatory legislation and customs**

- Many countries passed laws to enforce the isolation of people affected by leprosy. Often these laws remained long after any medical justification applied. The project intends to develop a global directory of such laws and the years they were in force.
- Information will also be gathered on social customs that discriminated against people affected by the disease, often in overtly cruel ways.

To promote these activities, a website has been constructed and will be used for gathering records of and linking historical resources. It is also expected that at least for the major leprosy countries today, the Project will work with local groups of interested people and organisations.

In summary, then, the Project will:

- *Identify* what historical resources exist
- *Encourage* others to keep records well
- *Link* resources and collections
- *Fill gaps*, for example verbal testimonies
- *Use* what we learn for education and current leprosy work

Please send us information about any historical resources (personal collections, libraries, museums, organisation records, etc.) of which you are aware.

Also, please make contact if you want to be involved and think you might be able to help with one or more of the Project activities.

Global Project on the History of Leprosy
45-47 Banbury Road
Wellcome Unit for the History of Medicine
Oxford University
Oxford
OX2 6PE

Tel: +44.(0)1865.284627 Fax: +44.(0)1865.274605
E-mail: leprosy.history@wuhmo.ox.ac.uk
Web URL:http://www.leprosyhistory.org

Project Steering Group Members

Dr Yo Yuasa (Chair of the Steering Group and President of the International Leprosy Association)
Anwei Law
Professor Michel Lechat
Dr S.K. Noordeen
Julia Sheppard
Paul Sommerfeld
Prof Bernardino Fantini
Dr Mark Harrison

In attendance at meetings

Kay Yamaguchi, Sasakawa Memorial Health Foundation

Project Staff

Jo Robertson	Research Officer
Debbie Emmitt	Administration Assistent

Can you help others discover the treasures to be found in the history of leprosy?

A network/database is being set up to create a pathway for future research into the rich but disappearing history of leprosy.

We are looking for information about:

Libraries/book collections/
Bibliographies
Medical records
Collections of personal testimonies
Hospitals
Private collections
Collections of photographs/paintings
Stamps
Museums
Maps
Leprosaria
Reports/Journals
Churches
Organisation archives
Coinage
Any other historical sources

PLEASE SEND THE FOLLOWING DETAILS TO:

Organisation name/owner

Relating to what period(s)

Location/address

Access/contact details

Summary of information/content

Any other helpful contacts

The last leprosy communities...and the people who call them home

Anwei Skinsnes Law
International Project Coordinator
The International Association for
Integration, Dignity and Economic Advancement
PO Box 133
Oak Hill
WV 25901
U.S.A.

Email: ideausa@inetone.net

Preface: Updating images and terminology

'If we may be permitted to concisely state our cry, it is that "Even if we have leprosy we are human beings"'

(Zen Kan Kyo National Hansen's Disease Patients' Association, 1952)

'In the old days we depended on charity.
Now our lives are based on human rights'.

(Humbert Willems, Suriname, 1999)

'To the souls of all who have already left this world
with deep resentment, we solemnly report our long awaited achievement and pray for the repose of their souls'.

(Michihiro Koh, Japan, 1999)

For more than three thousand years, individuals affected by leprosy have been dehumanized and labelled as "lepers" by society. Despite the advent of a cure for leprosy almost 50 years ago, individuals affected by this disease continue to struggle to regain identities lost to leprosy and society's mistreatment of them.

'To many of us worse than the very disease is the prejudice that comes along with it. Many of us stopped being called Francisco, Joe, Maria, and we started being called leprosy patients, "lepers" and recently Hansenites . . . .'

'I believe that our greatest challenge is to make sure that millions of people who have lost their identities will go back to being called by their own names'.

(Francisco A.V. Nunes, Brazil)

A major step in eliminating the stigma associated with leprosy is by acknowledging that it affects millions of *individuals* with unique personalities and real names who were denied family, community and personal identity because they had a disease that was feared and misunderstood.

'I have sometimes clasped in my arms the trees of the forest, praying God to animate them and give me a friend'.

(Guascoz, an individual with leprosy isolated in a tower outside the City of Aoste, France, where he died in 1803, as recorded by Xavier de Maistre)

Authors, historians, archaeologists, palaeontologists and the general public have an opportunity to be a part of the global "Quest for Dignity", a process through which individuals affected by leprosy are challenging old stereotypes by promoting a positive image of the disease, complete with personal accomplishments and participation in society.

Rather than perpetuate the offensive language of the past under the guise of "historical accuracy", it is far more humane and appropriate to start referring to individuals diagnosed with this disease by their own names, whether they are buried in a Medieval churchyard in Chichester, England, living in one of the world's last leprosy communities today, or being treated as an outpatient in India. If their names are unknown, it is far better to say a "person affected by leprosy" or "a person affected by Hansen's Disease." By refraining from using the word "leper", either historically or in modern times, we will be challenging the stigma, restoring the humanity of those who lived hundreds of years ago, and promoting the dignity of those individuals affected by this disease today.

The last leprosy communities

"Roots go deep when you live in one place for a long time. If I had to leave Kalaupapa, I would feel deserted and forever abandoned. When you don't have a husband or the support of an immediate family member or the government, where do you turn?"

(Olivia Breitha, Resident of Kalaupapa Settlement, Hawaii, for more than 60 years)

Fifty-eight years ago, on March 10th, 1941, Promin was first used at Carville, Louisiana, United States as a treatment for leprosy and this paved the way for a new era in the history of leprosy in which individuals could be treated as outpatients in the community. The discovery of the sulphone drugs as a cure for leprosy represented a medical milestone in the history of this disease. In the next few decades, many people throughout the world left communities such as Carville and reintegrated themselves into society. However, this exodus was not universal since social progress lagged far behind the medical advances.

For a variety of reasons, including forced loss of identity, separation from family, widespread prejudice and discrimination, the presence of disabilities, lack of training in job skills, and the reality of simply having nowhere to go, resulted in many individuals staying in these facilities which they had, in effect, transformed into homes.

'There is one primary concern - to be allowed to stay in the places that have become their homes . . . . All of us, as children, as adults, were taken away from our homes, our families, and confined, incarcerated, because of compulsory segregation laws that were designed to protect society. This emotional trauma has never left us. So, any decision concerning the closing of these places touches the very psyche of our being'.

(Bernard K. Punikai'a, Hawaii, separated from his mother at the age of six and a half)

As the population of people who contracted leprosy prior to the advent of a cure declines and all newly diagnosed individuals are treated as outpatients, the final remaining residents of leprosy communities throughout the world are now facing another crisis, the prospect of losing their homes. Outpatient treatment has rendered isolation facilities, such as leprosy hospitals, settlements and villages, obsolete (at least in terms of modern treatment). Yet their continued presence is of vital concern to their elderly residents. From St. Giles in England, with only five remaining residents who have had leprosy, to Rovisco Pais in Portugal with about 25 residents, to Kalaupapa with 50, Fontilles in Spain with about 100 and Kikuchi Keifu-en in Japan with a population of more than 800, the concern is the same: to be able to spend one's last years in the security of the place that has become home. The following are the words of Les Parker, one of the five remaining persons affected by Hansen's Disease at St. Giles, East Hanningfield, in England:

'With so much uncertainty about the future of Hansen's Disease centers throughout the world, the experience of St. Giles, the last Hansen's Disease hospital in England, and its five remaining residents, is an important example of how progress can lead to a feeling of abandonment...'

'In the mid-1980s, St. Giles ceased to be a hospital and was taken over by a housing association. Over time, we have been joined by 34 individuals, most of whom have mental disabilities. I sympathize with them, as they have to be housed somewhere due to similar closure of their own centers, but I question whether this population was best suited to join us at St. Giles . . . .'

'The atmosphere at St. Giles is nothing like it was when we all had the same problems and understood each other. In many ways, I feel more isolated and alone than ever...'

'It's like we're becoming second-class citizens in our own homes and this is going to happen all over the world, I think'.

'Hansen's Disease centers have been homes to thousands. As our numbers decrease, decisions about mixed-use for these centers cannot be made simply on financial grounds. Psychological, medical and social factors must be taken into consideration and discussed with those whose lives will be impacted by the decisions. It is essential that abandoned buildings don't result in abandoned lives, no matter how small our numbers may be'.

This situation has come about largely because social progress did not keep pace with medical advances. In a world that places primary emphasis on the physical aspects of leprosy and its cure, what is society's responsibility, and in particular the responsibility of individuals working in this field, to those individuals who lost their families and their freedom while waiting for the medical miracle to occur?

We are facing a real dilemma. How do we justify keeping these centres open when they no longer receive new patients and generally cost a great deal to maintain? How do we reconcile the treatment of the past with modern treatment? How do we justify the tremendous expense involved in these centres that house fewer and fewer individuals when other needs are so acute? At the same time, how do we ensure that the final chapters in these communities are just and humane?

These questions need to be addressed in order that everyone can share experiences and learn from what is being done in other places. Above all, it is essential these centres and their residents recognize that they are not alone and that others around the world are undergoing similar situations. Each place will need to develop creative ways to ensure that the residents of these communities do not live their last years feeling forgotten and alone and that their lives have served no meaningful purpose.

As people get older, three things become increasingly important: 1. Contact with family; 2. Being able to remain in one's home, and 3. Knowing that one's life has had meaning. Many members of the older generation of people affected by leprosy do not have close family ties, or any relationship with their families at all. Indeed, in Japan, even the ashes of individuals who have had leprosy cannot be returned home.

In many cases, the residents of these leprosy centres have created new families within these communities, which has added to the feeling of home. However, as these new family members die, which is inevitable considering their age, there is no new generation to take their place, and thus there is a void that cannot be filled. This adds an additional burden of loneliness for the final residents of these communities who watch as this, the only family that many of them have, dies out.

'I was 22 years old when they brought me here. That was in 1939. I was told that if I didn't come here, they would put a yellow flag on my house. I was heartbroken because I had to leave home. When I got here, I just cried all night'.

'My wish is to stay here where they brought me. I'm 80 now so I've been at Carville for 59 years. I don't think I could stand another move. Every morning I go to visit my husband, Waysey's grave, say my rosary, and tell him I love him. It breaks my heart to think about having to move away from him'.

(Mary Ruth Daigle)

When people look at the residents of these leprosy communities today, they often appear to be happy and well taken care of, simply elderly citizens like all others. It is easy to forget what they have experienced because many have appeared to survive their experiences so well. It is easy to forget that most of these individuals were denied their basic human rights simply because they had leprosy.

'Like convicted felons or Indians living on a reservation, Carville patients were denied the right to vote in state or National elections'.

(Stanley Stein, U.S.A.)

Carville citizens were also denied the right to marry and ride on public transport. Their mail was fumigated, as was the mail at Kalaupapa, and this was done entirely to make the public feel safe, without concern as to how the residents felt. A notice on Carville's well-known magazine, The STAR, stated:

'This paper and all outgoing mail are sterilized before leaving this hospital. This is done only as a gesture of respect to the unconvinced and not because there is any scientific necessity for it'.

Many people at Carville were encouraged to change their names, and this also occurred in other countries.

'In 1951, I was confined in Oshima leprosarium. At that time, I was encouraged to use a different name. In the towns and cities, no one understood the disease and therefore there was a lot of discrimination. I opted to use the name Kozaki Masao. From that moment I felt that my human rights and humanity had been denied'.

(Michihiro Koh, Japan)

In Japan, individuals were forcibly sterilized if they wanted to get married. Forced abortions also took place. There are still about 5,000 people living in the 15 leprosy sanatoria throughout Japan and they are literally a generation without children.

'In October of 1945, very soon after the end of the Second World War, I too had a vasectomy performed in order to marry. I agonized over whether to submit to the procedure but, in the end, love prevailed and it was performed'.

'But, when I found myself on the operating table, lying on my back and my legs spread, panic set in and I felt an urge to flee. This was a procedure to be performed on animals, not on a human being. When the procedure was over, I felt that I was a failure as a human being'.

(Kaoru Matsumoto, Japan)

It is important to note that individuals like Mr. Matsumoto have not only written books documenting the painful experiences of the past but have gone on to actively contribute to society. Mr. Matsumoto's story continues:

'I wasn't able to go to school. I am now giving scholarship funds to children of people with Hansen's Disease in Nepal. I have helped four people go to school with funds from my pension. In the future I want to be able to help 100 people. This is the greatest fun of my life and I want to meet them when possible . . . . They call me "Dad"'.

It is probably unrealistic to expect that Mr. Matsumoto and the residents of all of the world's last leprosy communities will be allowed to live out their lives in these, their homes. Some changes, like those taking place at Carville, are so radical that one wonders how changed a place has to become before it is no longer one's home? In addition, who would want to be the last person or the last ten people in one of these places, even if it was feasible economically? Thus, while I think we have a responsibility to try and come up with creative solutions so that people can continue to live quality lives in these communities that have become their homes, we need to do more, especially since, for some, relocation will be inevitable.

Thus, we come to the third thing that is increasingly important to people as they get older, having a sense that their lives had meaning; that they did not go through separation from families, various forms of social injustice, and denial of human rights simply for these to be forgotten or minimized by future generations. Some individuals have recorded their own history. This is especially true in Japan, where discrimination was so severe and continued by law until only a few years ago. Many, many people have written their autobiographies and some are also undertaking an oral history project. There are also at least two amazing museums in Japan which document the history of leprosy through photographs, artefacts including special coins used, various medicines administered, and items that simply represent everyday life in the "sanatoria". Much of the collection and interpretation has been done at these museums by individuals affected by leprosy themselves.

In the United States, fairly extensive oral histories have been collected and archived at Kalaupapa in Hawaii and Carville, and a few individuals have written their memoirs, but only the surface has been scratched.

'I wrote a book because I wanted people to know what I feel, what I felt. How I felt when I became a patient. How I felt during my time as a patient, what was done to me, how much I struggled, fighting the disease, fighting bureaucrats, trying not to be a statistic, trying to be a person.'

(Olivia Breitha)

Part of our responsibility to the older generation is to see that there is a concerted effort to record their oral histories. Some people have said that oral history may be the 20th century's substitute for the written memoir. Conducting an extensive oral history project with people affected by leprosy will not only provide us with a useful record of the history of leprosy from the perspective of those who were most affected, but the interviewing process can provide an additional degree of healing for those being interviewed. There has been an official recognition of the importance of recording oral histories among survivors of the Holocaust, Japanese-Americans who were sent to internment camps in the United States like Manzanar, slaves and their descendants, and even survivors of the Titanic. Eyewitness testimonies such as these provide personal insights into the history of these events that can be too easily lost if accounts are strictly based on statistics and data that can be researched in an archive.

The Yale Archive has set forth some of their principles regarding oral history that they have initiated with survivors of the Holocaust.

'We feel that the survivors should be given their own voice and that academic history, however important, should not speak for them at this point, though it can provide an extension and integration of what these first-hand accounts describe. Instead of statistics and data the testimonies provide living portraits and are the nearest we will come to the actual experiences of eyewitnesses'.

'The testimonies help restore the humanity of the persons . . . .'

Comparisons of human experiences are not necessary. Each history has its own integrity and its own substance. It is not important to try and ascertain who suffered the most. What is important is to acknowledge instances where dehumanization and a violation of human rights has occurred. People may flinch when these terms are used in relation to leprosy. We know that there are great instances of compassion towards people affected by this disease. However, we cannot ignore the fact that there were also many instances, even in this century, of extreme mistreatment which we have never really fully acknowledged. We have a responsibility to the older generation, and even to the younger generation, to acknowledge these and ensure that these events and those who lived through them will not be forgotten.

However, we must tell all sides of the story. Just as we must remember society's marginalization of people affected by leprosy because they had a disease, so must we also record and remember the achievements of those individuals who turned a diagnosis of leprosy into a personal challenge and ultimately a personal victory. We must remember Peter Greave in England, who became a famous author, and He De in China who was a famous actor of the Yue Opera in Guangzhou. He organized an opera troop in Xinzhou Leprosy Hospital and performed many operas to amuse the people and enrich their cultural life. We must remember Stanley Stein, who took on the task of radiating the light of truth about Hansen's Disease through his publication The STAR. We must remember Akashi Kaijin and all the other talented poets and artists in Japan. We must remember David and Annie Kupele who, in 1967, were the first couple at Kalaupapa to celebrate a 50th wedding anniversary, and we must remember Zhu Lan from China, who was an elderly lady who resided in a leprosy hospital in Guangzhou. She was greatly loved for her willingness to help everyone by cutting their hair or mending their clothes. We must remember too Mr. Toho Kumita and others in Japan who were so desperate for a connection with the outside world that they learned to read braille with their tongues, and we must remember so many other people who simply met the challenges of this disease with grace and dignity.

IDEA (International Association for Integrity, Dignity and Economic Advancement) has inaugurated a Banner of Honour, adapted from the idea of The Names Project AIDS Quilt, that will remember and honour exceptional individuals who turned a diagnosis of leprosy into a personal challenge and, ultimately, a personal victory. While we are remembering these individuals and making their memory known in different ways to the public, it is likely that we will also start seeing the stigma associated with leprosy start to disappear. This will happen in two ways: 1. By replacing the public's old images of disability and despair with new images of people and their abilities, and even more important, 2. By increasing the self-confidence and pride amongst those individuals who have had this disease, many of whom have led lives dominated by feelings of inferiority.

'It is a fact that institutional life tends to give one an inferiority complex'.

'Minority groups such as ours, must contend with the desire on the part of the majority groups to feel superior, who thus defend themselves from any contact with the adjudged inferior that would tend to equalize'.

(Stanley Stein, U.S.A.)

If we are to get rid of the stigma associated with leprosy, we must do things that equalize. We must value the memories and experiences of people who have had leprosy. We must discourage the use of labels such as "PAL" (people affected by leprosy) and resist calling people "patients" outside of the medical setting and once they are cured. We must acknowledge and record the full history of leprosy and how it impacted our fellow citizens, and we must remember that the residents of the last leprosy communities throughout the world gave up their families and their freedom in order to protect society. Most of them do not have children to whom they can pass down their history, so we must be the people who ask the questions and give them the assurance that what they have been through will be remembered, not only as a history of medical triumphs, but as a history of personal triumphs as well as a history of how fear and ignorance cause people to create outcasts out of human beings.

While we document the history of how leprosy impacted our fellow human beings, we should always keep in mind that leprosy is only one chapter in their lives and thus, we must make sure to record not only their experiences with leprosy, but their experiences as human beings. In the words of the poet, Haruko Tsuda:

'When I look back at the past, my soul is beckoned to the poems I wrote of my father. There are times when I think it would be more natural for me to write about my life as a leprosy sufferer, but my attitude of mind as a poet is inclined more to write about my existence as a 'human being.' I want each line and each stanza that I create to be the reflection of the real me, my heart and soul, that is not free from love and hatred'.

NB: The figures included with this paper illustrate some of the people today living with leprosy, and those working with them

Ms Wu Ze Geng, China. Photo: Pamela Parlapiano.

Sam and Theresa Wilson, Trinidad. Photo: Pamela Parlapiano.

Dr. P.R. Gopal and family, India. Photo: Pamela Parlapiano.

Phillip D'Souza, poultry farmer, India. Photo: Pamela Parlapiano.

Sam Wilson prepares for the 1st International Day of Dignity and Respect, March 11, 1999, Carville, Louisiana, with Kathryn O'Zniel looking on. Photo: Pamela Parlapiano.

Yasuji Hirasawa (center) speaks at IDEA workshop "The Last Leprosy Hospitals and the People who call them Home". Fontilles, Spain. Photo: Pamela Parlapiano.

Bernard Punikará and Jose Ramirez. Photo: Pamela Parlapiano.

Clinical Leprosy

Leprosy worldwide: current status

A. Colin McDougall
Department of Dermatology
The Churchill Hospital
Headington
Oxford OX3 7LJ
England

Address for correspondence
87 Lower Radley
Near Abingdon
Oxfordshire
OX14 3BA
England

Abstract
The problem of leprosy in particular parts of the world is described, and emphasis placed on the progress that has been made on the elimination of leprosy using multidrug therapy and improving case detection. However, in 24 countries leprosy frequency rates remain a public health problem and will continue to be until about 2003-2005. Inequality in access to health care services for some countries remains a problem that needs to be tackled.

Keywords

Leprosy, frequency rates, elimination, multidrug therapy, health care access

The current extent and seriousness of leprosy worldwide can be described in terms of prevalence, incidence, disability and child rates, bearing in mind, however, that these indices do not reflect the social, psychological and economic consequences of this diagnosis in many parts of the world. It can, however, be recorded that remarkable progress has been made in the control of leprosy in the community and in the treatment of individuals in recent years. This is largely due to the implementation of multiple drug therapy (MDT) for all cases, as recommended by the World Health Organisation (WHO) in 1982. Following the introduction of dapsone (DDS) in the 1940s, many hundreds of thousands of patients were treated successfully, and it became apparent, for the first time, that leprosy could be diagnosed and treated on an out-patient basis, as opposed to the previous policy of institutional segregation for virtually all cases, often on a life-long basis. Dapsone was, however, never implemented widely enough in endemic areas to have an epidemiological impact, and in the early 1970s drug resistance became increasingly widespread in many parts of the world. In late 1981, the WHO convened a meeting of experts in Geneva to develop recommendations for the treatment of all cases of leprosy with regimens of relatively short duration, using combinations of dapsone, rifampicin and clofazimine. These recommendations, now seen as historical in the fight against this disease, were published in 1982 (WHO 1982).

As can be seen in Tables 1 to 3, the worldwide leprosy situation has improved dramatically in the intervening years. By the beginning of 1998, more than 10.7 million cases had been cured by MDT; over 99% of all registered (known) cases were receiving MDT; relapses were extremely few; the number of countries with high prevalence rates had been reduced from 122 in 1985 to 32. In 1991, the World Health Assembly set a goal for the elimination of leprosy as a public health problem (less than 1 case per 10,000 of the population) by the year 2000. Among the 122 countries considered endemic in 1985, 98 have now reached the elimination goal and the prevalence of leprosy globally has been reduced by 86% (WHO, 2000a). The disease thus remains a public health problem in 24 countries. There are 672,596 registered cases and 67,086 newly detected cases in the top 11 countries, which accounts for 89% of the prevalence and 92% of the cases detected worldwide. India alone has 67% of the prevalence and 73% of the detection worldwide, the most heavily affected states being Bihar, Madhya Pradesh, Orissa Uttar Pradesh and West Bengal. The top 11 countries (India, Brazil, Myanamar, Indonesia, Nepal, Madagascar, Ethiopia, Mozambique, the Democratic Republic of Congo, The United Republic of Tanzania and Guinea) have a prevalence rate of 4.1 cases per 10,000 of the population compared to a global rate of 1.25. They have already missed the target date of the year 2000 and are now considered unlikely to achieve elimination until about 2003-2005.

While the prevalence rate is considered by the WHO and other agencies as being satisfactory as a main indicator of progress in leprosy control, concern has been expressed by others about the continuing increase in case detection figures for the past six years, despite enormous efforts to find cases and treat them with MDT. Through the WHO, the Sasakawa Memorial Health Foundation in Japan and Novartis in Switzerland, the drugs needed for MDT have been produced in blister calendar packs and are available completely free of charge to all leprosy patients worldwide. The increasing trend of case detection (the best available indicator we have of incidence) may be partially due to the extension of the geographical coverage of MDT, and the intensified health education and case finding activities, notably through "Leprosy Elimination Campaigns", as developed by the WHO. These were introduced in 1995 and have been carried out in 26 countries, covering an estimated population of 466 million, and resulting in the detection of 176,000 new cases (WHO, 2000b).

Apart from the WHO strategy to eliminate leprosy as a public health problem, considerable discussion has taken place on the possibility, at some, as yet ill-defined point in the future, of completely eradicating the disease from the face of the earth. A recent review of this subject examines in detail the other measures, apart from case finding and chemotherapy, which are available to aid eradication (Noordeen, 1999). These include immunoprophylaxis (vaccine development) and chemoprophylaxis (the use of one or more of the existing anti-leprosy drugs to prevent disease). The conclusion is that they present interesting, though limited, possibilities for the future and that, in general, eradication is not a practical strategy with the tools we currently have at hand.

In summary, the current world leprosy situation is a mixture of highly significant progress, particularly in case detection and the implementation of MDT regimes and reduction of prevalence. This is combined with inexplicably high detection rates of new cases, even in parts of the world where control programmes have been in operation for many decades, with good coverage of detected cases. It has also been recognised that our support and management of individuals already disabled and deformed by this disease is far from satisfactory and is not carried out on the scale needed.

A WHO Technical Advisory Group on Elimination has recently summarised the situation as follows (2000c):

'Leprosy elimination stands at a critical and extremely difficult juncture. This is partially because the commitment to eliminate leprosy in many epidemic countries is beginning to slacken (among decision-makers and in the field). Moreover, those areas that are easy to reach and to work in have been effectively covered. The residual problem is far more difficult – from all perspectives- and is further complicated by structural inadequacies in local health services. Even today, people in many areas do not have ready access to diagnosis and MDT (including those with long-standing disease). Therefore, achievements will no longer be sustainable if a significant number of hidden cases remain undetected and accessibility to treatment services remains difficult.

There is a need to critically review existing strategies and to develop pragmatic approaches, adapted to field realities, in order to facilitate the delivery of essential activities leading to the elimination of leprosy at the local level. Failure to do so could be misinterpreted as a failure of the current elimination strategy as well as the technology behind it'.

References

Noordeen, S.K. 1999 The Future of Leprosy Elimination. *International Journal of Leprosy.* 67 (4), supplement: S56-58.

World Health Organisation Study Group 1982 *Chemotherapy of leprosy for control programmes* W.H.O. Technical Report Series 675. Geneva: World Health Organisation.

World Health Organisation 2000a Weekly epidemiological record. *W.H.O.* 75:226-232

World Health Organisation 2000b Weekly epidemiological record. *W.H.O.* 75:361-368

World Health Organisation 2000c *Progress towards the elimination of leprosy: challenges and opportunities.* W.H.O. Technical Advisory Group for the Elimination of Leprosy. WHO/CDS/CPE/CEE. Geneva: W.H.O.

Note:

Up-dated information on the worldwide leprosy situation (March 2002) is available in Weekly Epidemiological Record No 1, 2002, **77**, 1-8, World Health Organisation, Geneva, Switzerland.

Table 1: Leprosy estimates: mid 1960's – 2001

Global estimates between mid 1960's – 1980's	10-12 million
Revised estimate 1991	5.5 million
Projected number of people with undiagnosed leprosy 1999-2001	1.5-2 million

Table 2: Leprosy global statistics: early 1999

Registered	834,988
MDT coverage (chemotherapy)	99%
New cases 1998-1999	714,876
Cured with MDT since 1985	10.7 million
Projected number of people with undiagnosed leprosy 1999-2001	1.5-2 million
Country with the highest number of patients	India: about 60% of the world total.

Table 3: Elimination prospects: 2000

Country	**Comment**
India Brazil Nepal Madagascar Mozambique Guinea Angola	Still a public health problem
Myannar Indonesia DR Congo	Possibly
Reasons:	
High prevalence Intense disease transmission Limited geographic coverage Civil strife Lack of infastructure	

Social reactions in the past and present of leprosy and the socio-economic rehabilitation of leprosy-cured persons

Jal Mehta
c/o Poona District Leprosy Committee
Manisha Terrace
2nd Floor Flat No. 35
2A Moledina Road
Pune 411001
India

Abstract

This paper discusses the social aspects affecting people with leprosy and their families, and the changes that have taken place in modern times. Despite the increasing success of treatment for curing leprosy, the stigma and low socio-economic status of people cured of, but suffering disabilities from, leprosy means that it must not be forgotten as a public health issue.

Keywords
Poverty, stigma, history, India, public health.

1. Introduction

Leprosy has not been eradicated, but instead, it seems to have been eliminated as a public health issue. However, in India, many new cases are still being reported and all varieties of treatment schedules are being prescribed. A new one-day single dose treatment has been advised and, every day, the statistics of leprosy prevalence keep improving. Nevertheless, India may still be producing a minimum of 100,000 new cases annually. For example, in Bihar, prevalence in many districts is 42 per 10,000 population but what about the millions of leprosy affected and leprosy-cured persons in the world to which we have not applied our minds.

2. The history of leprosy in India

Leprosy is an old disease and has been described in many ancient texts such as the Old Testament. It was chronicled in the Charak Samhita around 800 BC and studied by Sushruta Samhita (c.600BC) who clearly mentioned anaesthesia as a cardinal sign of leprosy (Dharmendra, 1967), thereby distinguishing it from other skin diseases which were classified together in the category of "leprosy". The Hindus more than two millennia ago knew that leprosy was not only a skin disease, but that it affected all the essential parts of the human body. It is significant that the brain is not mentioned, as we know that *M. leprae* does not affect the central nervous system (Jopling and McDougall, 1988).

Descriptions of the disease have appeared in the lay literature of many languages, instilling fear and misconceptions in the public mind. One observation of Vagbhat, that the disease has something to do with water, is interesting. He mentions that the waters of rivers arising in the Western Ghats and flowing eastwards may give rise to "Kushtha" (a term covering leprosy and other skin diseases) and it is a known fact that the prevalence of the disease in India increases rapidly as one goes eastwards. Does water play a part in the spread of the disease and could ingestion be a portal of entry? Recent research has shown that there is a significantly higher prevalence of leprosy in areas where the water used for bathing and washing is PCR-positive for *M.leprae* DNA (Matsuoka *et al.*, 1999). Here, we have an example of Vagbhat's observation and a belief of 800 BC about the spread of leprosy through water now being fortified and proven, to some extent, by today's scientists using modern tools of investigation. In the ritual exclusion of "lepers" from the general community, a person with leprosy is forbidden to wash his hands or his belongings in spring or stream water of any kind.

3. Modern history

The bacillus of leprosy was discovered by Hansen in Bergen in 1873 but his father-in-law, Danielssen, did not believe that leprosy was an infectious disease. He insisted that it was hereditary, and in this controversy Danielssen inoculated himself on several occasions with leprotic material but did not contract leprosy. Furthermore, Hansen in his enthusiasm inoculated leprotic material into the anterior chamber of the eye of a female leprosy patient, resulting in litigation and subsequent removal from his post (Nyfors, 1997). What I want to stress is the spirit of enquiry that Danielssen, Hansen, Looft, Lie, Neisser, Virchow and others of that time had with respect to this disease.

Mahatma Gandhi took a special interest in leprosy throughout his life. In 1881-82 at Porbandar, Ladha Maharaj of the Bileshvar temple to the Nature Cure Clinic was reciting the Ramayana and, one day, he detected a person suffering from leprosy and was so concerned about him that he sent him for treatment. It is recorded that the then Kondhawa leprosy home gave the necessary treatment to this person who was later admitted to the Veer Sanatorium. In 1947, at the prayer meeting in New Delhi, Gandhi mentioned leprosy on two successive days, October 23rd and 24th. He mentioned the necessity of working with the leprous in India, continued to be concerned about this disease, and wanted scientific and social action to be taken for the treatment of leprosy patients until his death.

3.1 Poverty, leprosy and stigma

Following the control of lepromatous leprosy in India, the effect of poverty on the spread and stigmatisation of this disease has become more apparent. Leprosy is an infectious disease caused by the bacillus *M.leprae*, but there is an additional causative factor in the spread of the disease, poverty, which is anthropogenic.

After the British came to India, considerable research was done in the field of leprosy and, in the 1870s in the Census Return, the Bengal Presidency showed the highest prevalence of leprosy at 5.2 per 10,000 population. In the 1950s, the prevalence had reached an average of 50 per 10,000 population for the whole country, following an increase in poverty and rapid population growth.

Recently, the Poona District Leprosy Committee (PDLC) received a Top Management Award from the Governor of Maharashtra. In his speech, the Governor referred to the Constitution of India, securing to all its citizens - Justice, Liberty, Equality, Fraternity - whereby every individual would be allowed to grow and flourish, and develop all their faculties, irrespective of caste, creed, race, religion or sex. The Governor also spoke about poverty. One-third of the one billion people in India has a per capita income of only 206 rupees in rural and 280 rupees in urban areas per month (which is equivalent to 5 or 6 US dollars). India is listed at 136th in the world poverty ranking, but out of 100 paise of Government funds meant for Social Welfare only 15 paise go to the poor. The remainder is "swallowed up" somewhere in administrative and other expenses. The Governor went to the extent of saying that, although the political will seems to be there, the management of the funds indicates inefficiency, dishonesty and corruption. A few days later, the Government of India issued a statement that the number of people living below the poverty line had fallen by 75%, reporting a much higher per capita income than previously suggested by the World Bank and the Indian Government. We, therefore, have a large proportion of the Indian population suffering from leprosy and living in poverty. It is relatively easy to treat leprosy, but almost impossible to treat poverty, and this of course applies to many of the infectious diseases in the world today.

We are enthused with words such as "free market economy" which, it is promised, will solve all our problems of poverty, but will it? The Nobel Laureate Amartya Sen has introduced the solution of the Social Security Net (SSN). However, this net would have to be almost equal to the size of the globe because the world population of poor people is growing at a rate that we may never be able to catch up with. One example of progress is "Operation Flood". India has become one of the largest producers of milk, ready to export milk powder even to Europe and, yet, millions of children in India get no milk during their growth period because the price is so high. Reduce the price of milk and the entire stock will be consumed within a matter of days! Our concern has always been humanitarian, and it is only through Christian compassion, religious and spiritual understanding, together with scientific and practical knowledge, that we will be able to tackle the problem of leprosy rehabilitation. However, in order to achieve this, leprosy-cured patients need to earn a living and be able to support themselves and their families in a healthy community.

4. Stigma and social acceptance

'The process of stigmatization can be divided into two stages. The first stage describes how certain cognitive dimensions of leprosy lead to a variety of effective responses towards the disease. The second stage involves how these effective responses contribute to social devaluation of the leprosy patient and, consequently, the adoption of negative behaviours towards them.'(Bainson and van den Borne, 1998: 341).

To this we have to add the stigma of poverty, making the situation more complex. In my experience, the most effective measure is economic betterment of the leprosy-cured person. With this economic power they are able to overcome the stigma of leprosy. However, disabilities, ignorance and stigma hinder the economic advancement of many people cured of this disease (Diffey *et al*., 2000). For example, a large hospital in Bombay regularly bought bandage cloth, surgical gauze, and hospital linen to make uniforms for the hospital in-patients, which were manufactured by leprosy-cured individuals. One day the hospital suddenly stopped purchasing our products on the plea that the uniforms were giving the patients an itching sensation all over their bodies! Another example of this type of stigma is illustrated in the case of Mr Datar.

I recommended Mr Datar, a highly educated but badly disabled patient at our hospital, for a National Award, for which he was rejected on two consecutive years. On the last occasion they took him in the category of the blind, as he had visual impairment due to iridocyclitis caused by leprosy. However, he still did not receive an award, the reason being given was that Datar had become blind in adult life and was not blind from birth. I wrote to the authorities and asked whether they were afraid to allow a leprosy disabled, but cured, person to enter the august Ashoka Hall of the Rashtrapati Bhavan (President's palace) in New Delhi, and to receive the award from the hands of the President of India. Within two weeks of sending this letter, the award was conferred on Mr. Datar who, to the best of my knowledge, became the first leprosy patient to ever receive this prestigious prize. Finally, I cite the example of a lady who worked as a secretary at the leprosy hospital in Kondhawa. When her neighbours discovered where she worked they refused to allow their children to play with her children, despite the fact that she never came into contact with the patients at the hospital. The woman did not have the courage to continue to work there.

The laws in India, such as the Indian Christian Marriage Act (1872), the Muslim Marriage Act (1939), the Motor Vehicle Act (1939), the Indian Railways Act (1890), Tenancy Acts, Election Rules and the Hindu Succession Act (1956) have all contributed to the propagation of leprosy stigma and its associated distress. Leprosy is so fertile in deformity that "lepers" are considered wicked and sinful but they are too poor to be wicked. Healthy children of leprosy parents suffer as do the relatives, and the doctors who treat leprosy patients. Stigma is perpetrated in subtle and devious ways. For example, a female leprosy patient's young and pretty daughter was forced into marriage with an old aged uncle, who threatened that he would reveal to the community that the young girl's mother was suffering from leprosy.

5. The Future

Today, some patients still refuse treatment for leprosy, even when free medicines and food are provided. However, stigma has been considerably reduced. This is largely the result of work carried out by the Co-operative Society, where patients carry out sophisticated engineering jobs for some of the major industries around Pune. This Society is a model for the rehabilitation of leprosy-cured persons in industrial centers, where considerable migration, including the leprosy affected, takes place from rural areas (Dambalkar *et al.*, 1995). One of the greatest achievements of this Co-operative Society is that leprosy-cured people have their own homes and facilities, and those who once relied on charity are now doing charity for others. They donate 25% of their income to our Charitable Trust, namely the Poona District Leprosy Committee for Leprosy Relief Work.

6. Conclusions

Let us not make the mistake made years ago, by concentrating only on the control of the infection and not on the physical and social rehabilitation of leprosy sufferers. It has been reported that there are now more than four million people who have been spared crippling deformities as a result of multi-drug therapy. However, it is difficult to reconcile these figures with the large numbers of people suffering from plantar ulcers and loss of sensation of their hands, in leprosy centres throughout India. I do not prescribe to the WHO Classification of Deformities. In loss of sensation of the hands and feet: proprioception, vibratory and thermal sensations, surface scanning, density judgment, humidity, superficial pain and other abilities are also lost. I, therefore, consider loss of sensation as loss of peripheral consciousness requiring the most stringent care. Finally, the fight against the spread and elimination of leprosy can only be won by also taking into account the social aspects of stigma and poverty that hinder the rehabilitation of leprosy cured patients into society.

References

Bainson, K.A. and van den Borne, B. 1998 Dimensions and process of stigmatization in leprosy. *Leprosy* Review 69 (4):341.

Dambalkar, K., Ramesh, V. and Vashist R.P. 1995 Problems due to migration of leprosy patients into urban areas. *Leprosy Review* 66 (4):326.

Dharmendra, R. 1967 *Notes on Leprosy*. Bombay: Samant and Co.

Diffey, B., Vaz, M., Soares, MJ., Jacob, A.J.W. and Piers, L.S. 2000 The effect of leprosy-induced deformity on the nutritional status of index cases and their household members in rural South India: a socio-economic perspective. *European Journal of Clinical Nutrition* 54 (8): 643-649.

Jopling, W.H. and McDougall, A.C. 1988 *Handbook of Leprosy*. Fourth Edition. Oxford: William Heinemann Medical Books.

Matsuoka, M., Izumi, S. and Saeki, K. 1999 *Mycobacterium leprae* DNA in daily using water as a possible *source of leprosy infection*. Indian Journal of Leprosy 71 (1):61.

Nyfors, A. 1997 Armauer Hansen (1841-1912): The life of the discoverer of the aetiology of leprosy. *Journal of the European Academy of Dermatology and Venereology* 9 (1001): 135.

Leprosy: a correctable model of immunological perturbation

John Lawson Stanford and Cynthia Ann Stanford
Department of Medical Microbiology
Windeyer Institute of Medical Sciences
Royal Free and University College Medical School
46 Cleveland Street
London W1T 4JF

E-mail: j.stanford@ucl.ac.uk

Abstract

Leprosy is placed in context with other mycobacterial diseases and some of the special aspects of the disease are described. The mycobacterial persistor, adjuvants, antigens and the immune responses they induce are discussed in relation to the modern presentation of the disease and inferences can be made about the likely parts they played in ancient disease. Modern treatment with chemotherapy and the on-going need for immunotherapy to correct immune perturbation are considered, as is the possibility of developing a vaccine that could prevent new cases of the disease, both directly and by elimination of subclinical infection. It is suggested that the recognition of subclinically infected individuals, without bony deformity, by the polymerase chain reaction on bone samples from the nasal cavity might give a better idea of the epidemiology of leprosy in past communities.

Key words

Mycobacteria, mycobacterioses, antigens, immunity, immunopathology

1. General introduction

Leprosy is often quoted as one of the most ancient of diseases, though with little evidence other than its frequent mention in the bible. The disease probably became endemic in Israel/Palestine between 300 and 100BC. Evidence from India takes it back to 600BC (Browne, 1970) and Chinese writings and pictures may take it back even further. This, however, is not particularly ancient for an infectious disease and others have similar pedigrees. The origin of leprosy bacilli is unknown but the most likely scenario is that they arose by mutation from an environmental species. Kazda *et al.* (1979) and Kazda (1981) found good epidemiological and bacteriological evidence that the precursor of *Mycobacterium leprae* in Norway lived in sphagnum vegetation. There is little similarity between leprosy and any of the other existant mycobacterial diseases of humans or other animals. Even "murine leprosy" caused by an organism called *M.lepramurium*, a difficult-to-grow variant of *M.avium*, has little similarity with human leprosy.

Without treatment, leprosy can be slowly but relentlessly progressive, disfiguring and disabling in most cases. Nonetheless, just as with tuberculosis, in some patients the disease appears to be self-limiting and arrested cases occurred. Prior to an understanding of infection, such a terrible disease was thought to be a divine punishment. The descriptions of early public health measures in Ecclesiastes show that the concept of contagion was understood, though whether with evil or with disease is uncertain. Methods of isolation advocated then remained the principal public health measure in Europe for 2000 years (Richards, 1977). Fear of the disease has only partially relented since the discovery of effective drugs to prevent it from progressing to deformity and to reduce infectivity.

Hansen and Danielson, working in Norway in the last half of the nineteenth century, recognised the infectious nature of leprosy and, in 1874, Hansen described the bacilli in the tissues (see Hansen and Looft, 1973). At first called "Hansen's bacillus" this was the first mycobacterium to be discovered, although it still has not been successfully cultured beyond a few logs of replication.

Within the genus *Mycobacterium* are two distinct subgenera of very ancient separation, known as the slow-growers and the rapid-, or fast-, growers (Goodfellow and Magee, 1998). About half of the 100 or so recognised species fall into each subgenus. Not only do mycobacteria cause major human and animal diseases, but they are very common in the environment and contact with them has probably played a major part in the evolution of the immune system.

2. The mycobacterial dseases: worldwide

2.1 Tuberculosis

Tuberculosis is easily the most important mycobacterial disease (mycobacteriosis) today with a third of the world's human population infected and eight to ten million new cases arising every year, two and a half to three million of whom die. However, these are not the great days of tuberculosis when, as the "Captain of all these men of death", it swept through Europe as an epidemic striking down members of every family and sometimes wiping out whole families. This occurred with the industrial revolution in the last half of the eighteenth and early nineteenth century, and since then the disease steadily declined in Europe for rather uncertain reasons. It could still behave in its old epidemic guise as was seen in the late nineteenth century when it was accidently introduced into native populations in Africa, Asia, Oceania and South America.

The so-called "human type" of tuberculosis, caused by *M.tuberculosis* itself, is primarily a disease of the lungs from which it can spread to any other part of the body. The "bovine type" of tuberculosis caused by the exceedingly similar organism called *M.bovis* is primarily a disease of cattle infecting humans predominantly via milk and the intestine, from which it too can spread to any other organ.

With the development of antituberculosis drugs in the 1950s the tail end of the tuberculosis epidemic was finally almost wiped out, although tuberculosis had started declining well before this time. Immunity lowered by infection with the human immunodeficiency virus (HIV) in the 1980s onwards has led to a recrudescence of tuberculosis, now sweeping Africa and Asia. Even in the Western world tuberculosis has reappeared as a new epidemic of the impoverished, the inner city itinerant, the drug addict and in prisons; this time of multi-drug-resistant tuberculosis (MDRTB) makes tuberculosis a very difficult disease to treat (Bloom, 1992).

Evidence from skin-testing with Tuberculin (a soluble extract of the cultured bacilli) suggests that up to one third of the entire human population has been infected with tubercle bacilli. The ability of tubercle bacilli to remain alive in human tissues as "persistors" (see below), apparently metabolising very slowly and not replicating, means that most of those once infected still carry living bacilli (Grange, 1992). That persistors may be capable of reactivation is one explanation of the association of HIV with tuberculosis (Festenstein and Grange, 1991). Mechanisms of reactivation of tuberculosis due to awakening of persistors are under intense investigation (Kell and Young, 2000).

From a palaeobacteriological viewpoint, pulmonary tuberculosis, or phthisis, often involved the pleura which could become grossly thickened in terminal disease. This provides a source of material for polymerase chain reaction (PCR) studies and this material survives in partially mummified remains for long after the lungs themselves have disappeared. "Surgical tuberculosis", or bone and joint disease, was also common in the past, leaving clearly recognisable bony lesions as another source of detectable DNA.

2.2 Leprosy

The worldwide burden of leprosy, the second most important mycobacterial disease, is difficult to evaluate. The introduction of multi-drug therapy (MDT) in the 1980s has led to far fewer patients being registered as under "active treatment". Today, treatment for many patients is completed within 2-3 years, whereas in the past when monotherapy with dapsone was the rule, treatment was often for a lifetime. This led to a fall in registered cases from some 10 million in 1985 to less than four million in 1998. Nonetheless, in 1998 more new cases of the disease were detected than in any previous year. This is partly the result of better case detection methods, of the more active case finding that always accompanies the introduction of an improved therapy, and of the recognition of very early cases that may not have progressed to the classical disease. Despite this, there has probably been a reduction in numbers of cases due to an improved standard of living, the protective capacity of the BCG (Bacillus of Calmette and Guérin) vaccine, the reduction in infectivity of diagnosed patients due to the greater efficacy of MDT, and the slow erosion of the social stigma that has always accompanied leprosy. The disease remains associated with the poor.

Infamous for its bony deformities, leprosy is one of the easiest diseases to recognise in skeletal remains. Nonetheless, many patients develop transient leprosy leaving no bony abnormalities and many people become infected but never develop the disease at all. Since many of these cases involve the nasal cavity, it may be possible to detect remnants of leprosy DNA dried onto the bony remnants of the nasal cavity in persons dying whilst they were infected which could considerably improve epidemiological knowledge of leprosy in the past.

2.3 Opportunist mycobacterial infections

These are caused by species that live freely in the environment but which are capable of giving rise to disease if offered a suitable opportunity to invade the tissues. Such diseases are usually associated with a particular situation and are not normally passed from one person to another. Some cause skin conditions, others cause lymphadenitis and still others cause lung disease. A list of the commonest mycobacterial opportunist pathogens is shown in Table 1.

1. Buruli ulcer is caused by infection with *Mycobacterium ulcerans* from the environment through skin abrasions or minor penetrating wounds, such as insect bites or grass spines. The disease was first observed in Uganda in the 1890s, but not described and attributed to mycobacteria until the 1940s when a few cases were recognised in Australia (MacCallum *et al.*, 1950). It was rediscovered in Zaire and Uganda some 10 years later, when environmentally associated epidemics were described. Today it is spreading very rapidly in the West African Republics of Benin, Cote d'Ivoire, Ghana and Togo, where it has replaced leprosy as the second most important mycobacterial disease. Rarely fatal, deep ulceration of the skin can occur, sometimes involving more than 25% of the entire skin area. Though there is often significant residual soft tissue deformity and disability, Buruli ulcer is usually self-healing as described in the work by Asiedu *et al.* (2000). A small proportion of cases experience spread to the long bones, usually the distal ends, and rarely the

disease spreads to the finger bones. A chronic osteomyelitis is established at these sites which is difficult to treat and could potentially leave palaeopathological evidence.

2. Swimming-pool/fish-tank granuloma is a much less severe disease causing superficial, unsightly granulomas of the skin at sites of penetration of *M.marinum*. The disease responds well to treatment with broad spectrum antibiotics.

3. Mycobacterial injection abscesses can follow the use of non-sterile solutions or of poorly sterilised equipment. These chronic abscesses at the sites of injection are usually caused by the fast-growing species *M.chelonei* or *M.fortuitum*.

4. Cervical lymphadenitis of childhood is the commonest opportunist mycobacterial disease in the United Kingdom, and occurs in 1 to 2% of children between the ages of 2 and 6 years. The causative species are *M.avium*, *M.intracellulare*, *M.malmoense* and *M.scrofulaceum* which enter via the tonsils or small mouth abrasions and drain to the lymph nodes in the neck. These become enlarged, tender, eventually fluctuent and may track to the skin causing indolent sinuses and unsightly scars.

5. Opportunist lung mycobacterioses are usually infections of middle-aged or elderly people with underlying lung damage from another cause. Common causes are *M.kansasii*, *M.intracellulare*, *M.avium*, *M.malmoense* and *M.xenopei* which, to become infectious, must be inhaled as an aerosol. They cause a disease similar to tuberculosis and can be difficult to cure because of drug resistance. These infections are often geographically disseminated according to the distribution of the causative organisms and the likely evolution of aerosols.

2.4 The environmental mycobacteria

Soil is the natural habitat of mycobacteria from which the pathogenic species *M.leprae* and *M.tuberculosis* undoubtedly arose (Kazda, 1983). Many common environmental species are opportunist pathogens as described above. Within these species there are often particular types that give rise to most cases of diseases, and other variants within the same species are virtually non-pathogenic, or pathogenic for different host species (Shield, 1983).

Taking *M.avium* as an example from amongst the slow-growers, tuberculosis in chickens is usually caused by agglutination serotype 2, whereas in ducks it is caused by serotype 1. In HIV (Human Immunodeficiency Virus) non-infected persons, lung infections with *M.avium* are most frequently with serotype 2. In HIV seropositive persons serotype 2 infections occur, though other serotypes predominate, varying geographically. Thus, in Britain and the U.S.A., serotypes 4 and 8 are common but in Sweden type 6 predominates (Yakrus and Good, 1990, Hoffner *et al.*, 1990). Amongst fast-growers, *M.fortuitum* in soil is often of immunodiffusion serotypes II/III/IV, but almost all human infections are with serotype I (Grange and Stanford, 1974). These examples are probably illustrative of the way in which *M.leprae* and *M.tuberculosis* arose and differentiated from environmental species.

Mycobacteria are ubiquitous in the soil, although very variable in numbers and species distribution (Beerwerth and Schurmann, 1969, Barker *et al.*, 1972), varying with simple physical conditions including temperature, moisture, pH and sunlight. The greatest numbers of bacilli and of species can be found at the edges of fresh water between pH 6 and 7.5 (Stanford and Paul, 1973). As the situation becomes drier slow-growing species predominate and, as ultraviolet light becomes stronger, pigmented species are selected. Thus, in lush meadows many species may be present, but under stones in deserts only slow-growing scotochromogens are found. Their distribution is also affected by husbandry methods and the application of artificial fertilisers and other chemicals as described by Donoghue *et al.* (1997). Some species colonise artificial environments such as coldwater tanks and taps (*M.kansasii* and *M.chelonei*) and even hotwater taps (*M.xenopi*) as described by McSwiggan and Collins (1974) and Collins and Grange 1984). Interesting studies by J.K. Kazda (pers. comm.) showed that extracts of certain plants appeared to inhibit the growth and survival of *M.avium,* and that the distribution of higher plants in water could be an important regulatory system for the distribution of freshwater mycobacteria.

2.5 The antigens and adjuvants of mycobacteria

Bacterial antigens are substances to which immunological responses can be made. Adjuvants enhance these responses and determine the type of response. The antigens of mycobacteria can be divided into four groups: Group 1 are shared by all species, Group 2 are restricted to slow-growing species, Group 3 are restricted to fast growing species, and Group 4 are the antigens specific to individual species as shown in Figure 1 (Stanford and Grange, 1974, Stanford, 1983). Antigens capable of inducing and eliciting cellular immune responses are those of Groups 1, 2 and 4 although the type of immune responses commonly made to them are qualitatively different and may have different immunopathological significance (Stanford *et al.*, 1981, Rook *et al.*, 1981).

All living things share some antigenic material, some of which may be merely chance molecular conformations or amino acid sequences giving no evidence of inter-relationships. Far more important are the common molecules that have come down through ontogeny, which are essential to successful life and survival. These extremely ancient molecules among the Group 1 mycobacterial antigens include the heat shock proteins (hsp) of bacteria and stress proteins of humans (Cohen and Young, 1991). They are essential to all life and part of the response to all challenges causing cellular damage. Some assist in the correct folding of proteins and others protect the nucleic acids of the genome from damage. Without these functions life could not exist. Amongst the best known and most investigated are the hsp65 and 70 kilo Dalton (kDa) molecular weight proteins of bacteria. Those mycobacteria share some 60% of amino acid sequence homology with the 60 and 70kDa stress proteins of humans respectively. Such homology is both a strength and weakness. It is a strength when immune response is prevented by tolerance induced by the bacterial homologues and a weakness since the immune system is particularly attuned to respond to antigens differing slightly from self. Young (1992) describes how potentially tissue damaging responses induced by contact with bacterial hsps can lead to immune attack against self - auto-immunity.

Within the cell envelopes of mycobacteria are several potent adjuvant substances, some lipid and others water-soluble which vary between species in their chemical structure and function (Stewart-Tull, 1983). These adjuvants both enhance immune responses to antigens and direct the kind of immune response that will be made. It is this capacity of mycobacterial adjuvants, combined with their antigens, that make the mycobacteria so important, both in the evolution of immunity and in the immunopathology of many different diseases.

3. Immunity and mycobacterial infections

3.1 Introduction

Humans learn through two systems, the brain and the immune system, both of which build from inherited instinct and from learning experiences. Born with innate immunity plus some inherited information retained from the maternal circulation, the immune system begins to learn from the environment of the birth canal onwards. Just as the untaught mind cannot adequately cope with the dangers of life, neither can the naive immune system. Important teachers of immunity are environmental mycobacteria whose effects are particularly important because they combine antigens with potent adjuvants, as described above, which enhance and manipulate the nature of the immune response (Rook and Stanford, 1998). The type of immunity established determines subsequent responses to other mycobacteria, to pollens and other allergens, and to self. Fine tuning of the nature of the immune response to allergens is a major factor determining whether the child develops atopic dermatitis or asthma. Responses to self determine susceptibility to autoimmune diseases and to cancers (Cohen and Young, 1991). The most important mycobacteria are not those that cause disease, but those that shape our ability to survive as long lived mammals (see Stanford and Rook, 1983).

3.2 Immune memory of mycobacteria

Although needing to be taught only once for the lessons to be maintained, external boosting is essential to jog the immune memory. Joliffe's studies on Ghurka soldiers using a simple measure of immune response - the tuberculin skin test (pers. comm.), showed that the immediate immune memory for skin-test responses to individual mycobacterial species is less than two years although, with stimulation by meeting that species again, memory is restored. The immune memory decays with age, and elderly people may succomb to endogenous reactivation of tuberculosis held latent as persistors since infection much earlier in their lives.

3.3 The concept of the persistor

Epidemiological studies of mycobacterial disease show that effective contact with the pathogen leads directly to disease, or to apparent protective immunity. If sufficient numbers of people with apparent immunity are followed for long enough, some develop the disease, perhaps years later, without requiring fresh exposure to the pathogen. The pathogen is held as a live persistor in the tissues, as a latent infection by immune responses developed from prior contact with environmental mycobacteria. This latent infection can reactivate whenever immunity is lowered, such as when infected with the human immunodeficiency virus (HIV) or with older age. Recent work with genetic fingerprinting of strains of tubercle bacilli suggests that persistors may directly reactivate, or their presence may increase susceptibility to newly encountered exogenous strains. Persistors occur in leprosy, tuberculosis, probably Buruli ulcer, and perhaps environmental mycobacteria which may help to maintain immune competence. The physical form of persistors remains unknown, but some think that they are scanty normally stainable bacilli and others think that they are in a granular, variably acid-fast form, as originally described by Much (1907) and developed further by Minchin (1927), Stanford (1987) and Khomenko (1987).

Molecular techniques such as PCR, sensitive enough to detect ancient DNA, can also detect persistors. Very recent studies with the *in situ* PCR have demonstrated *M.tuberculosis* genomic material in macrophages and Type 2 pneumocytes in post-mortem material from people without other evidence of tuberculosis (Hernandez-Pando *et al.*, 2000). In Ethiopia and Mexico, tissues from 30-40% of people dying violently or from diseases other than tuberculosis were found positive by this technique and latent infection with persistors seems the likely explanation. Soon it may be possible to visualise the organisms and determine their physical form (Grange, 1992). If persistors can be detected in the recent dead, why not in the anciently deceased? Already there is a report by Pap *et al.* (1999) of an Hungarian mummy with no evidence of tuberculosis other than PCR detection of ancient DNA. As those dying with latent infection become detectable, the true epidemiology of mycobacterial disease in past communities may be determined. PCR can be used to type strains of tuberculosis and soon it should be possible for leprosy bacilli, allowing another layer of epidemiological investigation.

3.4 The maturation of thymic (T) lymphocytes

T-lymphocytes are the orchestrators of cell-mediated immunity involved in both protection and pathology. B-lymphocytes lead to antibody production, essential for overcoming acute bacterial infections and for aspects of immunoregulation. Many B-cells require help from T-cells and are regulated by other T-cells. Helper T(h)-cells, which can influence other cell types, can often be identified by antigenic markers on their surfaces and are known as CD4+ T-cells. As described by Mossman *et al.* (1986), these mature from precursors along at least two pathways into Th1-cells and Th2-cells, producing different sets of chemical messengers, or cytokines. These are released in response to signals of tissue damage or bacterial invasion, and can activate another cell type, which is the macrophage, or a major effector arm of immunity. Some Th1 cells become cytotoxic, enabling them to kill bacilli-laden cells, exposing the bacilli to more effective phagocytes.

3.5 The immune response to mycobacteria

Macrophages that engulf an invading mycobacterium present its antigens on their surface in a form that can be read by adjacent Th-cells. These release cytokines that attract other T-cells and more macrophages to the site. This causes expansion of clones of cells that respond to the antigen presented on the macrophage, which start to orchestrate the eradication of the invader and the repair of tissue damage. It is at this point that the pathogen may perturb the response, making it less effective and allowing disease to develop, as described by Rook and Hernandez-Pando (1996).

The regulation of cellular immunity is effected partly from within, partly from the environment and partly in response to the adjuvants of the invading organisms. Some of these interactions between mycobacteria and the immune system are indicated in Figure 2, though the real situation is much more complicated than that shown.

Of the cerebral functions that control hormone production from the hypothalamus at the base of the brain, leading to the different cascades of endocrine function influencing T-cell maturation, depression may have the greatest effect. People affected by bereavement, by financial ruin or by imprisonment may be more likely to develop tuberculosis than at other times. Those living under poor conditions are more likely to develop tuberculosis or leprosy than are their richer neighbours.

Some environments are protective and others increase susceptibility to leprosy or tuberculosis which is an explanation of disease endemicity (Shield, 1983). This is indicated in Figure 2 as contact with environments with different distributions of the two subgenera of mycobacteria, showing them to have different effects dependent on their cell wall adjuvants, promoting either Th1 or Th2 maturation. Protective immunity from mycobacterial disease is mediated by Th1 responses to Group 1, and common antigens, and immunopathology is associated with a poorly regulated interaction between Th1 and Th2 responses to Groups 2 and 4 antigens (Wangoo *et al.*, 2001).

3.6 The spectrum of immunopathology in leprosy

Since early in the nineteenth century, leprosy has been recognised to occur at two extremes, that is tuberculoid and lepromatous, both most commonly affecting skin and nerves. Demonstration of leprosy bacilli in the tissues in the 1870s enabled a further division into bacilliferous, or multibacillary, and non-bacilliferous, or paucibacillary disease (Hansen and Looft, 1973). Patients with lepromatous leprosy have millions of acid-fast bacilli stainable in their tissues, whereas in those with tuberculoid disease it is hard to find a single bacillus. The same is seen in Buruli ulcer where millions of bacilli are present in the early, necrotising, phase of disease and almost none can be found in the later granulomatous phase (Lucas, 1989). The term tuberculoid leprosy is confusing since it suggests parallels with tuberculosis. In reality the cell-mediated mechanisms of tuberculoid leprosy and tuberculosis are different, and a closer parallel is found between tuberculosis and multibacillary leprosy on which immunosuppression has been superimposed.

The development of tuberculin by Robert Koch (1890, 1891) from cultured tubercle bacilli led to attempts to produce a similar skin test reagent for leprosy. The non-cultivability of the leprosy bacillus led to lepromin being prepared from sterilised suspensions of highly bacilliferous tissues of untreated lepromatous leprosy patients (Mitsuda, 1923). When injected into the skin of tuberculoid leprosy patients, lepromin induced a nodular granulomatous response after about two weeks, known as the "Mitsuda reaction", which did not occur in lepromatous patients, demonstrating a major immunological difference. Later it was reported by Fernandez (1939) that there was an earlier response to lepromin at 48-72 hours, analogous with the response to tuberculin. Together, these led to the concept that tuberculoid patients had immunity to leprosy bacilli which was too slow in development, or insufficient, to prevent disease (Godal *et al.*, 1974). Lepromatous patients were thought incapable of developing an immune response, perhaps for genetic reasons, or were immunosuppressed by the infecting bacilli. The development of antigenic preparations of leprosy bacilli allowed antibodies to be measured, showing that tuberculoid patients produced little antibody and lepromatous patients produced a lot. Later it was realised that there were more types of leprosy than the two extremes and in the mid-20th century, Ridley and Jopling (1966) introduced their concept of the immunopathological spectrum of leprosy (Figure 3), which includes forms of borderline disease, as further elaborated by Ridley (1988).

Striking features of leprosy are the ability of the bacilli to suppress immune responses so that thay can multiply unopposed, enter Schwann cells supporting peripheral nerves and interfere with their function, and enter nerve cells themselves (Lucas, 1989). They can also be found in smooth muscle cells (Rees and Waters, 1972). Immune responses vary from the completely inadequate antibody associated and progressive macrophage lepromas in which bacilli multiply unchecked, to the excessive and only partially effective tuberculoid cell-mediated response (Godal *et al.*, 1974). Both responses directly damage the tissues invaded by bacilli, impairing the blood supply to skin and disrupting nerve function. The sudden immunological changes of leprosy reactions (see below) lead to accelerated tissue damage which can be around infected tissues or focussed by autoimmunity on distant sites such as uninvolved skin and nerves, joints, testes and eyes. The result is a disease which, untreated, produces the soft tissue and bony deformity that we know as related to the biblical and medieval "leper". The pathognomonic bone changes have focussed palaeopathological interest on the disease.

3.7 Leprosy reactions

Sudden deteriorations in the clinical state of leprosy patients have been long recognised as leprosy reactions (Naafs, 1989). These reactions are of two common types. One is erythema nodosum leprosum (ENL) and the other is the reversal reaction. These occur predominantly in different types of patients. ENL is a generalised condition affecting skin, joints and nerves, not necessarily at the sites of leprosy lesions, and due to a relaxation of the immune tolerance/immunosuppression seen in multibacillary disease. Reversal reactions are also generalised and particularly affect nerves. They are associated with changing immunoregulation at the more tuberculoid end of the spectrum, reducing the efficacy of the cell-mediated response to control the disease. Together these two types of reaction are responsible for most of the tissue injury and deformity seen in leprosy, and both are painful and make patients ill. There are other less frequent types of reaction, one of which, Lucio's phenomenon, a haemorrhagic vasculitis occurring in multibacillary patients, chiefly occurs in central America (Lucio and Alvarado, 1852; Naafs, 1989).

3.8 Deformity in leprosy

Lepromatous to borderline tuberculoid disease can lead to bony deformities which do not occur in polar tuberculoid leprosy. Bony changes are associated with loss of innervation of bones and perturbation of the integrity of the balanced system of osteoblasts and osteoclasts which lay down and resorb bone, respectively. As a result, bones affected are partly resorbed and changed in shape. Vasculitis may be the principal lesion leading to ulceration, loss of fingers, toes, hands and feet, destruction of the eye from within, loss of nerve function and loss of testicular function in multibacillary leprosy. The mechanism responsible for the vasculitis is not entirely clear, but one possibility is an excessive antibody and complementary response to human stress proteins. Bacillary invasion of cutaneous nerves leads to loss of sensation, loss of tissue integrity and the resulting anaesthesia allows soft tissues to be accidentally damaged, adding to the effects of immunological attack on the tissues occurring during times of reaction. A combination of soft tissue destruction overlying bony changes, together with the thickening of superficial tissues due to enormous numbers of bacilli packed into cells as space-occupying lesions, results in the abnormal appearance of late leprosy patients (Ganapati and Revankar, 1989).

The characteristic bony changes of the disease have made leprosy one of the few diseases that can be confidently identified by the palaeopathologist but molecular methods will take palaeobacteriology far beyond bony changes (Rafi *et al.*, 1994).

4. Who catches leprosy?

4.1 Genetic aspects

Although there is no known genetic marker of predisposition to developing the disease, genes may dictate the place on the spectrum by determining the nature of the immune response to bacillary invasion. Thus, patients developing lepromatous disease tend to lack the human lymphocyte antigen (HLA) haplotype DR3, and those developing tuberculoid disease show a positive association with HLA-DR4, as found by van Eden *et al.* (1982).

4.2 Environmental aspects

Risk factors for leprosy include living in an area where the disease is highly endemic and within endemic countries there are regions where the disease is common and others where it is uncommon. A major cause for this is thought to be the unequal distribution of the two subgenera of mycobacteria between contiguous regions. Heavy exposure to slow-growing mycobacterial species may increase susceptibility to mycobacterial diseases, and similar exposure to fast-growers may be protective, as discussed above (Shield, 1983). Contact with environmental mycobacteria also determines the efficacy of BCG vaccine seen in the varied protective effects of this vaccine against both leprosy and tuberculosis (Table 2). Besides environmental mycobacterial aspects, in theory susceptibility will be increased by Th2 responses enhanced by chronic parasitic or viral diseases (Morrow, 1997).

4.3 Domiciliary aspects

Most new cases of leprosy develop in people living close to the soil, usually meaning those who are poor and of lower social class, though it is by no means rare for even the very rich to develop disease. The majority of new cases occur amongst those living in highly endemic regions but without known close contact with patients, yet leprosy is commonest in close contacts of patients (Badger, 1959).

The weak genetic associations mean that non-consanguineous contacts are just as susceptible as are blood relatives. From experience of ex-patriots working in leprosy-endemic regions, short exposure, however close, rarely led to disease. Exposure for many years was required before missionaries in the past developed leprosy. The reason for this is likely to be a combination of the time taken for the local environment to have an effect, and the length of immunological memory of less leprous situations. Studies have shown that skin-test memory for the range of mycobacterial species met in a particular environment is about two years (D.Joliffe, pers. comm.), although this may be an over-simplification.

4.4 Bacteriological aspects

Infection with leprosy bacilli probably occurs through the nasal mucosa where bacilli from inhaled droplet nuclei enter epithelial cells, start to multiply, and release bacterial products (Barton *et al.*, 1982). Individuals with effective immunity may rapidly kill these bacilli and the susceptible person develops disease of a type determined by the interaction between their immunity and bacterial products. Many others restrict the infection to transient involvement of the nose without progression to clinical disease. Such subclinical cases may become carriers, releasing bacilli into the air as infectious droplet nuclei, just as do patients with multibacillary disease. Thus, new cases of leprosy may come from contact with such subclinical and unrecognised carriers silently spreading the disease, as well as from contact with clinical leprosy patients, making the disease hard to eradicate.

4.5 Immunological aspects

In a highly leprosy endemic region of central India, none of a group of healthy people with positive reactions to skin tests with Leprosin A developed the disease during a ten-year follow-up, whereas a proportion of those who were negative to the test did develop leprosy ($p<0.03$, Fisher's exact test). Unlike Mitsuda's lepromin, a suspension of killed leprosy bacilli from human tissue, leprosin A is a soluble sonicate of bacilli extracted from the tissues of experimentally infected armadillos (Binford *et al.*, 1976). It produces a skin reaction after 72 hours rather than the 2-3 week Mitsuda reaction (Shield *et al.*, 1982, Shield and Stanford, 1982). Recent studies in Malawi by Fine *et al.* (2001) showed that people producing positive skin tests to a range of fast-growing mycobacteria were protected from leprosy. No such association was found for positivity to slow-growers. Responses to Leprosin A and to fast growers are principally to Group 1 antigens by a Th1 mechanism. Responses to slow growers, in those who are negative to fast-growers, tend to be to Group 2, or to Group 4 antigens by a predominantly Th2 mechanism (Figures 1 and 2).

It is highly likely that the above aspects of susceptibility to, and protection from, leprosy were just as operative in communities in the distant past as they are today. However, not only does the distribution of leprosy vary geographically, but so too does the distribution of the spectrum. Thus, the proportions of the different clinical presentations of leprosy are different between Africa, India and South America, and also between different areas of the same country. This variation is likely to have been the same in the past, and the distribution of disease types seen in a leprosy-endemic country today, may not be the same as that found in the same place in the past. This may reflect changing ways of life and the changing environment. In addition, it is possible that when the disease first spread from the environment to humans its clinical expression was different. It is now known that the genome of *M.leprae* is severely depleted (Cole *et al.*, 2001) and this depletion may have occurred in a series of stages, once the organism became established as a human pathogen. Now that the genome of the leprosy bacillus has been elucidated, it will be possible to find out whether all strains from different parts of the world, or causing different clinical presentations, are the same. Analysis of the short lengths of DNA found in palaeopathological specimens may allow the process to be examined historically.

5. Prophylaxis and treatment of leprosy

5.1 Prevention of leprosy

The biblical and mediaeval principles of isolating patients in leprosaria are unacceptable today. Although stigma remains because of the unfortunate appearance of many advanced cases, diagnosis is generally made much earlier, a course of modern chemotherapy soon makes the patient non-infectious, and much deformity is prevented. The problem of subclinical but infectious carriers remains an important source of new leprosy patients, although their number should slowly reduce as clinical cases are reduced. If there is an environmental niche for leprosy bacilli, suspected by many, eradication of the disease may be an impossible task. Thus an effective, cheap and easily applied vaccine is still needed for leprosy.

5.2 Leprosy vaccines

Although BCG may be a better vaccine for leprosy than it is for tuberculosis (Fine *et al.*, 1986), operating through its capacity to induce immune recognition of the Group 1 antigens shared by both leprosy bacilli and tubercle bacilli, its effect is not good enough. Recent studies in South India by Gupte *et al.* (1998) have shown that vaccination with a killed preparation of a selected strain of the slow-growing opportunist species *M.intracellulare*, the ICRC (Indian Cancer Research Centre) bacillus, is more effective in the South India setting than is BCG. Our own studies with a killed preparation of the fast-growing species *M.vaccae* suggest that it can improve the efficacy of BCG, either when the two are mixed, or when the *M.vaccae* is given to those in whom past BCG does not seem to have induced protection (Le Van Truoc *et al.*, 2001). Widespread investigation of these new approaches have yet to be carried out but such vaccines may offer improved protection (Table 3) and eliminate the subclinical carrier state.

5.3 Chemotherapy of leprosy

Until some 20 years ago, chemotherapy consisted of daily doses of the sulphone, dapsone, frequently given for life (Cochrane, 1959). With the inevitable development of dapsone resistance in leprosy bacilli, first appreciated in the late 1970s (Waters *et al.*, 1978), treatment with more than one drug was introduced as multi-drug therapy (MDT) in which rifampicin is added to dapsone treatment, often with a third drug, prothionamide or clofazimine (WHO 1977). This was more effective and much shorter courses can be used; six months for paucibacillary disease and two years for multibacillary disease. The treatment of reactions remains the chief clinical problem for leprologists (Ganapati and Revankar, 1989; Naafs, 1989). Thalidomide is very effective in ENL, although its use is restricted to men and women past the childbearing age (Waters, 1989). Reversal reactions are more difficult to control, a combination of increasing the anti-leprosy drugs and use of cortisone being the usual treatments (Naafs, *ibid.*). All these treatments have problems and an effective immunotherapeutic is badly needed to supplement drug treatment.

5.4 Immunotherapy for leprosy

5.4.1 In antiquity

It was recognised late in the nineteenth century that leprosy might be improved by immunotherapeutic measures, and the first to be used was Koch's old tuberculin (Abraham, 1890). This was followed by a series of attempts to treat patients with different preparations of organisms grown from leprous tissues, or patients' clothes, and believed to be leprosy bacilli. Subsequently, a preparation of *M.scrofulaceum* called "*Marianum* antigen" (Blanc *et al.*, 1953) was used in a series of studies at different centres. Some centres reported success, and others failure (reviewed by Stanford, 1998).

5.4.2 Today

Although MDT may rapidly kill most living leprosy bacilli, in many patients it fails to prevent progress to disability, or the continuing tissue damage due to autoimmune sequelae triggered by the adjuvant activity of the leprosy bacilli. Even with good bactericidal drugs, leprosy bacilli are only removed very slowly from the tissues, and the immunopathogenic effects of these bacilli persist long after they have been killed. One of the aims of immunotherapy has been to hasten the rate at which tissue clearance of visible bacilli occurs. Injections of BCG, BCG+ killed *M.leprae* (Convit *et al.*, 1982), killed ICRC bacilli (Deo *et al.*, 1981), the so-called "*Mycobacterium W*"(Chaudhuri *et al.*, 1983), and another preparation of *M.intracellulare*, or of *M.vaccae* (Ramu *et al.*, 1998) have been shown to act thus. Another major use of immunotherapy is to prevent leprosy reactions, limit the damage that they do and truncate persistant reactions. Again, the preparations described above have been shown to be effective against ENL, although their value has yet to be determined in reversal reactions or Lucio's reaction. Some of the immunotherapeutics of potential use in leprosy are shown in Table 4.

These beneficial clinical effects are thought to be achieved by improved immune regulation of T-helper-cell maturation. A surrogate marker for correction of residual immune defects in patients who had been fully treated for multibacillary disease is cellular immune response to the Group 1 common mycobacterial antigens which can be recognised by skin testing. Table 5 gives some results of a study carried out in Spain some years ago (Stanford *et al.*, 1987).

At the moment all these approaches towards immunotherapy are experimental, but in the relatively near future they may become part of the way that all leprosy patients are treated. Notably, the best way to deal with post-leprosy syndromes discussed below would be by immunotherapy.

6. Post-leprosy syndromes

Unfortunately for the patient, bacteriological cure does not always mean clinical cure. "Fully treated" patients may continue to suffer leprosy reactions and be left with deformity, progressive neuritis, uveitis of the eyes, arthritis and vasculitis resulting from adjuvant activity of the remaining bacillary products with reactions continuing to residual bacterial antigens and to cross-reactive self-antigens. Thus, correction of the immune perturbation leading to the basic lesion of these conditions, arteritis is required. Clinical observations of the use of immunotherapy with heat-killed *M.vaccae* in anterior uveitis in leprosy (Stanford *et al.*, 1994), Raynaud's disease and laser-Doppler flowmetry in peripheral vasculature of long-treated leprosy patients (Abbot *et al.*, in press for 2002), together with preclinical studies of the control of myointimal hyperplasia after angioplasty (Stansby *et al.*, 2002), all indicate an important place for immunotherapy in the future.

7. Conclusion

Leprosy is an extraordinary disease presenting in different forms over a spectrum covering almost all the possibilities of human immunopathology. Even when bacteriologically cured, the disease can continue to exert its immunoregulatory effects for years. In addition to the well known erythema nodosum leprosum and reversal reactions, much of the pathology seen in leprosy is due to autoimmune complications. When the homoeostatic mechanisms normally preventing the immune system from attacking self are broken down, autoimmune sequelae become the major complaint of the patient. Within leprosy there are parallels with many other chronic diseases from HIV to rheumatoid arthritis and atherosclerosis. A major part of treatment for leprosy in this new millenium is going to be a combination of active bactericidal drugs and regulation of the immune response, control of tissue damage and prevention of deformity.

References

Abbot, N.C., Beck, J.S., Feval, F., Weiss, F., Mobayen, M., Ghazi-Saidi, K., Dowlati, Y., Velayati, A.A. and Stanford, J.L. (in press for 2002) Immunotherapy with *Mycobacterium vaccae* and peripheral blood flow in long-treated leprosy patients, a blinded, placebo-controlled study. EJVEVS

Abraham, P.S. 1890 Leprosy and Professor Koch's treatment. *British Medical Journal* 2: 1438.

Asiedu, K., Scherpbeir, R. and Raviglione, M. (eds) 2000 *Buruli ulcer: Mycobacterium ulcerans infection.* Geneva: World Health Organisation

Badger, L.F. 1959 Epidemiology. In R.G.Cochrane and G. McRobert (eds): *Leprosy in Theory and Practice.* Bristol: John Wright and Sons Ltd., pp. 51-77

Barker, D.J.P., Clancey, J.K. and Rao, S.K. 1972 Mycobacteria on vegetation in Uganda. *East Africa Medical Journal* 49: 667-671.

Barton, R.P., Rees, R.J.W., McDougall, A.C. and Ellard, G.A. 1982 The nose in lepromatous leprosy; bacteriological and histopathological studies of patients treated with dapsone monotherapy for varying periods of time. *International Journal of Leprosy* 50: 58-67.

Beerwerth, W. and Schurmann, J. 1969 Zur Okologie der Mykobakterien. *Zentralblatt der Bakteriologie (Orig.)* 211: 58-69.

Binford, C.H., Storrs, E.E. and Walsh, G.P. 1976 Disseminated infection in the nine-banded armadillo (*Dasypus novemcinctus*) resulting from inoculation with *M. leprae*. Observations made on 15 animals studied at autopsy. *International Journal of Leprosy* 44: 80-83.

Blanc, M., Prost, M-T. and Sister Marie Suzanne 1953 Influence de l'injection d'une suspension d'un mycobacterium isole d'un cas de lèpre (souche chauvire). *Bulletin de la Société de Pathologie Exotique* 46: 1009-1015.

Bloom, B.R. 1992 Tuberculosis. Back to a frightening future. *Nature* 358: 538-539.

Browne, S.G. 1970 *Leprosy*. Documenta Geigy. Acta Clinica No. 11.

Chaudhuri, S., Fotedar, A. and Talwar, G.P. 1983 Lepromin conversion in repeatedly lepromin negative BL/LL patients after immunisation with autoclaved *Mycobacterium W. International Journal of Leprosy* 51: 159-168.

Cochrane, R.G. 1959 Therapy. In R.G.Cochrane and G. McRobert (eds): *Leprosy in Theory and Practice.* Bristol: John Wright and Sons Ltd., pp. 203-220.

Cohen, I.R. and Young, D.B. 1991 Autoimmunity, microbial immunity and the immunological homunculus. *Immunology Today* 12: 105-110.

Cole, S.T., Eiglmeier, K., Parkhill, J. and 41 other authors. 2001 Massive gene decay in the leprosy bacillus. *Nature* 409: 1007-1011.

Collins, C.H. and Grange, J.M. 1984 Mycobacteria in water. *J. Applied Bacteriology* 57: 193-211.

Convit, J., Aranzazu, N., Ulrich, M., Pinardi, Reyes, O. and Alvarado, J. 1982 Immunotherapy with a mixture of *Mycobacterium leprae* and BCG in different forms of leprosy and in Mitsuda negative contacts. *International Journal of Leprosy* 50: 415-424.

Deo, M.G., Bapat, C.V., Chulawala, R.G. and Bhatki, W.S. 1981 Potential anti-leprosy vaccine from killed ICRC bacilli - a clinico-pathological study. *Indian Journal of Medical Research* 74: 164-177.

Donoghue, H.D., Overend, E. and Stanford, J.L. 1997 A longitudinal study of environmental mycobacteria on a farm in south-west England. *Journal of Applied Microbiology* 82: 57-67.

van Eden, W., de Vries, R.R., D'Amaro, J., Schreuder, I., Leiker, D.L., and van Rood, J.J. 1982 HLA-DR-associated genetic control of the type of leprosy in a population from Suriname. *Human Immunology* 4: 343-50.

Fernandez, J.M.M. 1939 Estudio comparative de la reaccion de Mitsuda con las reacciones tuberculinicas. *Revista Argentina de Dermatosifilologia* 23: 425.

Festenstein, F. and Grange, J.M. 1991 Tuberculosis and the acquired immune deficiency syndrome. *J. Applied Bacteriology* 71: 19-30.

Fine, P.E. Ponnighaus, J.M., Maine, N., Clarkson, J.A. and Bliss, L.1986 Protective efficacy of BCG against leprosy in Northern Malawi. *Lancet* 2: 499-502.

Fine, P.E.M., Floyd, S., Stanford, J.L., Nkhosa, P., Kasunga, A., Chagulaka, S., Warndorff, D.K., Jenkins, P.A., Yates, M. and Ponnighaus, J.M. (2001) Environmental mycobacteria in Northern Malawi: implications for the epidemiology of tuberculosis and leprosy. *Epidemiology and Infection* 126:379-387

Ganapati, R. and Revankar, C.R. 1989 Clinical aspects of leprosy. In C. Ratledge, J. Stanford and J.M. Grange (eds): *The biology of the mycobacteria.* London: Academic Press, pp. 327-358

Godal, T., Myrvang, B., Stanford, J.L. and Samuel, D. 1974 Recent advances in the immunology of leprosy with special reference to new approaches in immunoprophylaxis. *Bulletin of the Institut Pasteur* 72: 273-310.

Goodfellow, M. and Magee, J. 1998 Taxonomy of mycobacteria. In P.R.J. Gangadharam and P.A.Jenkins (eds): *Mycobacteria.* New York: Chapman and Hall, pp. 1-71.

Grange, J.M. 1992 The mystery of the mycobacterial persistor. *Tubercle and Lung Disease* 73: 249-251.

Grange, J.M. and Stanford, J.L. 1974 Re-evaluation of *Mycobacterium fortuitum* (Synonym: *Mycobacterium ranae*). *International Journal of Systematic Bacteriology* 24: 320-329.

Gupte, M.D., Vallishayee, R.S., Anantharaman, D.S., Nagaraju, B., Sreevatsa, Balasubramanyam, S., de Britto, R.L., Elango, N., Uthayakumaran, N., Mahalingam, V.N., Lourdusamy ,G., Ramalingam, A., Kannan, S., Arokiasamy, J. 1998 Comparative leprosy vaccine trial in South India. *Indian Journal of Leprosy* 70: 369-88.

Hansen, G.A. and Looft, C. 1973 *Leprosy: in its clinical and pathological aspects.* (Reprint of 1895 first edition). Bristol: John Wright and Co.Ltd.

Hernandez-Pando, R., Jeyanathan, M., Mengistu, G., Leon, D., Orozco, H., Harboe, M., Rook, G.A.W. and Bjune, G. 2000 Persistence of DNA from *Mycobacterium tuberculosis* in superficially normal lung tissue during latent infection. *Lancet* 356: 2133-2138.

Hoffner, S.E., Kallenius, G., Petrini, B., Brennan, P.J. and Tsang, A.Y. 1990 Serovars of *Mycobacterium avium* complex isolated from patients in Sweden. *Journal of Clinical Microbiology* 28: 1105-1107.

Kazda, J.K. 1981 Occurrence of non-cultivable acid-fast bacilli in the environment and their relationship to *M.leprae*. *Leprosy Review* 52 (Supplement 1): 85-91.

Kazda, J.K. 1983 The principles of the ecology of mycobacteria. In C. Ratledge and J. Stanford (eds): *The biology of the mycobacteria.* London: Academic Press, pp. 323-341

Kazda, J.K., Muller, K. and Irgens, L.M. 1979 Cultivable mycobacteria in sphagnum vegetation on moors in South Sweden and coastal Norway. *Acta Pathologica et Microbiologica Scandinavica* 87: 97-101.

Kell, D.B. and Young, M. 2000 Bacterial dormancy and culturability: the role of autocrine growth factors. *Current Opinions in Microbiology* 3: 238-243.

Khomenko, A.G. 1987 The variability of *Mycobacterium tuberculosis* in patients with cavitary pulmonary tuberculosis in the course of chemotherapy. *Tubercle* 68: 243-253.

Koch, R. 1890 Weitere Mitteilungen uber ein Heilmittel gegen Tuberculose. *Deutche Medizinische Wochenschrift* 16: 1029

Koch, R. 1891 Weitere Mitteilungen uber ein Heilmittel gegen Tuberculose. *Deutche Medizinische Wochenschrift* 17: 101, 1189.

Le Van Truoc, Ho Minh Ly, Nguyen Kim Thuy, Dang Duc Trach, Stanford, C.A. and Stanford, J.L. (2001) Vaccination against leprosy at Ben San Leprosy Centre, Ho Chi Minh City, Vietnam. *Vaccine* 19:3451-3458.

Lucas, S.B. 1989 Mycobacteria and the tissues of man. In C. Ratledge, J. Stanford and J.M. Grange (eds): *The biology of the mycobacteria.* London: Academic Press, pp. 107-76

Lucio, R. and Alvarado, I. 1852 *Opusculo sobre el mal de San Lazaro e elefanciasis de los griegos.* M. Munguia y Cia

MacCallum, P., Tolhurst, J.C., Buckle, G. and Sissons, H.A. 1950 A new mycobacterial infection in man. *Journal of Pathology and Bacteriology* 60: 93-122.

McSwiggin, D.A. and Collins, C.H. 1974 The isolation of *M.kansasii* and *M.xenopi* from water. *Tubercle* B55: 291-297.

Minchin, W.C. 1927 *A study in tubercle virus, polymorphism and the treatment of tuberculosis and lupus with Oleum allii.* 3rd edition. London: Bailliere, Tindall and Cox.

Mitsuda, K. 1923 *Les lepreux maculo-nerveux, d'une part, les tubereux, d'autre part, se comportent differemment á la suite d'une inoculation d'emulsion de tubercle lepreux.* Proceedings of the Second International Leprosy Conference, Strasburg, p.219.

Morrow, A. 1997 Trouble with ticks. *African Farming* July/August: 23-25.

Mossman, T.R., Cherwinski, H., Bond, M.W., Giedlin, M.A. and Coffman, R.L. 1986 Two types of murine helper T cell clone. 1) Definition according to profiles of lymphokine activities and secreted proteins. *Journal of Immunology* 136: 2348-2357.

Much, H. 1907 Uber die granulare nach Ziehl nicht farbare Form des Tuberkulose virus. *Beitrage Klinisches Tuberkulose* 8: 85.

Naafs, B. 1989 Reactions in leprosy. In C. Ratledge, J. Stanford and J.M.Grange (eds): *The biology of the mycobacteria.* London, Academic Press, pp. 359-403

Pap, I., Jozsa, L., Repa, I., Bajzik, G., Lakhani, S.R., Donoghue, H.D. and Spigelman, M. 1999 18th to 19th century tuberculosis in naturally mummified individuals (Vac, Hungary). In G.Pálfi, O.Dutour, J.Deák, I.Hutás (eds): *Tuberculosis. Past and present.* Szeged, Hungary: Tuberculosis Foundation and Budapest: Golden Book Publisher Ltd., pp. 420-428

Rafi, A., Spigelman, M., Stanford, J.L., Lemma. E., Donoghue, H. and Zias, J. 1994 DNA of *Mycobacterium leprae* detected by PCR in ancient bones. *International Journal of Osteoarchaeology* 4: 287-290.

Ramu, G., Grange, J.M. and Stanford, J.L. 1998 Evaluation of adjunct immunotherapy with killed *Mycobacterium vaccae* in the treatment of multibacillary leprosy and erythema nodosum leprosum. In M. Casal(ed): *Clinical mycobacteriology.* Barcelona: Prous Science, pp. 393-402.

Rees, R.J.W. and Waters, M.F. 1972 Recent trends in leprosy research. *British Medical Bulletin* 28: 16-21.

Richards, P. 1977 *The Medieval leper and his northern heirs.* Cambridge: D.S.Brewer

Ridley, D.S. 1988 *Pathogenesis of leprosy and related diseases*. Cambridge: Butterworth and Co.

Ridley, D.S. and Jopling, W.H. 1966 Classification of leprosy according to immunity. A five group system. *International Journal of Leprosy* 34: 255-273.

Rook, G.A.W. and Hernandez-Pando, R. 1996 The pathogenesis of tuberculosis. *Annual Review of Microbiology* 50: 259-84.

Rook, G.A.W. and Stanford, J.L. 1998 Give us this day our daily germs. *Immunology Today* 19: 113-116.

Rook, G.A.W.. Bahr, G.M. and Stanford, J.L. 1981 The effect of two distinct forms of cell-mediated response to mycobacteria on the protective efficacy of BCG. *Tubercle* 62: 63-68.

Shield, M.J. 1983 The importance of immunologically effective contact with environmental mycobacteria. In C. Ratledge and J. Stanford (eds):*The biology of the mycobacteria*. London: Academic Press, pp. 343-415.

Shield, M.J. and Stanford, J.L. 1982 The epidemiological evaluation in Burma, of the skin test reagent LRA6; a cell-free extract from armadillo-derived *Mycobacterium leprae*. Part 2: Close contacts and non-contacts of bacilliferous leprosy patients. *International Journal of Leprosy* 50: 446-454.

Shield, M.J., Stanford, J.L., Garbajosa, G., Draper, P., Rees, R.J.W. 1982 The epidemiological evaluation in Burma, of the skin test reagent LRA6; a cell-free extract from armadillo-derived *Mycobacterium leprae*. Part 1: Leprosy patients. *International Journal of Leprosy* 50: 436-445.

Stanford, J.L. 1983 Immunologically important constituents of mycobacteria: Antigens. In C. Ratledge and J. Stanford (eds): *The biology of the mycobacteria*. London, Academic Press, pp. 85-127

Stanford, J.L. 1987 Much's granules revisited (editorial). *Tubercle* 68: 241-242.

Stanford, J.L. 1998 Some aspects of immunotherapy with bacterial products in leprosy. In M. Casal(ed): *Clinical mycobacteriology*. Barcelona: Prous Science, pp. 403-408.

Stanford, J.L. and Grange, J.M. 1974 The meaning and structure of species as applied to mycobacteria. *Tubercle* 55: 143-152.

Stanford, J.L. and Paul, R.C. 1973 A preliminary report on some studies of environmental mycobacteria. *Annals de la Société Belge de Médecine Tropicale* 53: 389-393.

Stanford, J.L. and Rook, G.A.W. 1983 Environmental mycobacteria and immunisation with BCG. In C.S.F.Easmon and J.Jeljaszewicz (eds): *Medical microbiology*. London: Academic Press, pp. 43-69.

Stanford, J.L., Shield, M.J. and Rook, G.A.W. 1981 How environmental mycobacteria may predetermine the protective efficacy of BCG. *Tubercle* 62: 55-62.

Stanford J.L., Stanford, C.A., Rook, G.A.W. and Grange J.M. 1994 Immunotherapy for tuberculosis - investigative and practical aspects. *Clinical Immunotherapy* 1: 430-440.

Stanford, J.L., Terencio de las Aguas, J., Torres, P., Gervasioni, B.O. and Ravioli, R. 1987 Studies on the effects of a potential immunotherapeutic agent in leprosy patients. *Health Cooperation Papers* 7 (Proceedings of the Fourth European Leprosy Symposium), pp. 201-206.

Stansby, G., Chan, Y.C., Berwanger, C.S., Shurey, S., Rook, G.A.W. and Stanford, J.L. 2002 Prevention of experimental myointimal hyperplasia by immunomodulation. *European Journal of Vascular and Endovascular Surgery* 23:23-28.

Stewart-Tull, D.E.S. 1983 Immunologically important constituents of mycobacteria: adjuvants. In C. Ratledge and J. Stanford (eds): *The biology of the mycobacteria*. London: Academic Press, pp. 3-84.

Wangoo, A., Sparer, T., Brown, I.N., Snewin, V., Janssen, R., Thole, J., Cook, H.T., Shaw, R.J. and Young, D.B. 2001 Contribution of Th1 and Th2 cells to protection and pathology in experimental models of granulomatous lung disease. *Journal of Immunology* 166: 3432-3439.

Waters, M.F.R. 1989 The chemotherapy of leprosy. In C. Ratledge and J. Stanford (eds): *The biology of the mycobacteria*. London: Academic Press, pp. 405-474.

Waters, M.F., Laing, A.B. and Rees, R.J.W. 1978 Proven primary dapsone resistance in leprosy - a case report. *Leprosy Review* 49: 127-130.

World Health Organisation 1977 *Fifth report of the WHO expert committee on leprosy*. Geneva: WHO Technical Report Series No. 607.

Yakrus, M.A. and Good, R.C. 1990 Geographic distribution, frequency and specimen source of *Mycobacterium avium* complex serotypes isolated from patients with acquired immunodeficiency syndrome. *Journal of Clinical Microbiology* 28: 926-929.

Young, D.B. 1992 Heat-shock proteins: immunity and autoimmunity. *Current Opinions in Immunology* 4: 396-400.

Table 1: Opportunist mycobacteria and the diseases they cause

	Skin	Lymph nodes	Lungs
Slow-growers			
M.avium		+	+
M.intracellulare		+	+
M.kansasii			+
M.malmoense		+	+
M.marinum	+		
M.scrofulaceum		+	+
M.simiae			+
M.szulgai		+	+
M.ulcerans	+		
M.xenopei			+
	Abscesses		
Fast-growers			
M.chelonei	+	+	+
M.fortuitum	+	+	+

Table 2: Some trials of BCG against leprosy and tuberculosis

	Against Leprosy	Against Tuberculosis	
Uganda	80%	United Kingdom	78%
Burma	18%	Georgia/Alabama	14%
South India	33%	South India	0%

Table 3: Vaccines

1) BCG is very effective in some places and in some age groups.

BCG possesses adjuvants that can enhance either Th1 or Th2, largely dependent on previous priming from the environment.

2) BCG+another killed mycobacterium. It should be possible to design a vaccine effective everywhere against both leprosy and tuberculosis.

The killed bacilli provide additional group i antigen and, providing a species with Th1 adjuvant activity is used, it should be more broadly applicable than BCG alone.

3) A killed mycobacterium, other than BCG. Use of killed *M.intracellulare* alone, or killed *M.vaccae* alone, have been successful in preliminary trials.

Table 4: Candidate Immunotherapeutics

1) Killed leprosy bacilli, with or without BCG. Some benefits have been reported, but theoretical disadvantages, production difficulties and effective alternatives rule it out.

2) and 3) Killed ICRC bacilli or killed "*M.w*", both strains of *M.intracellulare*. There is evidence for both of improved clearance of bacilli from multibacillary patients, Lepromin conversion in lepromatous patients and control of ENL reactions.

4) Killed *M.vaccae* possesses unique adjuvant properties, regulating immunity towards Th1. Clearance of bacilli from multibacillary patients is improved, Lepromin and Leprosin A conversion occurs in some lepromatous patients, ENL reactions are controlled and autoimmune damage to eyes and blood vessels in chronic leprosy patients is reduced. Some of these effects can still be detected 10 years after a single injection.

(Data taken from various sources)

Table 5: Conversion to Leprosin A Positivity

Subjects: 61 Spanish patients fully treated for lepromatous leprosy - all were skin test negative to Leprosin A at the start.

They were randomised to receive either:-
Placebo: an intradermal injection of saline, or
Immunotherapy: an intradermal injection of 1mg killed *M.vaccae*

Results of re-testing 1-2 years later:

Saline group	**1/29** skin-test positive **(p=0.001*)**
M.vaccae group	**12/32** skin-test positive

*Fisher's exact test.

(Data taken from Stanford *et al.*, 1987)

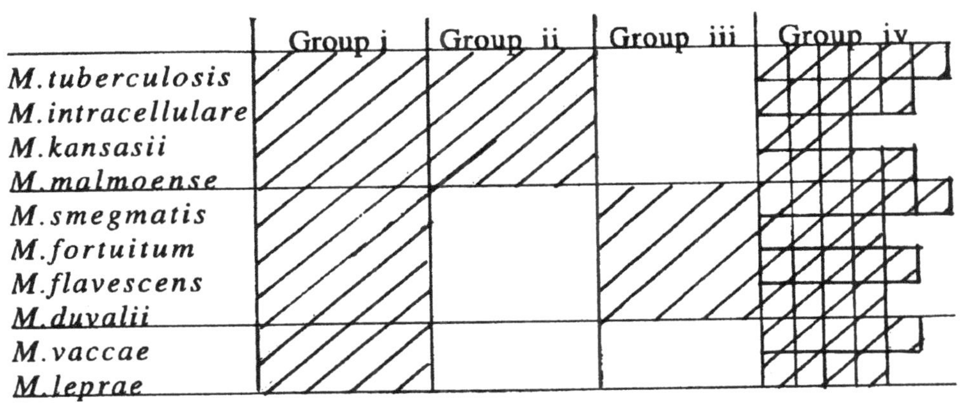

Figure 1: The antigens of mycobacteria (see elsewhere)
Figure showing the antigens demonstrable in sonicates of mycobacteria by precipitation with high titre rabbit antisera. The group i antigens are shared by all species, group ii are limited to slowly growing species and group iii are present in most fast growing species. Group iv antigens are specific to individual mycobacterial species. (Based on Stanford and Grange, 1974).

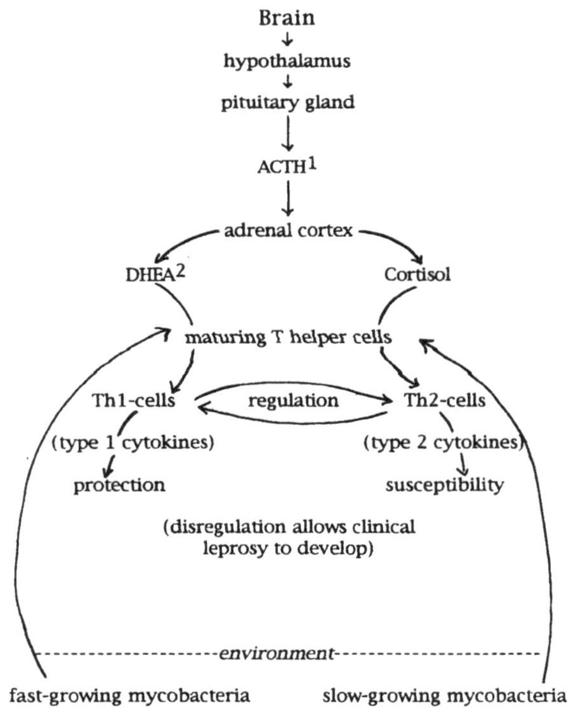

[1]adrenal cortical trophic hormone
[2]dehydroepiandrosterone

Figure 2: Immunity to mycobacteria in health and disease (showing some intrinsic and extrinsic influences on the maturation of T helper cells as part of cellular immunity, and the development of trends towards protective immunity or susceptibility to mycobacterial infectious challenge) – see elsewhere.

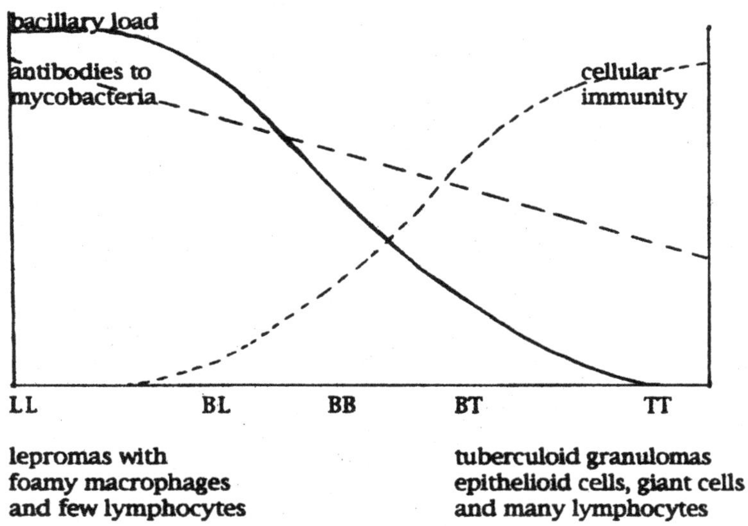

Figure 3: The immunopathological spectrum of leprosy (see elsewhere)
Figure illustrating some aspects of the immunopathological spectrum of leprosy. It can be seen that bacillary load and titres of antibodies to mycobacteria are high in lepromatous disease and low in tuberculoid disease, whereas cellular immunity is high in tuberculoid disease and absent in lepromatous leprosy. Salient histological features of the two ends of the spectrum are indicated below the diagram. (Based on Ridley, 1988).

Differential diagnosis at a leprosy referral clinic in Nepal

R.Butlin and M. Shah
Anandaban Leprosy Hospital
P.O.Box 151
Kathmandu
Nepal

Email: anandaban@mail.com.np

Abstract

This paper describes the experience of case findings of leprosy patients in Nepal over nearly a five year period. Forty per cent of 500 suspected cases of leprosy were genuine. The reasons suggested for such a high percentage of cases being diagnosed as non-leprous are explained and includes confusion of skin lesions for other conditions. It is suggested that better patient education regarding the signs and symptoms of leprosy, and better teaching of health workers (with access to diagnostic tools) would identify genuine cases of leprosy in the future.

Keywords

Leprosy, Nepal, diagnosis

1. Introduction

In Nepal there is an integrated leprosy control program and the main method of case finding over the past 10 years has been voluntary reporting. This relies on adequate awareness of leprosy amongst the public, a willingness to attend a clinic for examination when suggestive signs of leprosy are noticed, and the competence of health workers to make a clinical diagnosis (often without easy access to basic investigative tools such as a slit skin smear). Health workers given responsibility for leprosy diagnosis need to be familiar with the range of other conditions commonly presented by leprosy suspects.This study aims to throw light on the actual situation in recent years as reflected in a specialist leprosy clinic with open access to the public.

2. Methods

This study was carried out retrospectively at a leprosy referral clinic in the Kathmandu valley. The clinic is run by staff of the Anandaban Leprosy Hospital at the district general hospital in Patan City, once a week, on a day when no other out-patient clinics are open. This leprosy referral clinic serves primarily the 19 districts of the Central Development Region of Nepal. There are no charges for leprosy cases or for suspects.

Any person presenting themselves to the clinic for leprosy examination is initially screened by a senior paramedical worker or leprosy supervisor. Definite leprosy cases are then registered for treatment, whereas people in whom no diagnostic signs are found by preliminary physical examination are referred to the medical officer for further assessment as "suspects". Skin smear and biopsy facilities are available as well as simple blood tests and urinalysis.

Over a period of 4 years and 9 months (December 1993 to August 1998), 500 consecutive suspects were seen. Their notes have been reviewed to identify the range of presenting complaints, and common alternative diagnoses in these suspects. Complaints were classified as (a) skin lesions (b) sensory disturbance or motor paralysis (c) other (Appendix 1).

3. Results

Out of 1208 Voluntary Reporters seen for the first time at our clinic, 708 were immediately recognised as new leprosy cases and registered for treatment. These are not considered further. The remaining 500 people comprised of 311 males and 189 females. Amongst them 47 (10%) had originally been seen by another health worker elsewhere who simply advised them where to go for a leprosy examination. Household or family contact with a known leprosy patient was reported by 37 people.

3.1: The patients complaining of skin lesions were the largest group and consisted of 322 people (64.4%). In the majority of cases they mentioned a patch or patches, itching, blisters or papules. The most common diagnoses were fungal infections, eczema/ contact dermatitis, and psoriasis. Together these accounted for 46% of skin lesions and were characterized by some degree of itch. A very large number of other skin conditions occurred infrequently. In 21 cases no diagnosis was recorded apart from "not leprosy".

3.2: Symptoms related to the nervous system, i.e. sensory disturbance or motor paralysis, were represented by 117 people (23.4%). A surprisingly large number (19) of these presented with definite signs of spinal cord injury or spinal nerve root compression. A smaller group (9) gave a clear history of trauma (surgical or accidental) which had caused damage to peripheral or cutaneous nerves. Fifty-three people complained of generalised parasthesias, and in 46 of them no definite cause could be found with the range of investigations we had available.

3.3: "Other" complaints accounted for 61 people (12.2%). Of these 33 were chronic ulcers on hands or feet. These included 23 ulcers associated with Buerger's disease or other peripheral vascular diseases, and five "trophic ulcers" associated with spinal cord lesions or peripheral neuropathy. In addition, a variety of deformities and limb/joint pains and swellings of the extremities were presented to us. Connective tissue disorders and metabolic or endocrine diseases accounted for most of these problems. Two suspects clearly suffered from leprophobia, a delusional conviction of having leprosy in the absence of any objective signs.

3.4: Leprosy cases: in 142 cases the suspect had a skin smear done, and in some of these the requesting doctor had noted that it was "at the patient's request" or "for the patient's peace of mind", rather than for clinical reasons. Five skin smears were positive allowing a definite diagnosis of leprosy to be made in patients who did not have other cardinal signs.

In 23 cases a skin biopsy was performed to confirm a tentative diagnosis or to exclude leprosy. Of these, four were positive for leprosy. Seven people were particularly asked to come back for a follow-up examination - three had unexplained nerve lesions and four had "probably indeterminate leprosy". One of the latter did return much later, when the patch had developed anaesthesia and a diagnosis of tuberculoid leprosy could be made. Six people are currently lost to follow-up. Thus, another 10 leprosy cases were diagnosed amongst the suspects.

4. Discussion

The type of complaints presented by Voluntary Reporters reflects the public's belief concerning early signs and symptoms of leprosy. Amongst our cohort, about 30% came with itching lesions, none of which were due to leprosy, as typically leprosy lesions do not itch. There is a need to emphasize to the public that it is chronic, non-painful, <u>non-itching</u> patches, which are likely to be due to leprosy. Chronicity of lesions <u>does</u> seem to be recognised as a pointer to leprosy, although in this study we could not assess it, not having consistent data on duration of presenting complaint.

In Nepal, public health education material often mentions "parasthesiae" as an early symptom of leprosy. Generalised parasthesiae seems not to be useful as an indicator; it is common and non-specific, may be due to depression, anxiety, malnutrition, or various other disorders. Among 46 people who presented to us with generalised parasthesiae and no abnormal physical signs, nine had a personal reason to fear leprosy because they were contacts. On the other hand, localised parasthesia is, in our experience, usually found to be due to an identifiable disease.

In this series 36% of Voluntary Reporters had a non-dermatological problem, whereas teaching of "differential diagnosis" to health workers usually concentrates on skin lesions which mimic leprosy. Adequate attention needs to be given in leprosy courses to neurological differential diagnosis and to other causes of chronic ulcer.

If we had relied only on clinical examination on one occasion we would have missed 10 cases: five of these, diagnosed by skin smear, were multi-bacillary cases (now called by WHO "cases of consequence"). The four cases diagnosed by biopsy, and the one diagnosed on re-examination after an observation period, were tuberculoid and borderline tuberculoid types which are not a public health hazard, although some of them are at risk of disability if untreated.

5. Conclusion

Over a 4 year 9 month period only 40% of the people voluntarily reporting for leprosy examination at our clinic were genuine leprosy cases, but we would expect to find more cases by voluntary reporting if the public had a greater and more accurate knowledge concerning leprosy. Only if health workers are appropriately taught, and have the facilities to undertake skin smears, will they be able to correctly identify all genuine leprosy cases amongst a group of suspects.

Appendix 1: Range of conditions identified among leprosy suspects:

Group (a) Dermatological conditions

Acne, Allergy, Alopecia (senile and post febrile), Athlete's foot, Birthmark, Bacterial infection, Candida, Chromatomycosis, Contact dermatitis, Congenital ichthyosis, Congenital bullou-s disease, Darrier's disease, Dermatitis herpetiformis, Dermatomycosis, Dysdydrosis, Ecthyma, Eczema, Erythema dyschromia, Erythoderma, Exfoliative dermatitis, Fibrolipoma, Folliculitis, Fungal infection, Giant cell naevus, Granuloma annulare, Impetigo, Keloid, Leucoclastic vasculitis, Lichen planus, Lichen simplex, Lipomatosis, Madarosis of old age, Malnutrition, Melasma, Milaria, Morphoea, Mycetoma, Neurodermatitis, Photodermatitis, Pityriasis alba, Pityriasis rosea, Plantar fasciitis, Pompholyx, Porphyria, Post inflammatory hypo-pigmentation, Psoriasis, P.L.C., P..Y.L.E., P.L.V.E, Purpura, Ringworm, Scabies, Sear, Scleroderma, Sexually transmitted disease, Squamous cell carcinoma, Striae, Sun exposure, Systemic lupus erythematosis, Tinea capits, Tinea cruris, Tinea pedis, Tinea versicolor/ Pityriasis versicolor, Tuberculosis, Urticaria, Vitiligo, wart.

Group (b) Neurological conditions

Bell's palsy, Bernhardt's syndrome, Carpal tunnel syndrome, Cerebrovascular accident, Epilepsy, Guillain Barré syndrome, Hemiparesis, Meralgia parasthetica, paraplegia, peripheral neuropathy, Poliomyelitis, 'Saturday night palsy' (compression of brachial plexus), Sciatica, Scoliosis, Spina bifida, Spinal cord lesion, Spondylosis, Syringomyelia, Trauma to nerve, Trigeminal neuralgia, Ulnar nerve paralysis of unknown cause.

Group (c) Other conditions

Anaemia, Arterial embolism, Arterial thrombosis, Arteritis, Arthritis, Avascular necrosis (of diabetes), Anxiety, Buerger's disease, Diabetes mellitus (acute onset & complications of longstanding disease), Diabetic neuropathy, Erythema nodosum, Filariasis, Hypothyroidism, Lymphadenitis, Lymphoedema, Menopausal symptoms, Osteomyelitis (chronic), Peripheral vascular disease, Raynaud's syndrome, Rheumatoid arthritis, Tuberculosis, Varicose ulcer,Vitamin deficiency.

NB: Some overlap in diagnostic terms reflects usage by different Medical Officers

Acral bone resorption in multibacillary patients. A retrospective clinical study

X. Illarramendi [1], JA.C. Nery, L. M. M. Vieira and E. N. Sarno [2]
Instituto Oswaldo Cruz
Laboratório de Hanseníase
Av. Brazil 4365
Manguinhos
CEP. 21045-900
Rio de Janeiro
RJ Brasil

[1]E-mail: ximenill@gene.dbbm.fiocruz.br
[2]E-mail: esarno@gene.dbbm.fiocruz.br

Abstract

In order to evaluate the prevalence and the risks of bone resorption during multidrug therapy, clinical signs of bone resorption were evaluated in relation to epidemiological as well as to other clinical findings in adult multibacillary leprosy patients. A cohort of 93 newly-diagnosed patients admitted for treatment between 1990-1992 at the Outpatient Clinic of the National Reference Center for Leprosy was surveyed until 1999. Seven (7.53%) patients were found to have clinical acral absorption: two prior to treatment, four during multidrug therapy, and one after release from treatment. Bone resorption was found to be associated with age, degree of disability at diagnosis, the occurrence of three or more reactional episodes and the existence of other deformities. In addition to the well-known effect of neuropathy on bone resorption, as a result of this study it is suggested that immune mediators might also be exercising an effect on inducing osteolysis.

Keywords

Leprosy, bone resorption, impairment, acro-osteolysis.

1. Introduction

Since ancient times, leprosy has been regarded as a contagious, incurable, mutilating disease. Once the Global Strategy for the Elimination of Leprosy was implemented, with its emphasis on early diagnosis and the use of multidrug therapy (MDT), the disease became curable for the first time in history. The results have been so encouraging that there is now real hope that Hansen's disease will be eliminated from most of the world in the very near future. However, there remains a major unresolved concern which is the "human problem" (Srinivisan, 1995), the consequence of disabilities and handicap that prevent the afflicted from enjoying a normal, productive life which greatly limits their chances for self-fulfillment, not to mention the carrying out of everyday activities. Disabilities, and more specifically mutilations, are also responsible for the stigma and discrimination to which leprosy carriers continue to be subjected, not only within their families and communities, but also among health care professionals.

Throughout the world, the prevalence of disabilities in leprosy has been found to vary between 15 and 46.8% (Thappa *et al.*, 1990). According to the World Health Organisation (1998) in 1997, the disability prevalence rate in 32 endemic countries was 1.5 per 100,000 inhabitants. In spite of the sharp decrease in the frequency of disabilities over the last several years, from 9.3% in 1985 to 6.84% in 1997, disabilities still generate a myriad of unfavourable social and economic consequences. However, the current practice of releasing from control patients with impairment, which is especially true with the use of MDT, disguises the problem of disability in a community (Smith, 1992). The most seriously affected are the developing countries in which the disease is endemic. In some cases, the negative impact on local economies is substantial in that the disabled part of the population is generally unproductive and, consequently, has become a burden to society instead of a participant in the generation of wealth.

Brazil's share of leprosy cases in the Americas as a whole is an astounding 80%. It is second only to India in the number of cases (72,953) with a 0.83 per 10,000 detection rate of new cases among children under 15 years of age. This rate clearly constitutes high endemicity. The percentage of deformities produced by leprosy was reduced by 53% (up to 7%) during the period covering 1987-1997. However, 23,356 new cases with Grade of Disability (GD) II and III were diagnosed during that same period (Ministério da Saúde, 1998). Likewise, in clinical practice, the progression of deformities in treated patients, as well as the diagnosis of the disease in individuals with mutilations, continues to be frequently observed.

Bone and joint lesions have been classified according to pathogenic processes as: specific erosive and destructive changes caused by lepra reaction and *M. leprae* granulomas, non-specific changes as an end result of sensory loss, and disuse osteoporosis. Lesions that are non-specific to leprosy can be found in other diseases with neural involvement leading to loss of sensation and its consequences (Møller-Christensen, 1961, Paterson and Job, 1964, Carayon and Dharmendra, 1978).

Earlier studies of bone alterations in leprosy included patients quite different from those seen today. Quenau (1969) wrote about the difficulties in calculating the frequency of bone alterations in leprosy due to the variety of populations studied, the variability in terms of treatment initiation and the hardships involved in diagnosing new patients. Generally speaking, the prevalence of bone and articular lesions in leprosy has been reported to vary from 15 to 30%, although higher prevalence rates ranging from 60 to 80% have been reported in leprosaria (Carayon and Languillon, 1985). Paterson (1961) has reported incidence rates of 14.3% for specific leprous osteitis, 45% for non-specific bone changes and 10% for disuse osteoporosis in patients confined to a leprosarium, whereas only 2 to 3% were seen to have severe destructive alterations leading to deformity. More recently, Thappa *et al.* (1992) observed that 82.9% of the patients with disabilities had radiological bone changes. Quenau (1969), unlike other authors (Riordan, 1960), observed increased bone absorption in patients with multibacillary leprosy.

The development of bone loss in leprosy is a complex event and certainly involves different pathogenic processes. In spite of the known direct effect of *Mycobacterium leprae* on bone, some authors consider that sensory loss, together with successive trauma and superimposed infections, are the main cause of bone resorption in leprosy (Mallac, 1966, Carayon and Languillon, 1985). At the same time, others believe that the alterations have a neurovascular origin (Erickson and Johansen, 1948, Paterson, 1961, Paterson and Job, 1964, Skinses, 1982). More recently, both nervous pathology and bone damage have been attributed to inflammatory mediators released during immune episodes occurring during the disease (Sarno and Sampaio, 1996, Sampaio and Sarno, 1998).

In order to assess the risk of developing bone resorption, clinical signs of bone resorption in relation to epidemiological and other clinical findings were evaluated in multibacillary patients that had received at least 24 doses of MDT.

2. Patients and method

The patients were selected from the Souza Araújo Outpatient Clinic, National Reference Center for Leprosy, Department of Tropical Medicine of the Oswaldo Cruz Institute. All were residents of the State of Rio de Janeiro, which has a high leprosy prevalence rate (5.7 per 10,000 inhabitants) (Ministério da Saúde, 1998). The centre receives between 5000 and 6000 attendances per year and diagnoses a mean of 250 new cases of leprosy annually. Around 100 patients remain for treatment at the Center, nearly half of whom are MB. The rate of patient noncompliance is less than 2%.

The clinical records of 105 multibacillary patients (i.e. with positive slit-skin smear and/or with five or more skin lesions), over 15 years of age, admitted for treatment from January 1990 to December 1992 and under surveillance until the beginning of 1999, were evaluated. In view of the fact that the principal cause of non-specific bone changes is anaesthesia, patients with other possible causes of neuropathy but with a similar pattern of occurrence and progression of sensory involvement (Kemp *et al.*, 1986), namely diabetes, chronic alcoholism, and peripheral arteriopathy, were excluded.

The patients were evaluated by a dermatologist and a physiotherapist at the moment of diagnosis, when released from treatment, and annually during follow-up. They were also submitted to the routine proceedings of the Clinic i.e. neurological evaluation, slit-skin smears of six areas (ear lobes, elbows, skin lesion, and one knee), lepromin test, skin biopsy, radiographs of thorax and extremities, blood, urine and faeces analyses, and serum biochemistry. On their visit to the Center, all the patients were interviewed by a social worker and asked, among other things, about the approximate date they first perceived symptoms of the disease. The patients were treated with at least 24 monthly doses of 600 mg of rifampicin and 300 mg of clofazimine, together with 100 mg of dapsone and 50 mg clofazimine per day.

Patients were classified according to the Ridley and Jopling (1966) criteria as follows: 45 (48.4%) Borderline-Lepromatous (BL), 17 (18.3%) Mid-Borderline (BB), and 31 (33.3%) Lepromatous-Lepromatous (LL), based on the results of dermato-neurologic evaluation, bacilloscopy, Mitsuda reaction and biopsy of lesion. The GD was determined by an experienced physiotherapist at the beginning and end of treatment according to the WHO (1998) disability grading.

Bone resorption was clinically diagnosed according to the observation of loss of normal anatomical structure. It was classified as "discrete" if there was only loss of finger/toe tips or as "intense" if the loss exceeded one fifth of the segment (Ministério da Saúde, 1997).

Reactional states were diagnosed according to clinical and histopathological criteria in reversal reaction (RR), *erythema nodosum leprosum* (ENL), *erythema multiforme* (EM), neuritis and others (reactional hand, epididymo-orchitis, generalized oedema). In general, the drugs most frequently used to treat reactions were Thalidomide (300 to 100 mg/day) and/or Prednisone (1mg/Kg/day).

Data were analysed by means of a standard statistical program (EPI INFO 6.04 from the Center for Diseases Control, Atlanta, GA., U.S.A.). Simple frequencies and risk ratios were evaluated. An association was established with T and X^2 tests with a Confidence limit (CL) of 95%.

3. Results

From a total of 105 clinical records, 93 patients were included in the study. Clinical signs of acro-osteolysis were reported in seven individuals (7.53%). Bone resorption was observed in two patients at diagnosis, in three patients during specific treatment, in one patient during follow-up after treatment and in one both during and after MDT (Figure 1).

Alterations were reported in the hands and feet both together and separately. In three patients both the upper and lower extremities were absorbed, three other individuals had signs of bone resorption in one or both feet, and one case had absorption in only one hand. Five patients had discrete resorption whereas only two developed intense acro-osteolysis.

Despite the fact that the seven patients with acro-osteolysis were male, no statistical association with sex was observed due to the higher proportion of males (72) over females (21) in the sample. The majority of the patients (63.4%) were born in the State of Rio de Janeiro. Almost half of the patients were white (48% of the patients without resorption and 43% of those with absorption). Only 47% of the patients without acro-osteolysis, but most of the patients with absorption (71%), had a monthly income of less than 75 US$. More than half of the patients (71% with absorption and 52.3% without) had not completed the eight years of basic education. Although a higher proportion of the patients with absorption was LL and half of the patients without this impairment where BL, this difference was not found to be significant (Figure 1).

At diagnosis, patient ages varied between 16 and 76 years. However, clinical bone resorption was observed in patients who were diagnosed while under 45 years of age. Patient mean age when absorption was diagnosed was 27.9 (±6.32) years, while the mean age of the group without absorption was 42 (±16.73), which was statistically significant (2-tailed T-test $p<0.05$) (Figure 3). Although the average length of time taken for the disease to evolve, based on patient recall, was more than double in those with signs of resorption (2.9 years) in contrast to the group with no resorption (1.4 years), no statistical significance (Kruskal-Wallis $H=2.9$ $p=0.95$) was observed.

An association was seen between bone resorption and grade of disability (GD) both at diagnosis and upon release from treatment. No variation was seen in the GD of the group with absorption, while 25% of the patients without absorption had improvement (Figure 3). Among the deformities found to be associated with resorption in the hands and feet were motile clawing in both upper and lower extremities, ulcers in the former, and foot drop (Table 1).

None of the other clinical aspects evaluated, such as acrocyanosis, oedema, or the various neurological signs were associated with bone resorption. A reduction in muscular force was recorded in 40% of the patients but it resulted in no risk for bone resorption (RR=2.02, 95%CL 0.48-8.50). Loss of protective sensation and nerve thickening in upper and/or lower extremities were present in 70% of the patients (RR=1.08). Risk ratio did not differ in patients when sensory loss was analysed in the presence of preserved muscular force.
Osteomyelitis, secondary to chronic plantar ulcer, was reported in four cases and burn or electrical shock in two others. The one patient who developed absorption after MDT had recurrent and prolonged reactional episodes. Treatment duration was similar for both groups, an average of 2.2 years for patients with no absorption and 2.3 for those with acro-osteolysis.

Seventy-one percent of the patients had at least one reaction during MDT and/or at follow-up, but it was not found to be associated with absorption (Figure 4). ENL (51% of the patients) and RR (48%) were observed in a similar proportion in both groups. When considering the number of episodes during MDT and surveillance, association was observed with having three or more episodes ($X^2=4.02$, $df=1$, $p<0.05$) (Table 2).

4. Discussion

Leprosy is a disease that stands out among other contagious diseases, because it continues to be so difficult to accurately define. There is also no single reliable diagnostic technique available or a specific culture method for the causative agent. It is, therefore, almost impossible to evaluate the length of time the patient has been exposed to the mycobacteria, and the precise moment when the infection or the disease began (Van Beers *et al.*, 1996). As a consequence, when analyzing the length of time taken for the disease to evolve, it must be considered as a relative and only approximate variable, which depends on factors intrinsic to both the agent and the patient. Although no association between absorption and the duration of the disease was observed in this study, there have been reports of bone changes, impairments, disabilities and deformities being associated with a rather long period of evolution and a delay in diagnosis and treatment (Noordeen and Srinivisan, 1966, Thappa *et al.*, 1992, Rao *et al.*, 1994, Meima *et al.*, 1999). As a parameter, the duration of the disease may not be affecting this sample for various reasons. The patients are multibacillary and their low literacy may favour a low awareness of symptoms and therefore, a delay in perceiving the first symptoms of the disease. In addition, there is active case detection among patient contacts at the Center, which reduces the delay of diagnosis.

As mentioned above, active case detection, and early treatment and the use of MDT, has resulted in a reduction of the global prevalence of disabilities and deformities (Rao *et al.*, 1994). The prevalence of 7.53% of clinical bone resorption obtained in this study is not different from the general prevalence of leprosy disabilities in Brazil, which is 7% (Ministério de Saúde, 1998).

Several authors have reported the development of deformities, neural function impairment and bone alterations during MDT and surveillance after release from treatment (Hernandez *et al.,* 1982, Skinsnes, 1982, Rao *et al.*, 1994, Van Brakel and Khawas, 1994, de Rijk *et al.*, 1994, Carpintero *et al.*, 1998, Solomon *et al.*, 1998). In this sample, both the development and progression of bone resorption during and after treatment were observed. In the presence of neuropathy, vascular denervation and slowly progressive associated vascular sclerosis, and bone resorption, is ongoing despite the arrest and cure of the disease (Skinsnes, 1982). As the bone undergoes absorption, the soft tissues gradually shorten to adapt to the shortening of the supporting structure (Riordan, 1960). Accordingly, absorption is a very slow process, which is clinically evidenced after a long evaluation period.

The incidence of specific bone changes has been shown not to vary in the different forms of leprosy (Paterson, 1961, Thappa *et al.*, 1992). In this sample, of mostly non-specific bone resorption, no statistically significant variation was likewise observed. The risk of grade II impairment amongst impaired new cases has been found to be lowest among individuals of ages 45 and above (Meima *et al.,* 1999). Similarly, in this sample, clinical bone resorption was found in patients under 45 years of age. However, other authors (Paterson, 1961) have found that ageing increased the incidence of radiological non-specific and osteoporotic bone changes.

Deformities are produced in the hands and feet by a variety of different mechanisms. Alteration of pain perception contributes to the development of ulcers in both upper and lower extremities. Carayon *et al.* (1969) concluded that the main cause of mutilation in the hands is the persistence of motor activity, together with cutaneous anaesthesia and slow tissue metabolism of the segment. Sensory loss alone does not cause bone changes, but nerve lesions together with trauma and sepsis may lead to absorption. Bone lesions result, in the presence of sensory loss (sometimes even of subclinical degree) and of muscular imbalance, from unperceived minor trauma (Paterson and Job, 1964). Hence, successive trauma over areas with no protective sensation delay scar formation and allow bacteria to penetrate into deeper regions. Release of the necrotic areas produced during osteomyelitis leads to shortening of the segment. In the present study, these events led to absorption in four of the patients and has been demonstrated in other studies, in which a higher incidence of neurotrophic non-specific bone lesions was observed. (Hernandez *et al.*, 1982).

The most striking and permanently disabling bone changes can occur in the fingers during severe lepra reactions (Møller-Christensen, 1961). Paterson (1961) studied the pathogenesis of the bone destructive and erosive changes that take place in lepra reaction. The reactional arthritis in the hands and osteo-arthritis in the feet produce alterations at the site of penetration of nutrient arteries in the metaphysis. For other authors, during reactions, damage is produced by both the reduction of the innervation to the bones and the deficit of blood flow due to endarteritis of the nutrient vessels (Jopling and McDougall, 1988). Karat *et al.* (1968) reported that generalised transient osteoporosis takes place during the reactional states. In opposition to post-traumatic osteoporosis caused by increased activity of the sympathetic nervous system (Leriche, 1916) other mechanisms could explain the osteoporosis described in leprosy, as in the latter autonomic nerve impairment is common. The presence of immune mediators in bone tissue is likely to be involved in the physiopathology of bone changes in leprosy. Marks (1979) observed in leprosy patients, increased localised resorption of bone in some areas associated with osteoclast, osteolytic osteocytes and monocytic infiltration, and viewed osteolysis as an acceleration of the normal process. Marks suggested that bone loss might be related to the presence of mononuclear cells, and the skeletal effects of the disease could be associated more closely with the known immune changes that normally occur in leprosy. Several authors have recently described increased levels of tumor necrosis factor and interleukin-1 during reactional states, suggesting they could produce alteration of bone metabolism. One citocine was identified as the osteoclast activation factor (OAF) that acts synergistically with parathyroid hormone in bone remodelling (Stashenko *et al.*, 1987). Hence, regardless of the type of reaction, either recurrent episodes and/or a prolonged reactional state could allow the immune mediators produced to act on bone tissue to stimulate osteolysis.

5. Conclusions

Patient surveillance after release from MDT remains a necessary procedure specifically in individuals with neuropathy and recurrent or persistent reactions, in view of the fact that bone resorption continues to be observed when the patient is considered to be bacteriologically cured. Absorption in young adult patients will most probably lead to the development of some form of handicap at the most productive age of their lives.

Interestingly, in this study, a number of the risks factors linked to bone resorption that appear in the literature were not found to have statistical significance. The only exception was the presence of ulcers, previously described as an important cause of resorption. Deformities such as foot drop and claw hands resulting from nerve paralysis are acute events during the reactions. After the appropriate treatment, normal nerve function is usually restored (Naafs *et al.*, 1979, Sugumaran, 1997).

Association of clinical bone resorption with the occurrence of three or more reactional states during treatment might reflect the role immune mediators could be playing in acro-osteolysis. The irregularity in leprosy treatment that could be expected in patients with acro-osteolysis was not confirmed in this study in that most of the patients both with and without signs of bone resorption terminated their specifically-prescribed treatment within the allocated period of time. It is clear that the more fully acrodystrophy in leprosy is understood, the more effectively will the disabilities that so often follow impairment be prevented from occurring in the first place.

Acknowledgements

We are grateful to Ms. Judith Grevan for the correction of the manuscript.

References

Carayon, A. and Dharmendra 1978. Bone and joint changes in leprosy. In Dharmendra (ed): *Leprosy.* Bombay, Samant and Company, pp.872-885.

Carayon, A. and Languillon, J. 1985 Lésions ostéo-articulaires dans la lèpre (expérience 1960-1982). *Acta Leprol.* 3 (2): 133-153.

Carayon, A., Languillon, J. and Piquard, B. 1969 Recherches sur le mécanisme des mutilations et ostéolyses des extrémités du lepreux. *Bull. Soc Méd. Afr. Noire Lgue fr.* 14 (3):521-531.

Carpintero. P., Logroño, C., Carreto, A., Carrascal, A., and Lluch, C. 1998 Progression of bone lesions in cured leprosy patients. *Acta Leprol.* 11(2):21-4.

De Rijk, A.J., Gabre, S., Bypass, P. and Berhany, T. 1994 Field Evaluation of WHO-MDT of fixed duration at ALERT, Ethiopia: the AMFES project II. Reaction and neuritis during and after MDT in PB and MB leprosy patients. *Leprosy Review* 65:320-332.

Erickson, P.T. and Johansen, F.A. 1948 Bone changes in leprosy under sulfone therapy. *Memoria del V Congresso Internacional de la Lepra,* República de Cuba, 3rd to 11th of April 1948, pp. 193-206.

Hernandez, A.M., Fraguela, J.V., Fernandez, B.G., Sanabria, N.J., Sacerio, G.C. 1982 Hallazgos radiológicos en manos y piés clínicamente normales en enfermos de lepra. Informe preliminar. *Rev Cub Med Trop* 34:54-64.

Jopling, W.H. and McDougall, A.C. 1988 *Handbook of leprosy.* Fourth edition. London: William Heinemann.

Karat, S., Karat, A. and Foster, R. 1968 Radiological changes in bones of the limbs in leprosy. *Leprosy Review* 39:147-169.

Kemp, S.S., Dalinka, M.K. and Schumacher, H.R. 1986 Acro-osteolysis. Ethiologic and radiological considerations. *J. American Medical Association* 255:2058-2061

Leriche, R. 1916 Causalgia considered as an inflammation of the sympathetic nerves and its treatment by stripping and excising the periarterial neural plexus. *Presse Medical* 24:153.

Mallac, M. J. 1966 Onset and pattern of deformity in leprosy. *Leprosy Review* 37 (2): 71-91.

Marks, S.C. 1979 The cellular basis for extremity bone loss in leprosy. *Int. J Leprosy and Other Mycobacterial Diseases* 47:26-32.

Meima, A., Saunderson, P.R., Gebre, S., Desta, K., Van Dortmarssen, G.J. and Habbema, J.D.F. 1999 Factors associated with impairments in new leprosy patients: the AMFES cohort. *Leprosy Review* 70:183-203.

Ministério da Saúde 1997 *Manual de Prevenção de Incapacidades.* Brasilía: Fundação Nacional da Saúde.

Ministério da Saúde 1998 *Hanseníase no Brasil.* Progressos e Dificuldades em Relação à Eliminação. SPS/DGSP, CENEPI/FNS.

Møller-Christensen, V. 1961 *Bone Changes in Leprosy.* Copenhagen: Munkgaard.

Naafs, B., Pearson, J.M.H., Wheate, H.W. 1979 Reversal reactions: the prevention of nerve damage: comparison of short- and long-term steroid treatment. *Int. J. Leprosy and Other Mycobacterial Diseases* 47:7-12.

Noorden, S.K. and Srinivasan, H. 1966 Epidemiology of disability in leprosy. 1: A general study of disability among male patients above 15 years of age. *Int. J. Leprosy* 34:159-169

Paterson, D.E. 1961 Bone changes in leprosy, their incidence, progress, prevention and arrest. *Int. J Leprosy and Other Mycobacterial Diseases* 29:393-421.

Paterson, D.E. and Job, C.K. 1964 Bone changes and absorption in leprosy. A radiological, pathological and clinical study. In R.G.Cochrane and T.F. Davey (eds): *Leprosy in Theory and Practice.* Second edition. Baltimore: Williams and Wilkins, pp. 425-446.

Quenau, P. 1969 Lésions osseuses distales au cours de la lèpre á propos de 19 observations personelles. *Acta Lepr.* 3(2):113-141

Rao, P.S., Subramanian, M. and Subramanian, G. 1994 Deformity incidence in leprosy patients treated with multidrug therapy. *Indian J. Leprosy* 66:449-454.

Ridley, D.S. and Jopling, W.K. 1966 A classification of leprosy for research purposes. *Leprosy Review* 34: 119-128.

Riordan, D.C. 1960 The hand in leprosy. *J. Bone and Joint Surgery* 42-A:683-690.

Sampaio, E.P. and Sarno, E.N. 1998 Expression and cytokine secretion in the states of immune reactivation in leprosy. *Braz. J. Med. Biol. Res.* 31(1):69-76 .

Sarno, E.N. and Sampaio, E.P. 1996 The role of inflammatory cytokines in the tissue injury of leprosy. *Int. J. Leprosy and Other Mycobacterial Diseases* 64 (Suppl):S69-73.

Skinsnes, O.K. 1982 Infectious granulomas: exposit from the Leprosy Model. *Ann. Rev. Med.* 33:47-67.

Smith, W.C.S. 1992 The epidemiology of disability in leprosy including risk factors. *Leprosy Review* 63 (Suppl.): 23-30.

Solomon, S., Kurian, N., Ramadas, P. and Rao, P.S. 1998 Incidence of nerve damage in leprosy patients treated with MDT. *Int..J.Leprosy and Other Mycobacterial Diseases* 66(4):451-6.

Srinivasan, H. 1995 Deformities and disabilities - the unfinished agenda in leprosy work. *Leprosy Review* 66(3):193-200.

Stashenko, P., Dewhirst, F.E., Peros, W.J., Kent, R.L. and Ago, J.M. 1987 Synergistic interactions between interleukin 1, tumor necrosis factor and lymphotoxin in bone resorption. *J. Immunology*, 138:1464-1468.

Sugumarian, S.T. 1997 Steroid therapy for paralytic deformities in Leprosy. *Int. J. Leprosy and Other Mycobacterial Diseases* 65:337-344.

Thappa, D.M., Sharma, V.K. and Kaur, S. 1990 Disability index of hands and feet in patients attending an urban leprosy clinic. *Indian J. Leprosy* 62:328-337.

Thappa, D.M., Sharma, V.K., Kaur, S. and Suri, S. 1992 Radiological changes in hands and feet in disabled leprosy patients: a clinico-radiological correlation. *Indian J Leprosy* 64:58-67.

Thappa, D.M., Sharma, V.K., Kaur, S. and Suri, S. 1992 Radiological changes in hands and feet in disabled leprosy patients: a clinico-radiological correlation. *Ind. J. Leprosy* 64:58-67.

Van Beers, S.M., De Wit, M.Y.L. and Klaster, P.R. 1996 The epidemiology of *Mycobacterium leprae*: Recent insight. *FEMS (Federation of European Microbiology Societies) Letters* 130:221-230.

Van Brakel, W.H. and Khawas, I.B. 1994 Nerve damage in leprosy: an epidemiological and clinical study of 396 patients in West Nepal - Part 1. Definitions, methods and frequencies. *Leprosy Review* 65:204-221.

World Health Organization (1998) Elimination of leprosy as a public health problem. *Weekly Epidemiologic Report*, 73(40):308-311.

Table 1: Epidemiological, clinical, and laboratory data from seven patients with bone resorption (BR)

Rec. nº	Age at BR dx	Sex	Bone resorption		Clinical form	AFB Index		Other def.	Years to dx.	Years MDT	Chronic foot ulcer	EC	Edema	Sensorial alteration	Muscular force reduction	Type of Reactional episodes	Grade of disability		
			hands	feet		Initial	Final										At dx	MDT	f/up
2545	28	m	no	unilat	BB	1,16	0,00	both	0,2	3,0	yes	hands	hands	both	both	I, N	2	2	2
2493	48	m	unilat	no	LL	4,50	2,16	hands	4,5	2,3	no	both	both	both	both	No	2	2	2
2593	36	m	unilat	no	BL	2,83	0,00	both	4,1	2,3	no	both	feet	both	both	II	2	2	2
2609	23	m	no	unilat	BL	4,50	1,16	feet	0,8	2,2	no	hands	both	both	both	II, N	2	2	2
2328	34	m	no	unilat	LL	3,00	1,66	feet	0,5	2,3	yes	both	feet	both	no	No	1	1	2
2395	16	m	no	unilat	BL	3,50	3,00	feet	2,7	2,4	yes	hands	both	feet	feet	No	2	2	2
2351	23	m	no	unilat	LL	2,80	2,16	both	3,1	2,5	yes	both	both	both	hands	II	2	2	2

Rec=record
dx.= diagnosis
m=male
f=female
BB=mid-borderline
BL=borderline lepromatous
LL=lepromatous
MDT=multidrug therapy
AFB = acid fast bacilli
def=deformities
unilat=unilateral
bilat = bilateral
I = reversal reaction
N=neuritis
II = erythema nodosum leprosum and erythema polymorph
EC=palm-plantar erythro-cyanosis,
f/up= follow-up.

Table 2: Other deformities associated with bone resorption

	Other Deformities	With Resorption	No Resorption	Total
HANDS	ULCERS	2	6	8
	CLAW	2	4	6
	ABSENT	2	82	84
FEET	ULCERS	3	6	9
	CLAW	1	3	4
	FOOT DROP	3	0	3
	ABSENT	1	79	80

*NB. Some patients had more than one deformity.

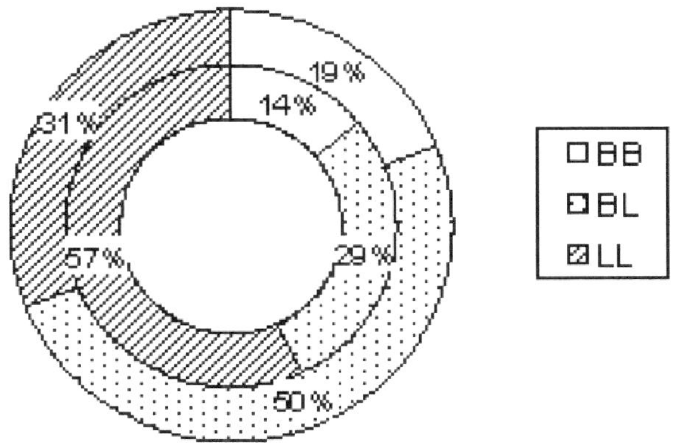

Figure 1: Bone resorption according to the clinical form of leprosy. Although more than 50% of the patients with absorption (inner circle) had lepromatous leprosy, no statistically significant association was observed between bone resorption and clinical form of leprosy.

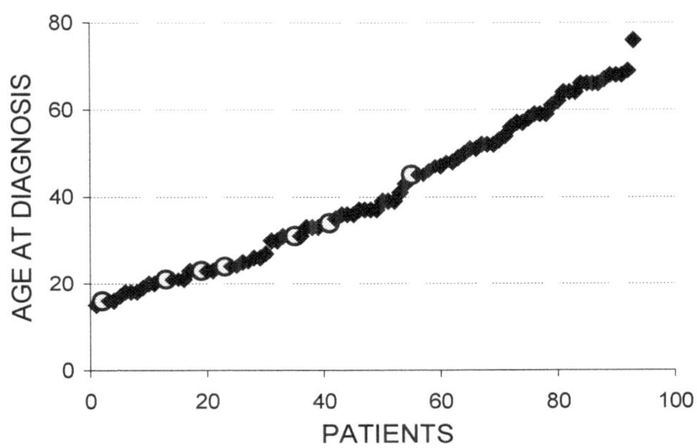

Figure 2: Age distribution of the patients. Patients with absorption had a mean age of 30 years and were all under 45 (open circles) when this impairment was diagnosed.

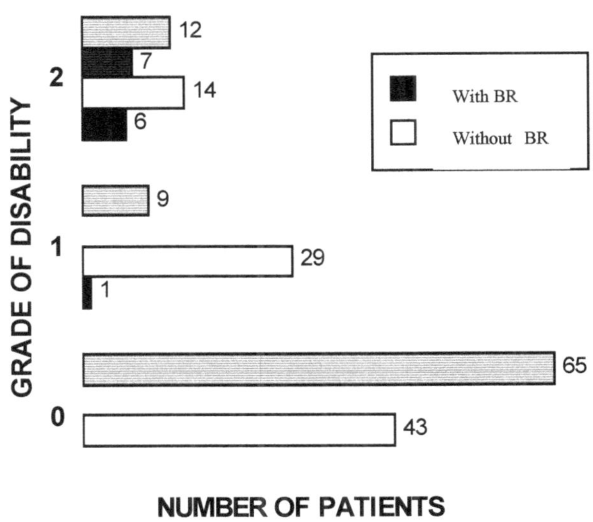

Figure 3: Progression of the grade of disability (GD) before (black bars) and after (grey bars) multidrug therapy. Improvement in the GD was only reported in the group of patients without absorption.

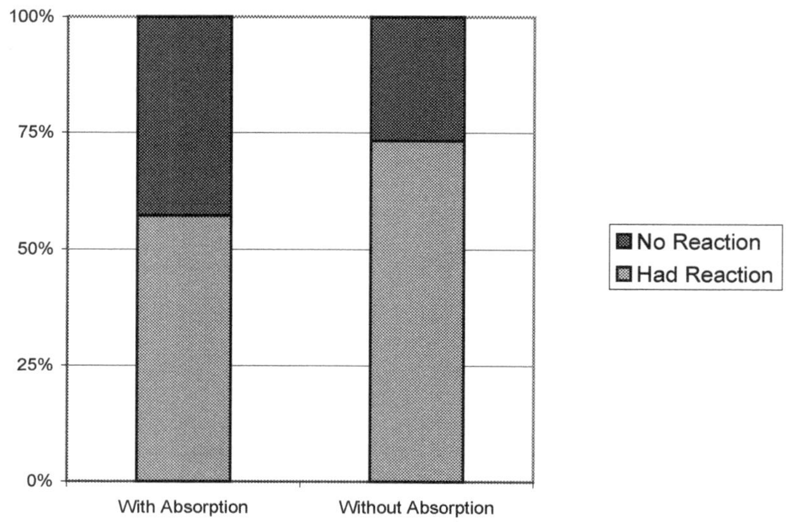

Figure 4: Occurrence of reactions in multibacillary patients. Approximately the same proportion of patients both with and without absorption had at least one reactional episode.

Progression of nerve damage after leprosy treatment

M.R. Jardim[1,2] , M. Skacel[1,2], S. Balassiano[2] , A.M. Sales[1] , M. Ferreira[1], E. N. Sarno[1]
[1]Leprosy Laboratory
Laboratório de Hanseníase
Department of Tropical Medicine
Oswaldo Cruz Institute
Av. Brazil 4365
Manguinhos
CEP: 21045-900
Rio de Janeiro
R.J. Brazil

[2]Departamento de Neurologia
Faculdade de Medicina
Universidade Estadual do
Rio de Janeiro
Brasil

Email: mmjardim@aonda.net
esarno@gene.dbbn.Fiocruz.br

Abstract

Neurological deficits may appear in around 25% of leprosy patients during the course of the disease. They are frequently related to neuritis, which may occur before, during and after specific leprosy treatment. The aim of this study was to evaluate 33 patients with persistent neurological impairments after one year of antileprosy treatment. The patients were from 16 to 72 years of age, and the great majority were males (72.7%). Neurological symptoms appeared prior to dermatological lesions in 43% of the patients. At first evaluation, neurological impairments were observed in 60% of the patients. Electrophysiological study carried out on 18 patients at diagnosis showed an axonal pattern. All of them had worsened when re-examined one year after discharge. The neurophysiological evaluation of those patients seen to evidence a mixed (demyelinating and axonal) pattern showed a higher rate of muscle weakness than those with an axonal pattern alone. This study suggests that, even in the absence of dermatological lesions, neurological impairments often remain for a long period of time after terminating multidrug therapy (MDT).

Keywords

Leprosy, peripheral neuropathy, electrophysiology, multibacillary

1. Introduction

Leprosy is one of the oldest chronic infectious diseases known to humans and is still subject to prejudice due to ignorance of the illness on the part not only of the general public but also of health care providers. Leprosy continues to be the major cause of neuropathy worldwide and deformity its overriding feature (Nations *et al.* 1998). In fact, more than one-fourth of the patients develop some form of disability, nearly half of whom become severely disabled for life (Scollard *et al.* 1994). At present, there are several million people in the world who are disabled due to leprosy (Tzourio *et al.* 1992).

It has been demonstrated that neuritis correlates significantly with disabilities both during and after treatment, and that neuritis frequently occurs in multibacillary patients (MB) (Pfalzgradd 1989). However, neurological impairments occur in practically all clinical forms of the disease. Nerve damage is known to develop in treated as well as in untreated patients, and even in those individuals who have completed antimycobacterial therapy (MDT) (Naafs 1996). After MDT has ended, the neurological impairments may remain stable or worsen, sometimes persisting for many years, particularly in MB patients. The patients that suffer from persistent nerve damage often become more and more disabled as time goes by (Britton and Lockwood 1997).

The objective of this study was to analyze the evolution of neurological impairments in a group of MB patients in an attempt to trace the development of continuing leprosy neuropathies in these patients. All of them presented neurological complaints one year or more after release from MDT, while the majority had nerve damage that became progressively more severe.

2. Patients and methods (study design)

A retrospective analysis of 33 multibacillary leprosy patients treated between 1989 and 1997 at the Leprosy Outpatient Clinic of the Oswaldo Cruz Foundation in Rio de Janeiro, R.J., Brazil, was carried out in an effort to trace their continuing leprosy neuropathies. The patients selected for this study had been subject to neurological examinations both at the outset and at the end of treatment, and it was found that their neurological impairments persisted after a minimum 12-month period had elapsed since the end of MDT. Moreover, at no time during the follow-up period did these particular MB patients show any clinical signs of relapse.

All the above-mentioned MB patients received the standard 24-dose MDT recommended by the World Heath Organization along with corticosteroids (1mg/weight) in the presence of reversal reactions (RR) and neuritis. The amount of corticosteroids was tapered off every fifteen days. Likewise, in the presence of *erythema nodosum leprosum* (ENL), thalidomide therapy (300mg/d) was administered until improvements were observed. The resulting clinical, neurophysiological, and epidemiological data can be seen in Table 1.

The neurological examinations undertaken on admission were carried out at the beginning of treatment and will be referred to as the "initial exam". In this text, the term "last exam" refers to the patients' last examination taken into account for the purposes of this study. All 33 patients have subsequently been subject to additional neurological exams, the results of which could not be included. At the initial examination, all of the patients were extensively questioned by a practicing neurologist as to the existence of any neurological complaints prior to the appearance of skin lesions. The neurological examination focused on the peripheral nerves, which were examined by palpation. Special attention was given to nerve consistency and the presence of nodules. Nerve pain was determined either when it occurred spontaneously or upon palpation. The palms and feet were examined for the presence of cyanosis, or erythroderma. Tactile sensation was tested by using anaesthesiometry, which is composed of six thin nylon filaments (the Semmes-Weinstein technique), while temperature sensation was tested by the use of cold or warm metal objects. A device was especially constructed from a compass (originally used for drawing, etc.) to enable the examiner to prick, but not perforate, the patient's skin in order to ascertain degree of pain sensation. Voluntary muscle strength was evaluated according to the Medical Research Council's (MRC) scale on "Aids to the examination of the peripheral nervous system", and muscle weakness was graded from 0 to 5.

Electrophysiological studies (E.S.) were alll undertaken by the same specialist on the Nihon-Koden – Neuropack 2. Room temperature was maintained at 20 to 23° C and skin temperature at 30 to 32° C. Motor and sensory nerve conduction was investigated by means of surface electrodes, and the electromyographies were performed with a coaxial concentric needle.

All the patients were subject to routine sensory nerve conduction studies on the median, radial, ulnar, and sural nerves. Digital filtering was between 20Hz to 3Hz, and sweep speed was constant at 1ms/div. Latency was calculated in milliseconds (ms) from the first stimulus artifact to the appearance of the first negative deflection. Amplitude was calculated in microvolts (mV) from the baseline to the first negative peak. Antidromic technique stimulation was used for the sural nerve studies and microvolts (mV) from the baseline to the first negative peak. Antidromic technique stimulation was applied to the sural nerve studies, and the orthodromic stimulation technique was used for the median, radial, and ulnar nerve studies. The conduction velocities of the median, radial, and ulnar nerves were performed by stimulation with ring electrodes around the first, third, and fifth fingers. The recording was obtained at the wrist with surface electrodes. For the sural nerve study, stimulation was applied to the lower third of the leg over the posterior aspect, and the recording electrode was placed around the lateral malleolus.

The motor nerve conduction study was performed on the median, ulnar, and common peroneal nerves. Digital filtering was maintained at 2OHz to 3Hz, and sweep speed was 5 ms/div. Latency was calculated from the first stimulus artefact to appearance of the first negative deflection (ms). Amplitude was calculated in millivolts (mV) from the baseline to the first negative peak. When the median nerve was studied, the active lead was placed at the thenar region, and the neutral lead was placed close to the metacarpophalangeal joint of the thumb. Stimulation was applied at the wrist, elbow, and the ERB point. For the ulnar nerve, the points of stimulation included: the wrist, above and below the elbow, the middle third of the arm, and the ERB point. The active lead was placed at the hypothenar region and the indifferent lead at the metacarpophalangeal joint of the fifth finger. The fibular nerve was stimulated at the ankle and below and above the head of the fibula. The active lead was placed over the *extensor digitorum brevis* muscle. The electromyography was performed with a coaxial concentric needle at the *abductor pollicis brevis, abductor digiti minimi*, and *extensor digitorum brevis* muscles. The electrodiagnostic examinations were carried out at no longer than one month intervals after the neurological examination and the median, ulnar, radial, sural and fibular nerves had been studied.

ES at the outset and end of treatment were not performed on all the patients. Only 19 of the 33 patients had had initial ES, whereas 31 were given ES at the end of treatment. Disability Grade (DG), as currently recommended by the World Health Organisation, was performed both at admission and at the end of treatment by either a practicing physiotherapist or occupational therapist. Facial examinations determined tear secretion (Schirmer's test), palpebral fissure, and corneal sensation (Corneal reflex). The extremities were evaluated by way of tactile sensation (the Semmes-Weinstein technique), nerve palpation, spontaneous pain and/or that caused by tapping, voluntary muscle strength, and the existence of deformities. Finally, DG was assessed on a scale from O to 2, as follows: anaesthesia = grade 1; muscle atrophy, lagophthalmos, footdrop, neuropathic plantar ulcer, bone absorption of distal phalanges, bone absorption of the middle and proximal phalanges, as well as wrist drop = grade 2.

3. Results

At the admissions interview, most patients did not initially complain of any neurological symptoms but, after some probing, the attending physician was able to learn that 43% (15) of the 33 patients referred to "previous neurological symptoms", such as paraesthesia, nerve pain, hyperaesthesia, hypoaesthesia, and/or cramps. This finding was more frequent among patients who reported loss of sensation at their first examination. The initial examination showed that 21% (7) of the patients had nerve pain, 18% (6) had paraesthesia, 42% (14) had erythroderma, 63% (21) had nerve enlargement, 54% (18) had sensation loss and 27% (9) had muscle weakness. At the last examination, 48% (16) of the 33 patients had nerve pain, 48% (16) paresthesia, 60% (20) erythroderma, 72% (24) nerve enlargement, 69% (23) loss in sensation, and 48% (16) showed muscle weakness. The percentages of the neurological parameters that were found when comparing the patients' "initial" and "final" exams are shown in Figure 1.

In the lepromatous leprosy (LL) group, the initial neurological examination demonstrated that nerve pain in association with muscle weakness was more frequently found (both 30%) than in the other two groups. Among the borderline leprosy patients, the most common neurological alteration seen was nerve enlargement (81%) which occurred much more often than in the other groups.

When the nerve trunks with alteration were compared with the neurological symptoms, it was seen that, in the initial examination, most paraesthetic complaints involved the ulnar nerve. Nonetheless, the neurological examination showed that the sural and not the ulnar nerve was the one most seriously impaired. On the other hand, at the last examination, paraesthetic complaints in the fibular, tibial, and sural regions increased. The ulnar nerve most frequently presented muscle weakness during the initial examination. At the last examination, however, it was revealed that the frequency of muscle weakness, as well as nerve pain, almost doubled in the peroneal nerve. The ulnar nerve was clearly the most often compromised nerve at the initial examination. By the last examination, however, similar findings were obtained for both the peroneal and ulnar nerves (see Figure 2).

All of the 18 patients on whom ES were performed at the outset were found to have axonal neuropathies. Nevertheless, the last examination showed that, of the 31 patients submitted to ES, two (6%) had normal results, 20 (64%) had an axonal pattern, one (3%) had demyelination, and eight (25%) had a mixed demyelinating and axonal pattern (Figure 3). When the electrodiagnostic pattern and the neurological symptoms were compared, it was observed that 57% of those with muscle weakness evidenced a demyelinating or mixed pattern, while loss in sensation was more closely associated with an axonal pattern.

4. Discussion

When specifically asked, most patients replied that their neurological symptoms occurred long before their first dermatological lesion was observed and that, at the time, these symptoms were in no way thought to be related to leprosy. Neurological symptoms may manifest themselves many years prior to a leprosy patient's diagnosis. In the present study, it was found that 43% of the multibacillary leprosy patients (MB) had experienced neurological symptoms, for the most part referring to changes in sensation, before any dermatological lesions were first observed. At the initial stages of the disease, insidious symptom development along with the absence of nerve pain may be a major contributing factor in the delay of diagnosis. Nerve enlargement was the principal finding at the initial examination, thereby meriting special attention as being a significant sign of leprosy.

In spite of regular treatment and close follow-up work, it is important to point out that (on an average, and regardless of slight improvements in nerve enlargement rates among the BL group) the frequency rates of the neurological data evaluated in each patient's most recent examination to date were actually higher than the percentages found at their initial examination (Becx-Bleumink 1990). One of the most worrying findings was that as many as half of all BL patients demonstrated a considerable worsening in motor weakness. Subsequently, in some patients, a progressive deterioration of peripheral neuropathy together with a bacteriological cure was observed. The clinical forms, however, did not seem to in any way influence the evolution of neuropathy.

At the beginning, most of the patients' complaints revolved around sensory and motor alternations in the ulnar nerve. However, at the last examination, an overall worsening in the condition of damaged nerves, particularly in the lower limbs, was demonstrated. In addition, the evaluation of the patients that began treatment with unquestionable evidence of sensory and motor alteration was clearly seen to have worsened when compared to the other patients in the study, principally among those who demonstrated muscle weakness at the initial examination. As mentioned above, at the end of the study, the patients' DG showed marked improvement, in sharp contrast to the results of their neurological examinations, which had clearly worsened.

The most prominent finding of this study was not the confirmation, as has been reported in other papers, of the existence of an axonal pattern in these patients, but that the neurological evaluation of those patients seen to evidence a mixed (demyelinating and axonal) pattern showed a higher rate of muscle weakness than those with an axonal pattern alone (McLeod et al 1997; Ramakrishanan and Srinvasan

1995). Perhaps the worsening of the evolution of the neuropathy was determined by the damage caused by secondary demyelination.

References

Becx-Bleumink, M. 1990 The management of nerve damage in the leprosy control services. *Leprosy Review* 61:1-11.

Britton, W.B. and Lockwood, D.J. 1997 Leprosy reactions: current and future approaches to management. *Baillières's clinical infectious diseases* 4 (1): 1-23.

McLeod, J.G., Hargrave, J.C., Walsh, J.C., Booth, R.S., Gye, R.S. and Barron, S.A. 1975 Nerve conduction studies in leprosy. *International Journal of Leprosy* 43 (1): 21– 31.

Naafs, B. 1996 Treatment of reaction and nerve damage. *International Journal of Leprosy* 64: S21 – S28.

Nations, S.P., Katz, J.S., Lyde, C.B. and Barohn, R.J. 1998 Leprous neuropathy: An american perspective. *Seminars in Neurology* 18 (1):113-123.

Pfaltzgraff, R.E. 1989 The management of reaction in leprosy. *International Journal of Leprosy* 57: 103-109.

Ramakrishanan, A.G. and Srinvasan, M.T. 1995 Electrophysiological correlates of hanseniasis. *International Journal of Leprosy* 63 (3):395-408.

Scollard, D.M., Smith T. and Bhoopat, L. 1994 Epidemiological characteristics of leprosy reactions. *International Journal of Leprosy* 62: 559-567.

Tzourio, C., Said G. and Millan, J. 1992 Asymptomatic nerve hyperthrophy in lepromatics leprosy: a clinical, electrophysiological and morphological study.*J. Neurology* 239(7): 367 – 374.

Table 1 – Clinical and electrodiagnostic data

Patient Clínical Data	Classification	G	R (N0.)	Age	Length attendance (years)	B.I.	DG		PNS	Nerve Pain		Paresthesia		Erythroderma		Nerve Enlargement		Changes in sensation		Muscle Weakness		ES	
							I	II		I	II	I	II	I	II	I	II	I	II	I	II	I	II
ABD	BL	M	R (1)	37	8	3.66	0	0	-	-	-	-	-	-	-	-	+	-	-	-	-		A
ALO	BL	M	RR (4)	59	8	4.16	1	1	-	+	+	+	+	+	+	+	-	+	+	-	-	A	M
AAFM	BL	M	RR (2)	34	6	3.66	1	1	+	-	+	-	+	+	+	+	+	+	+	+	+	A	A
CNS	BL	F	RR (1)	47	8	2.00	0	0	-	+	-	+	+	-	+	+	-	+	-	+	+	A	M
ESF	BL	M	RR (5)	46	8	1.16	0	0	+	-	+	-	+	+	+	+	+	-	+	-	+	A	M
HS	BL	M	R(1)	61	9	2.00	1	0	-	-	+	-	+	-	-	+	+	-	-	-	-	A	A
IGO	BL	M	N (2)	31	7	3.33	0	0	+	+	-	-	+	+	+	+	+	+	+	+	+		M
JCS	BL	M	RR (4)	46	10	2.16	1	0	-	-	+	-	-	-	+	-	+	-	-	-	+		M
JCB	BL	M	RR (1)	66	8	0.66	0	0	-	-	-	+	-	+	-	+	+	-	-	-	-		A
LCS	BL	M	RR (2)	18	12	1.30	0	0	-	-	-	-	-	-	-	-	-	-	+	-	-		D
MBM	BL	M	RR (1)	49	7	0.00	1	1	+	-	-	-	+	-	+	+	+	+	+	-	-		A
MJS	BL	M	RR (5)	46	8	1.00	1	1	+	-	+	+	-	-	-	+	-	+	+	-	+	A	A
NDS	BL	F	ENL (1)	48	7	4.83	1	0	+	-	-	-	+	-	-	+	-	+	+	-	-	A	
PSA	BL	M	ENL (5)	34	6	3.66	0	0	-	-	+	-	-	-	-	+	+	-	-	-	-	A	A
SNF	BL	M	RR (1)	16	8	3.50	0	0	-	-	-	-	-	+	+		+	+	+	+	+	A	A
TCSL	BL	F	N (2)	48	7	0.16	2	2	+	-	+	-	-	+	+	+	+	-	-	-	+	A	NAD
AJR	LL	M	N (2)	45	8	3.66	0	0	-	+	+	-	+	+	+	+	+	+	+	-	-	A	A
DEO	LL	M	ENL (5)	21	7	4.00	2	2	+	+	-	-	-	+	+	+	+	+	+	+	+	A	M
FFC	LL	M	N (1)	29	10	0.60	1	1	+	-	+	-	-	-	-	+	+	+	+	-	+	A	A
HPA	LL	M	N (2)	45	8	4.50	2	2	-	+	+	-	+	+	+	+	+	+	+	+	+	A	
MTV	LL	F	ENL (3)	16	8	2.83	1	0	-	-	-	+	+	-	+	+	+	+	+	+	+	A	A
MPM	LL	M	ENL (1)	23	7	4.50	2	2	+	-	-	-	+	-	+	-	+	-	-	+	+		M
MHMFL	LL	M	ENL (1)	35	6	3.66	2	0	+	-	-	-	-	-	+	-	-	-	+	-	+		A
MNG	LL	M	ENL (1)	23	8	2.66	1	0	+	+	+	+	-	+	+	+	+	+	+	-	+	A	A
MCCD	LL	F	ENL (1)	58	8	3.50	1	0	+	-	-	-	+	-	-	-	+	+	-	-	+		A
PSS	LL	M	RR (3) ENL (1)	28	6	4.50	0	0	-	-	-	-	-	-	+	-	+	-	-	-	-		A
RSS	LL	F	ENL (2)	35	8	4.33	0	0	-	-	-	-	+	-	-	-	-	-	+	-	-		A
STP	LL	M	ENL (9)	39	8	4.33	0	0	-	-	+	-	-	-	-	+	+	+	-	-	-		A
VLPN	LL	F	ENL (4)	25	11	3.50	1	0	-	-	+	-	+		-	-	-	-	+	-	+		A
ACAF	BB	F	ENL (1)	72	8	2.16	1	1	+	-	+	-	+	+	+	+	+	+	+	-	+	A	A
JCP	BB	M	ENL (4)	21	7	3.16	2	2	-	-	-	-	-	-	+	-	+	-	+	-	+		M
VLMP	BB	F	RR (1)	34	6	0.00	0	0	-	-	+	+	-	+	-	-	-	+	+	-	+		NAD
WLF	BB	M	RR (1)	64	9	1.00	1	1	+	-	-	-	-	+	+	-	+	+	+	+	-	A	A

G = Gender; R = Reactional episodes; (No.) = Number of reactional episodes; RR = Reversal Reactions; ENL = Erythema nodosum leprosum; N = Neuritis; B.I.= Bacillary Index; DG = Disability grade; PNS = Previous Neurological Symptoms; I = Initial exam; II = Last exam; ES = Electrophysiological study; A = Axonal; D = Demyelinating; M = Mixed; NAD = No abnormality detected

Figure 1 - Neurological symptoms evaluation

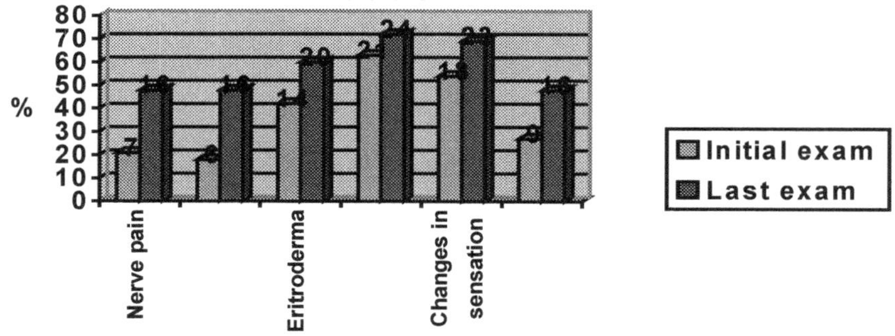

Figure 2 - Identification of the damaged nerves

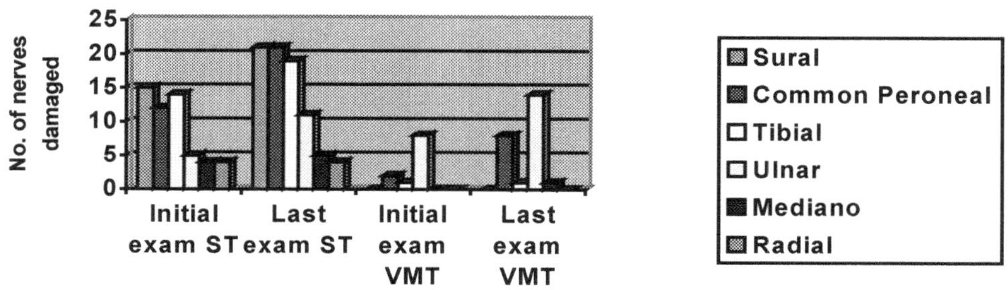

Figure 3 - ES date in the follow-up - 31 patients

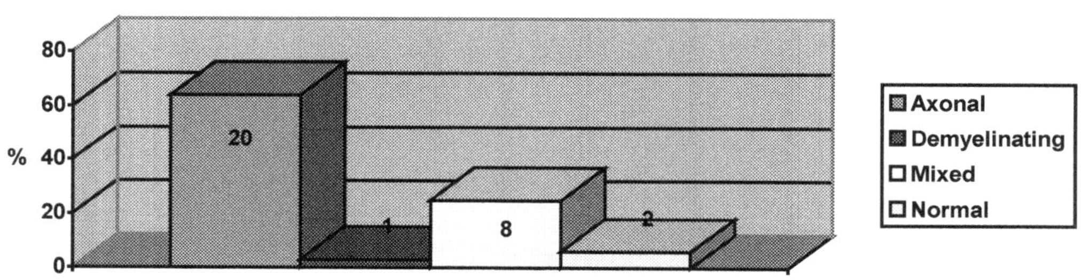

Hepatitis B and C infection among leprosy patients attending the sanatorium of Fontilles (Spain)

P.Torres[1], J.R.Gomez[2], V.Gimeno[1], J.J.Camarena[2], J.M.Nogueira[2], J.C.Navarro[2]
[1]Sanatorium San Francisco de Borja
Fontilles 03791
Alicante
Spain

Email: ptorres@fontilles.org

[2] Department of Microbiology
Dr.Peset University Hospital
Av. Gaspar Aguilar 90
46017 Valencia
Spain

Email: juan.camarena@uv.es

Abstract

Several scientific groups have proposed a possible association between infection by hepatitis B (HBV) and C (HCV) and leprosy. In this study, the available data is reviewed and most recent diagnostic tests applied to 214 institutionalized leprosy patients, and outpatients, controlled by The Sanatorium of Fontilles (Spain).

The prevalence of infection in the leprosy group, 100 / 214 patients or 46 % for hepatitis C markers, and 13 of 214 in the hepatitis B group, was significantly higher than the control groups of 3.5 % ($p < 0.001$) and 3.0% ($p< 0.05$), respectively. A history of blood transfusions prior to 1992 could be indicative of a high risk factor for hepatitis C infection.The results in this study indicate an increased risk of HBV and HCV infection for patients confined to leprosaria during a prolonged period of time, due to open skin lesions,blood transfusions and intravenous daily injections with non disposable syringes in the past.

Keywords

Leprosy, virus markers, risk factors, co-infection, confirmation tests, Spain

1. Introduction

A possible association between infection by hepatitis viruses B (HBV) and C (HCV) and leprosy has been proposed (Balakrishnan *et al*., 1983, Egawa *et al*., 1996). The data available is based mainly on most studies of HBV, discovered by Blumberg and Melartin (1967), on HBsAg (hepatitis B surface antigen) status alone, evaluated by different methods, although some recent reports include other serological markers (anti-HBs and HbeAg) (Yvonnet *et al*., 1989).In the case of HCV,its discovery by Choo *et al.* (1989) led to enzyme immunoassays (EIA) for serological diagnosis by detecting corresponding antibodies (anti – HCV).

Hepatitis B (HBV) and hepatitis C (HCV) viruses are transmitted by blood (transfusions, parenteral injections), possibly sexual contacts, and probably other unknown routes. They can cause chronic liver disease and some individuals may develop cirrhosis and hepatocellular carcinoma. Populations with increased risk of these viral infections,especially patients with hemophilia and on hemodialysis, have been identified.

Patients with leprosy possibly also form a high risk group because of skin lesions, blood transfusions and confinement in institutions during prolonged periods of time. HBV and HCV seroprevalence varies in different parts of the World; it is 2-3% by hepatitis B surface antigen (HbsAg) detection and 2% by RNA (ribonucleic acid) -HCV determination in the Spanish community of a Valencia population (3,000,000), respectively. In 1967 the prevalence of hepatitis B surface antigen in a group of patients suffering from lepromatous leprosy was found to be significantly higher then in a control group (Blumberg and Melartin, 1967). Subsequent studies have confirmed this finding.

The possibility of evaluating HCV antibodies, introduced by EIA tests (a type of immunoassay technique to detect antigens and antibodies), revealed HCV as the major cause of blood borne, non A or B, hepatitis worldwide. Some consider that the two polar forms of leprosy (tuberculoid and lepromatous) provide a model of interaction between cellular immunity and the hepatitis viruses.

In this study, the distribution of HBV and HCV virus markers were evaluated in 214 leprosy patients, mostly long term institutionalized, in The Sanatorium of Fontilles and compared with matched controls.

2. Patients and methods

2.1 Patients

Two hundred and fourteen institutionalized leprosy patients, and outpatients controlled by The Sanatorium of Fontilles and classified by the clinical and histological criteria of the Ridley-Jopling classification (1966), were studied. One hundred and ninety were lepromatous (88.7%) and 13 were tuberculoid (6.07%). The remaining 11 patients were of an intermediate boundary type (5.01%).

The age of the patients ranged from 30 to 83 years; 117 were male with an average age of 60.8 years and 97 were female, with an average age of 60.9 years. Serving as controls were randomly selected blood donors and people from the general population attending the Hospital Dr. Peset Aleixandre Valencia, with an age and sex distribution matching the studied patients.

2.2 Methods

Serum samples were collected during the study and stored at - 40°C and preserved with sodium azide until tested.The sera were used for testing of liver function abnormalities and markers of HCV and HBV infections.The routine liver function tests (alanine and aspartate aminotransferases level) and platelet counts were done in the Laboratory of The Sanatorium of Fontilles and all other tests at the Microbiology Department at the Hospital Dr.Peset Aleixandre, Valencia.

2.2.1 Anti-HBV and HCV Testing

Sera were put through the same protocol required for screening of blood donors. Initially, two third–generation microparticle enzyme immunoassays were undertaken by Axsym.Abbott., the HCV version 3.0 for the qualitative detection of antibody to hepatitis C virus (anti-HCV), and HbsAg (V2) for the qualitative detection of Hepatitis B surface antigen (HbsAg) both in human serum or plasma. Samples non-reactive by AxSYM HbsAg (V2) and AxSYM HCV 3.0 versions were considered negative for HbsAg and anti-HCV, respectively, and were not tested further.

2.2.2 HBsAg Reactive

A repeatedly reactive sample for AxSYM HBsAg (V2) was confirmed by the Digene Hybrid Capture HBV DNA Assay, This is a signal amplification, solution hybridization chemiluminescent detection system for the detection and quantification of hepatitis B viral DNA in serum. Specimens containing the target DNA hybridize with specific full length genomic HBV RNA probes.

2.2.3 Anti-HCV Reactive

Specimens which were reactive on retest by AxSYM HCV version 3.0 were considered reactive, and specimens on the grayzone range (sample rate / Cutoff rate : 0.80-0.99) were considered borderline reactive. We considered them as indeterminate and analysed them again by The Inno-Lia HCV Ab III update. The test uses strip technology in which a combination of synthetic peptides and recombinant proteins are applied as lines in an ordered manner. Each strip has six coated antigens and four control lines on nylon with a plastic backing. HCV RNA quantification by PCR was determined by The Amplicor Monitor. The level of RNA from HCV of the samples is determined by comparing the amount of absorbance from the standard.

2.2.4 Statistical analyses

Frequencies were compared using the X^2 – test and Fisher´s exact test. Group means were compared using Student´s t-test on a Minitab-release 12 version statistical computer package.

3. Results

3.1 Markers of HCV infection in leprosy patients

Anti-HCV was detected in 76 (35%) of 214 patients covering the entire immunological spectrum of leprosy. The number of infected individuals increased when the 53 cases of anti-HCV (+/-) (doubtful, on the grayzone) were confirmed by a supplementary more specific test (Inno-Lia HCV Ab III update). This was found to be reactive using an anti-HCV screening procedure, and 24 more were found positive, increasing the total number of infected individuals to 100 (46%). This prevalence rate was much higher than the one obtained (3.5%) from age and sex matched controls ($p<0.001$). Twenty-three doubtful anti-HCV results were confirmed negative by this test and six were indeterminate (Table 1).

Table 2 compares markers of HCV and HBV infections in leprous patients divided into age groups. The prevalence of antibodies detected by either test (anti-HCV or Line Immunoassay) was higher in patients aged 70 to 79 years (33 of 53 or 62%) than the other groups of younger or older patients, with the lowest prevalence being in the 30 to 39 year age group (2 of 11, or 18%).

Table 3 presents RNA-HCV positivity in anti-HCV (+) samples (46 of 55, or 83%). In the other test group, 14 of 24 Line Immunoassay positives were RNA-HCV tested and only 3 (21%) presented viremia, giving a total of 49 of 69 or 71% RNA –HCV (Anti-HCV or Line Immunoassay (+)). Sixteen of 23 Line Immunoassay negatives and 6 of 6 Line indeterminates RNA-HCV tested were all negative.

Table 4 expresses the influence of disease types on the prevalence of HCV markers, with a lower prevalence in the intermediate forms group. No influence of sex on HCV infection could be established (Table 5).

A history of blood transfusions could be traced for 45 individuals and 29 were found to be anti-HCV / Line Immunoassay (+) or 64%, and 25 of 29 presented viremia when RNA-HCV tested, indicating transfusions as a high risk factor (Table 6).

3.2 Markers of HBV infection in leprous patients

As shown in Table 2, 13 of 214 (or 6.0 %) of the leprosy patients presented HBsAg positivity, a prevalence rate significantly higher than the control population (3.0 %, $p < 0.05$). Only 1 of 13 (or 0.5 %) of the total number of patients presented HBV –DNA, and four cases had delta virus associated co-infection. Three of 13 (or 23%) HBsAg cases had a history of blood transfusions.

3.3 Liver function abnormalities in leprous patients infected with HCV or HBV

Patients with anti-HCV or Line Immunoassay positivity had elevated alanine aminotransferase levels (ALT) more frequently than those negative to the tests as presented in Table 7. The percentages are 48 of 100 (or 48 %) in the positive group and 20 of 114 (or 17 %) in the negative group ($p < 0.05$). The aspartate aminotransferase (ST) values between the two groups were not statistically significant.

The mean levels of the ALT values (normal $\leq$ 45 U / L) were also significantly elevated in the positive group (85 +/ - 25) compared to the negative group (24 +/-12, $p < 0.05$). Patients with HCV viremia (RNA-HCV (+)) present the highest ALT levels as a group with 39 of 49 (or 80%) above 45 U /L, together with the highest mean level (20 +/- 35) (Table 8).

Low platelet counts were obtained in 16 of 100 (or 19.7 %) of the positive patient group ($n \geq 1.5 \times 10^5$) and 4 of 114 (or 9.7 %) of the anti-HCV or Line Immunoassay (-) patients. The ALT values in the HBsAg group were elevated in 5 of 11 (or 45 %) and platelet counts were lower in 2 of 11 (or 18 %) of the group (Table 9). Only two patients presented infection with both viruses.

4. Discussion

Several groups of investigators have reported high prevalence rates of hepatitis B (HBV) and C (HCV) viruses in leprosy patients. The HBV infection studies are based mainly on HBsAg status alone, with a lack of consistency in the methods used due to varying degrees of sensitivity of HBsAg detection (Fakunle and Whittle, 1981; Moudgil and Irshad, 1988) offered by different investigators such as CIE, RPHA, RIA, ID and ELISA, and the results obtained have differed from group to group and remained inconclusive.

Recently, some groups have improved the detection methods used and results can be compared, although they vary from different parts of the world. Yvonnet *et al.* (1989) in a study conducted in Senegal on 987 leprosy patients and 6,187 healthy adults (controls), found HBsAg detected by RIA in approximately 25 % of the leprosy group in comparison to 15.2 % in the control group, with no significant difference for other serological markers (anti – HBc, anti – Hbe and HbeAg). There is also a significant difference for HBsAg carriage among LL compared with TT (29.7 % vs.15.2 %). Nuti *et al.* (1975) also found a significant difference among leprosy types in 223 Somalian patients with HBsAg present in 24.4 % of LL cases and 11.5 % of TT patients. Recently, Banerjee *et al.* (1994), in 41 patients in India, found 8 of 13 with lepromatous leprosy (or 61 %) to be positive for either HbsAg assayed by EIA and HBV-DNA analysis. In the high cell-mediated immune response category (BT and TT) two of 10 patients showed positivity, while in the control group positivity was approximately 10 %.

In our study HBsAg positivity was detected in 13 of 214 leprosy patients (or 6 %) and this was significantly higher than in blood donors (3%), all cases belonging to the LL category. This group represents 88 % of the patients in the study, a finding consistent with the results obtained in other studies that obtain a higher percentage of HBsAg carriage among patients of the LL category. Only one of 13 HBsAg was HBV – DNA (+). This situation has been reported when HBV DNA becomes integrated into hepatocellular chromosomes, and also in the case of acute viral hepatitis where HBsAg appears in the serum before HBV DNA (Brechot *et al.*, 1985). Three cases presented delta co-infection.

The prevalence rate of HCV markers of infection among the leprosy patients was 35% (76 of 214) for the anti-HCV test, and the number increased when 53 anti-HCV doubtful (+/-) results were confirmed by a more specific test providing 24 more positives. This raises the total number to 100 of 214 (or 46 %), a statistically significantly higher prevalence rate than the control group (3.5 %, $p < 0.001$). Fifty-five positive anti-HCV samples were tested for RNA-HCV and 46 presented viremia (83 %). Viremia was less frequent in the confirmatory test group as it was only detected in 3 of 14 (or 21 % of patients), giving a total of 49 of 65 (or 71 %) RNA-HCV. The prevalence increases with age and is highest in the 70–79 year age group. No significant differences in prevalence rates could be detected between sex, and the risk of HCV infection was different in the two polar forms of leprosy (LL: 46 % versus TT: 53 %), and only a moderate decrease was seen in the intermediate forms (27 %). Patients with anti-HCV or Line Immunoassay (+) tests had a higher frequency of elevated ALT values (48 of 100, or 48 %) compared with the group that tested negative (20 of 114, or 17%). This frequency continues to increase if we consider only the patients with viremia (39 of 49, or 80 %). Hence, HVC infection can induce severe hepatic disease that may result in hepatocellular carcinoma. Platelet counts were also reduced in the infected group (19 % vs 9.7%). Other groups of investigators have reported similar results with high prevalence rates of anti-HCV testing by EIA. In a survey conducted in six African countries, Denis *et al.* (1994) detected anti-HCV in 7.1 % of 1309 leprous patients, significantly higher than 2.6% of the 1560 control group. Egawa *et al.* (1996) also report a detection rate of anti-HCV by a second generation EIA in 68 (30 %) and RNA-HCV in 41 (18%) of 229 leprosy patients with characteristics very similar to this population studied at the Sanatorium of Fontilles (number, age detection methods employed, long term institutionalized etc.).

A history of blood transfusions could be traced in 45 individuals before screening for HCV was mandatory, and 25 of 29 individuals of this group (29 of 45 or 64 %) were anti-HCV. A supplementary test confirmed the presence of viraemia, indicating transfusions as a high risk factor for HCV infection, while three of 13 (or 23 %) of HbsAg patients had a history of blood transfusions.

Although studies on HBV and HCV infection among leprosy patients continue to produce contradictory results from different authors and areas investigated, Frommel *et al.* (1993) report prevalence rates on HBsAg carriage and anti-HCV infection similar to the control groups among Senegalese leprosy patients. In addition, Rosa *et al.* (in Tibbs *et al.*, 1991), in a study of 216 leprosy patients in Goiania, Central Brazil divided into outpatients and inpatients, found a prevalence rate of HCV not significantly different among leprosy patients and blood donors (no significant difference between outpatients and institutionalized). This was in contrast to a large difference between the HBV –leprosy and HBV blood donors (7.4 % versus 1.9 %). The results of the present study combined with others (Denis *et al.*, 1994, Egawa *et al.*, 1996, Frommel *et al.*, 1993) indicate an increased risk of HBV and HCV infection for patients confined in leprosaria during a prolonged period of time, due to open skin lesions, blood transfusions and intravenous daily injections with non-disposable syringes in the past. Studies in tropical countries should include confirmatory tests due to higher carriage rates in the general population (Tibbs *et al.*, 1991) and non-specific reactions may occur as a result of circulating autoantibodies in leprosy and malaria (Acetti *et al.*, 1990). The anergy, or decreased cell-mediated immunity of lepromatous leprosy, compared with increased delayed–type immune responses in the tuberculoid end of the leprosy spectrum, could only be observed in this study in the HBsAg carriage rate, a fact reported by Blumberg *et al.* (1967), but the question of the possible disorders of the immune system that accompany leprosy and their involvement with an increase in susceptibility to viral infections still remains open.

References

Acetti, A., Taliani, G., De Bac, C. and Sebastiani, A. 1990 Anti-HCV false positivity in malaria. *Lancet* 336:1442-1443.

Balakrishnan, S., Bhatia, V.N. and Harikrishnan, S. 1983 Hepatitis B surface antigen in leprosy patients. *Lepr.India* 55(1): 45-48.

Banerjee, K., Ghosh, S., Moudgil, K.D. and Khandekar, P. 1994 Status of HBV DNA and HBsAg in leprosy patients. *Leprosy in India* 62 (3): 444-446.

Blumberg B.S and Melartin L. 1967 Association between lepromatous leprosy and Australia antigen. *Lancet* 2:173-176.

Blumberg, B.S., Melartin, L., Lechat, M. and Guinto, R.S. 1967 Association between lepromatous leprosy and Australia antigen. *Lancet* 2:173-6.

Brechot, C., Degos, F., Lugassy, C., Thiers,V., Zafrani, S., Franco, D., Bismuth, H., Trepo, C., Benhamon, J.,Wands, J., Issenbacher, K., Tiollais, P. and Berthlot, P. 1985 Hepatitis B virus DNA in patients with chronic liver disease and negative tests for hepatitis B surface antigen. *New England J. Med.* 312:270-274.

Choo, QL., Kuo, G.,Weiner, A.J., Overby, L.R., Bradley, D.W. and Houghton, M. 1989 Isolation of a cDNA clone derived from a blood-borne non-A, non-B viral hepatitis genome. *Science* 244:359-62.

Denis, F., Aussel, L., Ranger, S., Martin, P., Itoua-N'Gaporo, A., Frommel, D., Teckle-Haimanot, R.T., San, A., M'Boup, S., Millan, J. and Qubati, A. 1994 Prevalence of antibodies to hepatitis C virus among patients with leprosy in seven African countries and the Yemen. *J.Med.Virol.* 43:1-4.

Egawa, K., Yakawa, T., Anakawa, S., Tanaka, T., Tsuda, F., Okamoto, H., Miyakawa, Y and Mayumi M. 1996 Hepatitis C virus antibody,viral RNA and genotypes in leprous patients in Japan. *Journal of Hepatology* 24: 397-402.

Fakunle, Y.M. and Whittle, H.C. 1981 Hepatitis B virus infection in patients with leprosy : a serological study in a leprosarium in northern Nigeria. *Trans. Royal Soc. Trop. Med. Hyg.* 75:623 – 625.

Frommel, D., Tekle, H.R., Berhe, N., Aussel, L.,Verdier, M., Preux, P.M. and Denis, F. 1993 A survey of antibodies to hepatitis C virus in Ethiopia. *Amer. J Trop.Med.Hyg.* 49: 435 –9.

Moudgil, K.D and Irshad, M. 1988 Global overview of the prevalence of hepatitis B virus markers (HBsAg and anti-HBs) in leprosy patients. *Trop.Gastroenterol.* 9:184-190.

Nuti, M., Tarabini, G., Palermo, P., Tarabini, G.L. and Thamer, G. 1975 Leprosy and Hepatitis B virus markers : Incidence of HBsAg and HbeAg in Somalian Patients. *International Journal of Leprosy* 47 (4):580-584.

Ridley, D.S. and Jopling, W.H. 1966 Classification of leprosy according to immunology. The five group system. *Int. J. Leprosy* 34:255-273

Tibbs, C.J., Palmer, S.J., Cocker, R., Clark, S.K., Parsons, Gm., Hojvat, S., Peterson D. and Barratvala, J.E. 1991 Prevalence of hepatitis C in tropical communities: the importance of confirmatory assays. *Journal of Medical Virology* 34:143-145.

Yvonnet B, Vincelot, P., Millan, J., Lesage, G., Denis, F., Languillon, J., Mboup, S., Coursaget, P., Diop Mar, I. and Chiron, J.P. 1989 Hepatitis B virus infection in lepromatous and tuberculoid patients from Senegal. *Acta Leprologica* 7 (1): 63 – 66.

Table 1: Reactivity to six HCV proteins measured by line immunoassay in doubtful anti-HCV results

	(Anti-HCV)±	**Line Immunoassay Positive**	**Line Immunoassay Negative**	**Line Immunoassay Indeterminate**
Number of samples	**53**	**24**	**23**	**6**

Table 2: Prevalence of HCV and HVB markers in patients with leprosy

Age	**N**	**Anti-HCV and Line Immunoassay (+)**	**HBs Ag**	**HBV-DNA**
30-39	11	1+1 (2), 18%	1 (9%)	0
40-49	20	2+2 (4), 20%	4 (20%)	0
50-59	49	11+5 (15), 32%	5 (10%)	1
60-69	74	33+6 (14), 52%	1 (1.3%)	0
70-79	53	25+8 (14), 62%	1 (14%)	0
>80	7	4+0 (4), 57%	1 (14%)	0
Total	**214**	**76+24 (100), 46%**	**13 (6%)**	**1 (0.5%)**
Controls	**100**	**3 (3%)**	**3 (3%)**	**0.4% (0.2%)**

Table 3: RNA and HCV positivity in the anti-HCV (+) and line immunoassay (+) samples tested

	Number of samples	**RNA-HCV (+)**
Anti-HCV (+)	55	46 (83%)
Line immunoassay(+)	14	3 (21%)
Total positive samples RNA-HCV tested	**69**	**49 (71%)**

Table 4: HCV infection in different forms of leprosy

	Number	**Anti-HCV/Line Immunoassay (+)**
Lepromatous	190	88 (46%)
Tuberculoid	13	7 (53%)
Intermediate	11	3 (27%)

Table 5: Sex distribution of HCV infection among leprosy patients

Patients	**Anti-HCV/Line Immunoassay (+) (Number and %)**
Males	57/117 (48%)
Females	41/97 (42%)

Table 6: Possible risk factor of hepatitis viral infections

	Transfusions (N=45)
Anti-HCV/Line Immunoassay (+)	29 (64%)
HCV-RNA (Anti-HCV/Line Immunoassay (+))	25 (55%)

Table 7: Liver function tests in patients with leprosy with or without markers of HCV infection

	Anti-HCV/Line Immunoassay (+)	**Anti-HCV/Line Immunoassay (-)**
Number	100	114
Alt	48 (48%)	20 (17%)
Mean value (U/L)	85 ± 25	24 ± 18
Platelets (<1.5 X 100,000)	16 (19.7%)	4 (9.7%)

Table 8: Alt liver values in the RNA HCV (+) group

Number of DNA-HCV (+) samples	*Alt values >45U/L*	**Mean alt value of the group**
49	39 (80%)	120 ±35

Table 9: Alt liver values and platelet counts in the HBs Ag (+) group

	HbsAg (+)
Number	11
Alt values >45U/L	5 (45%)
Platelets (<1.5 X 100,000)	2 (18%)

Comparison of PCR primers for detection of *M. leprae* in skin slit samples

Pedro Torres[1], John Holton[2], John L. Stanford[2] and Helen D. Donoghue[2]

[1]Sanatorio San Francisco de Borja
Fontilles
Alicante 03791
Spain

Email: ptorres@fontilles.org

[2] Department of Medical Microbiology
University College London
46, Cleveland Street
London W1T 4JF
England

E-mail:j.holton@ucl.ac.uk
j.stanford@ucl.ac.uk
h.donoghue@ucl.ac.uk

Abstract

There are several specific PCR-based methods in clinical use to detect *Mycobacterium leprae* DNA. However, the amplicons are quite large which may be a disadvantage when examining samples from treated patients in which the DNA is likely to be damaged and fragmented. Therefore, two sets of new *M.leprae*-specific nested primers were used to screen ten skin biopsies for the presence of *M.leprae* DNA. All were from patients at the Sanatorio San Francisco de Borja in Fontilles, Spain who had predominantly lepromatous leprosy. The new primers were more sensitive and enabled the *M.leprae* DNA load to be estimated in each sample. Three samples with a zero bacterial index were positive, but only with the most sensitive nested PCR. It was concluded that these new primers are of use in detecting very low levels of *M. leprae* DNA which may be of clinical significance.

Keywords

Mycobacterium leprae, PCR, clinical samples

1. Introduction

Mycobacterium leprae, the causative agent of leprosy, is one of the few organisms that cannot be cultivated *in vitro*. Diagnosis of the disease is based on clinical examination of the patient, histopathology and the demonstration of acid fast bacilli in slit-skin smears or biopsies. However, the detection limit based on microscopy is approximately 10^4/ml (Shepard and McCrae, 1968) In order to increase the sensitivity and specificity of the detection of *M. leprae*, several authors have described methods based on the polymerase chain reaction (PCR). These have used different DNA target sequences in the *M. leprae* genome and result in relatively large (320 - 530bp) amplicons compared with the 123bp amplicon produced by primers used routinely to detect *Mycobacterium tuberculosis* DNA (Eisenach *et al.*, 1990).

The primers designed by Hartskeerl *et al.* (1989), based on the 36kD antigen (Ag) gene, are still used by many laboratories. Although the amplicon is 531bp in size, the authors reported that their method enabled a single bacterium to be detected. This PCR has been used recently to monitor nasal carriage of *M.leprae* amongst patients at Fontilles and gives satisfactory results. Williams *et al.* (1990) used primers for the *M. leprae* 18kD antigen gene which yielded a 360bp product and could detect as few as 100 *M. leprae* in seeded uninfected human skin biopsy preparations. Pliyakis *et al.* (1990) devised a nested two-step PCR reaction which amplified a 578bp outer product and 347bp inner product from the *M. leprae groEL* (65kd antigen) gene. Their method could detect as little as 3fg of DNA which they equated with the amount found in a single bacillus. Woods and Cole (1989) described a PCR based on an *M. leprae*-specific repetitive element (RLEP), which was found by Yoon *et al.* (1993) to be capable of detecting *M. leprae* DNA in 73% of patients with a Bacterial Index of zero. Santos *et al.* (1993) developed this further by optimising the DNA extraction from clinical samples and including DNA hybridisation. They reported the ability to detect 100ag (10^{-16} g) of target DNA, equivalent to about one-tenth of the bacterial genome. The amplicon in this case was 372bp.

However, all these different *M. leprae*-specific PCR methods result in amplicons which are quite large. Treated patients may continue to harbour *M. leprae* DNA in their tissues, particularly in the case of older patients who received dapsone monotherapy (Rafi *et al.*, 1995). However, even treatment with rifampicin, which kills the organisms rapidly, results in slow DNA degradation occurring over a period of at least two to three months

(Fiallo *et al.*, 1992) and Rafi *et al.* (1995) suggested that this residual DNA, possibly degraded, may play a role in late reactions. Therefore, it is important that residual DNA can be detected.

Archaeological studies have shown that the smaller the DNA target sequence, the greater the likelihood of detecting DNA that has been damaged by the passage of time. Applying the same principle, it should be more productive to use PCR based on smaller target sequences for the detection of DNA from organisms that have been damaged by antimicrobial therapy in human hosts. Therefore, a small number of samples from patients who had initially presented with lepromatous (LL) or borderline (BL) leprosy, were examined for the presence of *M.leprae* DNA using new PCR primers, based on the RLEP sequence and 18kD Ag gene that had been designed to be both specific and more sensitive than those described previously (Donoghue *et al.*, 2001).

2. Materials and methods

2.1 Skin slit samples

Ten fresh skin slit samples were obtained from nine patients from the Sanatorio San Francisco de Borja in Fontilles, Spain and stored at –20°C until use. The samples dated from 1996 to 1998 and seven were obtained prior to treatment. The clinical data were supplied by the Pathology Laboratory of the Sanatorio San Francisco de Borja, Fontilles, Spain (Table 1). Seven patients had been initially diagnosed as having lepromatous leprosy and two with borderline leprosy. Most had been treated with the following regimen for at least two years and until they had a Bacterial Index of zero: rifampicin (600mg/month), chlorfazimine (100mg alternate days) and dapsone (100mg daily). However, two patients had received dapsone monotherapy only. One patient (SK2) had been discharged 10 years previously but had returned with a relapse of the disease. The second patient to have monotherapy had completed treatment 10 years earlier and returned to the hospital with a reversal reaction. One sample was taken at this time (SK9) and another (SK10) after the completion of the recommended World Health Organisation regimen of rifampicin, clofazimine and dapsone for six months and 20-30mg steroids/day.

2.2 DNA extraction

Stringent precautions were taken to avoid cross-contamination (Donoghue *et al.*, 2001 and in this volume). Each skin-slit sample (9 - 43 mg) was incubated in Proteinase K and EDTA at 56°C until completely disaggregated, incubated in lysis buffer L6 (Boom *et al.*, 1990) which contained guanidium thiocyanate, EDTA and Triton X-100, and the DNA captured onto silica. This was washed, dried and the DNA eluted into a small volume of water. The detailed method is described by Donoghue *et al.* (2001). Negative extraction controls were processed in parallel with sample tubes.

2.3 DNA amplification by PCR

Eight primers were devised, which comprise outer and inner nested pairs based on the *M. leprae*-specific 18kD Ag gene and the *M. leprae*-specific repetitive element (RLEP). Details of the product size and target sequences are shown in Table 2. Details of the PCR mix and experimental conditions were as described elsewhere (Donoghue *et al.*, this volume). Both single-stage and nested PCR was carried out. The level of *M.leprae* DNA was estimated by determining the maximum dilution in each sample that gave detectable PCR product by single-stage PCR with the 18kD Ag primers. The reciprocal of this maximum dilution was defined as the titre. This was compared with PCR of fresh dilutions of lepromin reagent. This is an autoclaved suspension of *M.leprae* bacilli extracted from armadillo tissues, which was supplied at an estimated bacterial load of 6×10^9 organisms ml^{-1}.

3. Results

3.1 Use of single stage PCR to detect M. leprae DNA in clinical samples

After 45 cycles of amplification, no positive results were obtained with either RLEP or 18kD Ag primers from the treated patient samples SK8-10. The remaining samples showed significant differences in *M.leprae*-specific DNA load, and some yielded a large amount of PCR product (Figure 1). Results are summarised in Table 1. The agreement between the microscopically-determined Bacterial Index and PCR titre was not good. There was no product from a DNA negative extraction control, nor from the PCR negative control.

3.2 Use of nested PCR to detect M. leprae DNA from clinical samples

After nested PCR with the 18kD Ag gene primers, a band of the expected size (110bp) was visible from eight of the ten samples. All samples were positive with the RLEP primers (99bp product). The first sample from the patient with a reversal reaction (SK9) was negative with the 18kDAg primers, but positive when the RLEP primers were used. Sample SK10, from the same patient obtained immediately after completion of the WHO treatment regimen, gave a very weak positive but only with the RLEP primers. Sample SK8 from a treated patient also with a Bacterial Index of zero, was positive with both the RLEP and 18kDAg primers (Figure 2).

4. Discussion

The new primers from both target sequences gave clear results from this small set of skin-slit samples. The titre of *M.*leprae DNA varied greatly between samples and some had a similar or greater level of *M.leprae DNA* than the lepromin reagent, i.e. $\geq 6 \times 10^{9}$ organisms ml^{-1}. Nested PCR increases the sensitivity of PCR as it enables a total of 50 amplification cycles to be used, but avoids over-amplification artefacts. It was only with nested PCR based on the RLEP primers that all three of the clinical samples with a Bacterial Index of zero were found to be positive. It is unsurprising that the RLEP PCR is more sensitive as this sequence is present at 29 copies/cell of *M.leprae*. The PCR based on the single-copy 18kDAg gene, was approximately 10 times less sensitive yet even so was approximately 100-fold more sensitive than PCR based on the Hartskeerl primers (Donoghue *et al.*, 2001).

The lack of agreement between Bacterial Index and PCR titre is most likely to be due to experimental error. However, it is also conceivable that naked *M. leprae* DNA is present in samples that would not be detected microscopically. In addition, the DNA may be fragmented in some cases. Now that the facility to detect such low levels of *M. leprae* DNA is available, it would be of interest to follow-up sufficient numbers of treated leprosy patients to determine whether or not such residual *M. leprae* DNA is of any clinical significance.

Acknowledgements

Thanks are due to Dr M.J. Colston (National Institute for Medical Research, London, England) who supplied the lepromin skin test reagent as a source of *Mycobacterium leprae*.

References

Boom, R., Sol, A., Salimans, M., Jansen, C., Wertheim-van Dillen, P. and van der Noordaa, J. 1990 Rapid and simple method for purification of nucleic acid. *J. Clin. Microbiol.* 28: 495-503.

Donoghue, H.D., Holton, J. and Spigelman, M. 2001 PCR primers that can detect low levels of *Mycobacterium leprae* DNA. *J. Med. Microbiol.* 50: 177-182.

Eisenach, K.D., Cave, M.D., Bates, J.H. and Crawford, J.T. 1990 Polymerase chain reaction amplification of a repetitive DNA sequence specific for *Mycobacterium tuberculosis* . *J. Infect. Dis.* 161: 977-981.

Fiallo, P., Williams, D.L., Chan, G.P. and Gillis, T.P. 1992 Effects of fixation on polymerase chain reaction detection of *Mycobacterium leprae*. *J. Clin. Microbiol.* 30: 3095-3098.

Hartskeerl, R.A., De Wit, M.Y.L. and Klatser, P.R. 1989 Polymerase chain reaction for the detection of *Mycobacterium leprae*. *J. Gen. Microbiol.* 135:2355-2364.

Plikayis, B.B., Gelber, R.H. and Shinnick, T.M. 1990 Rapid and sensitive detection of *Mycobacterium leprae* using a nested-primer gene amplification assay. *J. Clin. Microbiol.* 28: 1913-1927.

Rafi, A., Donoghue, H.D. and Stanford, J.L. 1995 Application of polymerase chain reaction for the detection of *Mycobacterium leprae* DNA in specimens from treated leprosy patients. *Int. J. Lepr.* 63: 42-47.

Santos, A.R., De Miranda, A.B., Sarno, E.N., Suffys, P.N. and Degrave, W.M. 1993 Use of PCR-mediated amplification of *Mycobacterium leprae* DNA in different types of clinical samples for the diagnosis of leprosy. *J. Med. Microbiol.* 39: 298-304.

Shepard,C.C. and McRae, D.H. 1968 A method for counting acid-fast bacteria. *Int. J. Lepr.* 36: 78-82.

Williams, D.L., Gillis. T.P., Booth, R.J., Looker, D.L. and Watson, J.D. 1990 The use of a specific DNA probe and polymerase chain reaction for the detection of *Mycobacterium leprae*. *J. Infect. Dis.* 162: 193-200.

Woods, S.A. and Cole, S.T. 1989 A rapid method for the detection of potentially viable *Mycobacterium leprae* in human biopsies: a novel application of PCR. *FEMS Microbiology Letters* 53: 305-309.

Yoon, K.-H., Cho, S.-N., Lee, M.-K, Abalos, R.M., Cellona, R.V., Fajardo, Jr., T.T., Guido, L.S., Dela Cruz, E.C., Walsh, G.P. and Kim, J.-D. 1993 Evaluation of polymerase chain reaction amplification of *Mycobacterium leprae*-specific repetitive sequence in biopsy specimens from leprosy patients. *J. Clin. Microbiol.* 31: 895-899.

Table 1: Clinical data relating to skin-slit samples from leprosy patients and results from single-stage PCR using 18kDAg primers

Sample	Disease state[1] and Bacterial Index[2]	Maximum dilution for PCR+	Subjective assessment of band strength	Estimated titre of *M.leprae* DNA
SK1	LL - 4+	1×10^{-6}	++	1,000,000
SK2	LL - 2+	1×10^{-5}	+	100,000
SK3	BL - 1+	1×10^{-5}	±	100,000[3]
SK4	LL - 2+	1×10^{-5}	(±)	100,000
SK5	LL - 3+	1×10^{-4}	±	10,000
SK6	LL - 4+	1×10^{-3}	±	1,000
SK7	LL - 3+	1×10^{-2}	+	100
SK8	LL - 0	Neat	-	<1
SK9	BL - 0	Neat	-	<1
SK10	BL - 0	Neat	-	<1

[1] LL Lepromatous leprosy; BL Borderline leprosy
[2] Bacterial Index based on microscopic examination of 100 fields:
0 = negative; 1+ = 1-10 bacilli in some fields; 2+ = 1-10 bacilli in every field; 3+ = 10-100 in every field; 4+ = 100-1000 bacilli in every field; 5 = >1000 bacilli in every field.
[3] equivalent to 1×10^{-5} dilution of lepromin reagent

Table 2: *M.leprae*-specific primer sequences[1]

Primer	Sequence (5' to 3')	Target	Product size
LP1	TGCATGTCATGGCCTTGAGG	RLEP outer	129bp
LP2	CACCGATACCAGCGGCAGAA	" "	"
LP3	TGAGGTGTCGGCGTGGTC	RLEP inner	99bp
LP4	CAGAAATGGTGCAAGGGA	" "	"
LP5	ATCGACTGTTGTTTGCGCAAC	18kD Ag inner	110bp
LP6	CCAGCAACCGAAATGTTCGGA	" "	"
LP7	TCATAGATGCCTAATCGACTG	18kD Ag outer	136bp
LP8	GGCACATCTGCGGCCAGCA	" "	"

[1] Donoghue *et al.*, this volume

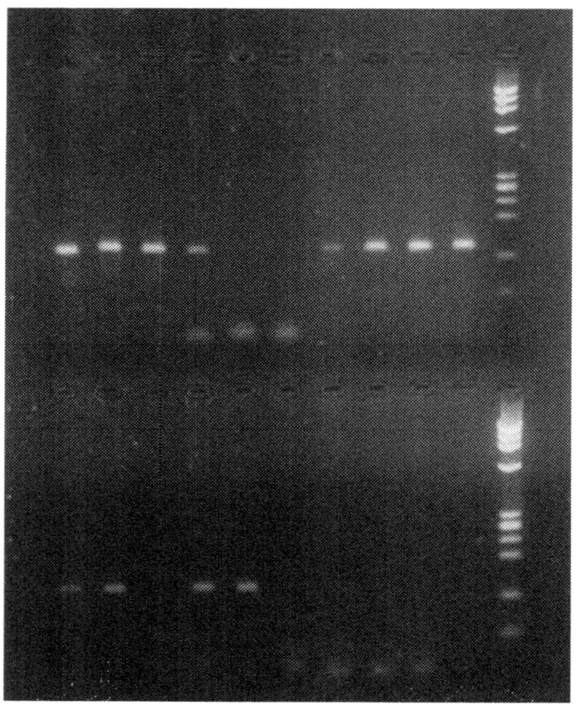

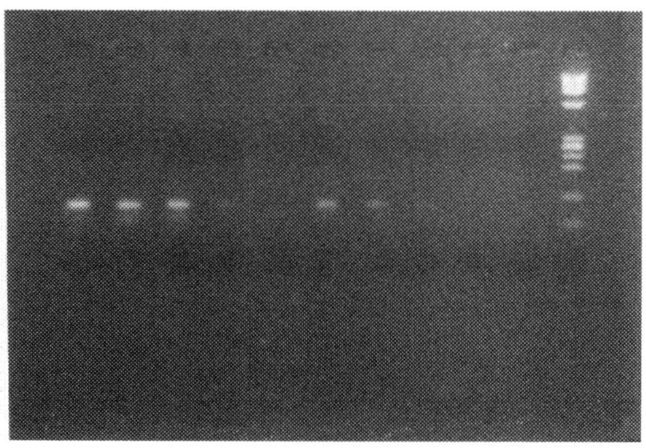

Figure 2: *M.leprae*-specific PCR after 2 x 25 cycles of amplification with outer then inner RLEP primers (99bp product). Left to right: lanes 1-3: *M.leprae* 10^{-4} - 10^{-6} dilutions; lanes 4-5: sample SK9 neat and 10^{-1} dilution; lanes 6-7: sample SK8 neat and 10^{-1} dilution; lanes 8-9 sample SK10 neat and 10^{-1} dilution; lane 10: PCR negative control; lane 11: ØX174*Hae*III digest molecular markers.

Figure 1: *M.leprae*-specific PCR after 45 cycles of amplification with outer primers for the 18kDAg gene (136bp product). Top row (left to right): lanes 1-6: sample SK2 10^{-2} to 10^{-7} dilutions; lanes 7-10: SK1 10^{-2} to 10^{-5} dilutions; lane 11: ØX174*Hae*III digest molecular markers.
Bottom row (left to right): lane 1: *M.leprae* 10^{-5} dilution; lanes 2-3: sample SK1 10^{-6} to 10^{-7} dilutions; lanes 4-9 sample: SK6 10^{-2} to 10^{-7} dilutions; lane 10: PCR negative control; lane 11: ØX174*Hae*III digest molecular markers.

Skeletal Diagnosis of Leprosy

Infective bones changes in leprosy

Keith Manchester
Department of Archaeological Sciences
University of Bradford
Bradford BD7 1DP
England

Abstract

The dual nature of inflammatory bone change in leprosy is discussed, and the concept that leprous neuropathy is the underlying pathological feature which allows the invasion of secondary pathogenic micro-organisms is stressed. The understanding of the pathogenesis of bone change in leprosy benefits from the input of clinicians, radiologists, and palaeopathologists. It is hoped that in leprosy, as in all other diseases manifest in palaeopathology, that such a multidisciplinary input will be of value to us all, and maybe also to the sufferer of this disease.

Keywords

Palaeopathology, inflammation, rhinomaxilla

The bone lesions in leprosy which are the result of the direct action of pathogenic bacteria are of two aetiological types. Although distinct and distinguishable in their individual characteristics, they may, nevertheless, be co-existent and part of the pathological syndrome in leprosy.

The least common osseous lesion is a small discrete spherical cortical or subcortical erosion in the peripheral appendicular skeleton or in the oronasal floor (Job, 1963 and Figure 1). Such lesions, although rare, are recognised in osteoarchaeological specimens but, of course, from such material there can be no possibility of histological verification of the diagnosis. The presumption is that these bony manifestations are the result of the presence of *Mycobacterium leprae* granulomatous lesions at the site. To my knowledge, no verification of these lesions has been made in clinical contexts. More common are those inflammatory lesions of peripheral bones and joints, and the oronasal area, which are ubiquitous in the skeletal manifestations of leprosy. Before examining these lesions it must be stressed, however, that the precise pathological nature of each affected bone does not differ from the change which may occur in infective bone change in non-leprous individuals. The pathognomonic feature of leprosy is the actual distribution of such infective lesions in the skeleton.

Let us then initially consider the pathological features themselves. They consist of cortical inflammatory changes of the palatine process of maxilla, diaphyseal cortical surface, and intra-articular cortical surface. There is inflammatory pitting with subsequent subperiosteal new bone formation (Andersen *et al.*, 1994 and Figure 2). This process consists of osteoclastic absorption resulting, microscopically, in Howship's lacunae. This is promoted by hyperaemia and high oxygen saturation, itself a feature of pyogenic inflammation and, probably, leprous autonomic neuropathy. Coincidentally, or rapidly sequentially, there is osteoblastic activity resulting in the deposition of new woven bone at the site. Organisation of this woven bone into lamellar bone ultimately produces the irregularly smooth outer cortical surface, which is so frequently noted in palaeopathological contexts. It is only if death occurs during an early phase of this infective process that the aggressive woven bone will be noted. Later, classical osteomyelitis may develop with intramedullary abscess and cloaca formation (Figure 3). In leprosy, these bone lesions develop not by haematogenous bacillary spread from a distant infective focus, but by direct bacterial invasion from an adjacent ulcerated and infected skin surface. In palaeopathological, and indeed modern clinical contexts, non-leprous osteomyelitis is predominantly in the metaphyses of the tibia and femur and, to a lesser extent, the long bones of the upper limb. Such lesions are due, in the main, to haematogenous bacillary dissemination. The bacteria responsible are non-specific pyogenic bacilli. In leprosy, *Mycobacterium leprae* causes peripheral neuropathy through leprous granulomatous reaction, with the consequent loss of all modalities of sensation distally. Hence, the skin and deep tissue anaesthesia facilitates skin ulceration with deep penetration to a periosteal surface. This then is a ready track for invasion by environmental pathogens. Non-specific inflammatory change develops throughout the invaded soft tissue and the subjacent skeleton. As noted, the autonomic nerve dysfunction in leprosy contributes to the sequence of pathological events. It is the distribution of multiple skeletal lesions which characterises leprosy in palaeopathological contexts, and which may hint at the immunological type of leprosy in the individual exhibiting the lesions. For example, the changes of the oral and nasal surfaces of the palatine process of the maxilla, which are part of the rhinomaxillary syndrome, which I will consider shortly, only develop in lepromatous or near lepromatous leprosy. Bilaterally symmetrical peripheral changes are also characteristic of lepromatous leprosy, while unilateral findings indicate the tuberculoid end of the immune spectrum. This is because the pattern of distribution of skeletal inflammatory lesions in each peripheral limb is governed by the integrity of the motor nerve supply to that limb.

Inflammatory lesions alone are not readily observed in the metacarpals and hand phalanges. This may be because superficial injury to the hand is immediately noticeable to the leprosy sufferer, whereas such damage to the anaesthetic plantar surface of the foot may be unnoticed for some considerable time. Therefore it is, in palaeopathology, the distribution of osseous inflammatory lesions in the lower limbs which allows us to deduce the degree of motor nerve damage, which is associated with the sensory loss. In a foot with minimal motor damage and, therefore, with normal anatomical architecture, the sites of maximal skin pressure and torsion are the plantar surfaces of the 1st and 5th metatarsophalangeal joints (Figure 4), and the infero-posterior aspect of the calcaneum. Both diaphyses and joint surfaces may be changed. For reasons unknown, inflammatory change of the calcaneum has not been reported in palaeopathology. If the transverse arch of the foot is lost, the trauma and subsequent pathological change will tend towards the mid forefoot (Figure 5). If the longitudinal arch is lost, then the change is mainly of the tarsus and proximal metatarsals.

Whilst considering the lower limb, there occurs, almost always in leprosy, an inflammatory reaction of absorption and repair of the distal half to two thirds of the tibia and fibula. Macroscopically there is gross irregularity of the cortical surface of these bones. Not infrequently there is evidence of active woven bone present. Dr. Johs. Andersen has stated (pers. comm.) that, in his experience, these tibiofibular inflammatory lesions are sterile. This suggests that the inflammatory lesions are due to the local response to toxins from bacterial infection of the feet. From my own clinical and radiological observations in Addis Ababa, Ethiopia it is noted that the tibiofibular lesions may occur in the presence of midfoot plantar soft tissue ulceration, even in the absence of subjacent bone change. Thus, osseous infection in the foot is not an invariable prerequisite for the tibiofibular change. Maybe for the reasons of earlier subjective awareness already proposed for the hand, inflammatory changes of the distal radius and ulna, whilst not unknown, are rare (see Lewis *et al.*, 1995).

The rhinomaxillary changes in leprosy (Andersen and Manchester, 1992), are those features described by Møller-Christensen as *facies leprosa* (1978). There is inflammatory change of pitting and new bone on the oral surface of the palatine process of the maxilla. This commences, and is maximal, in the zone around the palatine suture. The nasal surface is similarly affected and, with the oral change, may lead to perforation of the hard palate. Associated with the nasal surface change there may be inflammatory erosion of the conchae and the nasal septum. The changes occurring at the margins of the pyriform aperture have already been discussed by Ortner (this volume). There is absorption of the anterior nasal spine. There is also absorption of the alveolar process of the maxilla. This process is absolutely bilaterally symmetrical and restricted to the premaxilla. The surface change observed palaeopathologically is not, it is considered, comparable with local pyogenic infection.

In this paper the dual nature of inflammatory bone change in leprosy has been discussed, and the concept that leprous neuropathy is the underlying pathological feature which allows the invasion of secondary pathogenic micro-organisms has been stressed. The understanding of the pathogenesis of bone change in leprosy benefits from the input of clinicians, radiologists, and palaeopathologists. It is hoped that in leprosy, as in all other diseases manifest in palaeopathology, such a multidisciplinary input will be of value to us all, and maybe also to the sufferers of these diseases.

References

Andersen J.G. and Manchester, K. 1992 The rhinomaxillary syndrome in leprosy: a clinical, radiological and palaeopathological study. *International Journal of Osteoarchaeology* 2: 121-129.

Andersen J. G., Manchester K. and Roberts C.A. 1994 Septic bone changes in leprosy: a clinical, radiological and palaeopathological review. *International Journal of Osteoarchaeology* 4:24-30.

Job, C.K. 1963 Pathology of leprous osteomyelitis. *International Journal of Leprosy* 31(1): 26-33.

Lewis, M.E., Roberts, C.A. and Manchester, K. 1995 Inflammatory bone changes in leprous skeletons from the Medieval hospital of St James and St Mary Magdalene, Chichester, England. *Int. J. Leprosy* 63(1):77-85

Møller-Christensen, V. 1978 *Leprosy Changes of the Skull*. Odense: University Press.

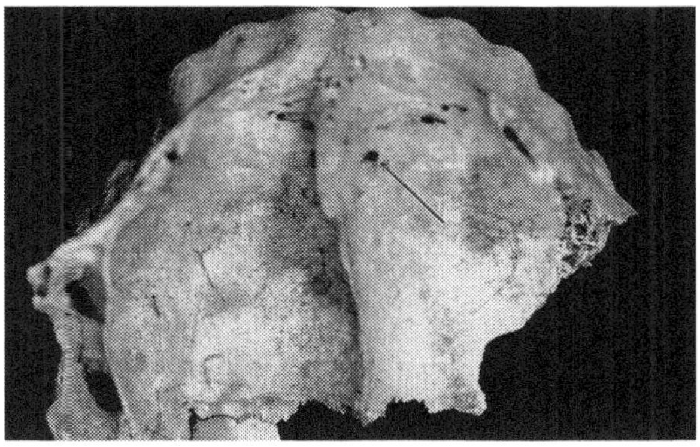

Figure 1: Leprosy. Nasal floor. Arrowed erosive granulomatous pit. Medieval. Hospital of St. James and St. Mary Magdalene, Chichester (Calvin Wells Laboratory, University of Bradford.).

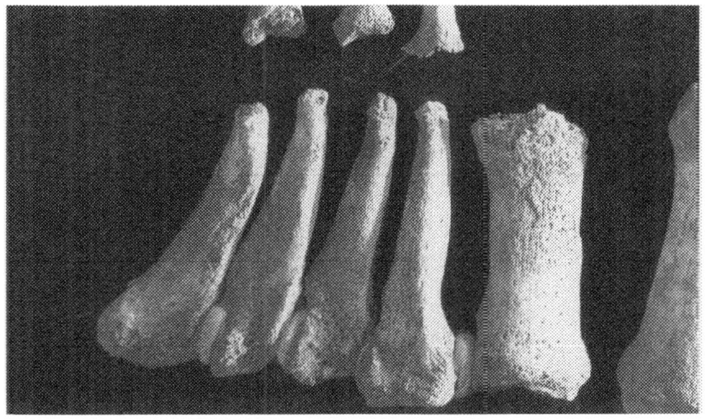

Figure 2: Leprosy. Metatarsals showing inflammatory changes of 1st metatarsal diaphysis. Medieval. Hospital of St. James and St. Mary Magdalene, Chichester (Calvin Wells Laboratory, University of Bradford.).

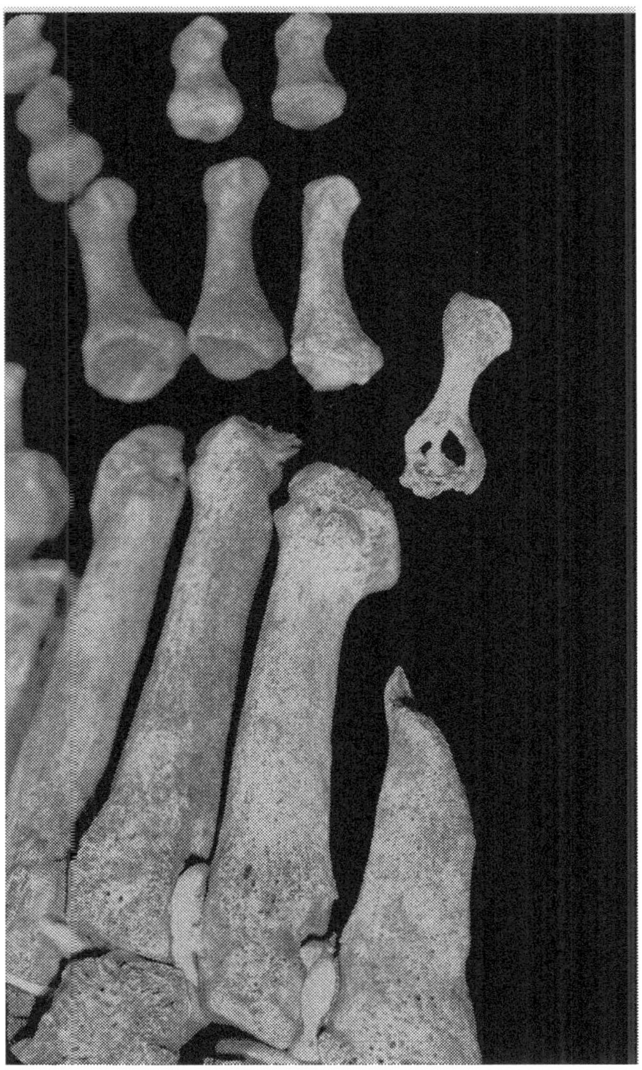

Figure 3: Leprosy. Metatarsals and phalanges. Osteomyelitic lesion in base of the 5th proximal phalanx. Medieval. Hospital of St. James and St. Mary Magdalene, Chichester (Calvin Wells Laboratory, University of Bradford.).

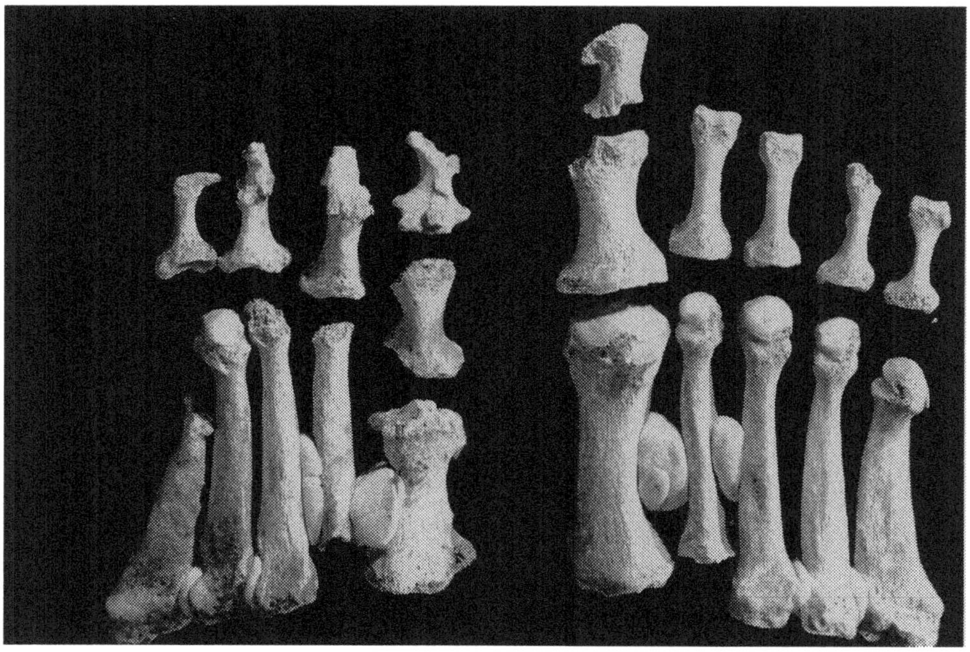

Figure 4: Leprosy. Metatarsals and phalanges. Maximal inflammatory change at 1st and 5th metatarsophalangeal joints. Medieval. Hospital of St. James and St. Mary Magdalene, Chichester (Calvin Wells Laboratory, University of Bradford.).

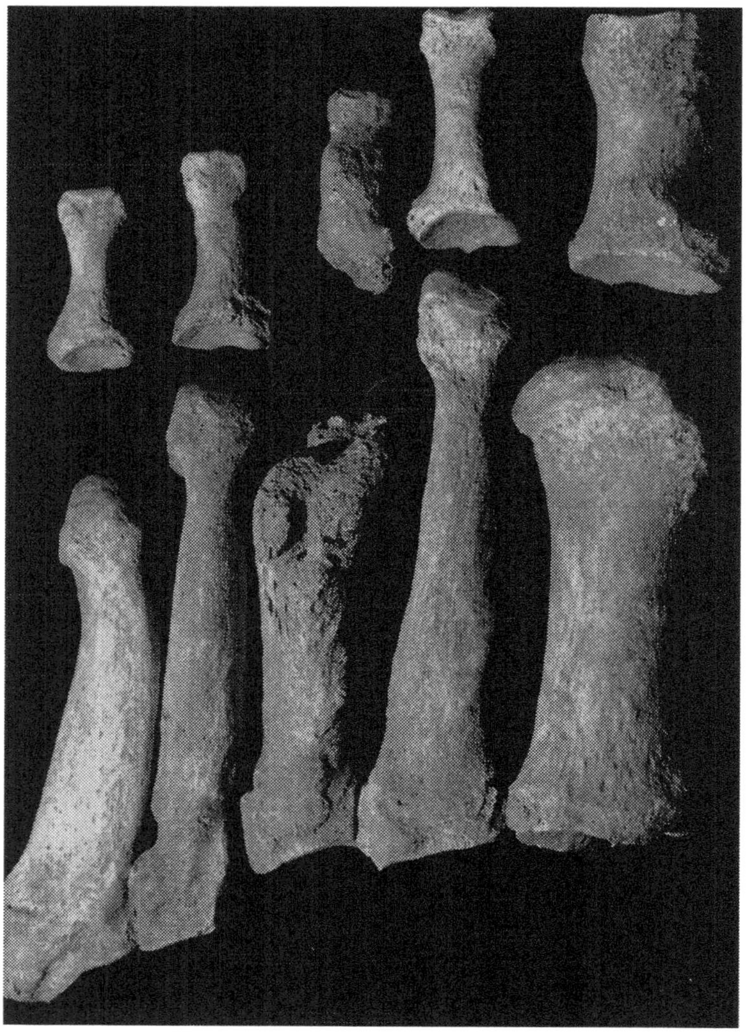

Figure 5: Leprosy. Metatarsals and phalanges. Maximal inflammatory change at 3rd metatarsophalangeal joint. Medieval. Hospital of St. James and St. Mary Magdalene, Chichester (Calvin Wells Laboratory, University of Bradford.).

Observations on the pathogenesis of skeletal disease in leprosy

Donald J. Ortner
Department of Anthropology
National Museum of Natural History
Smithsonian Institution
Washington, D.C. 20560
United States and

Department of Archaeological Sciences
University of Bradford
Bradford
West Yorkshire
BD7 1DP
England

Email: ortner.don@nmnh.si.edu

Abstract

Skeletal involvement in leprosy predilects the peripheral appendicular skeleton and the bones of the rhinomaxillary area of the skull. Bone changes in leprosy include remodelling most commonly associated with bone loss that typically affects the margins of the pyriform aperture, the anterior maxilla and the phalanges, metatarsals and metacarpals. Proliferative periostitis, which occurs most often in the lower extremities and particularly the distal tibiae and fibulae, probably is the result of a generalized inflammatory response to secondary infection. The dentition, particularly the formation of dentine, can be affected if disease onset occurs during dental development. The approximately 400 burials excavated from the medieval cemetery associated with the hospital of St. James and St. Mary Magdalene in Chichester, England provide important insight regarding the skeletal manifestations of many diseases, but particularly leprosy.

The presence of destructive and proliferative lesions in the same individual is a common condition in other chronic infectious diseases including treponematosis. However, the very slow destructive remodelling that is typical at some sites in leprous skeletons raises the question of how this bone destruction occurs. Typically, one sees the end result of bone loss, including the enlargement of the pyriform aperture, the resorption of the anterior nasal spine and the alveolar process in the anterior maxilla, and the reduction of diaphyseal diameter in the metatarsals, metacarpals, and phalanges. One rarely encounters an actively resorbing bone surface (osteoclastic depressions called Howship's lacunae) in an archaeological human skeleton affected by leprosy. Why is evidence of active resorption so rare? Is there a more active destructive phase perhaps early in the development of the disease, is the destruction slow and continuous, or does the destruction occur in multiple and perhaps brief periods of bone loss followed by rapid osteoblastic repair and extended periods of quiescence?

The Chichester skeletal sample contains two skulls that exhibit a more active destructive pathological remodelling of the nasal area. It seems likely that the bone lesions in these two burials reflect an inflammatory process associated with a more acute phase of infection, at least in the nasal region. Because of the chronic nature of leprosy we rarely see this type of lesion but it is important to be aware that bone destruction in the nasal, and possibly other, areas of the skeleton can be active in situations where the person happens to die during the brief period of active bone destruction.

Keywords

Rhinomaxillary pathological remodelling, periostitis

1. Introduction

Evidence of pathology in archaeological human skeletal remains provides important insight regarding the effects of leprosy on the human skeleton. Møller-Christensen's (1953) analysis of skeletons affected by leprosy from archaeological sites in Denmark defined pathological features of the skull associated with this disease. Subsequent research (Andersen and Manchester, 1988, 1992, Andersen *et al.*, 1992, 1994) has provided additional insight on skeletal manifestations of leprosy, particularly some of the effects that are secondary to other pathological conditions associated with this disease.

Palaeopathology also has the potential of contributing to our understanding of pathogenesis through careful analysis and interpretation of skeletal changes apparent in multiple cases of the same disease. There are significant conceptual hazards in doing this. Death, conceivably from causes unrelated to the disease resulting in skeletal pathology, can interrupt a skeletal disease process at almost any stage in development. Some of the stages may involve atypical skeletal manifestations that pose a particular challenge in differential diagnosis. However, there is a tendency for chronic infectious diseases to permit long-term survival and considerable pathological change in the human skeleton. This means that in archaeological examples of skeletal disease we are most often viewing the middle to end-stage of a disease process and not the more acute early phase of skeletal involvement. Some of the implications of the chronic nature of infectious skeletal disease have been explored in a previous paper (Ortner, 1998).

The bone changes of the rhinomaxillary region most typically involve a gradual loss of bone surrounding the pyriform aperture and the alveolar process of the anterior maxilla. There are, in addition, changes of the palatine process of maxilla, both the nasal and oral surfaces, which are mixed but predominantly irregularly osteogenic. Jopling (1978:25) calls the changes around the nasal area "atrophic rhinitis." In a clinical context, this is a useful term in describing soft tissue pathology. For skeletal palaeopathology, I prefer "pathological remodelling" since it accurately describes the osteological process without specifying the pathological process causing the bone changes.

In clinical cases of leprosy, the course of the disease may alternate between phases of acute activity and relative quiescence, depending on several factors including the patient's immune response, pregnancy, malnutrition, and intercurrent other infectious diseases (Jopling, 1978:66-68). Soft tissue nasal lesions are a common and early manifestation of leprosy (Jopling, 1978:25, Job, 1994: 211). They can ulcerate and this may lead to inflammatory stimulation of adjacent tissues. Recurring episodes of a localized infectious "flare-up" are probably associated with rhinomaxillary pathological remodelling seen in adjacent bone. Secondary infection with bone involvement is a common complication of nasal lesions (Pfaltzgraff and Ramu, 1994: 269) and both types of infection may contribute to the pathological remodelling seen around the nose.

Although age of onset for leprosy varies considerably, it tends to be acquired in childhood, and leprous skeletons of young people will be more likely to exhibit early stages of skeletal involvement than skeletons of older individuals. In the Chichester skeletal sample, the large proportion of burials affected by leprosy provides an important source of data on the range of bone changes associated with this disease. In the overwhelming majority of leprous skeletons in the Chichester sample, bone changes are indicative of a slow, chronic pathogenesis that took years to develop. In the pathological remodelling that occurs in the rhinomaxillary area, and in the phalanges, metatarsals, metacarpals, tarsals, and carpal bones, the destroyed surfaces of bone usually are remarkably smooth. In most cases there is no evidence of increased vascularity as indicated by an abnormal increase in vascular holes penetrating the cortical bone surfaces. Evidence of the roughened, finely-pitted, cortical surface, associated with osteoclastic activity, is also uncommon. Clearly, osteoclasts must have been active at some point for destructive pathological remodelling to occur. However, the fact that evidence of active bone destruction is so uncommon suggests that the destructive phase is brief and is quickly followed by slight repair that smooths the surface but normally adds little bone to the destroyed surface.

Since an acute phase is likely to be relatively brief, the probability of finding a skeleton in which the patient died during this phase is much less than the likelihood of finding skeletal changes associated with the chronic stages of an infectious disease. This means that evidence of an acute phase in chronic skeletal disease is likely to be rare, but does represent a condition that can be encountered and needs to be considered in differential diagnosis of pathological skeletal remains.

Two scenarios are offered as potential explanations for rhinomaxillary pathological remodelling. Bone destruction may occur during an early acute phase of the disease followed by surface repair by osteoblasts. A more likely possibility, and one that is congruent with the clinical presentation of nasal infection, is that multiple but relatively brief periods of bone destruction occur throughout the course of the disease. Most likely these destructive periods are a response to local episodes of infection by non-specific, pyogenic agents or by *Mycobacterium leprae* bacteria. These infectious agents stimulate a vascular response conducive to bone destruction (active hyperaemia). A contributing factor in this process may be leprous autonomic neuropathy. These multiple but relatively brief destructive phases are then followed by a different vascular condition (passive hyperaemia) and surface repair by osteoblasts. This is followed by an extended quiescent period in which no significant change occurs in bone. It is, of course, possible that in leprosy both of these scenarios, and perhaps others, occur at different stages in the progress of the disease. In the most probable scenario, the disease is intermittent but progressive. The longer a person survives with the disease the greater will be the pathological remodelling. In some cases of rhinomaxillary pathological remodelling the repair can result in a reduction of the pyriform aperture although the more common result is an enlargement of the aperture.

2. Review of relevant cases

The cases used in this report are from the archaeological site of the hospital of St. James and St. Mary Magdalene in Chichester, England. The skeletal sample from this site is curated and is being studied at the University of Bradford, England. Archaeological evidence indicates that the hospital, and its associated cemetery, were in use from the early part of the 12th century AD until the mid-17th century (Magilton and Lee, 1989). During the earlier history of the hospital its occupants were predominantly men diagnosed with leprosy. As leprosy began to decline in prevalence in the last half of the medieval period, the hospital accepted patients suffering from other diseases including chronic, debilitating diseases and mental deficiency. Toward the end of its history the hospital was used as an almshouse. During the later phases, females and children were included among the residents and their skeletons are found in the cemetery.

In the Chichester skeletal sample (N = ca. 350) 36 skeletons have a pattern of lesions for which leprosy was the probable cause. There are many more skeletons with lesions that could be the result of leprosy but diagnosis is less certain. Of the 36 probable cases of leprosy, 19 (53%) exhibit destructive remodelling of the rhinomaxillary area. Since completing my research on this material an additional 44 skeletons have been added to the sample. I have studied some of the cases of skeletal pathology in this material and will use one case to illustrate variation in early stage rhinomaxillary pathological remodelling. However, the statistical data are based on the initial sample of 350 skeletons.

Most of the examples of rhinomaxillary destructive remodelling studied thus far in the Chichester medieval skeletal sample appear to be chronic, with little or no evidence of active bone destruction at the time of death. These more typical and chronic manifestations are apparent in Chichester burial 34 (Figures 1 and 2). In this case the pyriform aperture is significantly enlarged and the normal sharp edges of the aperture are rounded. Both of these bone changes indicate that destructive pathological remodelling has occurred. However, the smooth cortical surface in the affected area (Figure 2) means that destruction was not active at the time of death. Chichester burial 48 (Figure 3) demonstrates a type of pathological remodelling in which the margins of the pyriform aperture have receded posteriorly but repair has been sufficient to reduce the size of the opening of the aperture. This rhinomaxillary pathology is less frequently encountered than the pathological process that enlarges the aperture.

An atypical expression of rhinomaxillary pathological remodelling is apparent in Chichester burial 187. This burial is that of a young male about eighteen years of age at the time of death and provides an important example of skeletal manifestations of infectious disease, in what appears to be an acute phase of rhinomaxillary remodelling associated with leprosy. The pathological manifestations of Chichester burial 187 have been described elsewhere (Ortner and Connell, 1996). Thus, only the pathological conditions relevant to this discussion are reviewed here. In burial 187, postmortem loss is limited largely to some of the phalanges and carpal bones. The bone tissue itself is very well preserved with very little postmortem destruction of bone surfaces. Some ribs are fragmentary and the cranium has been reconstructed. The major criteria for a diagnosis of leprosy in this case are seen in the bones of the feet and lower limb. The metatarsals and phalanges of the feet demonstrate pathological remodelling typical of leprosy with evidence of inflammatory remodelling of the distal tibiae and fibulae most probably due to the introduction of infectious pathogens through an injury to an anaesthetic foot (Møller-Christensen, 1953, Andersen *et al.*, 1994, Pfaltzgraff and Ramu, 1994: 281, Roberts and Manchester, 1995: 143-145) with subsequent bacterial or toxic extension along muscle and fascial planes. Periosteal bone formation reactive to an inflammatory stimulus extends through the lower ends of the tibia and fibula on both sides. Periosteal reactive bone formation occurs on the visceral surface of some ribs. The most common cause suggested for these rib lesions is tuberculosis (Kelley and Micozzi, 1984, Roberts *et al.*, 1994). This raises the possibility that this individual may also have had tuberculosis, which may have been the cause of death, as is so often the case in leprosy sufferers today.

The lesion of greatest interest occurs in the rhinomaxillary region. The cortical bone of the external surfaces of the frontal processes of the maxilla exhibit a transformation of the normally smooth compact bone, in which the surface is pitted by osteoclastic destruction with minimal, if any, evidence of reactive bone formation (Figures 4 and 5). Although the pitted surface bears a superficial resemblance to the porosity seen in the vascular response to chronic inflammation, the pits rarely penetrate through the cortical surface as is the case with the pathological vascular channels seen in anatomical sites of chronic inflammation in some cases of scurvy (Ortner and Ericksen, 1997). A small fragment of the inferior concha has been preserved and it is more porous and thicker than normal; this bone is often completely lost in the later stages of rhinomaxillary remodelling (Andersen and Manchester, 1992). The internal surface of the frontal process of the maxilla exhibits the pitting seen on the external surface (Figure 6). However, penetration through the cortex is more common. Porous lesions are also apparent on the nasal floor (Figure 7) indicating chronic inflammation in this area.

Burial 187 is unusual in having evidence of active destruction of the rhinomaxillary bone at the time of death (Ortner and Connell, 1996). This burial comes from the eastern end of the site and the dating of this area is more toward the 17th century although the exact date cannot be established with certainty. Leprosy had become a rare disease in England by the end of the 16th century (Manchester and Roberts, 1989) and a case dated in the 17th century would be unusual. While other diagnostic options for burial 187 exist, leprosy is the most likely cause of the lesions apparent in the lower extremities.

Chichester burial 354 demonstrates a more active phase of destructive pathological remodelling. The skeleton is that of an adult male about 27 years of age at the time of death. The margins of the pyriform aperture show the early stages of destructive remodelling (Figure 8). The right lateral interior surface of the pyriform aperture is abnormally porous which indicates the presence of chronic inflammation (Figure 9).

3. Discussion and conclusions

Differential diagnosis in palaeopathology is often a challenge. One of the reasons for this is that the skeletal response to an infectious disease may vary depending on the stage of the disease present when the person happens to die. This means that the palaeopathologist must be aware of, and take into consideration, the atypical manifestations of skeletal infectious disease that may be associated with a more acute phase in a disease whose skeletal manifestations are usually the result of chronic conditions. An acute phase of an infectious disease tends to be more destructive and can, thus, present surface change in bone that is quite different from the chronic skeletal manifestations of the same disease.

Developing the anatomical criteria for identifying these relatively rare manifestations of skeletal infectious disease means careful study of many cases of all the infectious conditions that can affect bone. My own research on the Chichester skeletal sample is instructive. I encountered Burial 187 well into the analysis of the sample. By the time I studied Burial 187 I had seen many examples of the more typical rhinomaxillary destructive remodelling. I had been puzzled by the lack of evidence of active osteoclastic activity in these cases where destruction had obviously occurred. Fortunately the lesions on the lower legs and feet of Burial 187 are virtually pathognomonic for leprosy. This diagnosis is complicated by the possibility that the young man may have had tuberculosis which can also result in destructive changes in the nasal area (*lupus vulgaris*). However, it seems most likely that the rhinomaxillary lesions seen in Burial 187 are caused by leprosy and that death, perhaps from tuberculosis, occurred when active destruction was taking place. Whether the observed bone change was due to the disease being in an early acute phase or in an intermittent "flare-up" of chronic disease is unclear. The bone changes apparent in the feet and lower legs suggest that the young man had the disease for some time despite his young age. This would be an argument for a more acute destructive phase coincident with, and possibly pathogenetically associated with, his terminal illness. Similarly, there is little doubt that the rhino-maxillary destructive pathological remodelling seen in Chichester burial 354 is also due to leprosy. In this case the destruction was perhaps in an even more acute phase at the time of death.

Acknowledgements

The author acknowledges with gratitude the assistance of Mr. Brian Connell, Museum of London, and the staff of the Calvin Wells Laboratory, Department of Archaeological Sciences, University of Bradford, UK. I also appreciate the suggestions made by Dr. Keith Manchester, also with the Department of Archaeological Sciences, University of Bradford, UK, that have materially improved the content of the manuscript. Mrs. Agnes Stix and Mrs. Marcia Bakry are responsible for preparing the figures and I am grateful for their contribution.

References

Andersen, J.G. and Manchester, K. 1988 Dorsal tarsal exostoses in leprosy: a palaeopathological and radiological study. *Journal of Archeological Science* 15:51-56.

Andersen, J.G. and Manchester, K. 1992 The rhinomaxillary syndrome in leprosy: a clinical, radiological and palaeopathological study. *International Journal of Osteoarchaeology* 2:121-129.

Andersen, J.G., Manchester, K. and Ali, R.S. 1992 Diaphyseal remodelling in leprosy: a radiological and palaeopathological study. *International Journal of Osteoarchaeology* 2:211-219.

Andersen, J.G., Manchester, K. and Roberts, C. 1994 Septic bone changes in leprosy: a clinical, radiological and palaeopathological review. *International Journal of Osteoarchaeology*. 4:21-30.

Job, C. K. 1994 Pathology of leprosy. In R. C. Hastings (ed.): *Leprosy*. Second Edition. Edinburgh: Churchill Livingstone, pp. 193-224.

Jopling, W.J. 1978 *Handbook of Leprosy*. Second edition. London: William Heinemann Medical Books Ltd.

Kelley, M.A. and Micozzi, M.S. 1984 Rib lesions and chronic pulmonary tuberculosis. *American Journal of Physical Anthropology* 65: 381-386.

Magilton, J.R. and Lee, F. 1989 The leper hospital of St. James and St. Mary Magdalene, Chichester. In C. A. Roberts, F. Lee and J. Bintliff (eds.): *Burial archaeology: current research, methods and developments*. Oxford: British Archaeological Reports, British Series 211, pp. 249-265.

Manchester, K. and Roberts, C. 1989 The palaeopathology of leprosy in Britain: a review. *World Archaeology* 21: 265-272.

Møller-Christensen, V. 1953 *Ten lepers from Naestved in Denmark: A study of skeletons from a Medieval Danish leper hospital*. Copenhagen: Danish Science Press Ltd.

Ortner, D.J. 1998 Male/female immune reactivity and its implications for interpreting evidence in human skeletal paleopathology. In A. L. Grauer and P. Stuart-Macadam (eds.): *Sex and gender in paleopathological perspective*. New York: Cambridge University Press, pp. 79-92.

Ortner, D. J. and Connell, B. 1996 Early inflammatory change in leprosy affecting the rhinomaxillary region. Case Report no. 19. *Paleopathology Newsletter* 96:1-11.

Ortner, D. J. and Ericksen, M.F. 1997 Bone changes in the human skull probably resulting from scurvy in infancy and childhood. *International Journal of Osteoarchaeology* 7:212-220.

Pfaltzgraff, R. E. and Ramu, G. 1994 Clinical leprosy. In R. C. Hastings (ed.): *Leprosy.* Second Edition. Edinburgh: Churchill Livingstone, pp. 237-287.

Roberts, C. and Manchester, K. 1995 *The archaeology of disease*, Second Edition. Ithaca, New York: Cornell University Press.

Roberts, C., Lucy, D. and Manchester, K. 1994 Inflammatory lesions of ribs: an analysis of the Terry Collection. *American Journal of Physical Anthropology* 95:169-182.

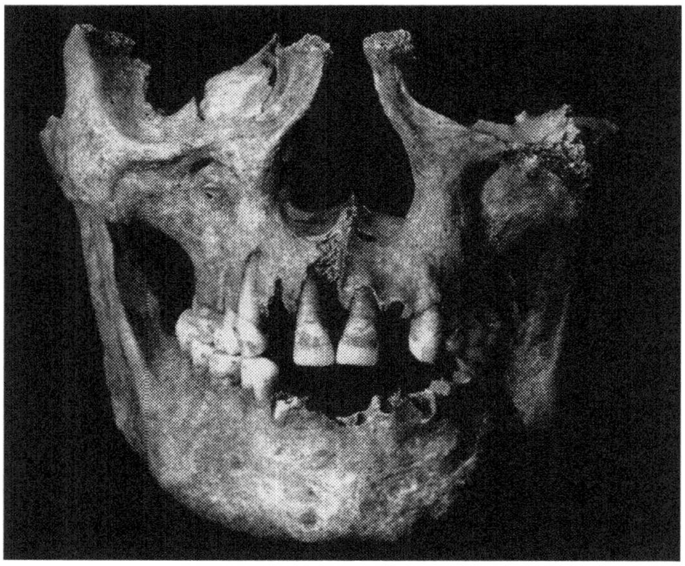

Figure 1: Anterior view of the maxilla of burial 34 from the cemetery of the medieval hospital of St. James and St. Mary Magdalene, Chichester, England. The remains are those of an adult male. The margins of the pyriform aperture are rounded and the alveolar process of maxilla exhibits destructive pathological remodelling. The compact bone surface at the margins of the pyriform aperture are smooth indicating the absence of active inflammation of the nasal area at the time of death. Note in particular that the alveolar bone associated with the central incisors is diminished and has exposed a considerable portion of the tooth root.

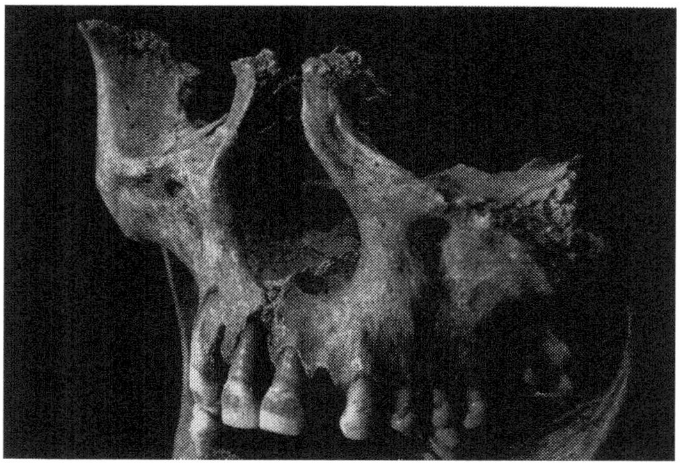

Figure 2: The margin of the right pyriform aperture showing the rounded anatomy and the smooth surface of the compact bone of burial 34 from Chichester.

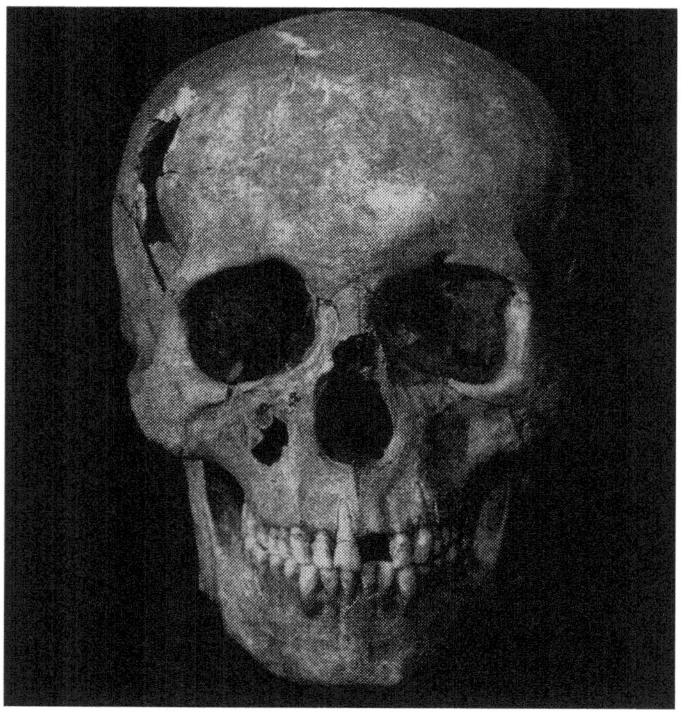

Figure 3: Anterior view of the skull of Chichester burial 48. Estimated age at death is 35 and the sex is male. Note the remodelling at the margins of the pyriform aperture that has resulted in a posterior migration of the margins. This would have been associated with a flattened nasal area in life. Although there has been considerable destructive pathological remodelling, the size of the aperture is abnormally narrow, particularly on the right side. This may indicate that remodelling resulting in loss of bone on the anterior surface of the margin also involved some bone formation on the medial margins.

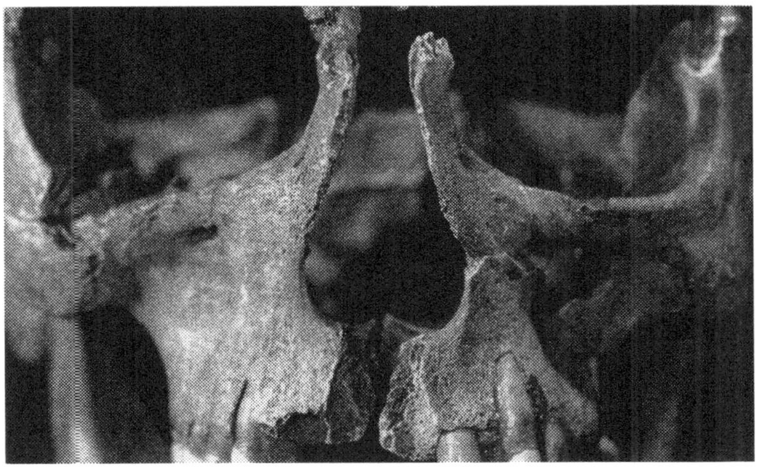

Figure 4: Anterior maxilla of burial 187 from the cemetery of the medieval hospital of St. James and St. Mary Magdalene, Chichester, England. The burial consists of the well-preserved skeleton of a male about 17-18 years of age at the time of death. Note the pitting and porosity of the compact bone adjacent to the pyriform aperture. This is indicative of active inflammation at the time of death associated with destructive remodelling of bone in the rhino-maxillary area.

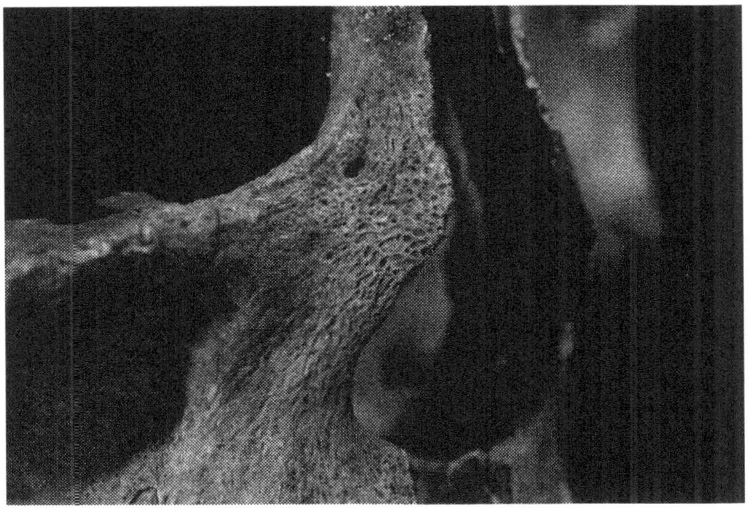

Figure 5: Detail of the right maxilla of Chichester burial 187 showing the pitting of the compact bone and the penetration of some of the pits through the compact bone surface.

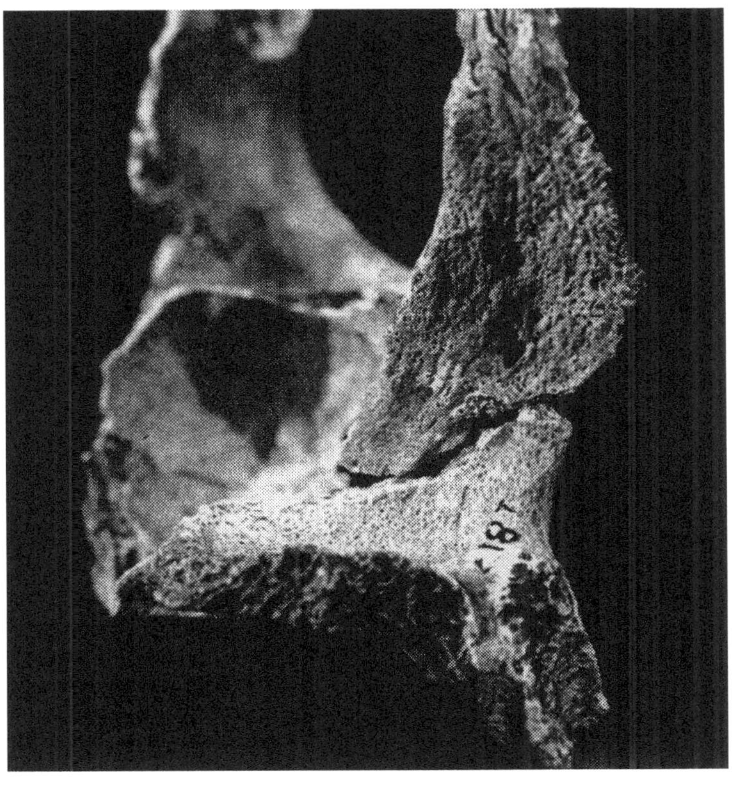

Figure 6: Detail of the compact bone surface forming the left lateral side of the pyriform aperture in Chichester burial 187. The pitting and porosity penetrating the bone indicates active inflammation of the overlying soft tissue at the time of death.

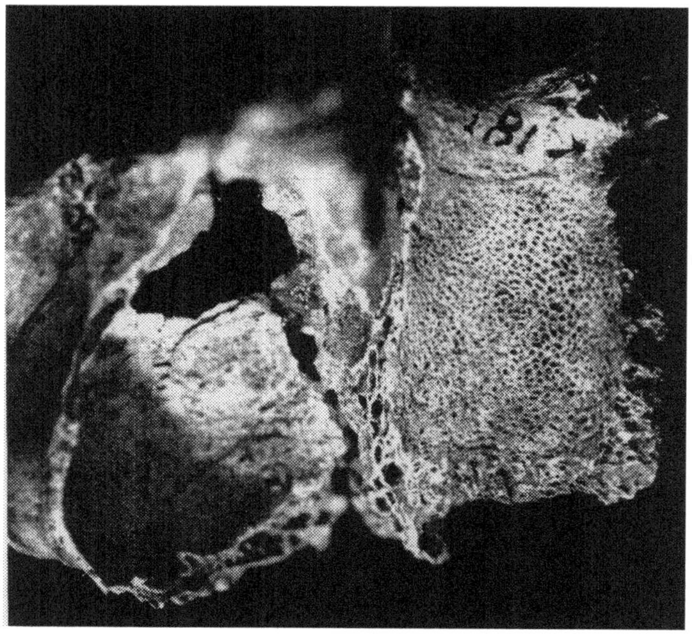

Figure 7: The left nasal floor of Chichester burial 187. The pitting and porosity of the compact bone indicates that this portion of the pyriform aperture was also affected by active inflammation.

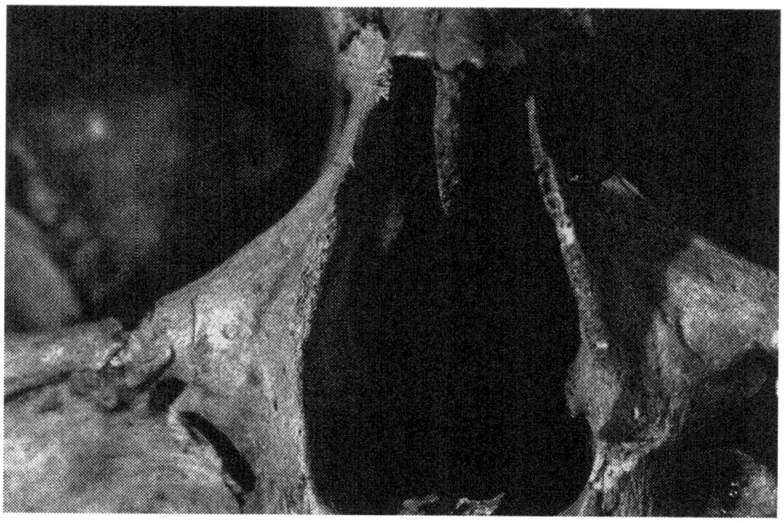

Figure 8: Anterior view of the nasal area of Chichester burial 354. The individual was an adult male about 27 years of age when he died. Note the irregular, porous edge of the pyriform aperture indicative of active inflammation and destructive pathological remodelling.

Figure 9: Interior surface of the right lateral pyriform aperture of Chichester burial 354. Note the fine porosity indicative of chronic inflammation.

Rhinomaxillary syndrome in the absence of leprosy: an exercise in differential diagnosis

Della Collins Cook
Department of Anthropology
Indiana University
Bloomington
IN 47405-6101
United States

Email: cook@indiana.edu

Abstract

Features of the rhinomaxillary syndrome can occur in isolation and in combination in certain conditions other than leprosy. Paleopathologists, epidemiologists, and medical historians have reached a consensus that leprosy was not present in the Precolumbian New World. Case studies from prehistoric North America are presented. In each a diagnosis other than leprosy is offered. This material suggests that the concept of loss of the anterior nasal spine as pathognomonic of leprosy by some researchers overstates the specificity of this lesion.

Keywords

Leprosy, treponematosis, trauma, Americas, palaeopathology

1. Introduction

As an Americanist and a paleopathologist, I have never encountered a plausible example of ancient leprosy in North America. Virtually everyone contributing to paleopathology today regards leprosy as absent from the Precolumbian New World (Merbs, 1992, Aufderheide and Rodriguez-Martin, 1998). This consensus has existed for the whole of the twentieth century. I will make use of this consensus to argue that features of the rhinomaxillary syndrome are less specific to leprosy than the Old World literature suggests. However, in the early days of paleopathology, there were a number of claims for the presence of leprosy in the Americas, most of them using art objects as evidence.

2. Recent origins of New World leprosy

The argument for ancient New World leprosy crystallized in a debate between Albert Ashmead and Rudolph Virchow in the 1890s. Ashmead attributed nasal and facial deformities in Olmec sculpture and Peruvian pottery to syphilis, tuberculosis, or leishmaniasis, arguing vigorously against leprosy as an appropriate diagnosis in these art objects. Rudolph Virchow, who had seen much leprosy among the living in Portugal and Norway, considered Precolumbian leprosy to be a possibility (Ashmead, 1896, Ackerknecht, 1953). Olmec and later Mesoamerican art representations of people with cleft or short upper lips are now interpreted as humans with jaguar attributes by more recent scholars (Verut, 1973, Covarrubias, 1957, Wells, 1964). Similarly, Mochica vessels with deformed noses and lips are now attributed to cleft lip, culturally patterned trauma, and leishmaniasis (Urteaga-Ballon, 1991, Pessoa and Martin, 1977). Treponematosis has also been proposed as a candidate for explaining nasal lesions in Peruvian pots (Tello, 1909). Kaufmann-Doig (1979) discusses historical evidence that mutilation of the nose, hands and feet was used as a punishment for adultery in protohistoric and early Colonial Peru. The Mochica huacos may thus represent the consequences of violating similar Precolumbian norms.

Evidence from historical accounts and from skeletal material was also advanced in the early part of this century, but none of this evidence has any currency (Wells, 1964; Merbs, 1992). The one case report that remains unchallenged in our literature is a 19th century skeleton from British Columbia in which Møller-Christensen diagnosed leprosy (Steinbock, 1976; Merbs (1992). It is of course irrelevant to the Precolumbian presence of the disease, and this specimen, if it is still extant, ought to be re-examined with a differential diagnosis in mind.

In contrast, Postcolumbian New World populations exhibit leprosy in a pattern that supports the view that the disease was introduced. In a study modelled on Friedlander's demonstration that the present distribution of leprosy in Mexico reflects the past immigration of African slaves, Mallory has shown that the prevalence of leprosy in Mexico corresponds closely to the proportion of African ancestry in the present-day population and to the former distribution of sugar estates that used African slaves as their labour force (1981). The rapidly declining focus of human leprosy in Louisiana is also thought to have been introduced with African slaves. Washburn (1950) reports the advent and decline of a leprosy focus among late 19th century Scandinavian immigrants to Minnesota. The Minnesota focus represents immigration to North America from the isolated rural communities that were the last remnants of leprosy in Northern Europe. These three examples illustrate the geographically restricted and transitory character of New World leprosy. The sylvatic leprosy in armadillos in the Southeastern United States may suggest a more complicated picture, but this is best seen as a relatively recent transmission from humans to one of the few animals that is prone to infection with the organism (see Merbs, 1992 for review). A recent report of leprosy in a New World monkey (Aufderheide and Rodríguez-Martin, 1998) is in error, because mangabeys are not New World monkeys. Most human leprosy cases in North America today occur in Cambodian and other Southeast Asian immigrants. This disjunct distribution argues strongly that leprosy is a recently introduced disease into this part of the world.

3. *Facies leprosa* and the anterior nasal spine

Møller-Christensen's concept of a triad of osseous lesions underlying *facies leprosa* is one of the monuments of paleopathology. Most of us hope that our work in paleopathology will prove relevant to health in the living, but Møller-Christensen's work provides one of the few examples of an advance in pathology that originated in the study of ancient remains. He recognized bony changes in crania from leprosaria before they were noticed underlying soft tissue lesions in living patients with Hansen's disease (Andersen and Manchester, 1992, Manchester, 1991, and contributors to the present volume). One of the three elements of Møller-Christensen's triad is destruction of the anterior nasal spine. This lesion has been viewed as pathognomonic of leprosy (Fornaiciari *et al.*, 1999), that is to say, the presence of this bone lesion is sufficient evidence to diagnose leprosy in an individual. Wells (1964) has suggested that bone loss around the maxillary incisors may be sufficient to diagnose leprosy. Roberts (1986) has similarly argued that in leprous odontodysplasia, resorption of the maxillary incisor roots is strong evidence for leprosy.

"Pathognomonic" is rather strong language for a field with the many limitations that we must recognize in paleopathology, but several contributors to the leprosy literature have argued that either destruction of the anterior nasal spine or the whole of Møller-Christensen's triad is indeed pathognomonic. If this is so, we should not expect to see destruction of the anterior nasal spine in the Precolumbian New World. This paper presents some very interesting New World noses that call into question the reliance on this lesion for diagnosing ancient leprosy.

4. Case studies

4.1 Illinois, Late Archaic: 11-C36-56

First, I want to resurrect Ashmead's concept of treponematosis as a cause for lesions mimicking *facies leprosa*. The case illustrated in Figure 1(a and b) is from the Pete Klunk Mounds in west-central Illinois, and dates to about 900 BC. In this young adult male, all margins of the nasal aperture are thickened and rounded. There are small nasal spurs. There is reactive new bone formation, particularly in the floor of the nasal cavity, on the oral surface of the bony palate, and in the lacrimal canals. The bony palate is thickened and there is a small palate perforation. The maxillary sinuses are not affected, and there is no evidence for dental lesions that might have contributed to these lesions. The maxillary incisors were not recovered, but the holes for the roots were very short. Maximum depth of the sockets ranges from 6 to 8 mm. from the alveolar ridge. The alveoli are conical, and their shape does not suggest that the roots were resorbed or constricted. The anterior nasal spine is somewhat damaged, but it is partially resorbed, broad and irregular (Figure 1b). The midline crest formed by the maxillary and palatine bones is elevated, porous, and irregular, particularly on the margins of the palate perforation. These lesions are consistent with treponematosis, and suggest the gangosa deformity in yaws and nasal and palatal lesions of congenital syphilis (Steinbock, 1976). Similar nasal lesions are common from the Archaic through Mississippian times in Eastern North America, and are commonly associated with generalized periosteal reaction on the long bones, sabre tibias, and lesions of the cranial vault. The enormous literature on treponematosis in the Americas provides many similar examples.

Short incisor roots in 11-C36-56 indicate that these lesions began in childhood. Short incisor roots have been reported in the clinical literature on congenital syphilis (McCormick, 1955). I have illustrated comparable treponemal lesions of the nasal aperture in other adults and children from west-central Illinois (Cook, 1994). One of these, Yokem Mounds Yo3-12, is a twelve-year old Mississippian child with sabre shins and nasal cavity lesions. This child has short anterior maxillary tooth roots, and the incisor roots (Figure 2) are constricted and irregular. These teeth are strikingly similar to those reported by Roberts in leprogenic odontodysplasia (1986), except that Yo3-12 is younger than her examples. An Arikara child, NMNH 382,929, from the Protohistoric Mobridge site has very similar constricted central incisor roots associated with reactive new bone formation in the bony palate and a healed injury to the mandible. This is interesting, because the Mobridge series shows little or no evidence for treponematosis. Both treponematosis and trauma may thus account for short, constricted incisor roots. In contrast to the destructive course of leprogenic odontodysplasia, the tooth root lesions reported here appear to be developmental in nature.

These points are not new. In his volume on Pecos Pueblo in New Mexico, Hooton presented two examples of nasal lesions that have resulted in complete loss of the anterior nasal spine. He suggested that these lesions were the result of syphilis (Hooton, 1930). The enormous literature on the treponematoses in the Americas includes many similar examples (for example, Hutchinson and Weaver, 1998). Manchester (1994) makes substantially this same point in his discussion of lesions resembling syphilis in Medieval leprosaria: nasal lesions of leprosy are easily mistaken for syphilis, and, we must add, vice versa.

4.2 Arizona Pueblo III FMNH 46684

Other infectious and inflammatory lesions can mimic *facies leprosa*. Figure 3a illustrates a young adult male, Burial 11, from Carter Ranch Pueblo, Arizona, dating to 1100-1225 A.D. (Danforth *et al.*, 1994). There is a healed, dense-walled destructive lesion of the maxillary alveolar bone extending from the alveolar ridge to the anterior nasal spine. The maxillary incisor roots are exposed throughout their anterior aspects in this lesion, but have been glued in place against the remaining posterior alveolar wall. Since they were recovered with the skeleton, they must have been supported by soft tissue in life. There is no discoloration that might suggest that these teeth had not been vital for a length of time in life. Margins of the lesion are rounded, and the anterior nasal spine is extensively resorbed (Figure 3b). What remains is a bifid, rounded base at the superior margin of the lesion. The crown of the right lateral incisor has been destroyed by caries. The remaining incisors are heavily worn and have exposed pulp cavities. Periodontosis secondary to caries or pulp exposure, alveolar cyst or tumour, and healed noma are plausible diagnoses here. Noma, or acute necrotizing gingivitis, has received little attention in palaeopathology (Pahl and Undeutsch, 1991), but cysts, ulcerations and other complicated sequelae of oral infections must have been more common in the past than our literature suggests.

4.3 Illinois Middle Woodland: TUI-7K

Developmental lesions can also mimic *facies leprosa*. Figure 4 shows a Middle Woodland skull from the Snyders Mound in Illinois. The excavation is not well-documented, but a plausible date is 150 B.C.—150 A.D. This small skull is probably male. Light dental wear and lack of suture obliteration suggest that this is a young adult. The skull is strikingly asymmetrical, with the asymmetry focused around the left mastoid, but visible in all planes. There is corresponding asymmetry of the mandible. The cranial base is deviated at least fifteen degrees from the midsagittal plane of the occipital. The left occipital condyle is elevated and saddle-shaped. This deformity, associated with fixation of rotation of the neck on the affected side, for example in individuals with an articular paramastoid tubercle, is an anomaly that is not present in this case.

The nasal bones are quite flat, and the nasal processes of maxilla are spurred. The inferior nasal margins are rounded, and there is no anterior nasal spine. The midline palatal suture is unfused on the facial surface of the maxilla and on the palate. Its margins form a small crest extending from prosthion well into the nasal aperture, where it joins a thick, extensive crest supporting the septum.

There are several plausible diagnoses here. Prechordal cranial base defects (Barnes, 1994), hemifacial microsomia (Akyuz *et al.*, 1998), and torticollis with pressure deformity of the external nose against the chest or shoulder are all possibilities. Post-cranial elements from this individual are fragmentary and mixed with remains of several other skeletons, making diagnosis difficult. However, it is clear that there has been a longstanding developmental disturbance. It is equally clear that this is not leprosy.

4.4 New Basketmaker: Cimarron

Soft-tissue lesions can affect the bone in ways that mimic *facies leprosa*. Figure 5 shows a Basketmaker mummy from New Mexico, that of a seven-year-old child. This child shows evidence for a large soft-tissue mass in the left side of the nasal cavity. The nasal septum is displaced to the right. The medial wall of the left maxillary sinus and the hard palate is thinned and defective, and the medial wall of the left orbit is displaced laterally. There is no reactive new bone formation, and the dentition is unremarkable. Ethan Braunstein and I are preparing this case for publication as an example of a nasal polyp.

The anterior nasal spine is absent. The midline crest that supports the vomer is so deviated to the right that the right nostril may have been closed, and we infer that the left nostril was occluded by the soft-tissue mass. The effect on the soft-tissue of the external nose is unclear, but the soft-tissue septum must have been displaced. Resorptive changes on the nasal sill as well as alveolar bone loss are visible in Figure 5. Pressure erosion can thus result in loss of the anterior nasal spine.

4.5 Arizona Pueblo III: FMNH 42678

Trauma can mimic *facies leprosa*. Figure 6a is an elderly female from Carter Ranch Pueblo, Arizona, with a healed, depressed nasal fracture, as well as a healed fracture of the left humerus and compression fractures of three vertebrae. There is a small, healed superficial compression fracture of the outer table of the frontal bone visible in Figure 6b. The posterior teeth were lost in life, and the remaining anterior teeth are very worn. The anterior nasal spine is reduced, and the midline crest behind it, within the nasal cavity, is expanded. A diagnosis of multiple, or perhaps repeated, trauma is warranted, and it is doubtful that anyone would call these lesions leprosy. However, this case illustrates the potential confusion that may arise from less extensive trauma to the nose that resulted in crushing of the septum or the external cartilaginous structures. I have seen comparable resorption of the anterior nasal spine associated with new bone formation and marked expansion of the vomer in a seven-year old modern child abuse victim from Indiana (Walker *et al.*, 1997). Clinicians considered tumour of the vomer as a possible diagnosis in a child with a flat external nose, but missed the significance of a lengthy history of other skeletal trauma. Relapsing polychondritis is a chronic inflammatory condition that can result in collapse of the bridge of the nose and thickening of the nasal septum that might occasionally account for features attributed here to trauma and treponematosis (Valesini *et al.*, 1995).

4.6 A small anterior nasal spine as a normal feature

Finally, a small nasal spine may be normal. There are substantial individual and population differences in the development of the nasal spine, and size alone cannot be a criterion for pathology. Projection of the nose is approximately three times the length of the nasal spine (Krogman, 1962:267). Hence, extremely reduced nasal spines are encountered in groups and individuals with relatively small noses (for example, Twiesselmann, 1942).

5. Discussion and conclusions

Is destruction of the nasal spine pathognomonic of leprosy? Clearly it is not in the New World. Perhaps we need to rethink the diagnosis of leprosy of such diverse historical personages as the Swabian-Italian king, Henry VII (Fornaciari *et al.*, 1999), or Robert the Bruce (Wells, 1964), as well as the many anonymous dead. Treponematosis, trauma, and developmental lesions may mimic leprosy if our focus is narrowed to loss of the anterior nasal spine (Manchester, 1994; see also Valentin *et al.* 1999).

I have selected cases for presentation here in which I can support an alternative diagnosis. The size of the several series from which these cases are drawn suggests that destruction of the anterior nasal spine can be expected in about one percent of Precolumbian North American skeletal material. No hand or foot lesions suggestive of leprosy have been reported in any of these series, and there is no compelling reason to argue that leprosy was present in any of them. A similar picture exists for anterior alveolar bone loss. We need to rethink the specificity of the rhinomaxillary syndrome as evidence for ancient leprosy.

Acknowledgements

I would like to thank the Field Museum of Natural History, the Evansville Museum, the Smithsonian Institution, and the Gilcrease Museum for access to several of the specimens discussed here.

References

Ackerknecht, E. 1953 *Rudolf Virchow: Doctor, Statesman, Anthropologist*. Madison: U. Wisconsin Press.

Akyuz, S., Mumcu, G., Atasu, M., Atalay, T., Eravuz, M., Kiziltan, M. and Cimilli, T. 1998 Hemifacial microsomia: oral, clinical, genetic and dermatoglyphic findings. *J. of Clinical and Pediatric Dentistry* 23:63-7.

Andersen, J.G. and Manchester, K. 1992 The rhinomaxillary syndrome in leprosy: a clinical, radiological and paleopathological study. *International J. of Osteoarchaeology* 2:121-30.

Ashmead, A.S. 1896 Pre-Columbian syphilis in Yucatan. *American Anthropologist* 9:106-9.

Aufderheide, A.C. and Rodríguez-Martin, C. 1998 *Cambridge encyclopedia of human paleopathology*. Cambridge: University Press.

Barnes, E. 1994 *Developmental defects of the axial skeleton in paleopathology*. University Press of Colorado: Niwot.

Cook, D.C. 1994 Dental evidence for congenital syphilis (and its absence) before and after the conquest of the New World. In O. Dutour, G. Pálfi, J. Berato and J-P. Brun (eds): *L'Origine de la syphilis en Europe: avant ou apres 1493?* Paris: Editions Errance, pp. 169-175.

Covarrubias, M.1957 *Indian Art of Mexico and Central America*. New York: Knopf.

Danforth, M.E., Cook, D.C. and Knick, S.G. 1994 The human remains from Carter Ranch Pueblo, Arizona: health in isolation. *American Antiquity* 59:88-101.

Fornaciari, G., Maliegni, F. and De Leo, P. 1999 The leprosy of Henry VIII: incarceration or isolation? *Lancet* 353:758.

Hooton, E.A. 1930 *The Indians of Pecos Pueblo*. New Haven: Yale University Press.

Hutchinson, D.L. and Weaver, D.S. 1998 Two cases of facial involvement on probable treponemal disease from Late Prehistoric Coastal North Carolina. *International J. of Osteoarchaeology* 8:444-453.

Kaufmann-Doig, F.1979 *Sexualverhalten im alten Peru*. Lima, Peru: Kompaktos.

Krogman, M.W. 1962 *The human skeleton in forensic medicine*. Springfield, Illinois: Charles C Thomas.

Mallory, L.D. 1981 La lepra en el Occidente de Mexico. *Pantoc Guadalajara* 1:17-36.

Manchester, K. 1991 Tuberculosis and leprosy: evidence of interaction of disease. ? In D. J. Ortner and A.C. Aufderheide (eds): *Human paleopathology. Current syntheses and future options*. Washington D.C.: Smithsonian Institution Press, pp. 23-35.

Manchester, K. 1994 Rhinomaxillary lesions in syphilis: differential diagnosis. In O. Dutour, G. Pálfi, J. Berato and J-P. Brun (eds): *L'Origine de la syphilis en Europe: avant ou apres 1493?* Paris: Editions Errance, pp. 79-80.

McCormick, J. 1955 Prenatal syphilis II--pathological sequence and clinical features. *Alabama Dental Review* 3:3-6.

Merbs, C.F.1992 A new world of infectious disease. *Yearbook of Physical Anthropology* 35:3-42.

Pahl, W.M., and Undeutsch W. 1991 Nomacancer aquaticus: first indication of the skin involving disease in ancient Egypt? In D. J. Ortner and A.C. Aufderheide (eds): *Human paleopathology. Current syntheses and future options*. Washington D.C.: Smithsonian Institution Press, pp. 297-304.

Pessôa, S.B. and Martin, A.V. 1977 *Pessôa parasitologica médica.* 10th edition. Rio de Janeiro: Guanabara: Koogen.

Roberts, C. 1986 Leprogenic odontodysplasia. In E. Cruwys and R. Foley (eds): *Dental anthropology.* British Archaeological Reports International Series 291. Oxford, pp. 137-147.

Steinbock, R.T. 1976 *Paleopathological diagnosis and interpretation.* Springfield, Illinois: Charles C. Thomas.

Tello, J.C. 1909 *La antiguedad de la sifilis en el Perú.* Lima, Peru: Sanmarti.

Twiesselmann, F. 1942 *Contribution a l'etude anthropologique des Pygmees de l'Afrique Occidentale.* Brussels, Belgium: Musee Royal d'Histoire Naturelle de Belgique.

Urteaga-Ballon, O. 1991 Medical ceramic representation of nasal leishmaniasis and surgical amputation in ancient Peruvian civilization. In D. J. Ortner and A.C. Aufderheide (eds): *Human paleopathology. Current syntheses and future options.* Washington D.C.: Smithsonian Institution Press, pp. 95-101.

Valentin, F., Herrscher, E. and Mafart, B-Y. 1999 Rhinomaxillary changes in a New Caledonia cranium. Palaeopathological differential diagnosis. *Int. J. Osteoarchaeology* 9:374-378.

Valesini, G., Priori, R. and Conti, F. 1995 Saddle nose. *New England J. of Medicine* 333(8):525-6.

Verut, D.D. 1973 *Precolumbian dermatology and cosmetology in Mexico.* New York: Schering (Chanticleer Press).

Walker, P.L., Cook, D.C. and Lambert, P.M. 1997 Skeletal evidence for child abuse: a physical anthropological perspective. *J. of Forensic Sciences* 43:196-207.

Washburn, W.L. 1950 Scandinavian settlers in the Upper Mississippi Valley, 1864-1932. *Bull. Hist. Med.* 24:113-48.

Wells, C. 1964 *Bones, bodies and disease.* New York: Praeger.

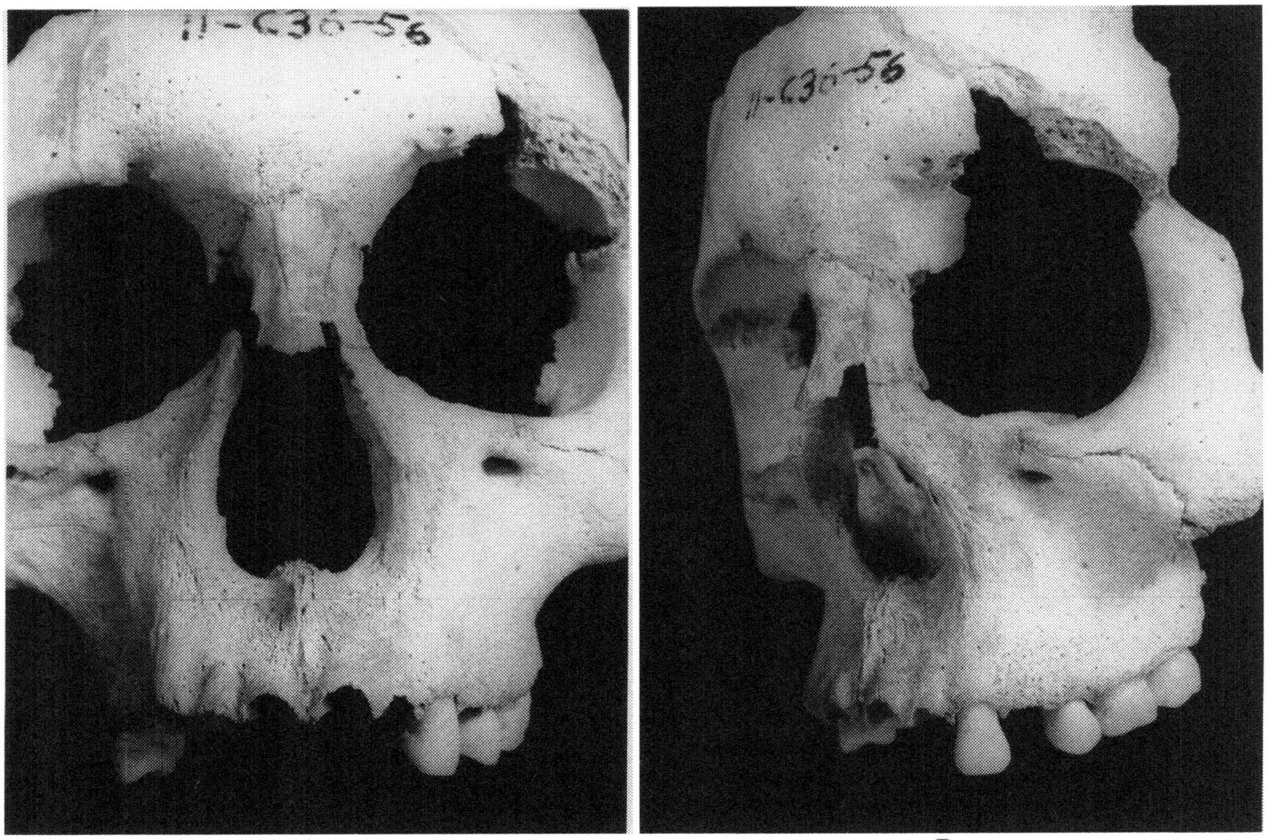

A B

Figure 1 A and B: Klunk (Archaic): treponematosis

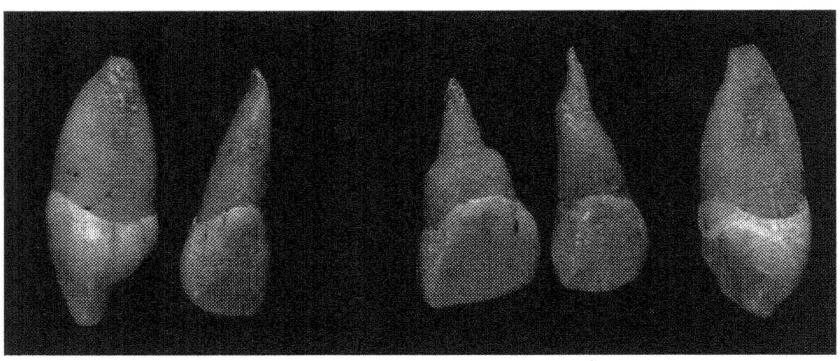

Figure 2: Yokem (Mississippian): dysplastic tooth roots in treponematosis

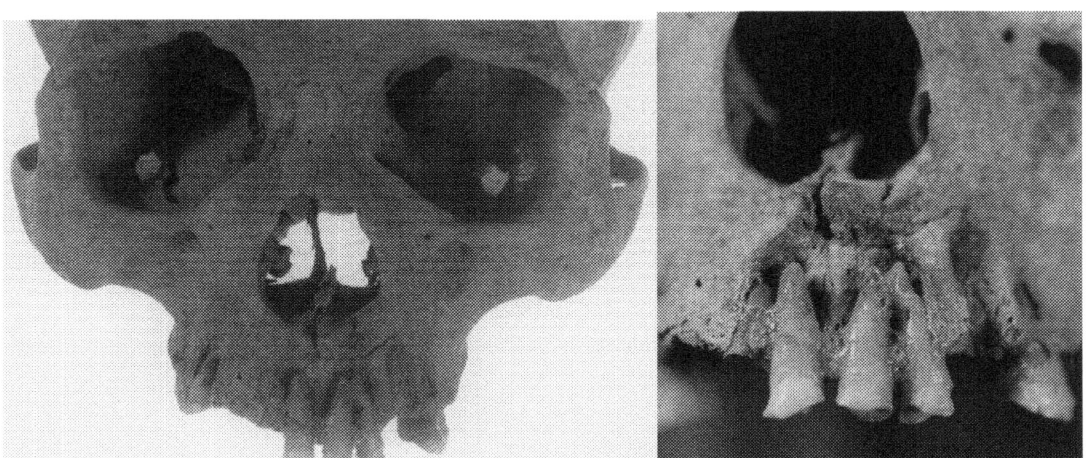

A B

Figure 3 A and B: Carter Ranch (Pueblo): destructive lesion of alveolar bone

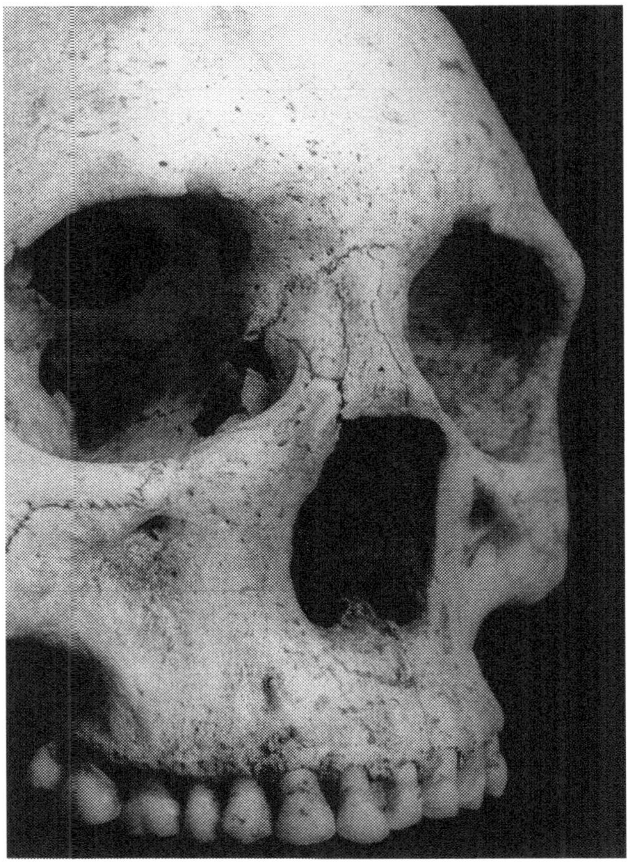

Figure 4: Snyders (Middle Woodland): disturbance of craniofacial development

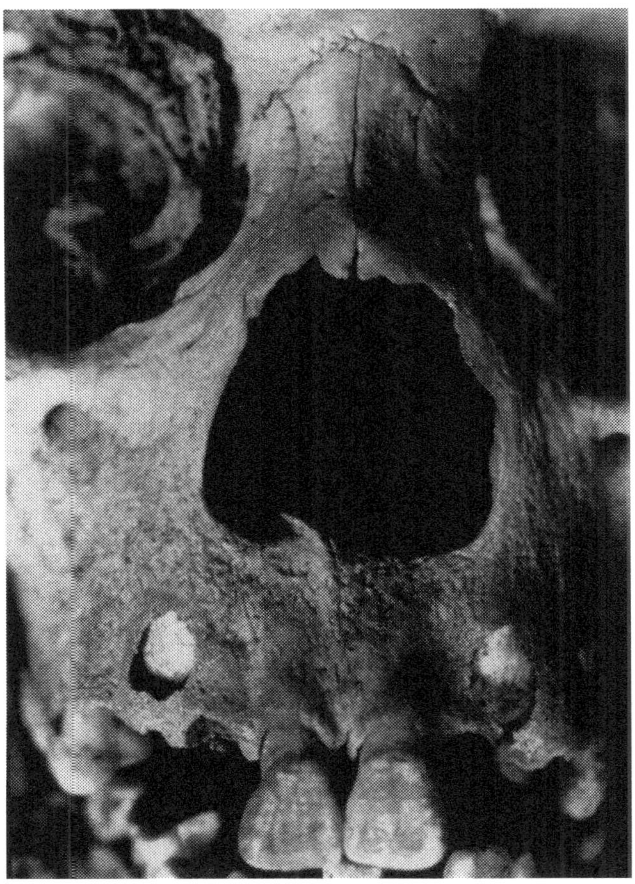

Figure 5: Cimarron (Basketmaker): pressure erosion from expansive soft-tissue lesion

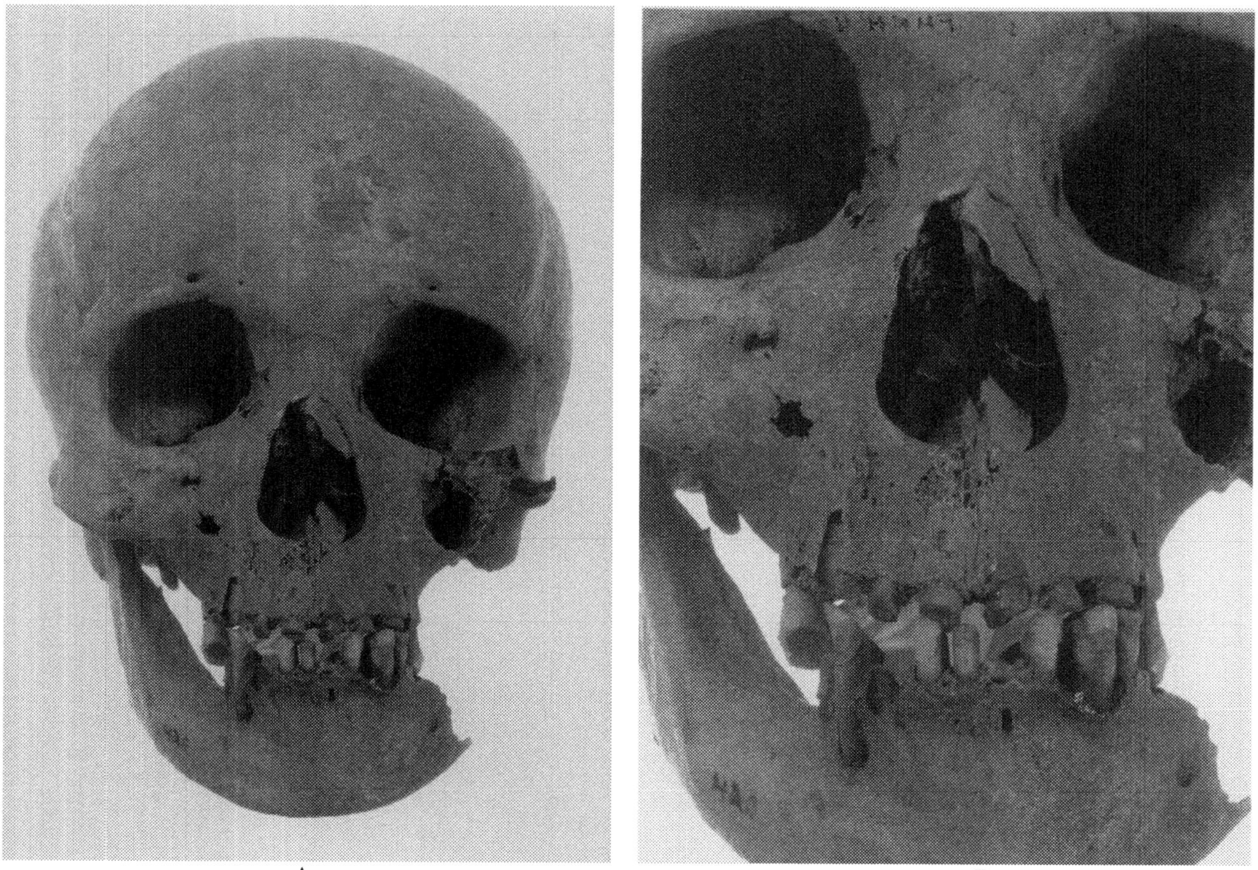

A B

Figure 6A and B: Carter Ranch (Pueblo): healed facial trauma

Diagnosis of leprosy in skeletons from an English later Medieval hospital using histological analysis

Michael Schultz[1] and Charlotte Roberts[2]

[1]Zentrum Anatomie
Georg-August-Universität
Kreuzbergring 36
D-37075 Göttingen
Germany

Email: mschult1@gwdg.de

[2]Department of Archaeology
University of Durham
South Road
Durham DH1 3LE
England

Email: C.A.Roberts@durham.ac.uk

Abstract

The diagnosis of leprosy in skeletal materials from archaeological sites can be problematic. Periostitis on the tibiae and fibulae is commonly seen in leprosy, in addition to other expected changes (e.g. alterations around the anterior bony aperture of the nose). However, periosteal new bone formation could be caused by a number of conditions. This paper describes the histological analysis of thin ground sections of tibiae with periosteal new bone formation from six individuals diagnosed with leprosy from a later Medieval hospital site in Chichester, England. In addition, one individual from a later medieval priory site on Lihou Island, Guernsey, with no characteristic lepromatous leprosy bone changes was examined. The aim was to determine, by light microscopic analysis of thin ground sections, whether the changes on the tibiae had been the result of leprosy or another condition. Results showed that there is a high probability that a specific histological appearance is characteristic of leprosy; for differential diagnoses, endemic syphilis is discussed.

Keywords

Leprosy, tibial periostitis, differential diagnosis, histology

1.Introduction

According to documentary sources, leprosy was a relatively common occurrence in later Medieval Europe (Richards, 1977). However, as with so many apparently frequent diseases, the number of cases diagnosed in archaeologically derived skeletal material from all areas of Europe, but particularly England, is low (see Roberts, this volume). There may be many reasons for this. Firstly, if people contract the high resistant (tuberculoid) form of the infection they may not develop any bone changes. Secondly, people with leprosy may have died before there was time for bone change to occur. In both these scenarios, the evidence would be absent skeletally (see Wood *et al.*, 1992 for discussion). Thirdly if, as suggested in contemporary sources, people with leprosy were isolated from the rest of the community into leprosy hospitals, then we may only expect to see cases in the cemeteries associated with those institutions. Few leprosy hospital cemeteries have been excavated in England (Roberts, 1994) although, paradoxically, lepromatous individuals have been identified in non-leprosy hospitals (Roberts, this volume). Fourthly, leprosy may not have been as common as suggested, and this may be related to misdiagnosis, i.e. people with another disease may have been labelled as leprous. Finally, the diagnosis of leprosy in skeletal material can often be hampered by poor preservation and excavation of facial, hand and foot bones, and lack of experience on the part of the observer. In addition, there may be several diagnoses that could be applied to the individual bone changes observed in leprosy.

When diagnosing leprosy, most researchers follow the guidance of Møller-Christensen (1961), Andersen and Manchester (1987, 1988, 1992), and Andersen *et al.* (1992, 1994). The focus is on the primary (or pathognomonic) changes of *facies leprosa* (Møller-Christensen, 1974) or the rhinomaxillary syndrome (Andersen and Manchester, 1992), seen in lepromatous or near lepromatous individuals. This consists of anterior nasal spine absorption, nasal aperture remodelling, inflammation of the palate, recession of the alveolar process of the maxilla, loss of the anterior teeth and sometimes lack of development of anterior maxillary tooth roots (leprogenic odontodysplasia – Danielsen, 1970). Secondarily, the hands, feet and lower legs may be affected as a result of bacterial invasion of the sensory, motor and autonomic nervous systems. This can result in: absorption of the distal ends of the terminal hand and foot phalanges, metacarpals and metatarsals, knife-edged remodelling of the metatarsals, concentric atrophy of the metacarpals, metatarsals and phalanges, "nicking" of the distal ends of the distal hand phalanges, periostitis, osteitis and osteomyelitis of hand and foot bones, septic arthritis of the foot and hand joints which may lead to fusion, cup and pencil deformities of the metacarpo-phalangeal and metatarso-phalangeal joints, associated with osteoporosis of the hand and foot bones, flexion deformities (indicated by "grooves") in the hand and foot phalanges, and dorsal tarsal bars of bone (indicating collapse of the arches of the feet and consequent strain on the ligaments). Additionally, the lower legs may develop a periosteal reaction in the form of new bone formation on the tibial and fibular shafts (Lewis *et al.*, 1995).

Because the skeleton can only react in a limited number of ways to disease, many disease processes (bone formation, bone destruction, or both) can produce similar changes in the skeleton, and patterning of the changes is key in attempting a diagnosis. There are a number of differential diagnoses that need to be considered for the changes thus described. Tuberculosis and the treponemal diseases of yaws and venereal syphilis may affect the rhinomaxillary area (Manchester, 1994). The changes in the hands and feet may also be caused by the joint diseases such as psoriatic and non-specific septic arthritis (Rogers and Waldron, 1995), and conditions that lead to flexion deformities of the hands and feet such as congenital claw-hand, and dorsal tarsal bars caused by flatfootedness (*pes planus*), pyogenic osteomyelitis, frostbite and diabetes mellitus (Aufderheide and Rodríguez Martín, 1998). The lower leg periostitis may also be the result of a number of conditions such as non-specific infection, which may be the result of "stress" (Goodman *et al.*, 1988), trauma (Ortner and Putschar, 1981:132), treponemal disease (yaws, endemic and venereal syphilis) - Steinbock (1976), primary and secondary hypertrophic osteoarthropathy (Resnick, 1995:4425), vitamin C deficiency (Aufderheide and Rodríguez Martín, 1998:311), overlying soft tissue infection (Ortner and Putschar, 1981:131), including venous stasis (Resnick, 1995:4425), and infantile cortical hyperostosis (*ibid.*). These problems of differential diagnosis emphasise the need to consider the distribution pattern of abnormal lesions in the skeleton, in addition to considering them together, rather than in isolation.

It is understood that the lower leg changes are usually a result of the transmission of secondary infection due to leprosy from the feet up to the lower legs. However, if the lower legs are affected, but not the rest of the skeleton, then a specific diagnosis is more difficult. This paper describes an attempt to diagnose leprosy purely based on the histological changes in sections of tibiae from five known leprosy cases, plus one individual with no evidence of a specific infection.

2. Material and Methods

Six skeletons were selected from the cemetery associated with the later Medieval hospital at Chichester, England, a hospital that was initially opened for leprosy sufferers in the 12th century AD (Lee and Magilton, 1989). All the skeletons showed diagnostic changes of leprosy, but also included tibial new bone formation. Based on accepted criteria (Buikstra and Ubelaker, 1994), five males and one probable female were identified, and all were adults ranging from young to older age; these data are summarised in Table 1.

All the skeletons were well preserved, although two (64, 202) were fragmentary. Diagnosis of leprosy was based on macroscopic criteria as outlined above and Table 2 summarises the features used.

In addition to the bone changes of leprosy, other pathological changes were noted, although they are not directly relevant to this study. The young adult male from burial 21 displayed cribra orbitalia, dental disease, Schmorl's nodes in the spine, and Harris' lines in the tibia. The mature adult male from burial 48 shows a number of dental diseases, Schmorl's nodes and spinal joint disease, and healed fractures of three ribs. The adolescent to young adult male from burial 64 presents dental disease and Schmorl's nodes. The young to middle adult male from burial 88 had evidence of dental disease and Schmorl's nodes in addition to maxillary sinusitis, non-specific infection of the radii, ulnae and metacarpals (possibly associated with leprosy – see Lewis *et al.*, 1995), while the young- to middle adult male from burial 148 similarly showed dental disease and Schmorl's nodes plus periostitis of the visceral surface of the ribs, and osteoarthritis in the hands and feet. Finally, the young- to middle adult probable female from burial 202 had evidence of dental disease, Schmorl's nodes, and osteoarthritis in the hip and rib joints.

Further to the six skeletons described above, another skeleton from a later medieval priory site on the Island of Lihou, Guernsey, was included in the analysis. This well preserved skeleton was from a male aged between 26 and 45 years at death who suffered from osteoarthritis of the spine, right clavicle and scapula, dental disease (enamel defects, calculus, caries and a maxillary torus), and some fused foot phalanges. Additionally, he had florid longstanding new bone formation on the tibiae and fibulae. The reason for including this individual in the analysis was because of the lack of any changes in the skeleton that would indicate a specific disease, including leprosy. With respect to leprosy, there were no changes to the facial, hand or foot bones, but lower leg periostitis was evident.
The tibiae that showed periosteal new bone formation on their surfaces were selected. The tibial shafts were cross sectioned and 6 to 10 mm thick slices were taken for

histological analysis. Thin ground sections of 50µm or 70µm were used. It is considered that morphological structures are only visible and detectable in polarised light when the section is 50µm or thicker. Of course, this is not the normal thickness used in histopathology when cells and other tissues still survive, and the section thickness is usually much reduced. Furthermore, decalcification of bone, as seen in modern histopathology, is not appropriate for archaeological bone for various reasons (e.g. see Schultz, 1986, 1997).

Only in the male from Lihou was a small sample (size 8 x 5 mm) taken for scanning-electron microscopy before embedding the rest of the bone section for light microscopy. For the individuals 64, 88, 148, and 202 from Chichester the right tibia was selected, and for individuals 21 and 48 from Chichester and the male from Lihou the left tibia was selected for sectioning. Figures 1-5 show the macroscopic and radiographic changes of the Chichester samples. Following sectioning of the bones, the samples were subjected to the embedding process and thin ground sections were produced by the techniques described by Schultz and Drommer (1983) and Schultz (1988). Histomorphological age was established by the methods of Kerley and Ubelaker (1978) and Wolf (1999), but also by accepted morphological methods described by Buikstra and Ubelaker (1994).

3. Results

3.1 Chichester burial 21

A complete cross section was taken from the middle of the left tibia shaft. From the proximal (Plate 1/A and Figure 7) and distal end of this section (Plate 1/B and Figure 6) thin ground sections were produced. More specifically, particularly on the external area of the shaft (external circumferential lamellae) of this well preserved tibia, bone collagen is present. In the other parts of the cross-section, almost no collagen has been preserved. Thus, the preservation is relatively good in parts of the bone. The histomorphological age determination suggests an individual age of approximately 20-25 years.

The external surface of the lateral aspect of the shaft, anterior to the interosseous margin, protrudes only slightly, whereas the surface of the medial aspect protrudes severely. The structure responsible for the "protrusion" consists of a relatively thin lamellar structure (maximum thickness is 1mm) which does not represent the original bone surface (= external circumferential lamellae), but rather a secondary layer of bone of periosteal origin caused by a pathological process. The layer which shows a very well organised stage of remodelling (Figure 6), covers the primary external circumferential lamellae and the original compact bone substance like the bark of a tree. Thus, there is a very fine borderline between the newly built bone and the original circumferential lamellae (Figure 7). This line does not represent the grenzlinie found in shafts of long bones affected by treponemal disease (cf. Schultz and Teschler-Nicola, 1987, Schultz, 1994). There are also no structures suspicious of polsters which are a criterion of treponemal diseases (cf. Schultz and Teschler-Nicola, 1987, Schultz, 1994). The morphology of the periosteal newly built bone indicates a slowly developing chronic inflammatory process from which the person probably suffered for several years. In the distal area of the shaft, the newly built bone was thicker (Figure 8). However, due to diagenesis, the layer is severely eroded. Nevertheless, the microscopic analysis reveals that this part of the new bone is of woven type which means that this layer is more recent than the one described before (Figure 6). Thus, the periosteal changes apparently represent a recurrent process.

3.2 Chichester burial 48

This bone sample represents a complete-cross section through the distal metaphysis of the left tibia. Macroscopically, the bone seems to be well preserved. However, at the microscopic level, the preservation is fair to poor. Only a few areas of bone collagen are visible. The age calculated by histomorphology is 60 years or more.

However, the osteoporotic change in the compact bone substance and of the cortical bone (maximum thickness is 2mm), as well as the extreme rarefaction of the spongy bone trabeculae, could also be caused by atrophy due to inactivity and not only ageing (Plate 1/C, Figure 9). Almost all of the circumferential bone surface is covered by a secondary layer of bone which represents periosteal new bone formation (Plate 1/C). These structures are secondarily well remodelled and changed by the osteoporotic process (Figure 10). Middle and high magnification microscopy (25x, 100x) reveals that the periosteal bone formation on the antero-lateral surfaces, and probably also on the dorsal face of the bone, was built up in at least two layers (Figure 10). This means that the pathological process, which was probably of inflammatory origin, was characterized by signs of recidivation.

3.3 Chichester burial 64

A small sample from the right tibia representing, in the cross section, only one fourth of the circumferential part of the shaft, probably from the medial face, was analysed microscopically (Plate 1/D). The preservation is poor, and no bone collagen was preserved. A histomorphological age determination was not possible because of the small size of the sample and poor preservation. The compact bone covering the spongy bone was very thin (maximum thickness is 3mm, minimum thickness is 1mm). This sample was probably taken from the metaphysis of the bone.

The external bone surface slightly "bulges". The changes are reminiscent of polster-like structures (Figure 11). Thus, there is only evidence of a very slight periosteal reaction as a result of an inflammatory process which is very well organised, i.e. remodelled. The thin newly built bone formation seems to have grown out of the original compact bone substance which was not affected in any way.

3.4 Chichester burial 88

A complete cross-section was taken from the middle of the shaft of the right tibia of this individual; it was relatively brittle and had been broken in several places. Microscopic analysis revealed extremely little evidence of bone collagen. Because of the poor preservation, histomorphological age determination gave only an approximate idea of the age of the individual which can be calculated as probably older than 30 or 35 years.

Low and middle magnification microscopy (10x, 25x) illustrate that the compact bone substance of the antero-medial face of the shaft is affected by osteoporosis which is very probably not due to the age, but caused by a pathological process of the medullary cavity (Plate 1/E). The rest of the cross-section of this tibia seems to be affected only very slightly by this process. The medullary cavity is enlarged which could be due to ageing, but also to the pathological process.

High magnification microscopy (100x) shows that the osteoporotic area was apparently caused by an inflammatory process, which affected not only the compact bone substance (osteitis), but also the medullary cavity (osteomyelitis) and the periosteum (periostitis). This periosteal new bone formation is mainly expressed on the external surface of the anterior half of the tibia, particularly on the medial aspect of the shaft (Plate 1/E), medially and laterally to the anterior margin. The changes, which are the product of a florid inflammatory process of the periosteum, are composed of woven bone and no bone remodelling is evident (Plate 1/E, Fig. 12). In the dorsal half, and also in the antero-lateral area of the shaft, the external surface of the tibia "bulges". In this region, the compact bone substance of the tibia is relatively thick (6mm). However, the external half of the compact bone substance (thickness is approximately 3mm) represents a remodelled area which was primarily an external, i.e. subperiosteal bony layer caused by a pathological process. Whether this last change described was the result of an inflammatory or a hemorrhagic process cannot be determined. This is because of remodelling of the original bone layer, which is now completely integrated into the compact bone substance, and the postmortem lack of the bone collagen. However, in middle and high magnification microscopy (25x, 100x), this remodelled layer on the external bone surface in the antero-lateral face of the shaft shows structures which resemble the polsters (Figure 13), characteristic in chronic treponemal disease (Schultz and Teschler-Nicola, 1987, Schultz, 1994). There is no visible demarcation between the periosteal new bone formation and the original compact bone, probably because of the poor preservation of the bone collagen. However, at the base of the new bone formation, some small and narrow blood vessel canals, which are orientated in a line parallel to the external bone surface, suggest a possible demarcation.

There are at least two suggestions for classifying the two different periosteal changes:

1) an acute (active) phase of an inflammatory process of the periosteum (periostitis) demonstrating rapid growth, which was apparently induced by an osteomyelitic-osteitic process

2) a chronic and mainly slowly changing process of probable inflammatory origin in the stage of remodelling which does not represent a typical florid haematogenous osteomyelitis

3) Theoretically, there is also a third possibility. Both processes could be due to the same disease. In this case, the second, relatively fresh periosteal reaction only represents the product of a recidivation. However, the different nature of these changes renders this not very probable.

3.5 Chichester burial 148

A complete cross-section from the middle of the shaft of the right tibia was taken for microscopic analysis. Macroscopically, the bone is well preserved whereas, on the microscopic level, preservation is poor because of the lack of bone collagen. Only in some "patch-like" structures in the middle of the compact bone substance, are there remains of bone collagen. The histomorphological age determination yields a result of between 40 and 55 years.

The external surface of the shaft "bulges" on its medial and lateral face (Plate 1/F). These areas are characterized by secondary new bone formation, which has an average thickness of 1.5mm and is well organised. On the lateral face, between the interosseous and anterior margins, this bone has a maximum thickness of 2.5mm, while all the compact bone, including the newly formed bone, measures only 4.6mm. The original periosteal layer presents in some areas small polster-like structures and is partly very well integrated into the original compact bone substance (Figure 14). In some other areas, however, there are remains of a very fine demarcation line between the original external bone surface and the periosteal new bone. This line is not comparable with the grenzlinie which is frequently found in chronic treponemal disease (cf. Schultz and Teschler-Nicola, 1987, Schultz, 1994). It seems that this inflammatory process was present for many years.

The original compact bone substance and the medullary cavity has not been affected by a pathological process. However, in the middle of the compact bone substance there are "field-like" areas of tangential lamellae which probably represent the result of inactivity atrophy (Figure 15).

3.6 Chichester burial 202

Macroscopically, the complete cross-section from the middle of the shaft of the right tibia, which is a relatively gracile bone, is fairly well preserved and shows no distinct vestiges of periosteal reaction (Plate 1/G). Microscopically, the bone is fairly well preserved. However, only very few remnants of bone collagen are observable. The histomorphological age determination suggests an individual of between 35 and 45 years.

On the medial aspect of this bone, there is a superficial small striated zone on the periphery of the original compact bone substance resembling a secondary layer. This structure is difficult to explain by analysis at low magnification analysis (10x). In this area, the original compact bone substance measures 3-3.5mm in thickness, whereas the external striation has a thickness of only approximately 1mm. However, middle and high magnification analysis (25x, 100x) reveals that this superficial layer, which has a bark-like character, is very well organised as lamellar bone and represents the vestiges of an extensive, remodelled periosteal reaction. Thus, the new bone formation was probably built up layer by layer over a relatively long time. A very similar structure is seen on the lateral aspect of the tibial shaft, anterior to the interosseous margin. Here, the border between the original external bone surface and the new bone formation, which is very well integrated into the compact bone substance, is still visible by the orientation of small, narrow and tangentially orientated blood vessel canals (Figure 16).

3.7 The male from Lihou

In the cross section, the left tibia is relatively normal in size. Microscopic analysis suggests that the preservation is poor to fair. There are relatively slight and diffuse vestiges of diagenesis (Plate 1/H). The destruction was apparently caused by algae and/or fungi (cf. Schultz 1986, 1997). The bone collagen is more or less totally destroyed by these postmortem changes; this makes diagnosis difficult. However, in some small areas of the original compact bone, remains of collagen are still seen. A reliable histomorphological age determination cannot be made, but a trend can be noted. In the original compact bone, the size of the Haversian canals and of the Haversian systems, and the distribution of the latter indicate an individual age of approximately 25 to 39 years. The medullary cavity is relatively large (Plate 1/H). On the dorsal face of the shaft, the compact bone measures 5.7mm in thickness, and on the lateral face only 3mm, excluding the new bone formed (Plate 1/H). There are also relatively broad tangential lamellae (Figure 17) and relatively wide resorption lacunae, which should not be diagnosed as due to ageing, but as characteristic features of inactivity atrophy, probably due to disuse of the leg because of the pathological process.

At low and middle magnification microscopy (10x, 25x), the bone demonstrates an obvious pathological feature. There is an irregularly structured, secondary bony layer of porotic character (Plate 1/H), which is responsible for the bulging of the external face of the shaft (Figures 18 and 19). On the medial face of the tibia, the layer is relatively thick (maximum thickness is 2.3mm) and represents a moderate stage of a periosteal reaction which was not very longstanding when the person died. This means that the layer was in an earlier stage of remodelling, whereas the newly formed bone on the lateral aspect of the shaft is only very thin and represents an old, very well organised, advanced stage of remodelling. Thus, the changes on the lateral aspect of the bone started at least several years before the death of this man. These newly built formations seem to be partly deposited on the original bone surface, and partly growing out of the compact bone substance around the periphery of tibia shaft (Figure 20). The remodelled layer shows structures which resemble the polsters (Figure 19) characteristic of chronic treponemal disease (Schultz and Teschler-Nicola, 1987, Schultz, 1994). Furthermore, and only visible by microscopic analysis, the original layer is partly demarcated against the original compact bone substance by a line of small and narrow blood vessel canals. A typical grenzlinie is not present, but in small areas of the section lamellar bone resembles this structure, characteristic of chronic treponemal disease. The character of the new bone formation suggests a slowly growing process, because there are no features of rapid bone growth such as small bone spicules (Plate 1/H). A recidivation of an inflammatory process is very probable.

The original compact bone substance apparently shows no convincing evidence of *intra vitam* changes such as bone remodelling (Plate 1/H). However, a slight affection of the compact bone cannot be completely excluded because, almost throughout the compact bone substance, there are two narrow cord-like structures which are rectangularly orientated to the external bone surface (Figure 21). These structures resemble changes seen in cicatrised metaplasia. There is no evidence of involvement of the medullary cavity such as is seen in inflammatory processes. Thus, a typical periostitis such as a florid inflammatory process, does not seem to be very probable.

4. Discussion

Although the sample of skeletons used for this study was small, and the features described are not pathognomonic to leprosy, this is the first attempt to describe histological changes of bone alteration associated with leprosy. The tibiae of the six individuals from Chichester, and of the male from Lihou, show characteristic signs of periosteal bone reaction which are macroscopically and microscopically very similar, or even identical, in all examined cases. Thus, there is a high probability that the changes were caused by similar, or the same, processes. However, this does not necessarily mean that the people suffered from the same diseases, because sometimes different diseases can produce similar or almost the same morphological changes in the bones (especially as bone can only react in a limited number of ways to inflammation). This is one of the main problems in paleopathology. As a rule, paleopathologists can only examine signs of ancient disease in macerated, i.e. dry bones. Thus, no soft tissues or cells, which play an important role in pathological investigations in the living, can be studied to establish a reliable diagnosis or for comparative purposes. This means that diagnostic criteria are sometimes relatively limited in paleopathology. On the other hand, reliable diagnoses can be established by using different characteristic signs which are not easy, or are even impossible to study in living patients or in recent-pathological specimens. At the microscopic level, there are such characteristics, such as faserfilz-osteon (e.g. in a primary bone tumour), polsters and/or grenzlinie (in treponemal disease).

Up to now, very little histopathological analysis of ancient leprous bones has been undertaken. It is not possible to diagnose leprosy by using only a bone section taken from the shaft of a tibia. However, it is possible to compare periosteal reactions at the microscopic level with the well-known characteristics of hematogenous osteomyelitis, treponemal disease, tuberculosis and even non-specific periosteal reactions such as inflammatory processes of the deep veins, primary and secondary hypertrophic osteoarthropathy (i.e. Bamberger-Marie disease, which could be caused by chronic heart-lung diseases) and scurvy in ancient skeletal material. Additionally, overlying soft tissue infections can produce a periosteal lesion. As a rule, it is not too difficult to differentiate between the bony products of an inflammatory or a hemorrhagic process (Schultz and Teschler-Nicola, 1987, Schultz, 1993). However, frequently both processes are mixed and this makes diagnoses again difficult (Schultz, 1993).

In recent literature describing the morphological changes in leprosy, emphasis is placed on the structural changes of the soft tissues (e.g. Waters, 1990), whereas macroscopic changes on bone surfaces and characteristic features of the micro-structure of dry bone specimens are neglected. Therefore, paleopathologists who intend to examine leprous skeletons are advised to study the reports on pathological investigations at the macroscopic (Hirschberg, 1923, Klingmüller, 1930), microscopic (e.g. Sawtschenko, 1891, Beitzke, 1934) and radiological (e.g. De La Camp, 1900, Deycke Pascha, 1906) levels carried out at the end of the 19th century, and into the first half of the 20th century. These papers are extremely helpful because of the sophisticated and detailed description of morphological features in leprous bones.

All thin ground sections produced from the seven samples were checked for characteristic features. Nature, frequencies and distribution of the microscopic bone changes and the characteristic features in the cross sections of these tibiae demonstrate an extensive similarity (Table 3). It is also important to bear in mind that all individuals from Chichester macroscopically showed the well-known features of leprosy in the skull. They also had changes characteristic of this disease in the postcranial skeleton, that is in the hands and/or the feet (Table 2). The probable female only has changes in the skull. The male from Lihou shows neither leprous changes in the skull nor in the hands and feet (Table 2).

All individuals examined in this study suffered from a periosteal reaction of the tibia. In all of these cases, the periosteal reaction was longstanding which had been well remodelled probably years before the death of the individuals (Figures 6, 9, 11, 13 and 16). In six of the seven cases, the new bone formation apparently grew relatively slowly. Only in one case (Chichester burial 48) can the nature of the growth not be estimated (Figures 9 and 10). Additionally, in two of the seven individuals, new bone formation could also be observed (Chichester burial 88 and the Lihou male). In the male from Lihou, some parts of the new bone formation were already in the early stage of lamellar organisation (Figure 19), whereas in the individual from Chichester (burial 88) the new bone consisted of woven bone (Figure 12). The changes in the latter case represent relatively rapid bone formation. Thus, in both cases, the nature and the type of bone formation was apparently different (cf. Plate 1/E and H). The new periosteal bone formation on the distal end of the tibia of the Chichester individual from burial 21 (Figure 8) represents a level of development which is comparable to that of the Lihou male (Figure 20). However, as described before, in the Lihou male the stage of development was not the same in all parts of the bone formed (compare Figure 19 with Figure 20). Furthermore, in three cases (Chichester burial 48 and probably also burial 21, as well as the male from Lihou) the periosteal bone formation showed vestiges of recidivation (Figure 10).

Only in one individual (Chichester burial 88), was the medullary cavity affected by the inflammatory process (Plate 1/E). In one of the individuals from Chichester (burial 48), the cause of the enlargement of the medullary cavity was osteoporosis due to age (Plate 1/C). In the male from Lihou, there was a slight enlargement of the medullary cavity which cannot be really interpreted (Plate 1/H). A possible cause could be disuse atrophy. In the individual from Chichester burial 88, the compact bone was affected in the same way as the medullary cavity. There are signs of an inflammatory process in the form of osteolysis, producing osteoporosis of the compact bone (Plate 1/E). In the individual from Chichester burial 48, the reduction of the compact bone of the tibial shaft was due to osteoporosis related to age (Plate 1/C). Slight changes in resorption, mainly in the endosteal area of the compact bone substance, are again probably due to disuse atrophy.

Special features indicating different stress markers, or even diseases, were observed at the microscopic level in the cross-sections of the seven individuals. The tangential lamellae which were found in the middle of the compact bone substance of the tibia of the individual from Chichester burial 148 (Figure 15), and of the Lihou male (Figure 17) are indicators of chronic inactivity atrophy

(Schultz, 1986, cf. Knese *et al.*, 1954). These structures need years for their development. In the Lihou male, these structures correlate with the reduction of the compact bone substance and the enlargement of the medullary cavity. In the individual from burial 48, tangential lamellae could not be found because of the extreme rarefaction of the compact bone substance due to old age involution.

Polsters are characteristic features at the microscopic level which are frequently found in chronic treponemal disease involving the long bones (Weber, 1927, Michaelis, 1930, Schultz and Teschler-Nicola, 1987, Schultz, 1994). In particular, well developed polsters are a good indicator of treponemal disease (Schultz, 1994). However, in three of the six cases from Chichester (burials 64, 88 and 148) which show the bony lesions of leprosy, and also in the skeleton of the male from Lihou, there are no typical polsters but only polster-like structures which are rudimentarily developed and relatively flat. At the microscopic level, polster-like structures were also detected by Blondiaux (pers. comm., cf. Blondiaux *et al.*, 1994, Blondiaux *et al.*, this volume) in medieval skeletons showing the bony signs of leprosy.

As a rule, a grenzlinie can also be observed in chronic treponemal disease which is not found in hematogenous osteomyelitis of long bones (Schultz and Teschler-Nicola, 1987, Schultz, 1994). Thus, a grenzlinie could also be a useful indicator for diagnosing chronic treponematosis using microscopy. In this investigation, faint signs of a possible grenzlinie were observed in at least two, or even three, of the six cases from Chichester (burials 21, 148 and possibly 88) as well as in the skeleton of the male from Lihou. However, the features were very subtle and only visible with difficulty. In contrast to these findings, in treponemal diseases including endemic syphilis, there are not only alterations in the subperiosteal bone, but additionally also osteoclastic changes in the endosteal bone and the bony trabeculae of the medullary cavity, as well as in the compact bone substance of the shaft of the affected long bones (Schultz and Teschler-Nicola, 1987, Schultz, 1994, Kuhnen *et al.*, 1999).

When we come to the differential diagnoses and the basic etiology of these lesions there are principally two possibilities:

1) We have to bear in mind that all six individuals from the Chichester cemetery suffered from chronic bone changes of leprosy, seen in the well-known macroscopic lesions of the face, and in the hands and/or feet. Additionally, all these individuals showed signs of a periosteal reaction on their tibiae which could be primarily characteristic of leprosy (cf. Møller-Christensen, 1974, Steinbock, 1976, Ortner and Putschar, 1981, Resnick, 1995, Aufderheide and Rodríguez Martin, 1998). The skeleton of the male from Lihou showed no signs of leprosy, but also presented periosteal new bone formation. This makes a diagnosis of leprosy also possible.

2) On the other hand, it could be argued from the morphological features resembling polsters and grenzlinie, that the periosteal reaction was caused by treponemal disease, probably endemic syphilis (cf. Schultz, 1994, Kuhnen *et al.*, 1999). Thus, all seven individuals could have been affected by this treponematosis, suffering additionally from leprosy which was proved by the bony changes in the face, the hands and/or the feet.

Indeed, with respect to the first possibility, at the microscopic level, the periosteal new bone formation in the tibiae are different from that observed in treponemal disease. Thus, these changes may primarily not be caused by treponematoses but by leprosy. Today, lepromatous leprosy induced periostitis is rarely seen (Waters, 1990, Resnick, 1995, Lewis *et al.*, 1995). Therefore, it does not seem very probable that all the individuals from Chichester who had suffered from leprosy would also have *M. leprae* induced periostitis. However, we do not know today in what way the disease was expressed or what the nature of the human immune system was some hundreds of years ago (although there is no evidence that it has changed in character). The changes in the individual from burial 88 from Chichester additionally showed signs of a florid osteomyelitis, combined with osteitis and periostitis. These relatively new changes were probably caused secondarily by infection of an ulceration of the skin of the lower leg, which led to a pyogenic osteomyelitis (cf. Waters, 1990, Resnick, 1995).

The second possibility also sounds plausible. As we know, patients in medieval leprosaria frequently suffered not only from leprosy but also from treponemal disease (Ackerknecht, 1963), although see Crane-Kramer (this volume). The microscopic features found in theses samples such as polsters and grenzlinie which are characteristic of treponemal disease, in their well developed stages, support this suggestion. Therefore, it is possible that the individuals described in this paper may have suffered additionally from treponematosis, probably endemic syphilis, which produced milder bony changes than those of venereal syphilis (cf. Schultz and Teschler-Nicola, 1987, Schultz, 1994, Kuhnen *et al.*, 1999). However, polsters and grenzlinie are only very poorly developed in these cases described above. Only in the case of Lihou male in whom, with the exception of the periosteal reaction, no signs of leprosy could be identified, could endemic syphilis be suggested as a diagnosis with a relatively high probability. Theoretically, this case from Lihou might indicate that all the six individuals from Chichester suffered additionally from endemic syphilis. However, this does not seem very likely.

Independent of the two possibilities mentioned above, osteoporosis (individual from Chichester burial 48), the enlargement of the medullary cavity and the rarefaction of the compact bone substance (individual from Lihou and probably also the individual from Chichester burial 48) and, particularly, the presence of tangential lamellae in the compact bone substance (individual from Chichester burial

148 and the male from Lihou) were very probably due to disuse atrophy and lack of use of the legs. In other cases, an advanced age could also be responsible for these changes (e.g. the individual from Chichester burial 48).

5. Conclusions

It can be assumed that in these seven cases, the signs of longstanding, i.e. remodelled, new bone formation, represents a primary periosteal inflammation. This inflammatory process was apparently relatively mild because there is no evidence that the compact bone substance was substantially affected. Additionally, there is also evidence of a florid, secondary periostitis (Chichester burial 88). In this case the compact bone substance, as well as the medullary cavity, was affected.

The microscopic analysis of the tibial samples from Chichester and Lihou shows that there is a special pattern of features which is associated with skeletal leprosy. The classification and interpretation of these morphological features, particularly the new built bone formation, is sometimes controversial. However, in summary, the probability is very high that the periosteal changes on the surfaces of the tibial shafts of the individuals from Chichester were caused by infections secondary to leprosy. The changes in the tibia of the male from Lihou are also suspicious of leprosy. However, it is still a possibility that other processes, such as endemic syphilis or inflammatory processes of the deep veins of the lower legs, resulted in these periosteal changes. Additional to the signs of inflammatory processes, there is also in at least two cases, evidence of limited function or even disuse of the affected leg. This also emphasises that the disease process was chronic.

Acknowledgements

The authors would like to thank Chichester District Council and John Magliton for the excavation of the cemetery at Chichester, Frances Lee for the initial analysis of the skeletons, Jason Maher of the Department of Archaeological Sciences, University of Bradford, for providing the sections, Jean Brown formerly of the Department of Department of Archaeological Sciences, University of Bradford, for producing the photographs and slides of the bones and radiographs, and Michael Brandt and Ingrid Hettwer-Steeger of the Zentrum Anatomie, University of Göttingen, for producing the thin ground sections and preparing the sample for scanning-electron microscopy.

References

Ackerknecht, E.H. 1963 *Geschichte und Geographie der wichtigsten Krankheiten.* Stuttgart: Ferdinand Enke Verlag

Andersen, J. and Manchester, K. 1987 Grooving of the proximal phalanx in leprosy: a palaeopathological study. *J. Archaeological Science* 14(1):77-82

Andersen, J. and Manchester, K. 1988 Dorsal tarsal exostosis in leprosy: a palaeopathological and radiological study. *J. Archaeological Science* 15:51-56

Andersen, J. and Manchester, K. 1992 The rhinomaxillary syndrome in leprosy: a clinical, radiological and palaeopathological study. *Int. J. Osteoarchaeology* 2:121-129

Andersen, J., Manchester, K. and Ali, R. 1992 Diaphyseal remodelling in leprosy: a radiological and palaeopathological study. *Int. J. Osteoarchaeology* 2:211-219

Andersen, J., Manchester, K. and Roberts, C.A. 1994 Septic bone changes in leprosy: a clinical, radiological and palaeopathological review. *Int. J.Osteoarchaeology* 4:21-30

Aufderheide, A.C. and Rodríguez Martin, C. 1998 *The Cambridge encyclopedia of human paleopathology.* Cambridge: University Press.

Beitzke, H. 1934 Erkrankungen der Knochen und Gelenke bei Lepra. In O. Lubarsch, F. Henke and R. Rössle (eds): *Handbuch der speziellen pathologischen Anatomie und Histologie*, 9 (2). Berlin: Julius Springer, pp. 594-611

Blondiaux, J., Duvette, J.-F., Vatteoni, S. and Eisenberg, L. 1994 Microradiographs of leprosy from an osteoarchaeological context. *Int. J.Osteo-archaeology* 4:13-20

Buikstra, J. and Ubelaker, D. (eds) 1994 *Standards for data collection from human skeletal remains.* Arkansas: Archeological Survey, Research Seminar Series 44.

Danielsen, K. 1970 Odontodysplasia leprosa in Danish Mediaeval skeletons. *Saertryk at Tandlaegebladet* 74:603-625

De La Camp, G. 1900 Periostitis bei Lepra. *Fortschr. Röntgenstr.* 4:36-48

Deycke Pascha, S. 1906 Knochenveränderungen bei Lepra tuberosa im Röntgengramm. *Fortschr. Röntgenstr.* 10:279-291

Goodman, A.H., Brooke Thomas, R., Swedlund, A.C. and Armelagos, G.J. 1988 Biocultural perspectives on stress in prehistoric, historical and contemporary population research. *Yearbook of Physical Anthropology* 31:169-202

Hirschberg, M. 1923 *La lèpre osseuse.* Strasbourg: 3rd International Leprosy Conference.

Kerley, E.R. and Ubelaker, D.H. 1978 Revision in the microscopic method of estimating age and death in human cortical bone. *Amer. J. Phys. Anthropol.* 49:545-546

Klingmüller, K. 1930 Die Lepra. *Handbuch der Haut und Geschlechtskrankheiten*, 10 (2):1-34

Knese, K.-H., Voges, D, and Ritschl, I. 1954 Untersuchungen über die Osteon- und Lamellenformen im Extremitätenskelett des Erwachsenen. *Z. Zellforsch.* 40:323-360

Kuhnen, C., Schultz, M., Bosse, A., Weber, A. Preuschoft, H. and Müller, K.-H. 1999 Endemische Syphilis an den rekognoszierten Reliquien des Gottfried von Cappenberg. *Pathologe* 20:292-296

Lee, F. and Magilton, J.R. 1989 The cemetery of the hospital of St James and St Mary Magdalene, Chichester – a case study. *World Archaeology* 21(2):273-282

Lewis, M.E., Roberts, C.A. and Manchester, K. 1995 Inflammatory bone changes in leprous skeletons from the Medieval hospital of St James and St Mary Magdalene, Chichester, England. *Int. J. of Leprosy* 63(1):77-85

Manchester, K. 1994 Rhinomaxillary lesions in syphilis: differential diagnosis. In Dutour, O., Palfí, G., Berato, J. and Brun, J-P (eds): *L'origine de la syphilis en Europe. Avant ou après 1493?* Paris: Editions Errance, pp. 79-80.

Michaelis, L. 1930 Vergleichende mikroskopische Untersuchungen an rezenten, historischen und fossilen menschlichen Knochen. *Veröff. Kriegs- und Konstitutionspath.* 6:1-92

Møller-Christensen, V. 1961 *Bone changes of leprosy.* Copenhagen: Munksgaard.

Møller-Christensen, V. 1974 Changes in the anterior nasal spine and alveolar process of maxilla in leprosy. *Int. J.Leprosy* 42:431-435

Ortner, D. and Putschar, W. 1981 *Identification of pathological conditions in human skeletal remains.* Washington DC: Smithsonian Institution Press.

Resnick, D. 1995 *Diagnosis of bone and joint disorders.* 3rd edition. London: W.B. Saunders Company.

Richards, P. 1977 *The Medieval leper and his northern heirs.* Cambridge: D.S. Brewer.

Roberts, C.A. 1994 Treponematosis in Gloucester, England: a theoretical and practical approach to the pre-Columbian theory. In Dutour, O., Palfí, G., Berato, J. and Brun, J-P (eds): *L'origine de la syphilis en Europe. Avant ou après 1493?* Paris: Editions Errance, pp. 101-108

Rogers, J. and Waldron, T. 1995 *A field guide to joint disease in archaeology.* Chichester: Wiley.

Sawtschenko, I. 1891 Zur Frage über die Veränderungen der Knochen beim Aussatz (Osteitis und Osteomyelitis leprosa). *Beitr. Path. Anat.* 9:241-265

Schultz, M. 1986 *Die mikroskopische Untersuchung prähistorischer Skeletfunde. Anwendung und Aussagemöglichkeiten der differential-diagnostischen Untersuchung in der Paläopathologie.* Archäologie und Museum 6. Liestal: Kanton Baselland, pp. 1-140

Schultz, M. 1988 Methoden der Licht- und Elektronenmikroskopie. In Knussmann, R. (ed) *Anthropologie. Handbuch der vergleichenden Biologie des Menschen 1 (1).* Stuttgart: Gustav Fischer, pp. 698-730

Schultz, M. 1993 Spuren unspezifischer Entzündungen an prähistorischen und historischen Schädeln. Ein Beitrag zur Paläopathologie. - Vestiges of non-specific inflammations in prehistoric and historic skull. A contribution to palaeopathology. In B. Kaufmann (Series Editor): *Anthropologische Beiträge Volumes 4 A and 4 B.* Aesch/Basel: Anthropologisches Forschungsinstitut Aesch and Anthropologische Gesellschaft Basel, pp. 81(A), 52 (B)

Schultz, M. 1994 Comparative histopathology of syphilitic lesions in prehistoric and historic human bones. In Dutour, O., Palfí, G., Berato, J. and Brun, J-P (eds): *L'origine de la syphilis en Europe. Avant ou après 1493?* Paris: Editions Errance, pp. 63-67

Schultz, M. 1997 Microscopic investigation of excavated skeletal remains: A contribution to paleopathology and forensic medicine. Haglund, W.D. and Sorg, M.H. (eds) In: *Forensic taphonomy. The postmortem fate of human remains.* Boca Raton/New York/London/Tokyo: CRC Press, pp. 201-222

Schultz, M. and Drommer, R 1983 Möglichkeiten der Präparateherstellung aus dem Gesichtsschädel-bereich für die makroskopische und mikroskopische Untersuchung unter Verwendung neuer Kunststofftechniken. In Hoppe, W.G. (ed) *Fortschritte der Kiefer- und Gesichts-Chirurgie* 28. *Experimentelle Mund-Kiefer-Gesichts-Chirurgie. Mikrochirurgische Eingriffe.* Stuttgart/New York: Gustav Thieme, pp. 95-97

Schultz, M. and Teschler-Nicola, M. 1987 Krankhafte Veränderungen an den Skelettfunden aus dem Karner der St.Martinskirche in Klosterneuburg, Niederösterreich. III. Entzündliche Veränderungen an den langen Röhrenknochen. *Annalen des Naturhistorischen Museums Wien* 89:252-296

Steinbock, R.T. 1976 *Paleopathological diagnosis and interpretation.* Springfield, Illinois: Charles C. Thomas.

Waters, M.F.R. 1990. Lepra. In D.A. Warrell (ed): *Infektionskrankheiten.* Weinheim, VCH, pp. 377-388

Weber, M. 1927 Schliffe von mazerierten Röhrenknochen und ihre Bedeutung für die Unterscheidung de Syphilis und Osteomyelitis von der Osteodystrophia fibrosa sowie für die Untersuchung fraglich syphilitischer, prähistorischer Knochen. *Beitr. Path. Anat. Allg. Path.* 78:441-511

Wolf, M. 1999 *Ergebnisse makro- und mikroskopischer Untersuchungen an den römischen Brandgräbern von Rheinzabern* (Rheinland-Pfalz). In M. Schultz (Series Editor): Beiträge zur Paläopathologie 3. Göttingen: Cuvillier

Wood, J.W., Milner, G.R., Harpending, H.C. and Weiss, K.M.1992 The osteological paradox. Problems of inferring health from skeletal samples. *Current Anthropology* 33(4):343-370

Table 1: Skeletons selected for analysis by age and sex

Number	Ch 21	Ch 48	Ch 64	Ch 88	Ch 148	Ch 202	Lihou
Sex	M	M	M	M	M	F>>M	M
Age	young adult (17-25 years)	mature adult (older than 45 years)	adolescent to young adult (16-20 years)	young to middle adult (26-45 years)	young to middle adult (26-45 years)	young to middle adult (30-40 years)	young to middle adult (26-45 years)

Table 2: Features used for the diagnosis of leprosy for the selected individuals from Chichester and Lihou

Number	Ch 21	Ch 48	Ch 64	Ch 88	Ch 148	Ch 202	Lihou
Face	+	+	+	+	+	+	-
Hands	+	-	-	-	+	-	-
Feet	+	+	+	+	+	-	-
Lower legs	+	+	+	+	+	+	+

+ = changes present
- = changes absent

Table 3: Bone changes and microscopic features of inflammatory processes in the six individuals from Chichester and the male from Lihou

Features	Chi 21	Chi 48	Chi 64	Chi 88	Chi 148	Chi 202	Lihou
Periosteum	+	+	+	+	+	+	+
Compact substance	-	[+]	-	+	-	-	(+)
Medullary cavity	-	[+]	-	+	-	-	[+]
Recidivation	(+)	+	-	-	-	-	+
Remodelling	+	+	+	+	+	+	+
Slowly growth	+	+/-	(+)	+	+	+	+
Osteoporosis	-	[+]	-	+	-	-	-
Polster-like structures	-	-	+	+	+	-	+
Grenzlinie-like structure	+	-	-	(+)	+	(+)	(+)
Tangential lamellae	-	-	-	-	+	-	+

+ = affected by an inflammatory process/feature present
(+) = not very pronounced, probably affected by an inflammatory process/very slight features probably developed
[+] = changes present, but probably not caused by an inflammatory process
- = not affected by an inflammatory process/feature not present
+/- = uncertain

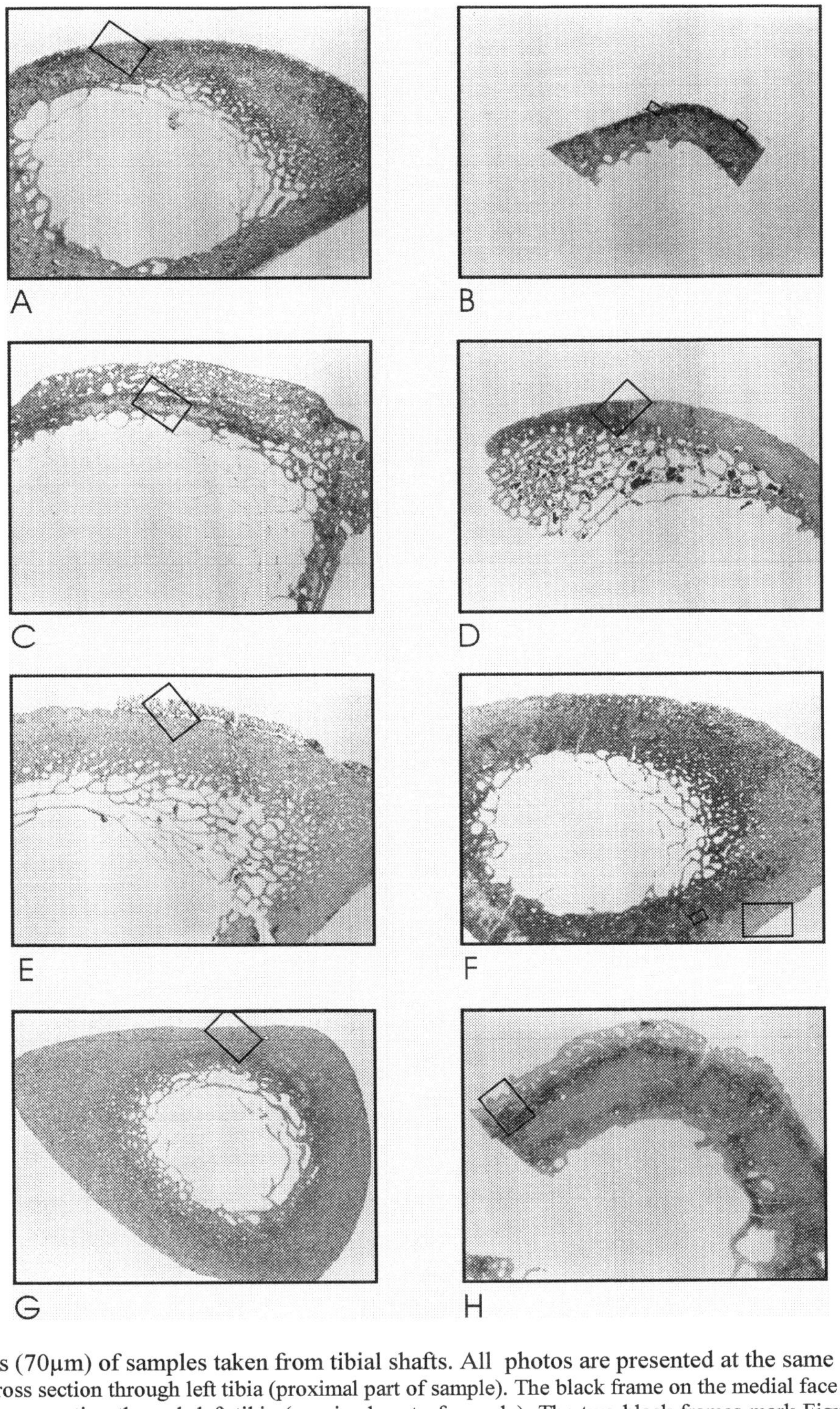

Plate 1:

Thin ground sections (70µm) of samples taken from tibial shafts. All photos are presented at the same scale (approx. 2x)

A) Chichester 21: Cross section through left tibia (proximal part of sample). The black frame on the medial face marks Figure 7.

B) Chichester 21: Cross section through left tibia (proximal part of sample). The two black frames mark Figure 6 (right frame) and Figure 8 (left frame).

C) Chichester 48: Cross section through left tibia. The black frame on the antero-lateral face marks Figure 9.

D) Chichester 64: Cross section through right tibia. The black frame marks Figure 11.

E) Chichester 88: Cross section through right tibia. The black frame on the medial face marks Figure 12.

F) Chichester 148: Cross section through right tibia. The large black frame on the antero-lateral face marks Figure 14; the small black frame in the compact bone substance marks Figure 15.

G) Chichester 202: Cross section through right tibia. The black frame on the lateral face marks Figure 16.

H) Lihou male: Cross section through left tibia. The black frame on the medial face marks Figure 19. From the gap in the compact bone substance (left) the sample for scanning-electron microscopy was taken (Fig. 18).

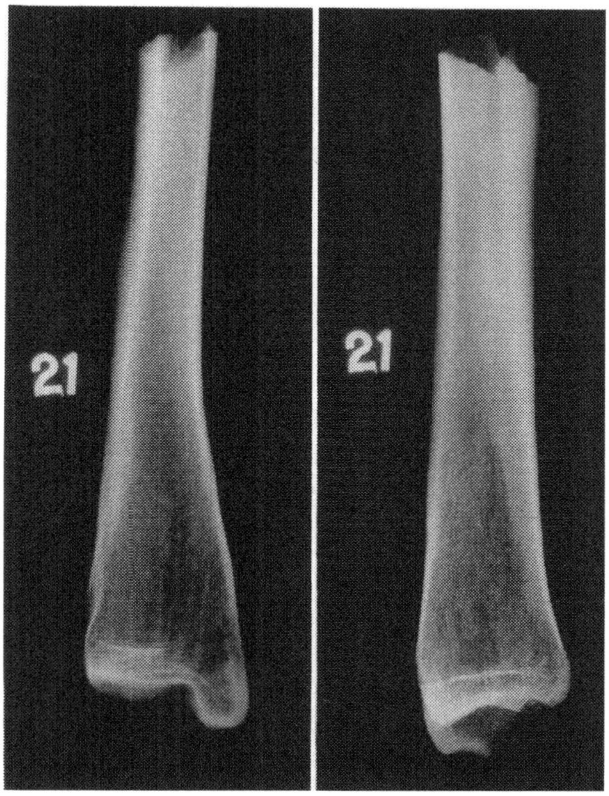

Figure 1: Chichester 21: Radiographic images of periostitis on tibia

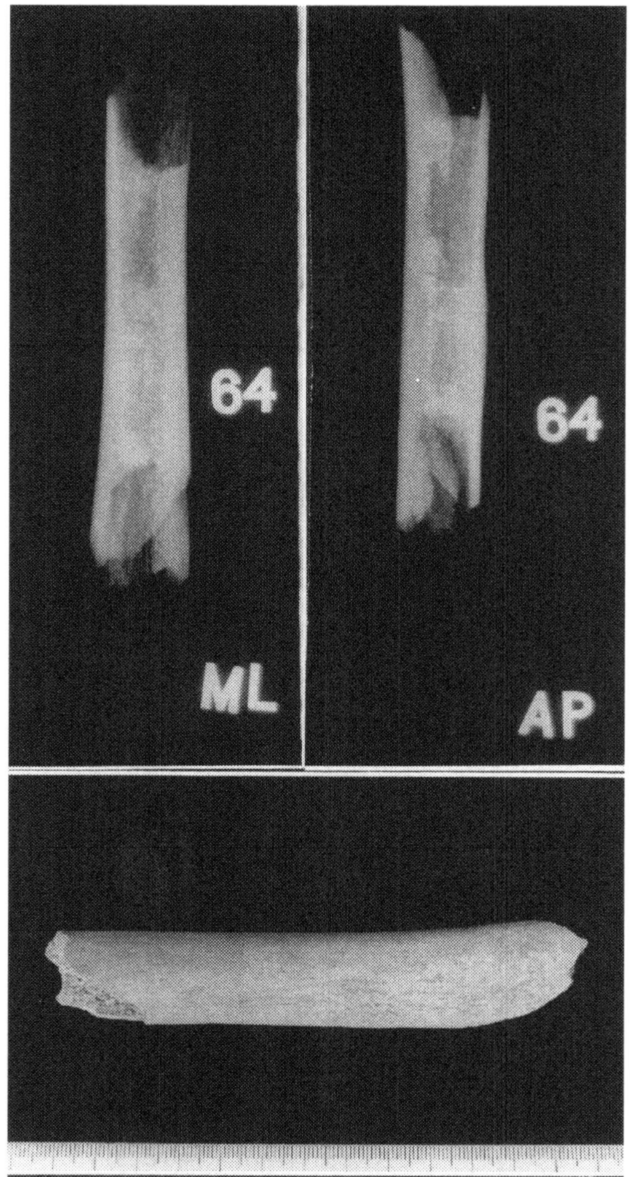

Figure 2: Chichester 64: Macroscopic and radiographic images of periostitis on a fragment of shaft of tibia

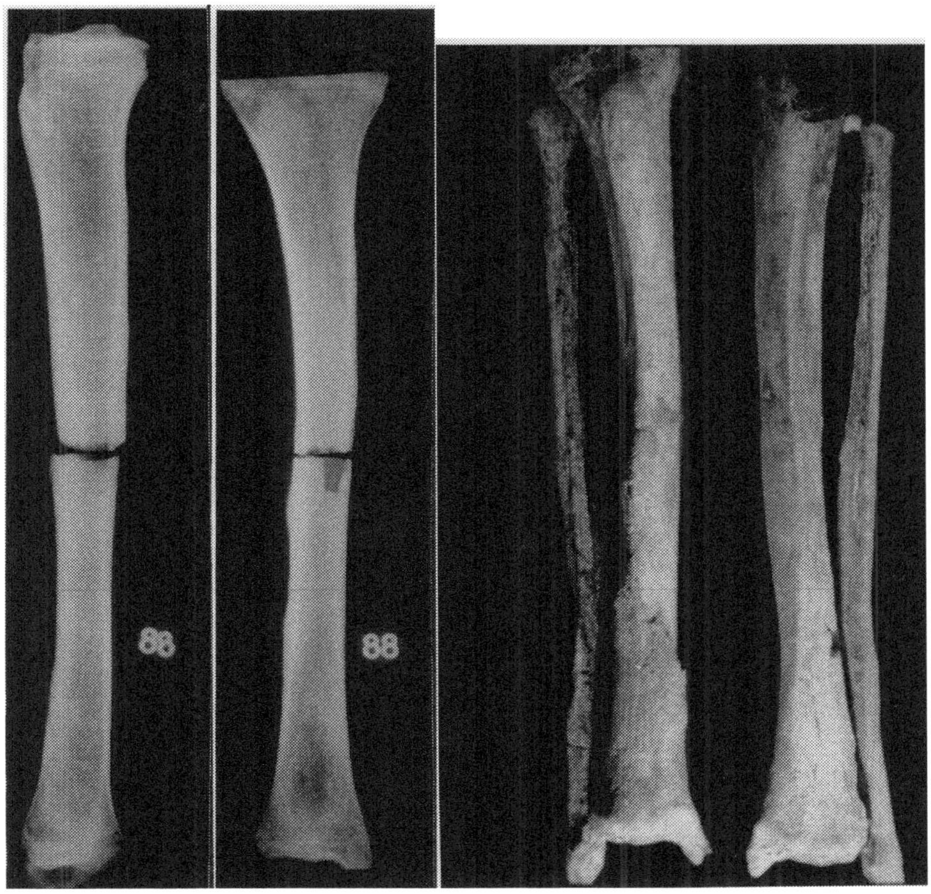

Figure 3: Chichester 88: Macroscopic image of both tibiae and fibulae showing extensive periostitis, and radiographic images of the tibia analysed

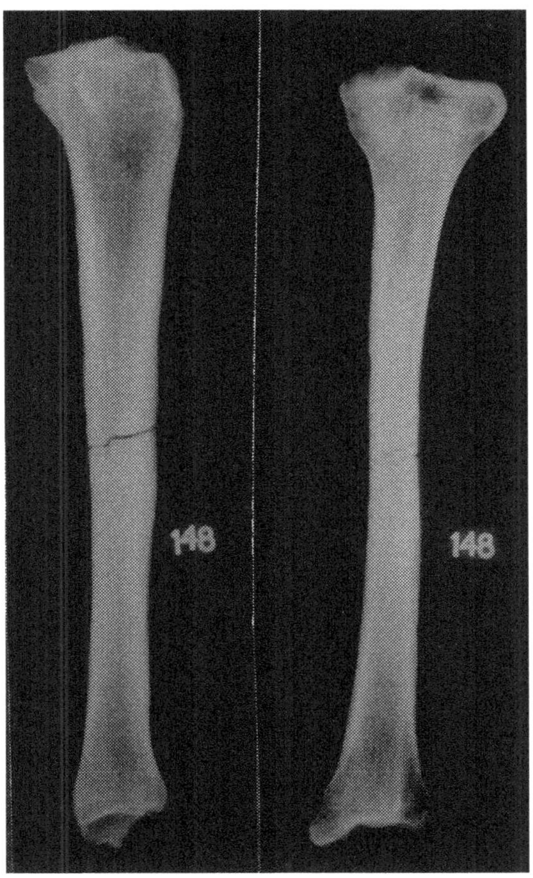

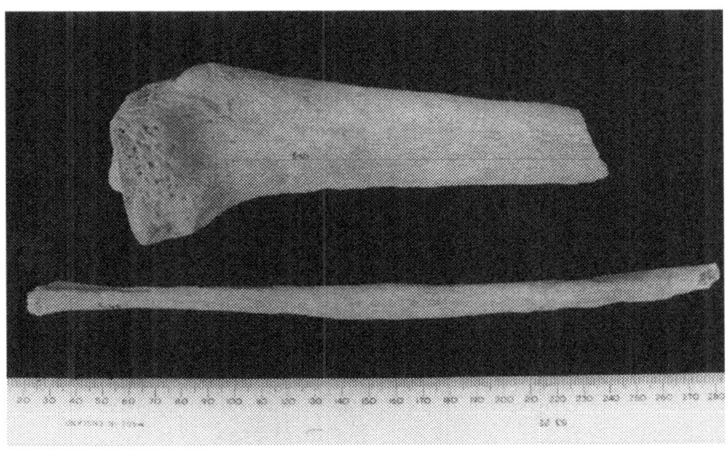

Figure 4: Chichester 148: Macroscopic image of tibia and fibula with periostitis used for analysis, and radiographic image of tibia

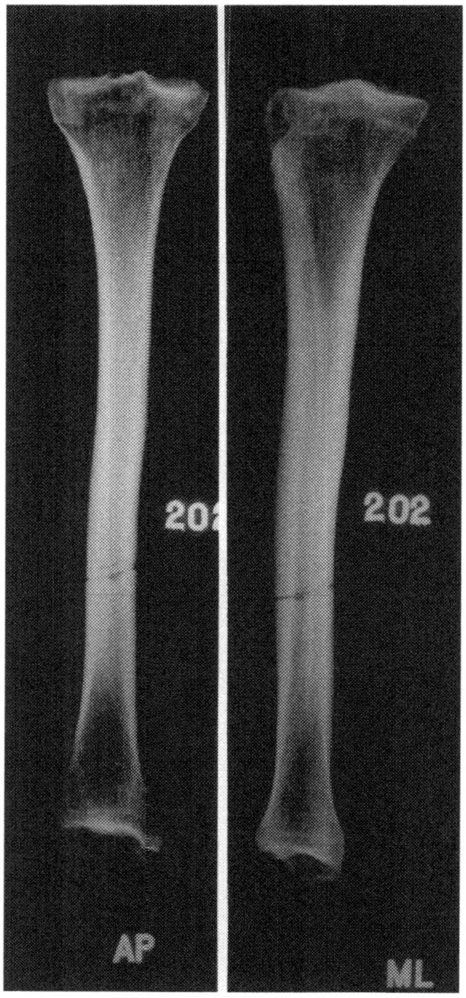

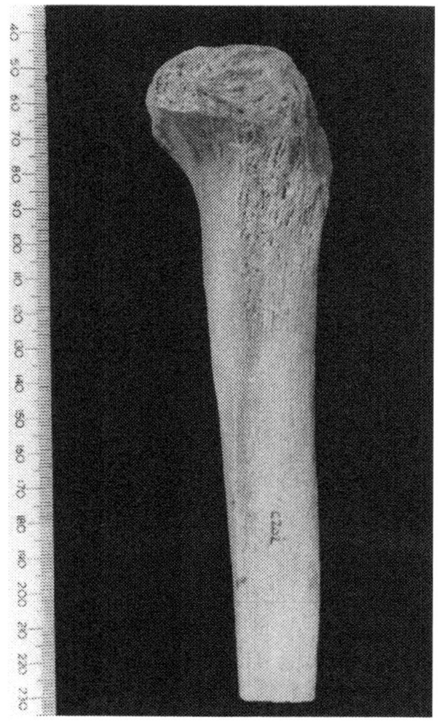

Figure 5: Chichester 202: Macroscopic image of tibia with periostitis used for analysis, and radiographic image of same bone

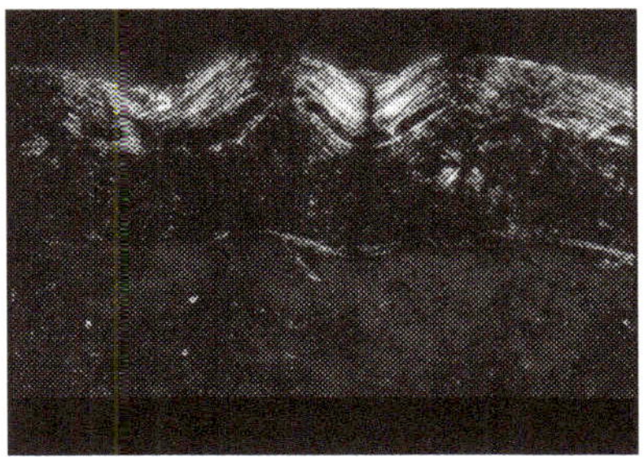

Figure 6: Chichester 21 (distal part of sample). Thin ground section (70µm) viewed by microscope in polarised light. Periosteal bone surface. Well organised secondary layer showing blood vessel impressions. Magnification 100x.

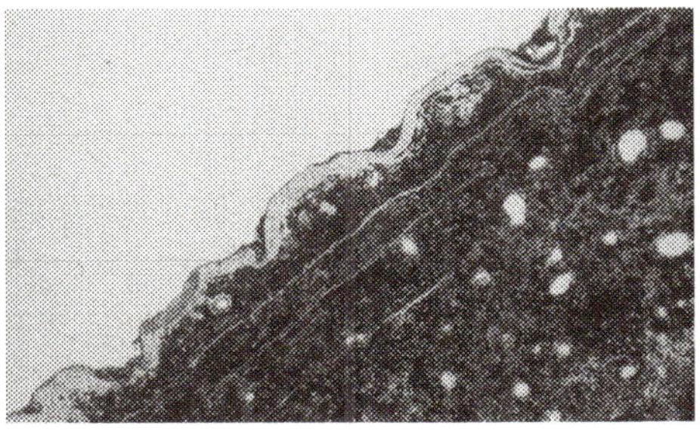

Figure 7: Chichester 21 (proximal part of sample). Thin ground section (70µm) viewed by microscope in plane light. Periosteal bone surface. Well organised secondary layer showing blood vessel impressions and fine border line resembling a grenzlinie. Magnification 25x.

Figure 8: Chichester 21 (distal part of sample). Thin ground section (70µm) viewed by microscope in polarised light using a hilfsobject red 1st order (quartz) as compensator. Periosteal bone surface. Relatively fresh newly built bone formation (woven bone) on the original external bone surface. Magnification 100x.

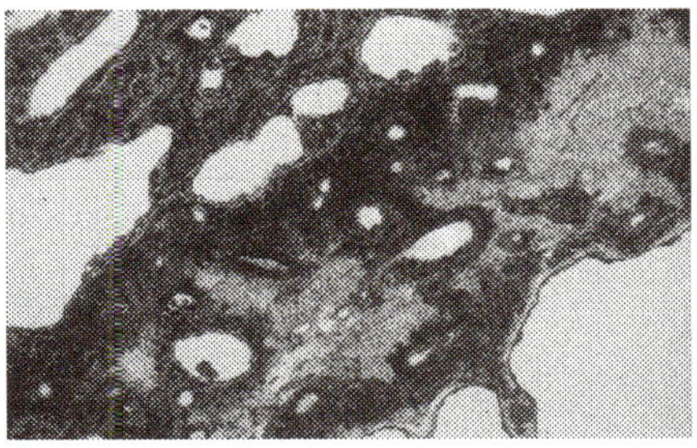

Figure 9: Chichester 48. Thin ground section (70µm) viewed by microscope in plane light. Secondary layer, compact bone substance and endosteal bone surface. Extreme reduction of compact bone substance demonstrating osteoporosis. Magnification 25x.

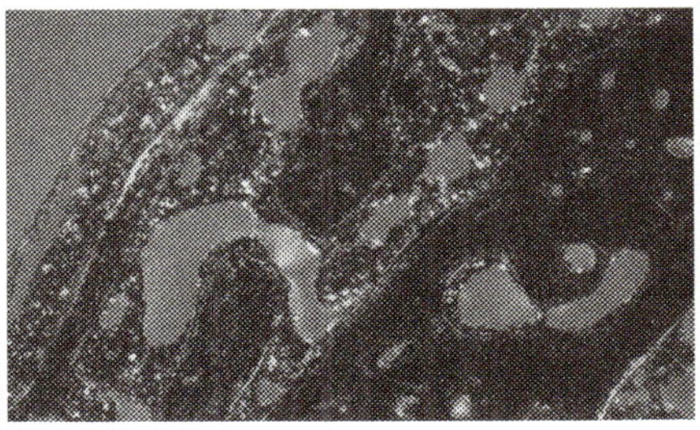

Figure 10: Chichester 48. Thin ground section (70µm) viewed by microscope in polarised light using a hilfsobject red 1st order (quartz) as compensator. Compact bone substance and periosteal bone surface representing two strata of newly built bone formation. Secondary osteoporosis. Magnification 25x.

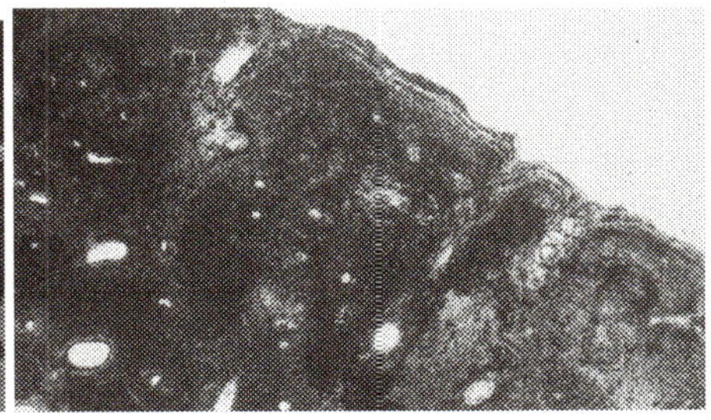

Figure 11: Chichester 64. Thin ground section (70µm) viewed by microscope in plane light. Periosteal bone surface. Very well organised periosteal bone surface showing polster-like structures. Magnification 25x.

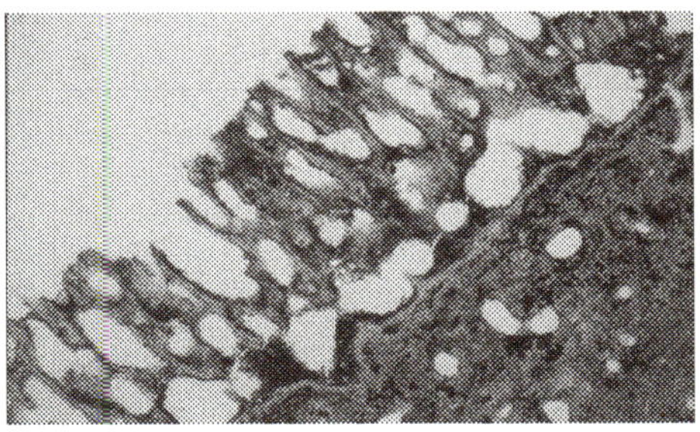

Figure 12: Chichester 88. Thin ground section (70µm) viewed by microscope in plane light. Periosteal bone surface. Product of periostitis (woven bone) on bone surface. Magnification 25x.

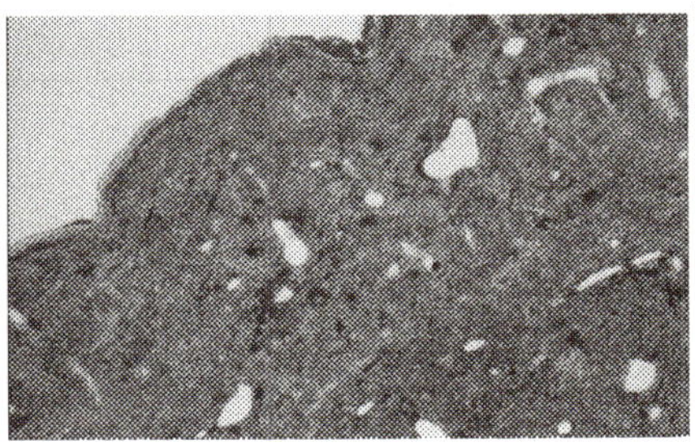

Figure 13: Chichester 88. Thin ground section (70µm) viewed by microscope in plane light. Periosteal bone surface. Very well organised periosteal bone surface showing polster-like structures. Magnification 25x.

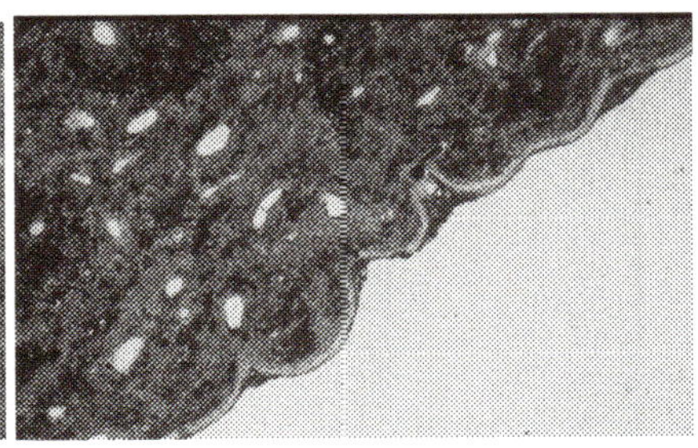

Figure 14: Chichester 148. Thin ground section (70µm) viewed by microscope in plane light. Periosteal bone surface. Well organised secondary layer showing blood vessel impressions and early polster-like structures. Magnification 25x.

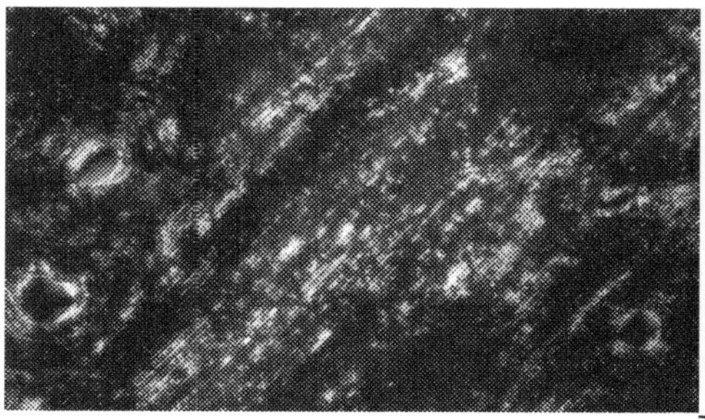

Figure 15: Chichester 148. Thin ground section (70µm) viewed by microscope in polarised light. Compact bone substance. Tangential lamellae suggesting inactivity atrophy. Magnification 100x.

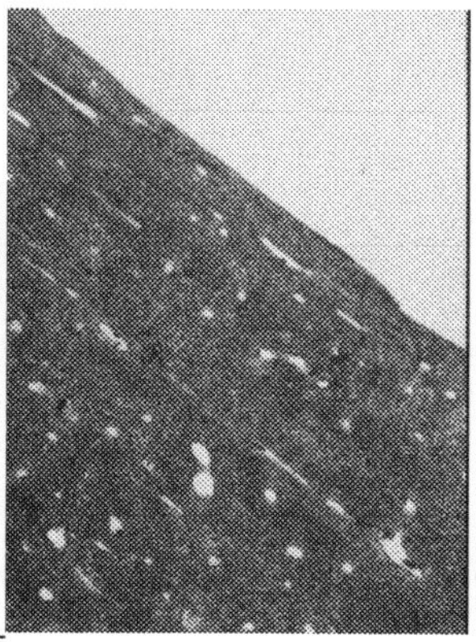

Figure 16: Chichester 202. Thin ground section (70µm) viewed by microscope in plane light. Periosteal bone surface. Between a relatively thick, very well organised secondary layer and the compact bone substance, there is a border line built up by very small, tangentially orientated blood vessel canals. Magnification 25x.

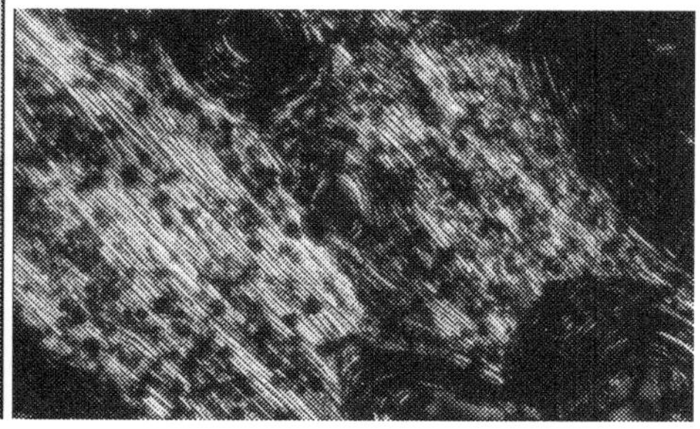

Figure 17: Lihou male. Thin ground section (50µm) viewed by microscope in polarised light. Compact bone substance. Tangential lamellae suggest disuse atrophy. Magnification 100x.

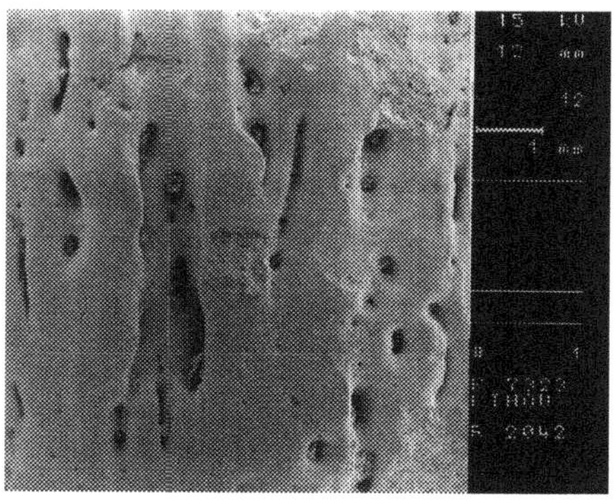

Figure 18: Lihou male. Scanning-electron microscopic image. External surface of the new bone formation. Magnification 12x.

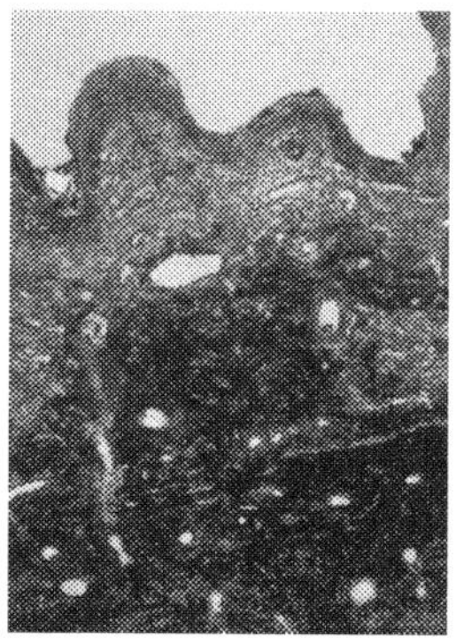

Figure 19: Lihou male. Thin ground section (70µm) viewed by microscope in plane light. Periosteal bone surface. Organised periosteal bone surface presenting polster-like structures as well as remains of a border line resembling a grenzlinie. Magnification 25x.

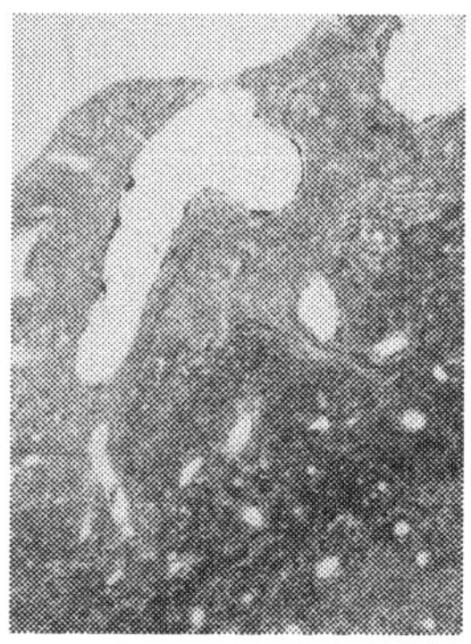

Figure 20: Lihou male. Thin ground section (50µm) viewed by microscope in plane light. Periosteal bone surface. The relatively fresh new bone formation (woven bone) is the product of a periosteal inflammation. Magnification 25x.

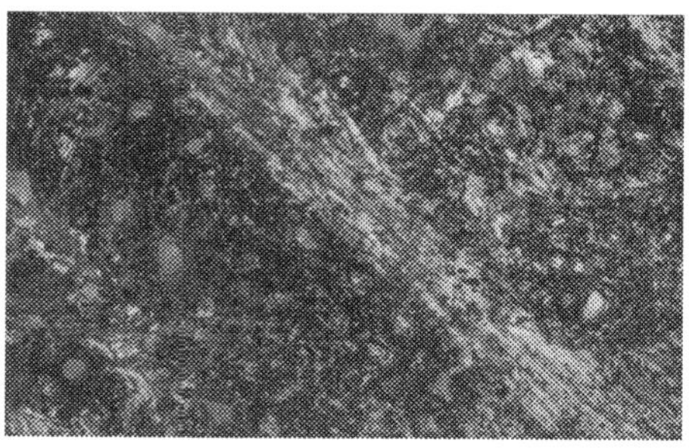

Figure 21: Lihou male. Thin ground section (50µm) viewed by microscope in polarised light using a hilfsobject red 1st order (quartz) as compensator. Cord-like structures in the interior area of the compact bone substance. Magnification 25x.

Microscopic study and X-ray analysis of two 5th century cases of leprosy: palaeoepidemiological inferences

Joël Blondiaux [1], Jean Dürr [2], Lahcen Khouchaf [2], Leslie E. Eisenberg [3]
[1]Centre d'Etudes Paléopathologiques du Nord
36 rue Jules Ferry
59127 Walincourt-Selvigny
France

Email : jblondiaux@nordnet.fr

[2]Laboratory of Physical Analyses
Centre de Recherche de l'Ecole des Mines
941 rue Charles Bourseul BP 838
59500 Douai
France

Email : khouchaf@ensm-douai.fr

[3]Burial Sites Preservation
State Historical Society of Wisconsin
816 State Street
Madison
Wisconsin
United States

Email: leeisenberg:mail.shsw.wisc.edu

Abstract

Applying modern techniques to palaeopathological samples, and by combining microscopy and crystallography, we conclude that cortical lesions due to leprosy and treponemal diseases on the tibia and fibula are identical. In addition, we confirm that bone destructive lesions due to tuberculosis and leprosy, although both are mycobacterial diseases, have different pathological processes and cannot be compared.

Keywords

Microscopy, X-ray analysis, 5th century AD, France, palaeoepidemiology

1. Introduction

Some 30 years ago, Coutelier and Renders (1969) and Coutelier (1971) collected samples of bone from leprosy patients who had undergone amputation, and described the different microscopic features of leprosy remodelling, both in acro-osteolysis of the phalanges and metatarsal bones, and in some the changes to diaphyses of the long bones of the leg. At the time microradiographic techniques were primarily used. Coutelier (1971) discussed simple destruction, patterns of new bone apposition (both slow lamellar, and fast irregular types), and the relationship between peripheral erosion of compact bone and osseous deposition in the medullary cavity. Coutelier and Render's 1969 study, the first of its kind, selected microscopic features characteristic of leprosy. Based on this work, we have inferred that microscopy might help the paleopathological diagnosis of leprosy (Blondiaux *et al.*, 1993, 1994). Here, we will not reintroduce the question of the acro-osteolysic patterns already studied, but will re-examine cortical remodelling of tibiae and fibulae.

Two years ago, a preliminary study of X-ray analyses of samples of tuberculous bone lesions was presented and then compared to normal bone and non-tuberculous osteomyelitic bone samples (Blondiaux *et al.*, 1998, 1999). This study demonstrated that tuberculous lesions in osteoarchaeological samples showed high levels of calcium carbonate deposits, and high levels of magnesium, compared to normal and osteomyelitic bone. Because leprosy and tuberculosis are both mycobacterial diseases, and both present with slow bone destructive patterns, we wanted to test the use of X-ray analysis on acro-osteolytic leprous bones.

2. Material and methods

2.1 Materials

Bone samples were obtained from two human skeletons excavated from sites in northern (Neuville-sur-Escaut) and southern France (Vaison-la-Romaine). They are believed to represent the earliest known cases of leprosy excavated anywhere in France. The following describe the samples:

2.1.1 Vaison La Romaine Burial (c. 500 AD)

This is a mature male skeleton excavated from an urban cemetery of 131 burials located at Vaison-La-Romaine (Vaucluse) in southern France. Archaeological material was scarce and, as a result, no detailed dating evidence was available. Based on anthropological and pathological data, it is believed that the individuals buried in this cemetery were derived from a low socio-economic class. The paleopathological examination of burial 23 yielded the following cranial and post-cranial evidence of disease:

Cranium

Discrete symmetrical absorption of the alveolar bone superior to the maxillary incisors (three teeth are missing postmortem); the alveolar process as well as the margins of the nasal aperture exhibit no remodelling. A small area to the left of the intermaxillary suture, and the frontal and medial surfaces of the palatine processes, exhibit pitting. The lower nasal conchae are missing antemortem.

Postcranium

Hands: linear grooved absorption of the distal articular surface of the first right metacarpal, acro-osteolysis of the fifth proximal phalanx, and distal and proximal palmar grooves of the proximal phalanges (evidence of clawed paralytic hands)
Lower limbs: irregular periosteal deposition on the distal halves and lateral surfaces of both tibiae and fibulae (Figure 1), loss of the foot phalanges bilaterally, with knife-edged remodelling of metatarsals and absorption of the dorsal surfaces of the tarsals; absorption of the distal extremity of a first phalanx (Figure 2).
Although the macroscopic features of concentric diaphyseal remodelling of phalanges, and knife-edged remodelling of the metatarsals supports a diagnosis of leprosy, the tibial and fibular macroscopic changes are, by themselves, not pathognomonic of leprosy.

2.1.2. Neuville Sur Escaut Burial 67 (c. 500 AD)

Excavations at the site of Neuville-sur-Escaut (Nord), located in Northern France, yielded approximately 100 burials for analysis. The site is situated on the bank of the Escaut river at an ancient strategic ford, belonging to the early Frankish kingdoms (5th to 6th centuries AD). The archaeological and anthropological study of the mortuary patterns and grave associations indicate an alamanic origin (southern Germany) for the cemetery (i.e. cremation, horse burial, female jewellry). The postcranial pathology observed on the skeleton of Burial 67, an adult female was documented, as follows :

1. Both tibiae exhibit periosteal deposition on their lateral and distal halves as well as on the medial aspect of the fibulae. The posterior edges of the bony appositions are sinuous, but are not incorporated into the underlying cortical bone, thus lying on the surface.
2. A thick periosteal deposit envelops the right calcaneum; the left side is devoid of any pathological change.
3. The left fifth metatarsal exhibits knife-edged remodelling and its right counterpart exhibits concentric diaphyseal remodelling. According to Andersen *et al.* (1992), Burial 67 from Neuville-sur-Escaut exhibits the earliest osteoarchaeological evidence of leprosy in France.

2.2 Methods

The following methods of analysis were utilised to investigate any changes specific to leprosy

1. *Light microscopy*: diaphyseal tibial sections taken from both skeletons were embedded in methylmethacrylate. Fifty micron-thick transverse sections were then examined using light and polarised light microscopy at X 200 amplification.
2. *X-ray diffraction:* half of the knife-edged metatarsal bone from Vaison-la-Romaine was ground in an agate mortar and studied by X-ray diffraction (Siemens D 5000 apparatus) for crystal species diagnosis.
3. *Environmental scanning electron microscopy equipped with microanalysis.* The other half of the same metatarsal bone from Vaison-la-Romaine was transected, polished and examined by an environmental scanning electron microscope (Electroscan) equipped with micro-analysis (Energy Dispersion Spectrometry).

3. Results

3.1 Microradiography

The microradiographs of the Vaison-la-Romaine tibial samples (Figure 3) and the light microscopy appearance of the Neuville-sur-Escaut and Vaison-la-Romaine tibial sections, while resembling Coutelier's description, are ubiquitous compared to the microradiographic patterns of sectioned acro-osteolytic metatarsals from both individuals (Blondiaux *et al.,* 1993). Compared to the description of osteo-periosteal remodelling of treponematosis by Schultz (1994), our leprosy samples demonstrated superimposable features (Figure 4) as follows:
1. The "Polster" structure, or slow lamellar sub-periosteal deposition, described as "polipous" or "villous" proliferations of a highly thickened periosteal layer (consisting of very dense, parallel lamellar structures), and subperiosteal new cortical bone with sinuous lacunae bordered by circular lamellar structures (Figure 5)
2. The "Grenzstreifen" (boundary line of lamellar bone separating the newly formed periosteal bone layer and the original cortical bone)
3. The mosaics at the edges of the "Grenzstreifen" (Figure 5). It is therefore possible to suggest that the slow periosteal apposition of bone, and cortical remodelling, on the tibiae is not only characteristic of treponemal disease but also of tibial remodelling due to Hansen's disease.

3.2 X-Ray Analyses

No evidence of intra-lamellar calcium carbonate deposits were seen in the Vaison-la-Romaine acro-osteolytic samples, either using X-ray diffraction or the environmental SEM equipped with micro-analysis.

4. Discussion

Based on the results of analyses of these two early skeletons discovered in France, we posited a theory of the spread of leprosy in the Old World during the first millennium AD. It appears the *Mycobacterium leprae* spread via the shipping and commercial routes from the south (i.e. India and Egypt) into northern Europe. We can trace a chronological route starting from where literature and osteoarchaeology converge for the most ancient evidence of leprosy in the Nile Valley.

If the earliest foundations of leprosaria are followed in western Europe, they are found in the Rhône, Saône, Rhine and Meuse valleys from the fifth to the tenth centuries AD. Roberts (1986) also traces the early distribution of leprosy hospitals, founded during the twelfth century in England, along the coast and some of the main rivers (e.g. the Severn, Thames and Humber). The osteoarchaeological discoveries follow the same south-to-north distribution in France (Vaison-la-Romaine, Neuville-sur-Escaut) and they present an east-west distribution in England. The same waterways also channelled many epidemics of the plague, beginning in the fourth century, from Alexandria to Marseilles and Treves. If these epidemics had a short duration of one to two years, the leprosy epidemic lasted much longer and most likely followed the merchants and armies associated with the Roman conquest. The correlation of early osteoarchaeological evidence and the foundation of leprosy hospitals along the same mercantile routes is noteworthy; this may be more than coincidence and should be considered. When considering the site of Neuville-sur-Escaut, on a Danubian trade route, this fits well with the concentration of leprosy hospitals from the sixth to the tenth centuries at the intersections of rivers and valleys in northern Europe. Another epidemiological variable to consider may be the soil-borne transmission of the disease (Chakrabarty and Dastidar, 1989). Populated areas of river banks and ports frequented by barefoot lower-class people working on damp soil could have been favourable to the spread of local endemic disease.

5. Conclusions

The utility of employing microscopic techniques has been demonstrated. Periosteal apposition of new bone and cortical remodelling of the tibia and fibula in leprosy are not microscopically distinct from those changes seen in the treponemal diseases. Consequently, the past attribution of periosteal new bone apposition on the tibiae and fibulae in leprosy samples should be re-examined. The effects of ascending infection from distal ulcerations, although possible in some cases, must be compared with immunological, vascular or neurological processes resulting in new and complex bone structures different from osteomyelitic features.

Additional studies are needed to confirm the patterns observed using X-ray analyses but the calcifications seen in bone lesions due to tuberculosis are not expected to be present in bone lesions resulting from Hansen's disease. In the former, the calcium deposit is due to the exacerbation of a process involving the tuberculous granuloma and its topographic connection with the degradation of *M. tuberculosis* (Baud and Lagier, 1999). Such a process is not encountered in the development of bone changes due to leprosy. Using modern techniques on palaeopathological samples, and by combining microscopy and crystallography, it is concluded that the cortical lesions resulting from leprosy and treponemal diseases on the tibia and fibula are identical. However, destructive lesions of tuberculosis and leprosy, both mycobacterial diseases, have different pathological processes and cannot be compared. Finally, plague and leprosy seem to have been transmitted by the same water routes at different periods. The relationship of humanity to bacterial pests, in the past and today, have many kinds of expressions that vary depending on social, economic and political variables. Palaeopathology is one window into those relationships.

References

Andersen, J.G., Manchester, K. and Shahzady, A. R. (1992) Diaphyseal remodelling in leprosy: a radiological and palaeopathological study. *International Journal of Osteoarchaeology* 2(3) 211-219.

Baud, C-A and Lagier, R. 1999 Calcification of tuberculous lesions : Its interest in general pathology and paleopathology. In G. Pálfí, O. Dutour, J. Deák, I. Hutás (eds): *Tuberculosis, Past and Present, Budapest*. Szeged, Hungary: TB Foundation and Buda[est, Hungary: Golden Book Publisher Ltd, pp. 319-322.

Blondiaux, J., Baud, C-A., Dürr J. and Flipo, R-M. 1998 L'ostéite infectieuse en paléopathologie : Intérêt des analyses par rayons X. *Journal de Physique* 4 (8) PR5: 393-398.

Blondiaux, J., Duvette, J.F., Vatteoni, S. and Eisenberg, L.E. 1993 *Two early cases of leprosy from northern and southern France*. Poster presented to the twentieth Annual Meeting of the Paleopathology Association, Toronto, Ontario, Canada.

Blondiaux, J., Duvette, J-F, Vatteoni, S. and Eisenberg, L.E. 1994 Microradiographs of leprosy from an osteoarcheological context. *International J. Osteoarcheology* 4: 13-20.

Blondiaux, J., Hedouin, V., Flipo, R-M., Dürr, J. and Khouchaf, L. 1999 Tuberculous bone and joint lesions in paleopathology: an X-Ray analysis study. In G. Pálfí, O. Dutour, J. Deák, I. Hutás (eds): *Tuberculosis, Past and Present, Budapest.* Szeged, Hungary: TB Foundation and Budapest, Hungary: Golden Book Publisher Ltd, pp. 267-271.

Chakrabarty, A.N. and Dastidar, S.G. 1989 Correlation between occurrence of leprosy and fossil fuels: role of fossil fuel bacteria in the origin and global epidemiology of leprosy. *Indian J. Experimental Biology* 27:483-496

Coutelier, L. 1971 The bone lesions in leprosy: a study based on microradiography and fluorescence microscopy. *International Journal of Leprosy* 39(2): 231-243.

Coutelier, L. and Renders, A. 1969 La microradiographie et la microscopie en fluorescence appliquées à l'étude des lésions osseuses de la lèpre. *Annales de la Société Belge de Médecine Tropicale* 49(5): 427-456.

Roberts, C. 1986 Leprosy and leprosaria in Medieval Britain. *M.A.S.C.A. Journal* 4 (1): 15-21.

Schultz, M. 1994 Comparative histology of syphilitic lesions in prehistoric and historic human bones. In O. Dutour, G. Pálfi, J. Berato and J-P Brun (eds): *L'Origine de la syphilis en Europe, avant ou après 1493?* Paris: Editions Errance, pp. 63-67.

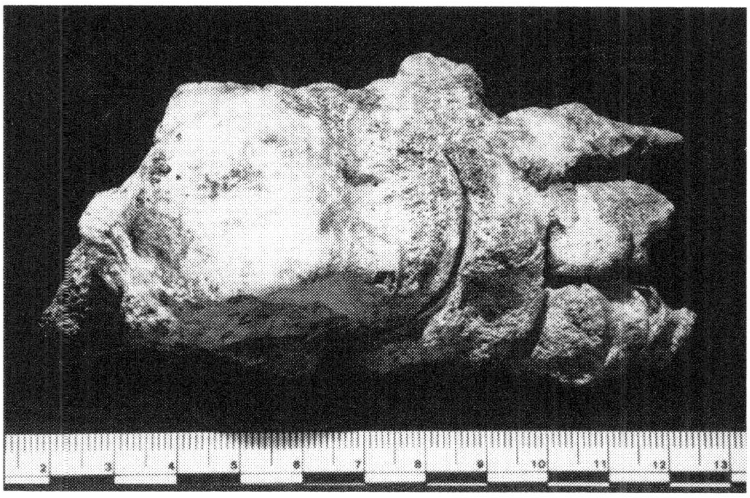

Figure 1: Vaison la Romaine 23: pathognomonic resorption of the forefoot

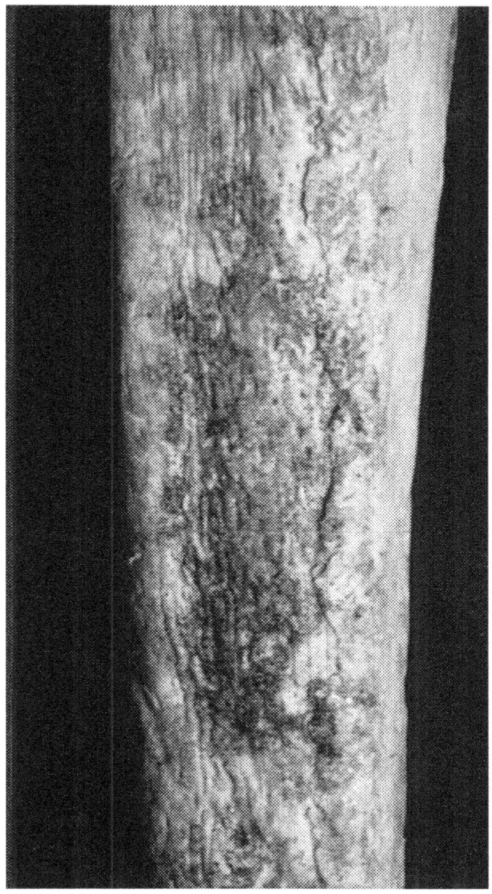

Figure 2: Vaison la Romaine 23: periosteal deposition on the external surface of the right tibia

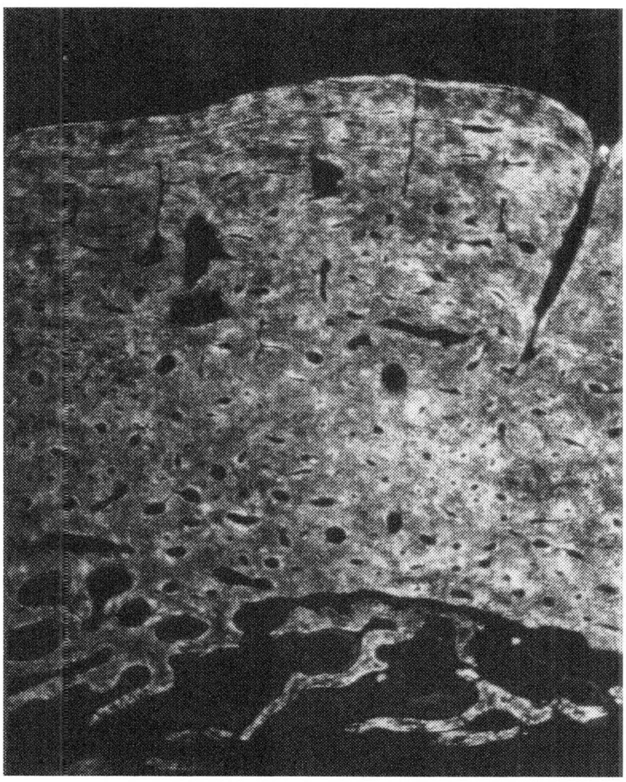

Figure 3: Vaison la Romaine 23: microradiography of tibial section (x 100); "Polster" structure.

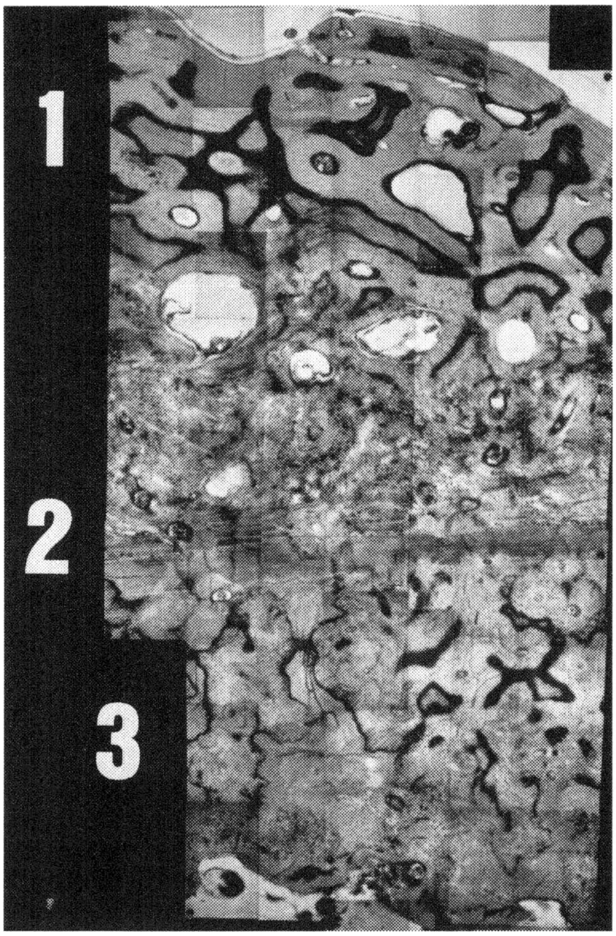

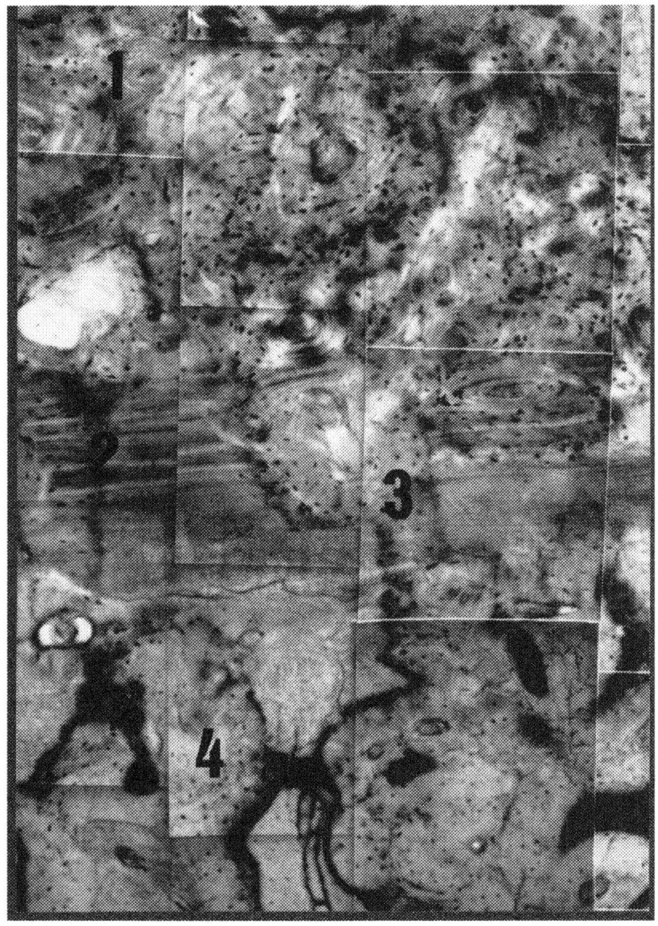

Figure 4: Neuville-sur-Escaut 67: light microscopy (x 100) of tibial section: 1: "Polster" structure and new vascular bone, 2: Greinstreifen, 3: Normal cortical bone and endosteal edges.

Figure 5: Neuville sur Escaut 67: light microscopy (x 200) details of: 1: Structure of new lamellar bone, 2: "Grensztreifen", 3: Mosaic, 4: Normal cortical bone .

Was there a Medieval diagnostic confusion between leprosy and syphilis? An examination of the skeletal evidence

Gillian M. M. Crane-Kramer
Department of Anthropology
State University of New York at Plattsburgh
101 Broad Street
Plattsburgh
New York
U.S.A. 12901-2681

E-mail: kramer@westelcom.com

Abstract

For several decades, a major tenet of the pre-Columbian hypothesis has been the possible Medieval diagnostic confusion between leprosy and acquired venereal syphilis. If this confusion can be demonstrated, then it is possible to suggest the pre-Columbian presence of venereal syphilis in European populations. Supporters of this theoretical position have argued that evidence for this confusion can be identified in historic documents that contain passages suggestive of a venereal transmission for leprosy. If this assertion is correct, one would expect this "literary confusion" to manifest itself in the skeletal record. Alternatively, the Columbian hypothesis argues for the absence of the condition in Europe prior to the voyages of Columbus at the climax of the 15th century. The aim of this paper is to present the results of a pathological examination of over 600 individuals from several leprous and non-leprous cemeteries from England and Denmark, extending from the Viking Period to the late Middle Ages. On the basis of this analysis it is suggested that, at present, a Medieval diagnostic confusion cannot be supported by the skeletal evidence.

Keywords

Treponematosis, leprosy, Pre-Columbian, diagnostic confusion, Northern Europe

For several centuries, scholars have debated the question of the origins and antiquity of treponemal infection. In recent times, three hypotheses have been presented to account for the epidemiological history of the genus *Treponema.* A vast majority of this work has concerned itself with an investigation of the evolutionary history of venereal syphilis. This paper will address specific assertions of the Pre-Columbian hypothesis, and the implications that these have on our understanding of disease processes in past populations. Proponents of the Pre-Columbian hypothesis assert that venereal syphilis was present in Europe prior to the return of Columbus and his crew from the New World at the climax of the 15th Century (Hackett 1963, 1967, Holcomb 1934, 1935, Hudson 1963, 1968, Kampmeier 1984). The lack of recognition of this condition prior to the European epidemic of 1500 AD resulted from the inability of classical and medieval diagnosticians to differentiate between venereal syphilis and leprosy. In the absence of substantial skeletal evidence for venereal syphilis, palaeopathologists and medical historians have relied on ancient and medieval written texts in order to establish the antiquity of the disease in the Old World (Baker 1985, cited in Baker and Armelagos 1988). In order to establish this diagnostic confusion, they have identified passages in historic texts which they argue present cases of venereal transmission for conditions grouped under the term "leprosy".

Supporters of the Pre-Columbian hypothesis have suggested that references to venereal syphilis and leprosy can be identified as early as in the Old Testament of the Bible written between the 8th and 2nd centuries BC (Holcomb, 1935, Willcox, 1949, Hudson, 1961, 1963, Brown *et al.*, 1970, Rosebury, 1971, 1992, Baker and Armelagos, 1988). However, confusion has surrounded the interpretation of the Hebrew word "tsara'ath" which was translated into Greek and Latin under the terms "λεπρα" and "lepra" respectively (Andersen, 1969). Scholarly review of Biblical texts suggest that the term "tsara'ath" refers to a number of diseases resulting in ritual uncleanliness, and cannot be held to represent solely descriptions of true lepromatous leprosy. In addition, the archaeological record has failed to reveal any evidence of true leprosy from Biblical times. With the exception of Zias' (1985, 1991) recovery of leprous skeletal material from Israel dating to 614 AD, and the 2nd century BC Ptolemaic and Coptic Christian material from Dakhleh Oasis, Egypt described independently by Dzierzykray-Rogalski (1980) and Molto (1998), there has been no other skeletal evidence of the disease in all of Israel, Palestine, or Egypt. In addition, in an examination of 600 individuals from the burial site of Lachish, Møller-Christensen (1967) could find no evidence of leprosy. This lack of confirmation in the archaeological record suggests that the terms in the Old Testament that were used to express a condition which was later associated with leprosy, were generic terms referring to a plethora of different diseases. Therefore, if one cannot establish the presence of clinical leprosy in Biblical times, by extension one cannot argue for a diagnostic confusion between leprosy and venereal syphilis.

Many of the concepts which contributed to the medieval understanding of leprosy were described by early Roman, Arabic, and particularly, Greek physicians. A considerable portion of this medical knowledge was preserved by the Byzantine and Arabic kingdoms and reintroduced into Western Europe in the early Middle Ages (Hollister, 1994). In the Middle Ages, there was a heavy reliance upon the work of these former physicians with little scrutiny or scientific enquiry. A number of Greek, Byzantine and Roman historians and physicians such as Pliny, Hippocrates, Celsus and Martial describe the occurrence of genital lesions resulting from intercourse (Hudson, 1961, 1963, Rosebury, 1971, Kampmeier, 1984). Martial and Pliny in the 1st Century AD refer to a condition known as "mentagra" which was a term that was euphemistic for the pubic region, and was generally used to refer to lesions resulting from sexual activity (Hudson 1961, Kampmeier 1984).

Interestingly, Greek physicians as early as the Alexandrian school of the 3rd century BC referred to cases of lepromatous leprosy under the term "elephantiasis" because of the thickening of the skin in this condition (Richards, 1977, Andersen, 1969). Although all the original records have been lost, in the 1st century AD, Rufus from Ephesos quotes the Alexandrian physician Straton as recording an accurate depiction of a low resistant form of leprosy (Andersen, 1969). It is clear that the medical practitioners of the Classical world did recognize and accurately diagnose a condition which can be related to advanced lepromatous leprosy, which they referred to as "elephantiasis". There was a distinct differentiation between "elephantiasis" and the disease described as "lepra", which was considered to represent a variety of skin conditions, one of which may have been leprosy.

In the ancient literature of China and India, Pre-Columbian scholars have identified passages that they argue demonstrate unmistakable evidence for the ancient connection between genital lesions and sexual intercourse (Campbell, 1934, Kampmeier, 1984). However, Wong and Wu (1936) and Lu and Needham (1967) both assert that no ancient Chinese author has ever described a condition resembling syphilis. Kampmeier (1984) suggests that the Indian Sanskrit Veda (c.1400 BC) contains multiple references to genital ailments, which some have interpreted as representing venereal syphilis. However, clinical descriptions of leprosy in India and China from as early as the 6th century BC present accurate clinical descriptions of lepromatous leprosy (Rogers and Muir, 1940, Dharmendra, 1967, Cochrane, 1964, Andersen, 1969). It is clear that ancient practitioners were capable of establishing the diagnostic criteria for leprosy in considerable detail. Therefore, it is difficult to argue that physicians in antiquity were unable to differentiate this condition from venereal syphilis.

Pre-Columbian theorists have also investigated medieval texts in order to establish that syphilis was included with other conditions under the heading of "leprosy". It appears that the 8th to the 9th century Arabian physician John of Damascus seems to have been the first to use the term lepra for what is now called lepromatous leprosy, and have so started the process which led to 'elephantiasis graecorum being called leprosy'(Hulse, 1972: 1024). Thus, the ancient generic term "lepra" was mistakenly translated into the Medieval Vulgate Bible as the term representing leprosy exclusively. Supporters of the Pre-Columbian theory argue that the diagnostic confusion between leprosy and venereal syphilis can be demonstrated in the medieval literature in two ways. Firstly, the appropriation of the ancient word "lepra" to refer to leprosy could suggest that medieval physicians continued to use this term as a catchall phrase for multiple diseases, as in ancient times (Andersen, 1969, Dols ,1979, Richards, 1977). Therefore, cases of venereal syphilis could have been contained within the group of conditions labelled as "lepra". Secondly, theorists point to a number of examples in the medieval literature that appear to suggest a confusion in the understanding of the mode of transmission for leprosy (Holcomb, 1935, Hudson, 1968, 1972, Brody, 1974, Rubin, 1974, Kampmeier, 1984, Baker and Armelagos, 1988). Numerous medieval writers state that leprosy can be transmitted through sexual intercourse with a "leper", contracted as an hereditary condition, or through the breast milk of a leprous nurse. In addition, there are abundant references to "leper whore", "venereal leprosy", and "hereditary leprosy", as well as descriptions of lesions on the genitals (Baker and Armelagos, 1988). It has been well established that leprosy is not transmitted through sexual contact and is not hereditary (but see Lewis, this volume). Although the disease does show a significant family incidence, transmission of the disease between spouses occurs in less than 5% of couples (Richards, 1977). Interestingly, modern research has demonstrated that viable leprous mycobacteria can be transmitted to infants via breast milk (Pedley, 1967).

While at first glance these descriptions could be used to argue for a medieval diagnostic confusion, there is another avenue of investigation that has not previously been fully considered. The biblical word "tsara'ath" used to refer to conditions resulting in ritual uncleanliness was translated into the Greek as the term "λεπρα". With the adoption of the Greek, and then Roman, word "lepra" into the Byzantine Latin Vulgate translation of the Bible, there developed an association between ritual uncleanliness or moral impurity and the disease leprosy. Therefore, as Richards (1977) points out, a previously well-defined disease of no religious significance became associated with the biblical notion of impurity, and culminated in the stigma attached to the word leprosy in modern times. In a period dominated by the doctrines of the Catholic Church, leprosy was associated with all forms of sin from avarice to murder. Importantly, one of the common notions in the Middle Ages, which can be found as early as 200 AD in the work of the physician Aretaios, was the idea that the "leper" was extremely lustful, and had a distinctly over-active sexual appetite (Dols, 1979, Andersen, 1969). As lust is one of the seven deadly sins, this aspect of the character of a person suffering from leprosy would have been particularly emphasized by the Catholic Church. Modern research has demonstrated that there is no indication that individuals suffering from leprosy are characterized by elevated sexual activity or desire. However, it is possible that the use of terms which imply sexuality to refer to leprous individuals stemmed from this pervading religious notion of sexual deviance, and does not indicate a diagnostic confusion between leprosy and venereal syphilis (Zias, 1989). Therefore, the presence of references such as "leper whore" may simply reflect the association between leprosy and moral impurity, and not a lack of diagnostic competence.

In addition to the written work of ancient and medieval scholars, proponents and critics of the Pre-Columbian theory have concentrated on the appearance of edicts against syphilis (Holcomb, 1934, 1935, Sudhoff, 1926, Harrison, 1959, Temkin, 1966, Kampmeier, 1984, Russell, 1989, Rosebury, 1992, Luger, 1993). The edicts on syphilis that began to appear in the late fifteenth century have been considered to be valuable sources of information concerning the history of syphilis in Europe. The Columbian school has argued that the timing of the edicts indicates the appearance of a new and virulent disease within European populations at the climax of the fifteenth century (Baker and Armelagos, 1988). Alternately, the Pre-Columbian contingent has suggested that the edicts simply represent the recognition of syphilis as a distinct disease at this time in Europe. Unfortunately, a considerable amount of confusion in the literature has resulted in errors in the establishment of accurate dates for particular edicts. A thorough discussion of the European edicts is beyond the scope of this paper. However, this paper will concentrate on the edict of Paris, which has been the foundation of the pre-Columbian position. The mainstay of the Pre-Columbian hypothesis presented by Holcomb (1935) rests upon an edict issued in Paris which bans people with "grosse verole" (syphilis) from the city limits. Holcomb states that 'on March 25, 1493, or ten days after Columbus arrived at Palos upon return from the first voyage, and more than a month before he could have arrived at Barcelona, an edict was issued in the faraway city of Paris driving syphilitics from that city on pain of being drowned in the river' (Holcomb 1935: 298). Holcomb may have retrieved this date from the work of Sudhoff (1926) who stated that 'as early as March 1493, when Columbus was returning from the first voyage, an ordinance against the great pox ("grosse verole") was being issued in Paris'(in Rosebury, 1992: 64). On the basis of this date Holcomb (1934, 1935) argues that the disease could not have been transported by Columbus from the New World, and that it must have been present in Europe prior to his return from the first voyage. Therefore, it was not until the late fifteenth century that medieval diagnosticians were able to differentiate between syphilis and leprosy. Harrison (1959) investigated the historic records of Paris which revealed that due to a series of mistakes in later French laws, the date had actually been misinterpreted. He asserted that this ordinance was actually issued on June 25, 1498. This research was verified by the work of Haustein (1930 in Temkin, 1966: 33). As a result of an exhaustive investigation of the French archives, Harrison stated that the earliest edict 'relating to syphilis in Paris which he had found was the Arret du Parlement de Paris dated March 6, 1497' (Harrison, 1959: 4). The decree in English translation is as follows, 'Today, the sixth March, since in this city of Paris there are many sick of a certain contagious disease, named "la grosse verole", which has become prevalent during the past two years...'(Harrison, 1959: 4). Therefore, the earliest edict of Paris dates to almost four years after the return of Columbus from his first voyage to the New World. Interestingly, in addition the edict states that the disease had only been prevalent in France for the two years preceding the edict, which is significant for two reasons. First, it suggests that the French did not perceive the disease as having a long history amongst their population. Second, it suggests that the French had no knowledge of the disease before the campaign of Naples in 1494-5, and that it was not identified in Charles VIII's army before its entry into Italy. Naturally, the Arret du Parlement de Paris does not provide definitive proof for the absence of treponemal disease in Europe prior to the return of Columbus from the New World. However, if treponemal infection had been present in significant numbers of people in France prior to this date, it is intriguing that a cosmopolitan city like Paris did not feel the need to release an edict before March 1497 AD It is significant that the original edict of Paris dates not to the year 1493, but to the year 1497, at which time venereal syphilis was well known in many European countries. Thus, it is impossible for the Pre-Columbianist position to argue that it represents recognition of the diagnostic differentiation between leprosy and venereal syphilis in 1493.

Proponents of the pre-Columbian theory have relied heavily on ancient and medieval literature as proof of a diagnostic confusion between leprosy and venereal syphilis. However, an evaluation of this material suggests that these literary references are ambiguous, and can therefore be interpreted in alternative ways. In addition, no evidence has ever been presented which demonstrates the existence of edicts against syphilis prior to the return of Columbus. Therefore, it is not possible to establish that this line of evidence represents a diagnostic recognition of venereal syphilis as distinct from leprosy.

In light of the ambiguity of the historic literature, the logical avenue for resolution of this question is the examination of skeletal material from European leprosy and non-leprosy hospital cemeteries. If in fact a diagnostic confusion between leprosy and venereal syphilis did exist in the ancient and medieval world, one should be able to identify individuals suffering from venereal syphilis in a leprosarium burial ground and *vice versa*. In reality, this is not the case. The remainder of this paper will concern itself with a presentation of the results of the author's doctoral research. The research involved the examination of skeletal material from three leprosy hospital cemeteries in England. In addition, individuals were examined from several non-leprosy hospital cemetery contexts. Table 1 presents the skeletal collections examined in this study, including the approximate dates of the cemeteries and number of overall burials. Figure 1 presents a general location map of the eight sites. An entire examination of the collections were conducted by the author with the exception of the cemeteries of Rivenhall, Ipswich and Wharram Percy in which only selected individuals were examined. The analysis of these complete collections was conducted by Mays (1991) at English Heritage, who kindly permitted examination of the skeletal material relevant to the research.

The pathogenic processes responsible for osseous change in venereal syphilis and leprosy have been well established in recent years. In this investigation, the identification of leprosy in skeletal material was based upon the pathological indicators established by the work of Møller-Christensen (1953a and b, 1961, 1969, 1974, 1978), Danielsen (1970), Manchester (1981, 1984, 1989, 1992, 1994), Roberts (1986), Andersen and Manchester (1987) and Andersen (1969). Diagnostic osseous changes indicative of treponemal infection, in particular venereal syphilis, have been presented in the work of Hackett (1946, 1963, 1967, 1976, 1981, 1983), Hutchinson (1857, 1861, 1909), Bradlaw (1953) and Ortner and Putschar (1985). Tables 2 and 3 present a summary of these sources for the major cranial and infracranial osseous diagnostic criteria used for leprosy and treponemal infection. In terms of the identification of leprosy in the skeletal collections, individuals categorized as those with definite leprosy were those skeletons that demonstrated clear evidence of *facies leprosa*, in addition to observable changes in the postcranial skeleton. Those individuals who were included in the possible leprosy category were further subdivided into two groups. The first group included all individuals who demonstrated both cranial and postcranial changes that were suggestive of leprosy, but not extensive enough for a definite diagnosis. The second group included individuals who demonstrated considerable postcranial change for the disease, but for whom the crania were not present for observation.

Results of the osteological analysis presented a clear demarcation of the two diseases in all included skeletal samples. In each of the cemeteries from leprosy hospital contexts, several individuals demonstrating clear evidence of lepromatous leprosy were identified. In no cases were osseous changes observed in either sub-adult or adult individuals that were indicative of treponemal infection. In an examination of the non-leprosy hospital cemeteries, three individuals were identified as manifesting osseous change consistent with a diagnosis of treponemal infection. In both the case from Blackfriars, Gloucester, and the individual from Ipswich, Suffolk, cranial changes representing stages of the "caries sicca" sequence were observed, in addition to considerable postcranial change (Roberts, 1994, Mays, 1991). The final case from Rivenhall, Essex demonstrates only minimal cranial changes. However, the postcranial changes in this individual are extensive and consistent with those observed in treponemal infection. A detailed discussion of the treponemal cases from Blackfriars Friary (Ipswich) and Rivenhall (Essex) will be presented in a future paper (Mays and Crane-Kramer, in review).

Individuals manifesting changes consistent with a diagnosis of leprosy were absent from all non-leprosy hospital cemeteries, with two exceptions. The site of Wharram Percy represents the remains of a deserted medieval village in North Yorkshire (Beresford and Hurst, 1990). The burials are primarily of "ordinary" rural people who lived either in Wharram Percy itself or elsewhere in the parish. In the churchyard cemetery, a single leprous non-adult skeleton has been identified in the burial population. However, importantly there is no evidence of treponemal infection in the remaining 686 individuals interred at this site. The only cemetery population in which individuals expressing osseous change for both leprosy and treponemal infection comes from the Blackfriars Friary in Ipswich. This burial ground contained a single definite case of treponemal infection, in addition to four leprous individuals (Mays, 1991). The burial ground was in use between AD 1263 and 1538 and contained remains of both the friars and friary benefactors (Page, 1975). A large number of the graves in this cemetery were disturbed by later interments, and it was not possible to establish precise dating for individual burials based on standard archaeological techniques. However, the treponemal case has been AMS radiocarbon dated (including a delta C13 factor) and revealed a date of 380 years ± 18 years b.p. (Mays, 1991). Of the leprous individuals, three of the four were young-middle adult males, the fourth being a mature female individual. As the cemetery is associated with the medieval friary, it is clear that the female individual represents either a church benefactor or a servant attached to the friary. It is impossible to state with any confidence whether the three male individuals represent friars or lay-benefactors. The treponemal case represents a mature female individual which one must assume also represents either a church benefactor or one of their immediate family, or a servant engaged at the friary itself. The burial of individuals demonstrating changes consistent with a diagnosis of leprosy and treponemal infection in the same friary cemetery does not indicate a diagnostic confusion between the two diseases. In the same cemetery individuals have been identified representing a number of pathological conditions such as tuberculosis, osteomyelitis, Paget's disease, trauma, and osteoarthritis. Individuals suffering from specific conditions were not segregated within the burial grounds, and it appears that the cemetery population represents individuals suffering from a number of pathological conditions. As the friary was not established as a leprosy hospital, it is not surprising to find a plethora of conditions found in the general population during the Medieval period. Therefore, one cannot argue for a diagnostic confusion existing between leprosy and treponemal disease in this burial population any more than argue for a diagnostic confusion between either of these conditions and tuberculosis or Paget's disease.

In light of the results of this research, it is extremely difficult to argue for the existence of a diagnostic confusion in the Medieval period in England. Not a single case of treponemal infection was identified in any of the three leprosy hospital cemeteries included in this study. In addition, with the exception of the single case from Wharram Percy, and Blackfriars Friary, not a single case of leprosy has been identified in the non-leprosy hospital cemetery populations (although see Roberts, this volume). As multiple cases of leprosy and treponemal infection have been identified in the entire sample, it is equally unacceptable to argue that the diagnostic criteria for the two diseases are not adequate to identify specific individuals with the conditions. As the majority of observed material comes from cemeteries within England, it is not possible to state that a diagnostic confusion did not exist in other countries in Europe. However, an examination of the Early Roman, Roman and Viking Period material from Stengade II in Denmark demonstrated no evidence of either leprosy or treponemal infection. This may in part be due to the fact that the preservation of this material was generally quite poor. It is also possible that these populations had no familiarity with either condition at this time. On the basis of her extensive examination of Danish skeletal material, Bennike (1985) has observed that no case of treponemal infection or leprosy has ever been identified in Danish prehistoric or Viking material. As a large number of cases of leprosy have been described by Møller-Christensen (1953a, 1953b, 1958, 1961, 1978), Møller-Christensen and Faber (1952), and Møller-Christensen and Weiss (1971) in both the material from St. Jørgen's gard, Naestved, and Aebelholt Monastery in Denmark, there is no question that, as in contemporary England, leprosy was well established in Denmark in the medieval period. However, in the hundreds of skeletons observed in these collections, Møller-Christensen was unable to identify a single case of treponemal infection. In addition, no dated cases of treponemal infection have been identified in Denmark in the Middle Ages prior to the 16th Century when the disease was well established throughout Europe. This lack of evidence amongst the examined material in Denmark supports the findings observed in English populations.

In conclusion, it appears that the Pre-Columbian assertion that a diagnostic confusion existed between leprosy and syphilis in the Medieval period in Northern Europe (and perhaps by extension other regions of the British Isles and mainland Europe), cannot be supported by the osteological evidence considered in this study. Therefore, it may not be tenable any more to use this line of enquiry in an examination of the Pre-Columbian existence of treponemal infection in European populations.

References

Andersen, J.G. 1969 *Studies in the Mediaeval diagnosis of leprosy in Denmark.* Copenhagen: Costers Bogtrykkeri.

Andersen, J.G., and Manchester, K. 1987 Grooving of the proximal phalanx in leprosy: A palaeopathological and radiological study. *Journal of Archaeological Science* 15:51-6.

Baker, B.J. 1985 Use of written documentation in diagnoses of Pre-Columbian syphilis. In J.B. Gregg (moderator): Section 2. Contributed papers. In Eve Cockburn (ed): *Papers on paleopathology presented at the twelfth annual meeting of the Paleopathology Association, Knoxville, Tennessee, 9-10 April 1985.* Detroit, Paleopathology Association, p.6.

Baker, B.J. and Armelagos, G.J. 1988 The origin and antiquity of syphilis: paleopathological diagnosis and interpretation. *Current Anthropology* 29:703-37.

Bennike, P. 1985 *Paleopathology of Danish skeletons: A comparative study of demography, disease, and injury.* Copenhagen: Akademisk Forlag.

Beresford, M., and Hurst, J. 1990 *Wharram Percy: deserted Medieval village.* New York: Yale University Press.

Bradlaw, R.V. 1953 The dental stigmata of prenatal syphilis. *Oral Surgery, Oral Medicine, and Oral Pathology* 6:147-58.

Brody, S. 1974 *The disease of the soul: leprosy in Medieval literature.* Ithaca: Cornell University Press.

Brown, W.J., Donohue, J.F., Axnick, N.W., Blount, J.H., Ewen, N.H., and Jones, O.G. 1970 *Syphilis and other venereal diseases.* Cambridge: Harvard University Press.

Campbell, H.S. 1934 Concerning the history of syphilis. *American Journal of Surgery* 26:402-12.

Cochrane, R.G. 1964 The history of leprosy and its spread throughout the world. In R.G. Cochrane and T.F. Davey (eds): *Leprosy in theory and practice.* 2nd edition. Bristol: J. Wright and Sons, pp. 1-12.

Danielsen, K. 1970 Odontodysplasia leprosa in Danish Mediaeval skeletons. *Saertrykaf Tandlaegebladet* 74:603-25.

Dharmendra, R. 1967 *History of the spread and decline of leprosy.* India: Ministry of Health.

Dols, M.W. 1979 Leprosy in Medieval Arabic medicine. *Journal of the History of Medicine and Allied Sciences* 34:314-33.

Dzierzykray-Rogalski, T. 1980 Paleopathology of the Ptolemaic inhabitants of the Dakhleh Oasis (Egypt). *Journal of Human Evolution* 9:71-4.

Hackett, C.J. 1946 A review of references to the bone lesions of yaws. *Tropical Diseases Bulletin* 43:1091-1104.

Hackett, C.J. 1963 On the origin of the human treponematoses. *Bulletin of the World Health Organization* 29:7-41.

Hackett, C.J. 1967 The human treponematoses. In D. Brothwell and A.T. Sandison (eds): *Diseases in antiquity*. Illinois, Springfield: Charles C. Thomas, pp.152-169.

Hackett, C.J. 1976 *Diagnostic criteria of syphilis, yaws and treponarid (treponematoses) and of some other diseases in dry bones*. Sitzungsberichte der Heidelberger Akademie der Wissenschaften, Mathematisch-Naturwissenschaftliche Klasse 4. Berlin and New York: Springer-Verlag.

Hackett, C.J. 1981 Development of caries sicca in a dry calvaria. *Virchow's Archives (Pathology and Anatomy)* 391:53-79.

Hackett, C.J. 1983 Problems in the paleopathology of the human treponematoses. In G.D. Hart (ed): *Disease in ancient man*. Toronto: Clarke Irwin, pp. 106-28.

Harrison, L.W. 1959 The origin of syphilis. *British Journal of Venereal Diseases* 35:1-7.

Haustein, H. 1930 Die frühgeschichte der syphilis, 1495-98. Historich kritsche untersuchung auf grund von archivalien und staatsdokumenten. *Archiv fuer Dermatologie und Syphilis* 161: 255-388.

Holcomb, R.C. 1934 Christopher Columbus and the American origin of syphilis. *United States Naval Medical Bulletin* 32:401-30.

Holcomb, R.C. 1935 The antiquity of syphilis. *Medical Life* 42:275-325.

Hollister, C.W. 1994 *Medieval Europe: A short history*. 7th edition. New York: McGraw-Hill.

Hudson, E.H. 1961 Historical approach to the terminology of syphilis. *Archives of Dermatology* 84:545-62.

Hudson, E.H. 1963 On the origin of the human treponematoses. *Bulletin of the World Health Organization* 29:7-41.

Hudson, E.H. 1968 Christopher Columbus and the history of syphilis. *Acta Tropica* 25:1-16.

Hudson, E.H. 1972 Diagnosing a case of venereal disease in fifteenth century Scotland. *British Journal of Venereal Diseases* 48(2):146-53.

Hulse, E.V. 1972 Leprosy and ancient Egypt. *Lancet* 2: 1024.

Hutchinson, J. 1857 Report on effects of infantile syphilis in marring the development of the teeth. *Transactions of the Pathological Society of London* 9:449.

Hutchinson, J. 1861 Heredito-syphilitic strums: and on the teeth as a means of diagnosis. *British Medical Journal* 1:515-18.

Hutchinson, J. 1909 *Syphilis*. London: Cassell and Company.

Kampmeier, R.H. 1984 Early development of knowledge of sexually transmitted diseases. In K.K. Holmes, P.A. Mardh, P.F. Sparling, and P.J. Weisner (eds): *Sexually transmitted diseases*. New York: McGraw-Hill, pp. 19-29.

Lu, G.D., and Needham, J. 1967 Records of diseases in ancient China. In D. Brothwell and A.T. Sandison (eds): *Diseases in antiquity*. Springfield, Illinois: Charles C. Thomas, pp. 222-37.

Luger, A. 1993 The origin of syphilis: clinical and epidemiologic considerations on the Columbian theory. *Sexually Transmitted Diseases* 20(2):110-17.

Manchester, K. 1981 A leprous skeleton of the 7th century from Eccles, Kent, and the present evidence of leprosy in early Britain. *Journal of Archaeological Science* 8(2):205-209.

Manchester, K. 1984 Tuberculosis and leprosy in antiquity: an interpretation. *Medical History* 28(2):162-73.

Manchester, K.1989 *Bone changes in leprosy: pathogenesis and paleopathology*. Bradford, England: Calvin Wells Laboratory, Department of Archaeological Sciences, University of Bradford. Unpublished.

Manchester, K. 1992 Leprosy: the origin and development of the disease in antiquity. In D. Gourevitch (ed): *Maladie et maladies: histoire et conceptualisation*. Geneva: Librairie Droz SA., pp. 31-49

Manchester, K. 1994 Rhinomaxillary lesions in syphilis: differential diagnosis. In O. Dutour, G. Pálfi, J. Berato, and J-P. Brun (eds): *L'origine de la syphilis en Europe: avant ou apres 1493?* Paris: Editions Errance, pp. 79-80.

Mays, S.A. 1991 *The Medieval burials from the Blackfriars Friary, School Street, Ipswich, Suffolk*. English Heritage, Ancient Monuments Laboratory Report 16/91, Unpublished.

Mays, S.A. and Crane-Kramer, G. (in review) Two probable cases of treponemal disease of Mediaeval date from England. *Amer. J. Phys. Anthrop.*

Møller-Christensen, V. 1953a *Ten lepers from Naestved in Denmark: a study of skeletons from a Medieval Danish leper hospital*. Translated from Danish by Hans Andersen. Copenhagen: Danish Science Press Ltd.

Møller-Christensen, V. 1953b Changes in the maxillary bone in leprosy. *International Journal of Leprosy* 21: 617-620.

Møller-Christensen, V. 1958 *Bogen om Æbelholt Kloster* (The Book about the Monastery of Äbelholt). Copenhagen: Dansk Videnskabs Forlag.

Møller-Christensen, V. 1961 *Bone changes in leprosy*. Copenhagen: Munksgaard.

Møller-Christensen, V. 1967 Evidence of leprosy in earlier peoples. In D. Brothwell and A.T. Sandison (eds): *Diseases in antiquity*. Illinois, Springfield: Charles C. Thomas, pp. 295-306.

Møller-Christensen, V. 1969 The history of syphilis and leprosy- an osteoarchaeological approach. *Abbottempo* (English Edition), 1:20-25.

Møller-Christensen, V. 1974 Changes in the anterior nasal spine and the alveolar process of the maxillae in leprosy: a clinical examination. *International Journal of Leprosy* 42(4):431-35.

Møller-Christensen, V. 1978 *Leprosy changes of the skull*. Odense: University Press.

Møller-Christensen, V., and Faber, B. 1952 Leprous changes in material of Medieval skeletons from the St. George's Court, Naestved. *Acta Radiologica* 37:308-17.

Møller-Christensen, V. and Weiss, D.L. 1971 One of the oldest datable skeletons with leprous bone changes from the Naestved leprosy hospital churchyard in Denmark. *International Journal of Leprosy* 39(2):172-82.

Molto, J.E. 1998 *Disease history from the Dakhleh Oasis, Egypt*. Paper presented at the Society for the Study of Egyptian Antiquities Annual Symposium, Scholar's Day, November 8th, 1998. Toronto, Canada.

Ortner, D.J., and Putschar, W.G.J. 1985 *Identification of pathological conditions in human skeletal remains*. Washington D.C.: Smithsonian Institution Press.

Page, W. 1975 *The Victoria history of the counties of England: Suffolk*. Volume 2. London: University of London Institute of Historical Research.

Pedley, J.C. 1967 The presence of *Mycobacterium leprae* in human milk. *Leprosy Review* 38:239-42.

Richards, P. 1977 *The Medieval leper and his northern heirs*. Cambridge: D.S. Brewer Ltd.

Roberts, C. 1986 Leprogenic odontodysplasia. In E. Cruwys and R. Foley (eds): *Dental Anthropology*. British Archaeological Reports. International Series 291, Oxford, pp.137-47.

Roberts, C. 1994 Treponematosis in Gloucester, England: A theoretical and practical approach to the Pre-Columbian theory. In O. Dutour, G. Pálfi, J. Berato and J-P. Brun (eds): *L'origine de la syphilis en Europe: avant ou apres 1493?* Paris: Editions Errance, pp. 101-108.

Rogers, L. and Muir, E. 1940 *Leprosy*. 2nd edition. Bristol: J. Wright and Sons Ltd.

Rosebury, T. 1971 *Microbes and Morals*. New York: Viking.

Rosebury, T. 1992 Columbus and the Indians. *Monthly Review (New York)* 44:61-70.

Rubin, S. 1974 *Medieval English medicine*. New York: Barnes and Noble.

Russell, P.A. 1989 Syphilis, God's scourge or nature's vengeance? *Archive for Reformation History* 80:286-307.

Sudhoff, K. 1926 Essays in the history of medicine. In F.H. Garrison (ed): *An introduction to the history of medicine*. New York: Medical Life Press, pages not given.

Temkin, O. 1966 Discussion. In S. Jarcho (ed): *Human paleopathology*. New Haven: Yale University Press, pp. 30-35.

Willcox, R.R. 1949 Venereal disease in the Bible. *British Journal of Venereal Diseases* 25:28-33.

Wong, K.C. and Wu, L.T. 1936 *History of Chinese medicine*. 2nd edition. Shanghai: National Quarantine Service.

Zias, J. 1985 Leprosy in the Byzantine monasteries of the Judean Desert. *Koroth* 9(1-2):242-48.

Zias, J. 1989 Lust and leprosy: confusion or correlation? *Bulletin of the American Schools of Oriental Research* 275:27-31.

Zias, J. 1991 Leprosy and tuberculosis in the Byzantine monasteries of the Judean Desert. In D. Ortner and A.C. Aufderheide (eds): *Human paleopathology. Current syntheses and future options*. Washington D.C.: Smithsonian Institution Press, pp. 197-99.

Table 1: Observed skeletal collections

Leprosy hospital cemeteries	**Date**	**No. Burials**
St. James and St Mary Magdalene, Chichester, England	11th-17th century AD	351
St. Mary and St. Thomas Hospital, Ilford, Essex	11th-17th century AD	25
St. Margaret's Hospital, High Wycombe, Buckinghamshire	13th-16th century AD	12
Non-leprosy hospital cemeteries		
Dominican Friary of Blackfriars, Gloucester	13th-16th century AD	140
Blackfriars Friary, Ipswich, Suffolk	13th-16th century AD	250
Wharram Percy, North Yorkshire	10th-19th century AD	687
Rivenhall, Essex	8th-19th century AD	229
Stengade II, Langelands, Denmark	0-2nd century AD 2nd-5th century AD 9th-11th century AD	8 2 61
Total		**1765**

Table 2: Cranial osseous diagnostic criteria for leprosy and treponemal infection

Leprosy	**Treponemal infection**
(Cranial: *facies leprosa*)	(Cranial: *caries sicca*)
Atrophy of anterior nasal spine and pyriform aperture	1. Initial sequence (clustered and confluent pits)
Rounding of nasal aperture	2. Discrete sequence (focal superficial and circumvallate cavitation, radialo scars)
Atrophy and recession of alveolar process of maxilla	3. Contiguous sequence (serpiginous and nodular cavitation, *caries sicca*)
Endonasal inflammatory changes of hard palate	Inflammation and/or absorption of oro-nasal regions
Inflammation of nasal antra	Mulberry molars
Inflammation and/or absorption of turbinates, vomer or ethmoid	Hutchinson's incisors
Leprogenic odontodysplasia	

Table 3: Infracranial osseous diagnostic criteria for leprosy and treponemal infection

Leprosy	**Treponemal infection**
Hands	*Non-gummatous lesions*
Diaphyseal concentric remodelling	Periostitis
Acro-osteolysis	Osteitis
Palmar grooves	Osteo-periostitis
Ankylosis	
Claw hand deformity	
Juxta-articular dorsal inflammatory change	
Septic arthritis (pyogenic sepsis)	
Subluxation and dislocation	
Cup and peg deformity	
Feet	*Gummatous osteo-periostitis*
True pancircumferential concentric remodelling	Gumma formation
Knife edge remodelling	Osteomyelitis
Pencilling	Sequestra/cloaca formation
Dislocation and subluxation	Pathological fracture
Septic arthritis (pyogenic sepsis)	
Ankylosis	*Joint involvement*
Dorsal tarsal bars	
Cup and peg deformity	Gummatous arthritis
Tarsal disorganisation	Clutton joints
Periostitis	Charcot joints (*tabes dorsalis*)
Osteitis	
Vascular grooves	
Osteoporosis	Osteochondritis
Pathological fracture	Wimberger's sign
Enlarged nutrient foramina	Sabre shin deformity
Erosive joint lesions	Dactylitis
Bone cysts	

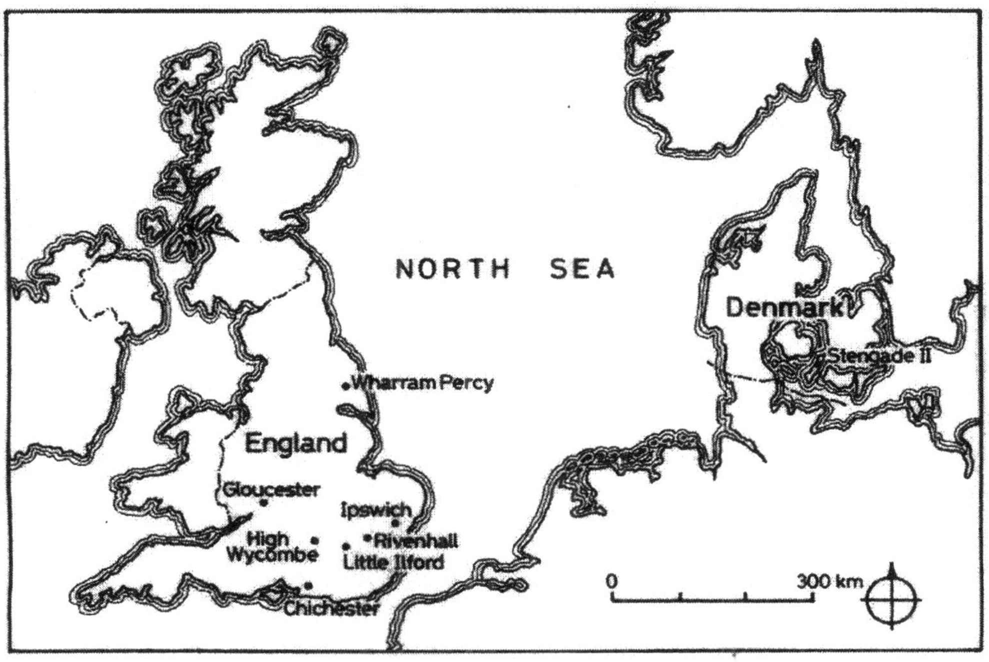

Figure 1: Location map for the eight archaeological sites included in this study

History and Palaeopathology of Leprosy Worldwide

The geography of leprosy in the Russian Empire: historical evidence for the dissemination of the disease

Alexandra Buzhilova
Department of Physical Anthropology
Institute of Archaeology of Russian Academy of Sciences
Dm. Ulianova str.19
Moscow
Russia 117036

E-mail: albu_pa@mail.ru

Abstract

The distribution of leprosy in the territory of the Russian Empire is analysed through medical historical documents. The data is compared with historical sources from the end of the 19th and the beginning of the 20th century. Computer generated geographical distribution maps estimate the frequency of leprosy in various regions using non-discrete lines. Three endemic areas of leprosy are identified. The first region is in the Baltic Sea circle. The second is noted in the northern Caucasus and in the Caspian Sea region. The third area is in the Far East of the continent. The geographical distribution of the zones of leprosy, and analysis of historical migrations of Eurasian populations, allows the estimation of the possible pathways for the dissemination of the disease.

Keywords

Leprosy, Russia, digital map models.

1. Introduction

Until the end of 19th century, the distribution of leprosy in the Russian Empire had not been investigated in detail, although approximate areas of endemic centres were known. At that time, Grebentchikov (1897) wrote that leprous people were frequently seen in the Baltic provinces, in the zone of Don Cossacks, in the Caucasus, in the Astrakhan province and near Baikal lake.

In 1887, the reporting of cases of the disease from separate districts began in Russia, and has allowed us to present statistical data on the geographical diffusion of leprosy. Nine centres of leprosy were identified (Table 1). The highest level of frequency of the disease was registered in the Baltic provinces, although Grebentchikov (1897) notes that it is the result of mistakes in collecting the data, and argues that the higher levels might be in the regions of Eastern Siberia, rather than in the Baltic provinces.

As a result of statistical analysis in the Baltic area, most cases of leprosy were found in the districts of Liphland, Kurland, Eastland and Petersburg. The coastal regions of the Arctic and Great Oceans within the limits of Siberia present a slightly lower frequency than in the Baltic lands (Table 1). It is necessary to remember that, due to the density of the population of Siberia, and possible mistakes in reporting, the Baltic region was classed by many researchers as the most active zone of diffusion of the disease (Torsuev, 1958). However, other areas endemic for leprosy share an identical prevalence. The frequencies for these areas are approximately 10 times below the prevalence noted in the Baltic and East Siberia. In provinces of the European part of central regions of Russia the disease is sporadic in nature, and therefore the frequency of leprosy did not exceed the minimal threshold and so is recorded as zero (Table 1).

More than half of the patients registered in the Russian Empire had the lepromatous form of the disease, and about a quarter of them were recorded as suffering from the tuberculoid form. The distribution of the different forms of leprosy had the same characteristics in both males and females (Figure 1). The peak prevalence of the disease was in males aged between 20 and 30 years and in females between 30 and 40 years. The lepromatous type was recorded most often in those aged 20-30, but tuberculoid and borderline presentations were more common in those aged 30-40 years. It was noted that people with tuberculoid forms of the disease lived on average 10 years longer, while only a quarter of lepromatous cases lived for more than 10 years (Torsuev, 1958).

Torsuev (1958) describes some geographical features. The cases of the tuberculoid form of leprosy were predominant in one of the districts of Liphland province (Baltic) and in the Viluyi district of Yakutia (Siberia). In almost all provinces among the patients there was a predominance of men over women with the Baltic seacoast, Terek River and Far East regions being the only exceptions. In Caspian areas, the patients were in the younger age categories (between 10 and 30 years), while in the Baltic provinces patients were older (30-50 years). In other zones, the average age of people with leprosy lay between these two (Figure 2). It is important to note that, while more than half of the patients lived with their families, 14 % were isolated from them, and only 31 % were patients in a leprosarium (Torsuev, 1958).

Subsequent studies on leprosy in the Russian Empire have given similar results, and have provided details on the diffusion of the disease. The number of patients gradually increased, despite a mortality rate of about 1.4 % per year. It is necessary to emphasise that, as Torsuev (1958) repeatedly specifies, the disease was not treated effectively and in many districts the segregation of patients was not enforced at all. People with leprosy were frequently abandoned due to unskilled health services in some areas, and a lack of money to maintain leprosaria.

On account of this, the dissemination of leprosy in the Russian Empire approached a natural model of development of infectious disease. This also explains how leprosy spread, as noted for the first time in Russia by Peterson (1891) when, 'each more appreciable attack of leprosy is surrounded by a peripheral girdle, in which there are separate cases of the illness' (Torsuev, 1958:40).

Using this information, an estimation of the geographical diffusion of leprosy was made by constructing computer models of factual material. Proximity of results to the natural mechanism of development of an infection, without active intervention by humans, has framed conditions for a more objective reconstruction of the reasons for the spread of the disease.

2. Material and methods

Summary data received from various sources was used for mapping early material. The file uses data from Ivanov and Mamonov (1923), and records the number of patients in various geographical regions from 1889 to 1910 (Torsuev, 1958). It was complemented by additional data compiled by Torsuev (1958) from historical medical sources, and represents both synchronous and later periods of diffusion of leprosy down to the year 1916. In addition, information from a report by Trzhtcyak (1913) describing evidence for the diffusion of leprosy at the end of the 19th century in Western Europe was used. Thus, data on the number of patients in 203 regions in the period from the end of the 19th to the beginning of the 20th centuries were analysed.

It is necessary to emphasise that the information in some cases was generalised, such as the frequency of leprosy in a certain district or province. In such case, the geographic point chosen was arbitrary and was placed on a map in the centre of the district or province. In other cases, the frequency of the disease corresponded to discrete geographic points (Figure 3).

In order to use the data correctly, all data were counted in percentiles (1:10,000) in relation to the density of the population. For this purpose, general consensus data compiled by Brokgaus and Efron (1991:106-128) from January 28th 1897 from the Russian Empire was used.

Map generation was carried out using an original computer programme for geographical and statistical analysis, developed by Balanovskaya *et al.* (1994) on the basis of a package MAG, or International Library Programme (Department of Cartography, Moscow State University; supervisor S.M.Koshel). The software was created specifically for geographical analysis of a gene pool (Balanovskaya *et al.*, 1990). Recently, this technology has been used in various anthropological and archaeological research projects in Russia.

A digital model, represented by a bi-dimensional matrix with a key and units of a uniform cartographic grid, were imposed on a map. Increased prevalence of the disease is shown by raised intensity of colouring of the geographical loci. The statistical parameters are indicated on the legends on the maps, where "K" is the number of basic points appropriate to empirical data (*), and "N" is the number of units of a regular grid (that is the number of meanings of a matrix received by an interpolation procedure). In the legend the maximum, minimum and average figures of an attribute, with reduction of the standard deviation, are also specified. On the constructed analytical maps it is possible to estimate the degree of variability of an attribute in view of the formation of local features (Figure 4-6).

Except for the analytical maps depicting the geographical analysis, the trend maps were constructed by a method of averaging of the means of an attribute on the areas at regular intervals (Figure 5-8). Developed by Balanovskaya and Nurbaev (1995), the methods are simplified to the following procedure. The average data for all units of a grid of a digital matrix on the area of a rectangular of the given size is calculated with the aid of the chosen algorithm. The result of averaging is appropriated to the central unit of each rectangle (window). Its centre then moves to the next unit of the grid and the previous procedure of averaging repeats is applied. As a result, a trend map is created on which there are only basic fluctuations of an attribute, without the detailed account of local features. This has allowed a concentration on a generalised analysis. Two geographical scales were used, the territory of Europe and the territory of Eurasia.

3. Results and discussion

3.1 Leprosy in the European part of the Russian Empire

The highest frequency of leprosy is seen in two regions of Europe. The first concentrates in the Baltic, and the second is near the Azov Sea, in the Northern Caucasus, near the Caspian Sea and in Transcaucasia (Figure 4).

3.1.1 The Baltic Region

In the Baltic the highest prevalence of leprosy was noted in the Eastland provinces in the Ezel and Revel districts (Table 2). It is necessary to explain that the Ezel district demonstrates similar high rates compared to Ezel Island (now known as Saarema, Estonia), which had a leprosarium during the Middle Ages. In the Liphland province, leprosy is seen mainly in the Phellin, Pernov and the Volmar districts. In Kurland leprosy is registered in the Vindava district. All points form a basic centre of illness determined by the empirical data. The high levels in the Baltic are comparable with Scandinavia by an interpolation procedure (Figure 4), and the result does not contradict the known historical data (Trzhtcyak, 1913).

The trend map shows that the greatest prevalence of leprosy was in the north-east coastal areas of the Baltic Sea (Figure 5). This feature suggests the possible mechanisms of transfer of the disease. Travel by sea, perhaps at the time of the Vikings (793-1066), and the introduction of leprosy into the seaports of the Baltic could have triggered the beginning of the diffusion of the infection in the coastal zone.

The Viking Epoch is characterised by a significant expansion of Scandinavians across Europe. In Europe they were known as Norsepeople, and in Russia as Varangians. Through migration Norsepeople have occupied the large islands of the Northern Atlantic, and have founded settlements in Britain, Northern France, Southern Italy and Sicily. Varangians served as bodyguards in Russia and Byzantium, and Norsepeople extended into the Volga region and Baghdad caliphate. These fishermen, who were brave seafarers and skilful ship builders, did not avoid long distance navigation in search of places to settle. Different northern German tribes took part in campaigns, but it is known that Swedes went to Eastern Europe (Svanidze, 1992). Perhaps when people came to the Baltic they included participants of other campaigns who had visited natural centres of leprosy. It is possible that infected people took part in these campaigns and came to the Baltic. This might theoretically have led to the spread of leprosy to the region.

The next phase of active migration to the Baltic zone occurred with German knights. At the end of the 12th century, crusaders from Germany formed the order of the Teutonic knights, who undertook many aggressive campaigns into the territory of modern Estonia, Latvia and Lithuania. After the defeat of this Order in the middle of the 13th century, the Order of Livonia was created and continued with aggressive politics. As a consequence, these knights won the territory of modern Estonia and Latvia from the 13th century until 1561, referred to as Livonia. By the end of 16th century, as a result of the Livonian war, Livonia ceased to exist. In two centuries this territory repeatedly changed its borders by falling under the control of stronger neighbouring countries (Latvian, 1938).

The historical medical data confirms that leprosy was known in the region as early as the 13th century, compatible with the time of the Crusader invasion. Some have argued that the spread of leprosy could have occurred much earlier, but evidence supporting this hypothesis is limited. Nevertheless, it is possible that Norse people and German crusaders could have been responsible for the introduction of leprosy into the Baltic. In the Middle Ages, leprosy began to spread and became more common. Indirect evidence of the high density of this disease at this time is found in Bergmann, with about 100 leprosaria functioning in Livonia in the 14th and 15th centuries (Torsuev, 1958, Derums, 1970). Thus, in the territories of the West-European states, where leprosy was introduced much earlier, the spread of the disease in the Baltic may be explained by migrations in the Middle Ages. The devastating gains of the German crusades, and rigid economic dependence of the local population, created favourable conditions for the spread of leprosy between people. However, despite a systematic search for excavated human skeletons from the villages and town cemeteries in the Baltic, no cases of leprosy have been identified (Derums, 1970). As the illness was visually recognisable, it is possible that, the patients were stigmatised, and hence, were not buried in local ordinary cemeteries.

3.1.2. The Volga Region, Northern Caucasus and Transcaucasia

The second centre of leprosy in Europe unites some regions with a high frequency of the infection. It is remarkable that, in the majority of cases, these are the locations of the Cossack forces from the Don, Cuban and Terek regions, Astrakhan, and also the territory of modern Armenia and Azerbaijan (Figure 4). The greatest frequency of leprosy patients was found in the zone of the Don river in the Cossack village of Elizavetinskaya, in the Astrakhan zone in the town of Krasnyi Yar, in Astrakhan, in the Terek region in the village of Blagodarnenskoe, and in Transcaucasia in the towns of Yerevan and Baku (Table 3). It is important to note, that the highest frequency is noted in three Cossacks areas of Don, Astrakhan and Terek. In Transcaucasia the prevalence of leprosy was double that elsewhere.

The first official document in which leprosy in Don Cossacks is mentioned is the report of W.F. Lautenberg from January 24th 1759 to the State Medical Board. The next two decades are characterised by the accumulation in the Board of similar reports not only from physicians, but also from the Orthodox Church. In 1776 the Voronezh department of the Orthodox Church recorded that among the Don Cossacks many people asked for divorce, as the husband or wife had “Crimean illness”. This raises the question as to what this illness was, whether it was infectious and whether it was treatable. To address the problem the State Medical Board sent the physicians Gaetano Mira and Andrei Yagodin to the Don Cossacks. From the report of A.Yagodin it is clear, that the illness was infectious and ‘... this extremely serious disease requires treatment; however it responds poorly to medicines... if the black icterus has joined, the chronic illness becomes incurable, infectious and soon comes to an end by death’ (Torsuev, 1958:8). The detailed and eloquent description of illnesses provided by Yagodin raises the possibility that this is leprosy, though the physician does not name it as such. From the report of another physician, Mikhail Kotelnitcky, from the town of Cherkassk (nowadays Starocherkasskaya) it is known that the main reason for the "Crimean illness" was the development of scurvy, '... which in this population... is almost like hereditary ...' (Torsuev, *ibid*: 9). Mikhail Kotelnitcky, as well as many at that time, links the occurrence of illness with the environment conditions and lifestyle of the Don population (*ibid*).

The medical data shows that leprosy from time to time is recorded in others places of southern Russia (Rikhter, 1814). In the subsequent 20 years reports about the wide distribution of leprosy in the Ural Cossacks are found (Pallas, 1809). The doctor Petr Simontovsky spent almost one year in the Astrakhan battalion. He felt that leprosy was brought there from Asia through the Astrakhan port. In fact, leprosy could have come from Asia via the Volga river, and from Persia via the Caspian Sea route. Thus, the territory of the Northern Caucasus and Transcaucasia can be viewed as an initial zone of diffusion of the illness to other districts of European Russia.

It is possible to discuss the occurrence of leprosy, at least in the 18th century, as endemic in the south of Russia. The geographical distribution of the illness is rather characteristic and it is has been known for more than 100 years. In the Northern Caucasus the coast of the Azov Sea, basin of Don, Cuban, and Terek, and the northern and eastern parts of the Stavropol region are most affected. In Transcaucasia the areas involved are Armenia and Azerbaijan, while Daghestan and Georgia are affected to a lesser degree. It is interesting to think how these centres could be formed. The trend map shows that the centres of leprosy in the Northern Caucasus and in Transcaucasia have a line of demarcation, such as the Caucasus mountains, and with another one they meet by the Caspian coast (Figure 5).

The history of the formation of Cossacks forces could provide us with information to understand the spread of leprosy. The population that settled in the 15th century near Azov is recognised by the Russian government as Don Cossacks at the end of the 16th century. Prior to the beginning of the 18th century, Cossacks did not receive the status of the non-regular Russian army, and the population was engaged in hunting and fishing. With the purpose of capturing military equipment, Don Cossacks made plundering raids on Turkish possessions on the Azov and Black seas coasts and also into the Caspian Sea. The official date of the organisation of Terek Cossacks is 1577. Their remote ancestors were Cossacks from the river Sunzha and from Don. In the middle of the 18th century the Astrakhan Cossacks force was organised (Popko, 1880).

Thus, the time period over which the Cossacks were formed, is limited to a few hundred years from the 15th to the 18th centuries. It is necessary to point out that Cossacks districts were established in areas which vary due to active migrations of the population both from west and east over a thousand years. In this area there were settlements of Greeks, and the land lay on branches of the Great Silk Route, which connected the natural centres of leprosy in China, India and Central Asian countries. Nevertheless, the coincidence of endemic foci of leprosy in the homelands of the Cossacks suggests late transmission of leprosy, not earlier than the 15th and 16th centuries AD. In this connection, the Don Cossacks could be seen as the initial centre of leprosy among Cossacks. First, it is the earliest Cossacks settlement, and secondly there is evidence for the involvement of Don Cossacks at the formation of the Terek, Kizlyar and Astrakhan forces. This suggests the transfer of disease to other territories (Popko, 1880). As it was only in middle of the 18th century that Cossacks became officially supervised by the government, information about infections could only be recorded from this time. However, it is not possible to be certain about the time of introduction of leprosy among Don Cossacks.

The occurrence of leprosy among Cossacks could have been at various times. The first option for the spread of leprosy could be contact with local inhabitants. For example, Cossacks forces were meeting people from Armenia or Georgia. It is known that the Terek force was comprised of Grebensky and Don Cossacks and the Armenians and Georgians, which joined in service with their families (Popko, 1880). On the other hand, the spread of leprosy could have been a consequence of contacts between Cossacks and the population of Asia Minor, Near East and Central Asia during the military campaigns organised both by the government, and also under their own initiative. For example, the Persian campaign of the Russian army and fleet to the Caspian lands of Persia, under the command of Piter the Great in 1722-1723, was one possible way of introducing leprosy to the Cossack forces. It is known that in the Russian army about 40 000 men where Cossacks. They reached southern Daghestan by sea and land and occupied Derbent by force. Russian troops stayed in the area for one year, fighting further battles (Knyazkov, 1914). It is important to highlight that, since the 16th century, Derbent was part of Persia, one of the most ancient centres of leprosy. Therefore, the Russian army was in an area endemic for leprosy that could have affected the Cossacks forces. In support of this, it is possible to show data from Pallas (1809) which specifies that many Yaik Cossacks (Ural region) believed that the "Crimean illness" developed from contact with former Astrakhan soldiers of the Persian campaign. It is necessary to note that, throughout their history, the Cossacks regularly split up and settled in territories within the limits of a designated "Cossack area of leprosy" (Figure 4). In later times it is possible to note that in Siberia the centres of leprosy, as well as in Europe, coincide with locations of Cossacks.

The name of a "Crimean illness" similar to leprosy and widespread in the region can be connected to the epoch of the Crimean kaghanate (15th to 18th centuries AD), when the right bank of the River Volga was named Crimean territory. Leprosy could have been introduced here by various routes, including from Transcaucasia. The sources suggest that in Transcaucasia leprosy had probably been present for a long time. Armenia, being a bridge from Asia to Europe, was a route for trade and also a bridgehead for various military actions. Taking into account the proximity of India, Persia and Turkey, where leprosy was widespread, the possibility of an early endemic centre of leprosy in Armenia becomes more plausible. Oganesyan (1946) writes that, in 260-270 AD near the Arbenut springs, the wife of the Armenian feudal lord Suren Salauni, named princess of Agvid, opened a leprosarium for 35 people. From Agatangelos it is known that in the 4th century leprosy in Armenia was widespread. Byuzand noticed that in the time of Nerses nobody saw poor people within the limits of Armenia. Patients stayed in leprosaria, where people brought them all necessary things for a comfortable life. However after the death of Nerses, the leprosaria were closed and burnt down. Moisei Khorensky, a historian of the 5th century AD, testifies that people with leprosy were expelled, as under the law they were considered "unhealthy" because it was possible to get the infection from them. They lived in the desert and other remote places. In response to this, the work of Stepan Orbelyan (14th century AD) tells us that in the 5th century two hermits, one of which was the son of a Persian king, constructed a monastery and leprosarium on the coast of Araks for those expelled from the leprosaria (Oganesyan, 1946). Oganesyan (1946) also suggests that in ancient times in Armenia the number of leprosy patients was greater than in more recent times; similarly, the area endemic for leprosy was larger than in the 19th century. Thus, it appears that the countries of Transcaucasia could be one of the sources of leprosy which later transferred to the Cossack areas. In addition, it is necessary to remember that the Turkish campaigns of the Cossacks and the Russian army. During the expansion of the Turkish Empire from the middle of the 17th century, this country ruled as vassal territories the Azov and Black Sea coasts of the Northern Caucasus. The time of the Iranian and Turkish wars (16th to 17th centuries) is characterised by the strong influence of the Turkish Empire in Transcaucasia (Miller, 1948). Obviously, all these territories were associated with the presence of Cossacks forces.

3.1.3 The Polarity of Two Areas of Leprosy

If we are to assess the centres of leprosy in the European part of the Russian empire, it is necessary to pay attention to the polarity of two areas, namely the Baltic and Caucasian regions (Figures 4-5). Perhaps the mechanism of formation and history of leprosy centres in these territories is markedly different. The numerous historical sources and geography of diffusion of the disease testify to this possibility. However, this is in contrast to the situation in the western part of Europe where the disease appears to have spread from the Middle East. It was distributed over the continent to northern France and then to England, to the countries of Scandinavia and later to Central and Eastern Europe. A similar cycle of diffusion of an infection is noticed for the pandemic of plague in the 14th century. In this connection the centres of leprosy in the European part of the Russian empire could be seen as "the fragments" of the cycle of diffusion of leprosy on the continent.

3.2 Leprosy in the Asian part of the Russian Empire

3.2.1. Middle and Central Asia

The change of geography allows us to look at the European centres of leprosy in a Eurasian context (Figure 6). The Caucasian centre near the Caspian zone could be associated with the Transcaspian, Central Asian and near Aral zone centres of leprosy (Table 4). On a map it is quite obvious to see two areas covering the north and the south coasts of Caspian. The history of these regions provides evidence to suggest that the regions were united by caravan routes of the Great Silk Route (Figure 7). Trading started in the 2nd century BC and acted as a bridge between China and Western Europe. It started in Xian, went through Lanchjou, and came to Dunkhuan, where the route split. A northern road passed through Turphan, then it crossed the Pamirs and went to Fergana and the Kazakh steppes. A southern route passed Lobnor lake and went to the southern part of the Taklamakan desert through Yarkend and the southern part of the Pamirs, to Baktria, and from there to Parphia, India and the Near East (Lubo-Lesnichenko, 1994). Northern and southern branches, crossing in Transcaucasia, went through Asia Minor to the countries of Western Europe (Figure 7). This observation does not lessen at all the influence of the Persian centre of leprosy, discussed above. Research by Rubin, which Torsuev (1958) described, showed that on the Turkmenian coast of the Caspian Sea, leprosy probably appeared with migrants from Iran in the 18th century. Iomuds, the tribe who came to Turkmenia, display a high level of leprosy.

Thus, the Caspian zone and the districts connected to it, as well as regions of Middle Asia and the Kazakh steppes, relate both to the Chinese and Persian leprosy centres. Concluding the analysis of the diffusion of the infection in the Middle and Central Asian districts, it is necessary to add that the natural features of this territory - mountainous districts and desert - promoted the isolation of the centres and prevented much transfer of the disease. Even long before the expansion of Russia, the local population was familiar with the symptoms of "makhay", an illness which may have been leprosy, and they practised isolation of patients. They were expelled and compelled to live in small settlements, or "kishlak-makhay", where they eked out a miserable existence, supporting themselves by begging, being deprived of any medical care (Torsuev, 1958:135). However, the illness was restricted to these districts by natural boundaries up until the present day and the prevalence of the disease is close to the minimal threshold.

3.2.2 Northern and eastern Asia

The following foci for the diffusion of leprosy in the Russian empire are found in the Krasnoyarsk area on the Yenisei river, in Transbaikalia, Yakutia, Primorye Territory and in Kamchatka (Table 5, Figure 6). It is necessary at once to point out that some centres, namely Krasnoyarsk, Barguzin (Transbaikalia) and Amur, are associated with the geographical location of Cossacks forces, formed in middle of the 19th century such as Yenisei, Transbaikalian and Amur Cossacks. This has been already discussed above, and only now is it possible to add that Cossacks have played an appreciable role in colonisation of sparsely populated areas of Russia.

Obviously, the most extensive centre of leprosy with a high prevalence of illness is the territory of modern Yakutia, including Kolyma and East Siberia, the coast of the Pacific from Kamchatka and the basin of Amur (Figure 6). The details of people with leprosy in Yakutia are described by the English researcher Marsden (1892).

Marsden tried to find leprosy patients in settlements which were widely scattered. Often she had no exact information as to where they could be, so she had to go on a journey into forests on the off-chance of finding somebody. Frequently the Yakuts did not invite her to the settlements because she had contact with leprosy patients (Marsden, 1892). Patients were evicted and moved 40-50 kms from the habitation of others. As Marsden wrote (1892), it is possible to imagine small yurts, for two people, where 5-10 leprosy patients actually lived. The yurts were made from thin logs and turf, but were unable to keep out wind, rain and snow reliably. They could not be repaired by the residents because of the absence of arm-hands. Inhabitants of the nearby settlements did not help because of their phobia of contracting the infection. The temperatures were often 50-60°C below zero and around them lived wild animals. Occasionally meals were brought by kinsmen so that people with leprosy did not come to the settlements. Their clothes were in tatters, and some of them were practically naked. Taking into account the climatic conditions, the patients did not leave the dwellings for eight to nine months in any one year. The Yakuts with leprosy lived in intolerable conditions. It is necessary to praise the energy of Marsden and her extraordinary selflessness. She spoke no Russian, and had to overcome the extreme difficulties of travel in Siberia to reach the people of Yakuts with leprosy, to examine them and to describe the disease in detail. She then returned and motivated the Russian public to provide suitable care and support for these outcasts (Torsuev, 1958:192). Other researchers of leprosy in Yakutia note that the disease was probably known in the region for a long time. The population distinguished this illness from other diseases, considering leprosy infectious, and consequently applied the most ruthless measures to isolate the patients.

The geographic analysis of leprosy suggests two variants for the spread of the disease to Yakutia (Figure 6). On the one hand, people with leprosy could have come to Yakutia from the coast of the Sea of Okhotsk. Alternatively, the potential for spread of the disease with migrations to the Baikal region through the Mongolian steppes from China in antiquity cannot be excluded.

Preservation of ancient centres of leprosy in Siberia is suggested also by observation of the centres in the 19th century. Despite identical climatic, economic and social conditions leprosy became endemic in some parts of the Vilyui, Kolyma and Verkhoyansk districts, but not in the Olekminsk region. A similar phenomenon is noted in districts near Baikal and the Pacific (Torsuev, 1958). Many physicians, studying the spread of leprosy on the coast noted the disease occurred there as well as in Kamchatka in the middle of the 19th century. The time corresponds with active colonisation of the territory resulting from the politics of the Russian government. Apart from Don and Ural Cossacks, there were migrants from Baku and Odessas, parts of the Northern Caucasus, and also settled Koreans. Interestingly, a local physician of Kamchatka named Dr. Natshokin noted that leprosy in Kamchatka was similar to so-called "northern leprosy", which was found in other northern countries such as Iceland, Norway and Scotland (Torsuev, 1958:165).

In this way, leprosy on the coast and in the north, including Kamchatka, seems to have been recently acquired. Its rapid spread can be caused both by low standards of living in local populations, and the various population migrations in the region. The "trend map" constructed for Eurasia suggests two centres for leprosy (Figure 8). The first centre is the continental zone in Yakutia, and the second is the north part of the Pacific coast. They are joined by intermediate zones where the illness was present with a lower frequency. It seems that population migrations and colonisation were the principle methods for the spread of leprosy to areas of the Far East and Siberia.

4. Conclusions

In summary, geographical analysis shows that, despite the various historical explanations for the introduction of leprosy into European and Asian parts of the Russian Empire, they are united by some common principles. The occurrence of regions endemic for leprosy in the Russian territory are characterised by migrations connected with military activity and aggressive expansion of territories in Europe, and with colonisation, particularly in the Far East and in Siberia. Therefore, leprosy in the European and Asian parts of the Russian Empire show similar mechanisms in the formation of endemic centres, which are a consequence of the activities of humans rather than natural factors.

Acknowledgements

The author is grateful to Prof. E. Balanovskaya and Mr. O. Balanovsky, Research Centre for Medical Genetics, Moscow for providing the original computer programmes for mapping the data. The author would like to thank the anonymous referees for their valuable comments on an earlier draft. Special thanks go to the reviewer for his helpful comments and alterations to make the English more understandable. The research has been partially supported by the Bioanthropology Foundation (UK).

References

Balanovskaya, E. V., Batsuur Zh., Belkovskii, A.N., Rychkov, A.V., Rychkov Yu., G. 1990 Genogeography of population: digital atlas. *J. Genetic.* 26 (5):925-935 (in Russian)

Balanovskaya, E.V., Nurbaev, S.D. and Rychkov Yu., G. 1994 Computer technology of genogeographic analysis of gene pool: II. Statistical transformation of maps. *J. Genetic.* 30 (11):1538-1555 (in Russian)

Balanovskaya, E.V. and S.D. Nurbaev. 1995 Computer technology of genogeographic analysis of gene pool: III. Derivation of trend surfaces. *J. Genetic.* 31 (4):536-559 (in Russian)

Brokgaus F.A. and Efron, I.A. 1898 *Encyclopaedic dictionary of Russia* 5: 54-55 (copy press in 1991:106-128 - in Russian)

Derums, V.Ya. 1970 *Diseases and treatment in the ancient Baltic.* Riga: Zinatne (in Russian)

Grebentchikov, V. N. 1897 Leprosy in Russia from published data entered on the 8th of April, 1895 and their registration on maps. *Bulletin of Public Hygiene, Judicial and Practical Medicine* 78:3-28 (in Russian)

Ivanov and Mamanov 1923 (cited in Torsuev, 1958 and no full reference given)

Knyzkov, S. 1914 *Sketches from a history of Great Piter and his time.* St. Petersburg: Publisher not stated (in Russian)

Latvian, H. 1938 *The chronicle of Livonia.* Moscow-Leningrad: Medkniga (in Russian)

Lubo-Lesnichenko, E.I. 1994 *China on a silk way.* Moscow: Medkniga (in Russian)

Marsden, K. 1892 *Travel of Ms. Marsden in the Yakut area.* Moscow: Nauka (in Russian)

Miller, A.F. 1948 *A brief history of Turkey.* Moscow: No publisher stated (in Russian)

Oganesyan, L.A. 1946 *A history of medicine in Armenia.* Volumes 1 and 2. Yerevan: Medkniga (in Russian)

Pallas, P.S. 1809 *Travel in different provinces of the Russian state.* St.Petersburg: No publisher stated (in Russian)

Peterson 1891 (cited in Torsuev, 1958 and no full reference given)

Popko, I. 1880 *Cossacks from Terek on ancient times.* St. Petersburg: No publisher stated (in Russian)

Rikhter, V.M. 1814 *A history of medicine in Russia.* Moscow: No publisher stated (in Russian)

Svanidze, A.A. 1992 Northwest Europe in the early middle ages. In A.O. Chubarian (ed): *History of Europe.* Volume 2 (Medieval Europe). Moscow: Nauka, pp.171-186 (in Russian)

Torsuev, N.A. 1958 *Leprosy in the Russian empire.* Rostov-na-Donu: Medkniga (in Russian)

Trzhtcyak, S. 1913 *Leprosy in the biblical era and now, with special reference to leprosy in Russia.* St. Petersburg: No publisher stated (in Russian)

Table 1: Distribution of leprosy in the 1895-1897 in Russian empire (data from Grebenshikov, 1897)

Region:	%
Baltic provinces	1.3
Coastline of Arctic Ocean	1.2
Seacoast of Caspian Sea	0.5
Seacoast of Aral Sea	0.2
Seacoasts of Black Sea and Sea of Azov	0.2
The Caucasus	0.2
Samarkand and Fergana provinces	0.2
Coastline of the Baikal Lake	0.2
Regions of European Russian provinces	0.0
Average	0.1

Table 2: Regions with the highest rates of leprosy in the Baltic

Region:	%
Ezel	32.2
Revel	22.2
Fellin	18.4
Pernov	17.4
Volmar	13.5
Vindava	9.0

Table 3: Regions with the highest rates of leprosy in the South of Russia

Region:	%
Elizavetinskaya, Don Cossacks	15.0
Azov, Don Cossacks	6.9
Krasnyi Yar, Asrtakhan Cossacks	13.9
Astrakchan, Astrakhan Cossacks	8.8
Chernyi Yar, Astrakhan Cossacks	5.8
Blagodarnenskoe, Stavropol	11.1
Erivan	8.6
Novobayazet, Erivan province	6.1
Baku	6.1

Table 4: Regions with the highest rates of leprosy in Central Asia

Region:	%
Eastern Caspian region	5.9
Turkestan, Caspian part	4.7
Karakalpakiya	3.6
Katta-Kurgan, Samarkand province	1.8
Pendjekent, Samarkand province	5.5
Kurgan-Tuybe, Samarkand province	4.4

Table 5: Regions with the highest rates of leprosy in Siberia and the Far East

Region:	%
Krasnoyarsk, Enisei region	2.1
Bargusin, Baikal region	15.2
Vilyuisk, Yakutia	46.7
Kolyma, Yakutia	23.5
Khabarovsk, Pacific Ocean	6.2
Pacific Ocean seacoast	9.0
Nikolaevsk-na-Amure, Amur region	68.8
Okhotsk, Pacific Ocean	42.0
Kamchatka	49.3

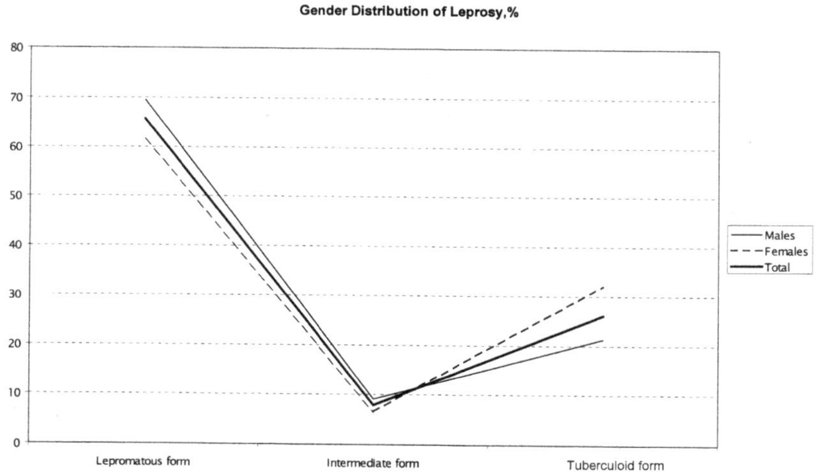

Figure1: Sex distribution (%) of leprosy patients

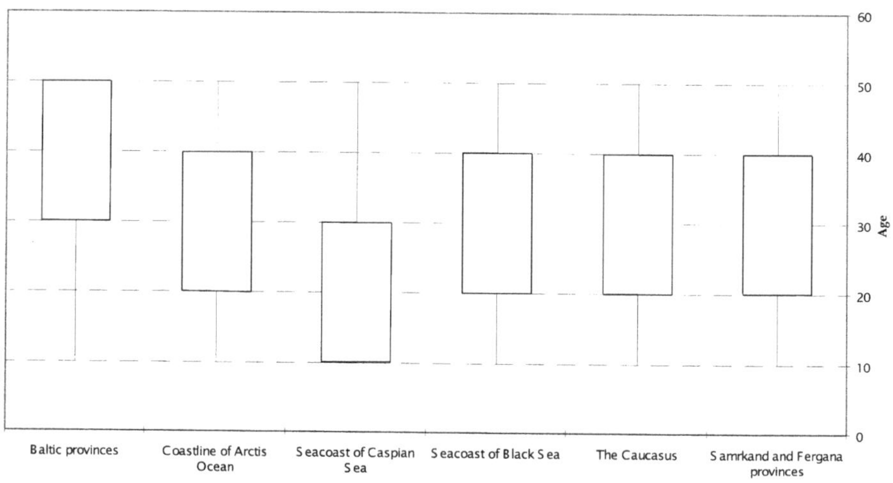

Figure 2: Age distribution of leprosy patients in different regions

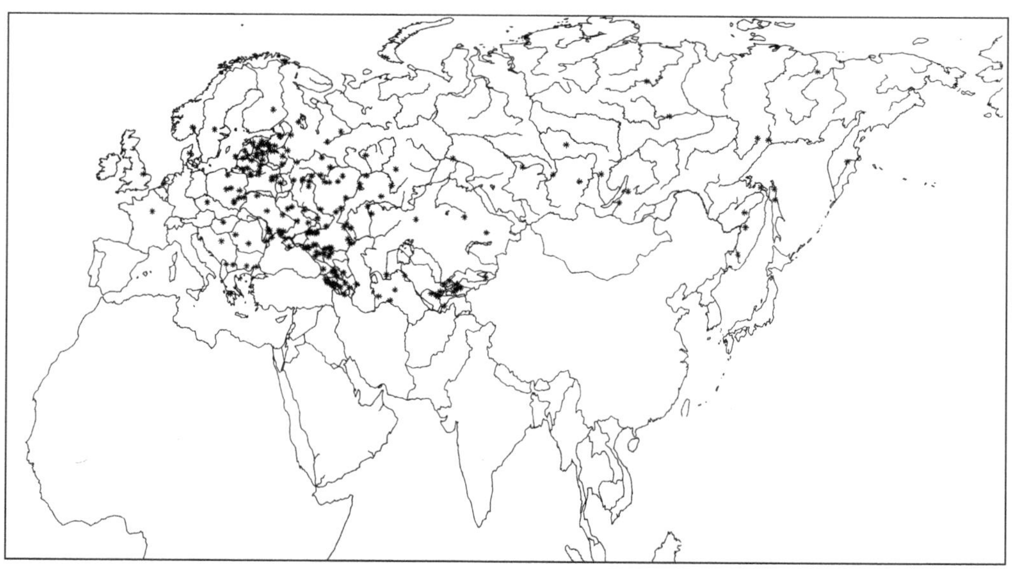

Figure 3: Geographical distribution of investigated data

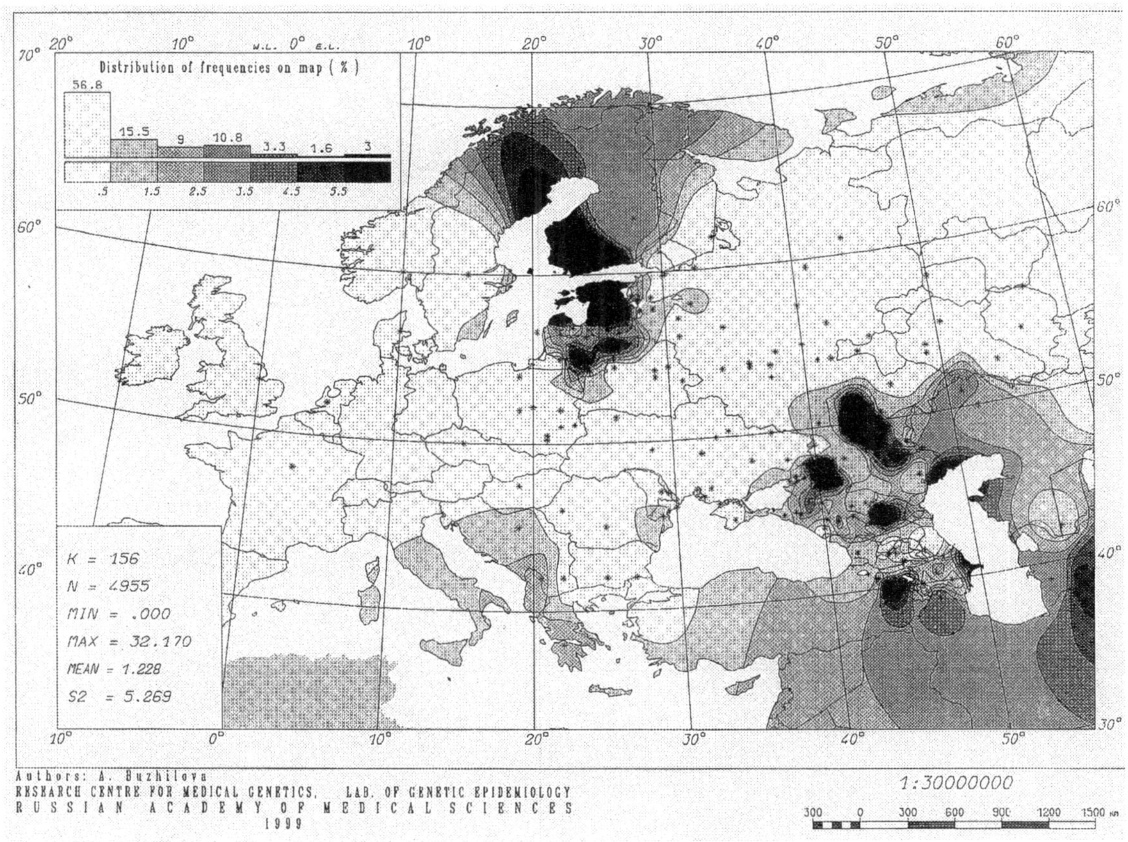

Figure 4: Analytical map of leprosy in Europe at the beginning of the 20th century

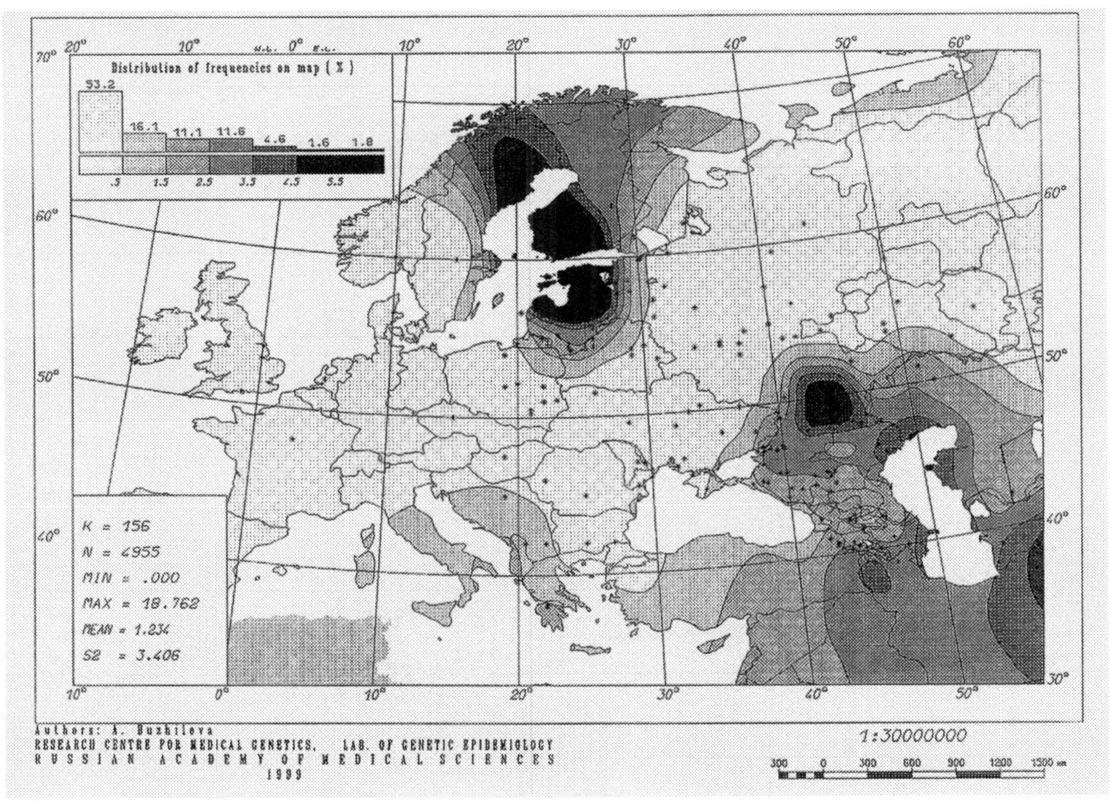

Figure 5: Trend for leprosy in Europe at the beginning of the 20th century

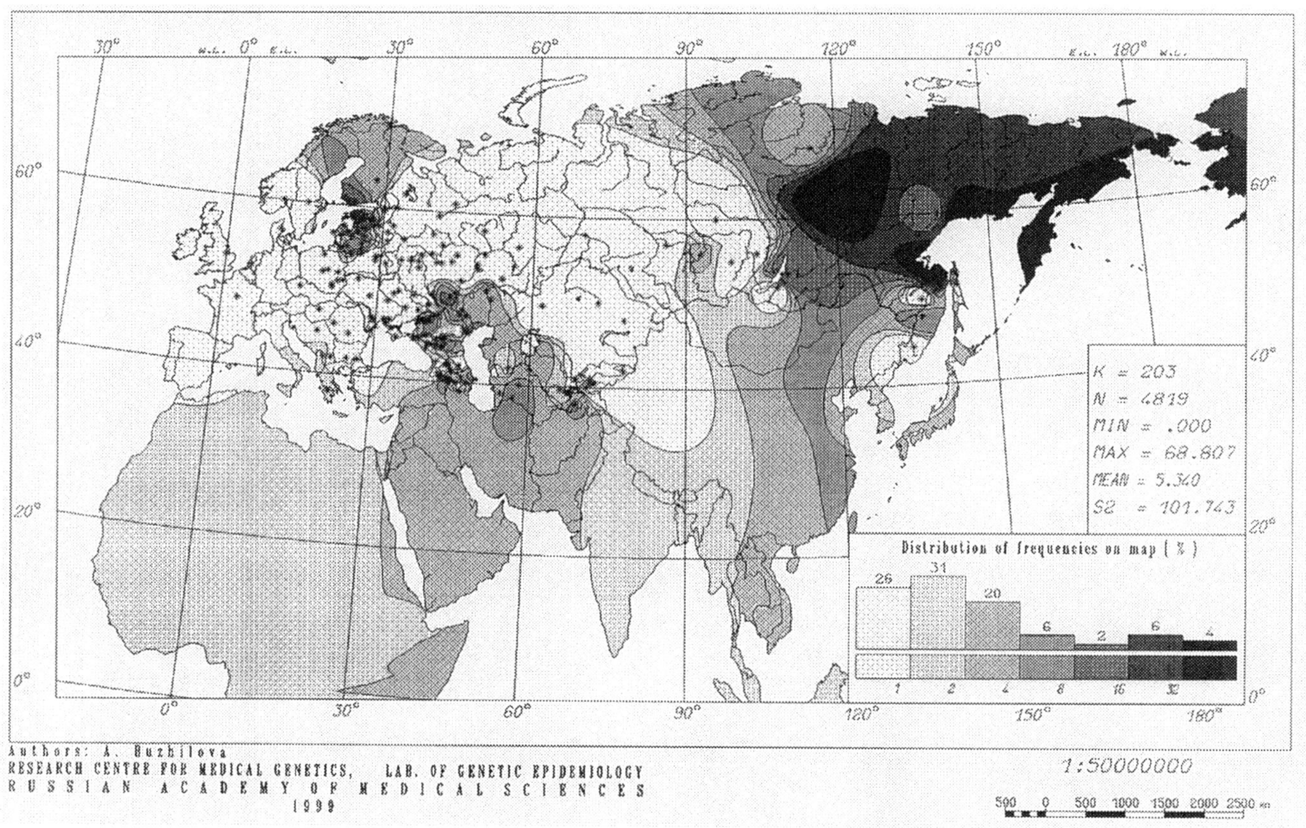

Figure 6: Analytical map of leprosy in Eurasia at the beginning of the 20th century

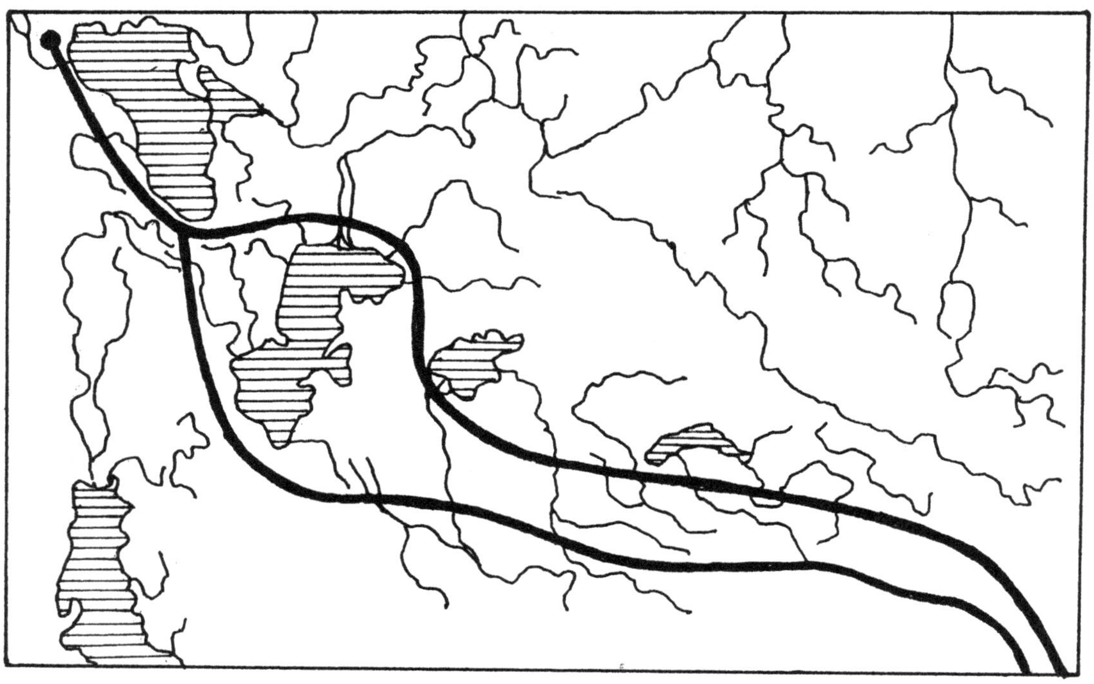

Figure 7: Part of the Great Silk Route (Lubo-Lesnichenko, 1994)

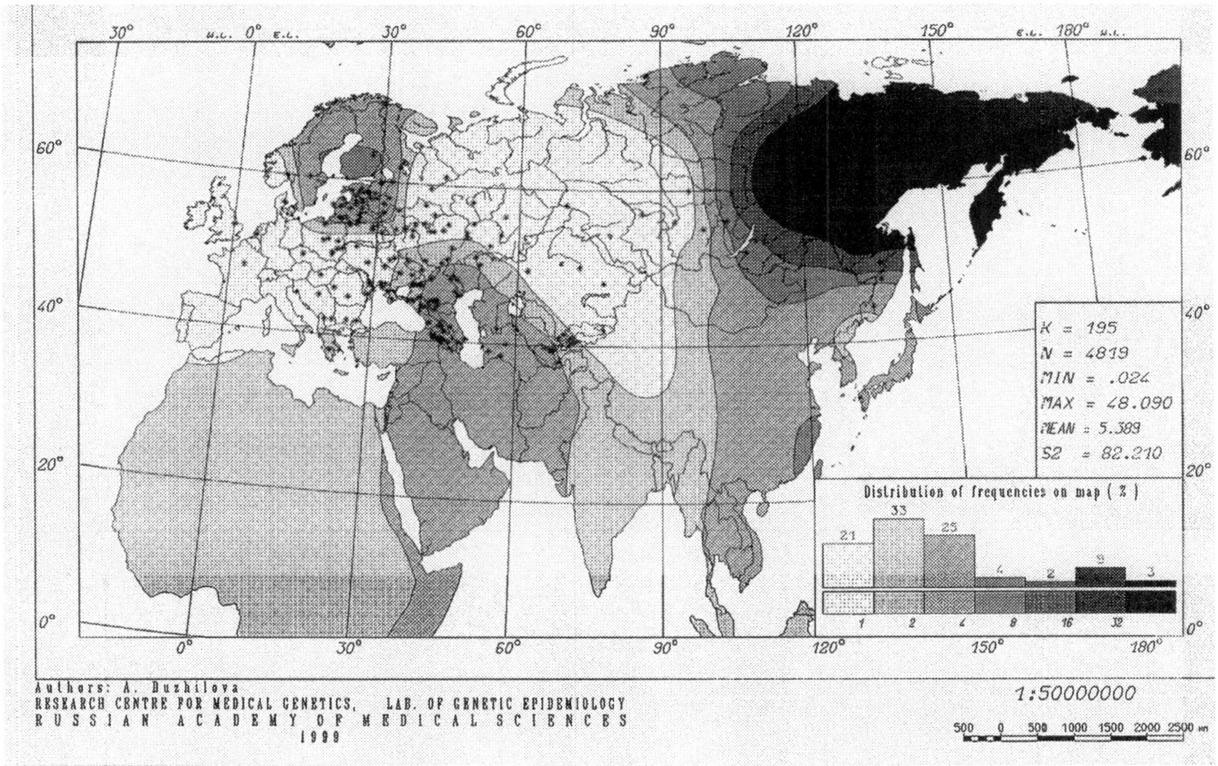

Figure 8: Trends for leprosy in Eurasia at the beginning of the 20th century

Vilhelm Møller-Christensen: his work and legacy

Pia Bennike
Laboratory of Biological Anthropology
University of Copenhagen
Blegdamsvej 3
DK-2200 Copenhagen N
Denmark

E-mail bennike@antrolab.ku.dk

Abstract

Those who met Vilhelm Møller-Christensen either during a visit to "his" leprosy museum in Denmark or at an international conference on palaeopathology, medical history or leprosy, would probably agree that he had a very impressive and enthusiastic attitude towards the study of palaeopathology. Indeed, his energy and direct approach seemed at times almost overwhelming. The same can be said about what he left behind: an enormous correspondence, illustrations including hundreds of slides and photographs, not to mention the skeletal samples from several sites, such as Æbelholt monastery and the famous leprosy cemetery at Næstved (Aaderup). After Møller-Christensen died in 1988, I was honoured to take over the responsibility for this material at the Museum of Medical History, University of Copenhagen and for preserving it for future studies. The skeletal samples were all documented in extensive files, and the most important cases were exhibited in the two small museums that he established.

Keywords

Leprosy, history, palaeopathology, medical history, Møller-Christensen

There are many reasons for writing a biography of a particular person of academic interest be it short or long. One purpose is to elucidate the historical development of, and changing theories in, a certain field of science through the merits of this person. Another purpose is to illustrate the ethics and policies which are reflected in the dimensions of the scientific and social life of the person. Furthermore, in a personal developmental process a description of a respected and admired scientist may also be an inspiration for one's own career. Rather than listing numerous events and publications, a biography therefore seems to be a relevant way of describing the development of a specific field and the theories involved through a person's scientific life. Questions raised include: how was a certain study planned and carried out, how were ideas, theories and methods born, and developed, and which new conclusions did they help to verify? In a small and multidisciplinary field such as palaeopathology the answers to such questions may prove to be quite significant. The purpose of the following short biography of Vilhelm Møller-Christensen is to give an impression of one man's career and its influence on the study of historical leprosy and palaeopathology, not only in Denmark, but on an international level.

The most important events of Vilhelm Møller-Christensen's professional life will be highlighted in the following: how he navigated through difficult times and sometimes dealt with massive scepticism from colleagues and archaeologists, and how he reached his goals and completed the work he had planned and started. The results of his skeletal studies on leprosy changes in bones are probably well known to those interested in leprosy and history, but the strategies leading to the improvement of the study of osteoarchaeology and palaeopathology will also be described.

Vilhelm Møller-Christensen was 85 years old when he died in 1988. After his death, I was given the honour of inheriting the responsibility for the skeletal material at the Museum of Medical History, University of Copenhagen and for preserving it for future studies. Those who met Vilhelm Møller-Christensen either during a visit to "his" leprosy museum in Denmark (Figure 1) or at an international conference on palaeopathology or leprosy, would probably agree that he was endowed with an extremely impressive and enthusiastic attitude towards the field of palaeopathology. Indeed, his energy and direct approach seemed at times almost overwhelming. The same can be said of what he left behind: an enormous correspondence, illustrations including hundreds of slides and photographs, not to mention the skeletal samples from several sites, such as Æbelholt monastery, the famous leprosy cemetery at Næstved (Aaderup) and the less famous leprosy cemeteries on the Islands of Bornholm and Møn. The skeletal samples were all documented in extensive files, and the most important cases were exhibited in the two small museum buildings that he established (Figure 2).

Møller-Christensen will mainly be remembered as the pioneer who initiated the systematic study of leprosy changes in medieval skeletons. Many questions may be asked concerning his activities. How did he manage to accomplish so much in science? To this one can only answer that he possessed enormous willpower which very few people or any obstacles could defy. At times it presented itself when he argued for his scientific theories. It was not always easy or agreeable to disagree with him. He put enormous, personal effort into solving any problem to reach his many goals. When one had been achieved, a new one was waiting to be fulfilled. He was a meticulous planner, and had a great talent for engaging and inspiring other people as volunteers, as research assistants, and as financial supporters for his many projects. The word "fundraising" was probably little known at the time, but he was very good at raising money, getting grants and support even from the Danish Royal Family. His scientific results speak clearly for themselves.

Vilhelm Møller-Christensen was born in 1903 in Haslev, a small town in southern Zealand, less than 20 km from the leprosy cemetery at Næstved which he later made internationally known to historians and leprologists. His parents were school teachers, and his father was the head of a school. This may well have meant that he was brought up in a home where education, studying, asking questions and getting answers was part of daily life. His parents' professional background may have influenced his wish to study, and later his very impressive ability to pass on knowledge through his many contributions at conferences, exhibitions, and through lectures, tutorials and several hundred publications.

Møller-Christensen graduated from medical school at the University of Copenhagen in 1928. During his internship at various hospitals and later as a physician in Hillerød, Møller-Christensen worked on his dissertation "The History of the Forceps" which was published in 1938. The dissertation clearly showed that Møller-Christensen had a unique, ambitious and untraditional approach to the theme. Later, this approach served him well when he became interested in the largely unexplored science, palaeopathology.

Møller-Christensen's first few publications were all concerned with historical topics related to medicine. According to him, his early interest in medical history was aroused and stimulated by lectures given by Professor Vilhelm Maar on the history of medicine in 1923. Neither in his very first publications nor in his dissertation on the history of the forceps did he ever mention skeletons or skeletal studies. His first publication on skeletal remains appeared in 1943, when he was 40 years old (Møller & Møller-Christensen 1943). However, during his search for forceps at the ruins of Æbelholt monastery in 1935, he became aware of the many buried skeletons. His interest in palaeopathology was sharpened when the first palaeopathological publication on Danish skeletal material appeared in 1936. It dealt with a study of 613 skeletons from Øm monastery at Øm in Jylland by another physician, Kristian Isager (1936).

The excavation of the cemetery at Æbelholt began in 1935 and revealed the remains of more than a thousand skeletons. When Møller-Christensen joined the excavation, he soon became aware of the many potential skeletal observations which were being lost due to the excavation techniques of that time. He therefore developed his own excavation procedure for skeletons, the so-called "osteoarchaeological technique" (Møller-Christensen, 1953a, 1973, and Figure 3).

It was the dentition of the skeletons that first attracted Møller-Christensen´s attention, and he soon realized that studies of dental conditions depended heavily on how well the teeth were preserved. He divided the skeletons into groups depending on the grade of tooth preservation. Together with dentists, he invented a so-called "archaeological dental index" on the basis of a specific calculation system (Brinch & Møller-Christensen, 1949). Møller-Christensen subsequently combined his interest in teeth with the occurrence of pathological changes of the spine. This led to the development of a theory that the occurrence of dental abscesses was related to pathological changes found in the cervical intervertebral joints. He named the correlation of these conditions *Morbus Æbelholt* (Møller-Christensen & Brinch, 1948, Ingelmark *et al.*, 1958). This theory was rather controversial and was met with some scepticism but, despite this, Møller-Christensen managed to persuade specialists in the field to study the spine (Ingelmark 1956, Sager, 1969).

Later, however, studying all signs of pathological conditions on skeletal remains became his true passion. After his first excavation, he studied the c. 750 skeletons from the monastery at Æbelholt. His results were not only published in several scientific and popular papers, but also as a widely read monograph which unfortunately was never translated into English (Møller-Christensen, 1958). Many scientists were invited to study the material which resulted in a large number of publications on dental conditions (among others Jørgensen 1956, Lundström & Lysell 1963, Lunt 1969, 1978), and he himself published a number of special cases (Møller & Møller-Christensen 1951, 1952, Møller-Christensen, 1969b). He helped to establish a small but fascinating museum at Æbelholt which still attracts many visitors during the summer months. Through his many activities, Møller-Christensen was very successful in passing on his interest in past populations and the study of palaeopathology to a number of people, an interest which proved to be very important for his studies on leprosy in the Middle Ages.

In the skeletal collection from Æbelholt monastery, Vilhelm found part of one female skeleton with specific pathological changes that he could not diagnose (Figure 4). The lower part of the skeleton was missing due to poor preservation conditions or excavation techniques, and the skull was normal, whereas the upper extremities were flexed maximally. Almost all the bones of the upper extremities and the scapulae bore traces of pathological conditions, especially the finger bones which were seriously afflicted with what seemed to be symmetrical nerve lesions resulting in necrosis. As neither Møller-Christensen nor his medical colleagues knew of any disease that would result in such skeletal changes, they believed that they were looking at a case of leprosy, with which none of them was familiar (Møller-Christensen & Møller, 1946). When the Second World War was over

Møller-Christensen was able to take the bones to a hospital in Bergen, Norway which still treated leprosy patients. The skeleton was studied by two pathologists familiar with leprosy. One of them, Prof. Erik Waaler, would not exclude that the changes might be due to leprosy, whereas Dr. Reidar Melsom clearly stated that they were not.

During this visit, Møller-Christensen realized that the only leprous skeletal material available in Bergen was one hand and one foot of a deceased patient. Not only was he confronted with a lack of skeletal material, but he soon discovered that information on leprous changes of bones in the literature was very scarce. He therefore decided to look for a mediaeval cemetery attached to a St. George's Hospital in order to excavate and study the skeletons buried there. His plan was to compile a systematic description of all bone changes, a plan and strategy which later turned out to be invaluable!

In 1898, Edvard Ehlers published details on the location of the sites of 19 medieval leprosy hospitals called St. George in Denmark. Much later, twelve more were added to the list (Møller-Christensen, 1961/62 - Figure 5). One of these sites lay close to the town of Næstved in southern Zealand. In 1948, Møller-Christensen went to the area during his spare time and, at the site where the hospital was supposed to have been, he asked the local farmers whether they had seen any human bones exposed while working in the fields. After several attempts, and after having been suspected of being a policeman looking for a murder victim or of being insane, he finally found a farmer who had seen many skeletons in his field and even in his yard. Shortly thereafter an excavation of the site commenced (Figure 6). However, work was constantly delayed due to problems with the archaeological authorities, among others, and the excavation was only completed in 1968, about 20 years after it had started.

Each and every skeletal change on the many well preserved skeletons was registered and catalogued, revealing a pattern of pathological changes (Møller-Christensen, 1951a, 1951b, 1952, 1953c, Møller-Christensen & Faber, 1952). It became clear to Møller-Christensen that the central parts of the upper jaw had atrophied in many of the skeletons. After this discovery Møller-Christensen returned to the leprosy hospital in Bergen, Norway to look for the same signs in the living patients. Together with two leprologists he found the same changes in the patients as in the Danish medieval skeletons. Møller-Christensen named this condition *facies leprosa* when seen on the skeletons, whereas he suggested that the clinical changes be named "The Bergen syndrome" in patients, as it was in Bergen that this syndrome was discovered (Møller-Christensen and Faber, 1952). In 1974, Professor Schmitz-Cliever suggested the names be changed to *"facies leprosa nasalis"* and *"facies leprosa maxillaris"*, or both. Later, Johannes G. Andersen and Keith Manchester suggested that the complex of changes be called the "the rhino-maxillary syndrome" (1992). In 1953, Møller-Christensen's first publication appeared as a monograph "Ten Lepers from Næstved in Denmark" (Møller-Christensen, 1953b). At the time, the term "leper" did not carry the stigma it does today. He described the bone changes involved in the term *"facies Leprosa"* which were 1) atrophy of the nasal spine, 2) atrophy and recession of the alveolar process of the anterior maxilla, and 3) endonasal inflammation changes. That same year Møller-Christensen went to the 6th International Congress of Leprosy in Madrid together with the two Norwegian pathologists. They drove south in a car together with ten leprous skeletons that Møller-Christensen wanted to show at the Congress (Figure 7). This must have impressed several participants who were mainly leprologists, for Møller-Christensen was later described in Graham Greene's book "A Burnt-Out Case" (Greene, 1960) as an old Danish doctor who had become so interested in leprosy that he brought skeletons with him to international conferences in his car! Graham Greene got the information from his friend Michel Lechat, a famous leprologist and the director of a leprosarium in the Congo at the time. Professor Lechat, at the conference on leprosy in Bradford in 1999 handed the author the book dedicated to him by Graham Greene, so that his role would not be forgotten. Møller-Christensen was of course proud of being so famous, but rather annoyed about being called an old man (in 1953 he was 50 years old). True to his nature, he contacted Graham Greene once he was in London to tell him that they were the same age!

The very untraditional contribution at the Madrid conference (the demonstration of medieval skeletons with pathological changes due to leprosy), was repeated with great success at many other international conferences on leprosy, anthropology (1955), rheumatology (1956), pathology (1960 and 1965 - Figure 8) and medical history (1964), just to mention a few. He often combined his attendance at conferences with visits to places where he could study leprosy patients or skeletal remains, alone or with colleagues. Together they published studies on two skeletons from Great Britain, one from Nubia (Møller-Christensen & Hughes, 1962, 1966), and on skeletal material from the Department of Anatomy at the University of Edinburgh (Møller-Christensen & Inkster, 1965). He visited the University of Glasgow in order to study the occurrence of cribra orbitalia which he had observed very frequently (63.1 %) in adult Danish leprous skeletons (Figure 9, Møller-Christensen, 1961, 1978, Møller-Christensen and Sandison, 1963). He paid a special visit to the Museé de l'Homme in Paris (Figure 10).

In 1959, Møller-Christensen began a comparative study of the changes he had found in the medieval skeletons with those in leprosy patients. However, as he himself had said in an interview, 'it was impossible to find money for trips to the Far East to study leprosy in those days'. He therefore signed on as a ship's doctor. 'Whenever they docked at a habour with a leprosy hospital nearby, they gave me shore leave' (Larsen, 1973: 260). This is how he made his way to Thailand and the Far East that year, where he studied patients at leprosy hospitals in Penang, Kuala Lumpur, Bangkok and other cities. Many years later in 1972, he returned to the Far East with financial support from the research council, together with Knud Danielsen, the dentist who described Møller-Christensen's observation of the very unusual occurrence of short roots in subadult leprous skeletons (Danielsen, 1968, 1970 and Figure 11). This time they also wanted to study changes in the maxillary bone. Together, Møller-Christensen and Danielsen studied and photographed 333 patients at the Prapradaeng Leprosarium in Bangkok. Here, Vilhelm met one of the patients he had seen in 1959. His own description in an interview was as follows (Larsen 1973: 261): 'She was now 47 years old, one eye blind, her nose destroyed and both hands and feet severely mutilated. However, when I met her again in 1972 she was newly married –for the fourth time! Her central incisors were lost and the lateral ones were loose. We offered her dentures in return for the loose teeth which we brought home to study. No doubt, she was very pleased with the exchange!' (Figure 12).

In 1958, a book was finally published on the Æbelholt skeletons that Møller-Christensen had started to study 20 years earlier, and a small museum was established in 1957 at the site of the ruins of the monastery. In 1961, he published a second monograph on leprosy (Møller-Christensen, 1961). He continued his innumerable activities (Møller-Christensen, 1963, 1965, 1967, 1969a, 1969c, Weiss and Møller-Christensen, 1971a, 1971b, Møller-Christensen and Weiss, 1971), and initiated many interesting research projects working together with other dental or medical scientists. Professor. W.H. Jopling, a well known English leprologist, collaborated with Møller-Christensen on a study of skeletons in the Catacombs of Paris in search of leprous bone changes (Møller-Christensen and Jopling, 1964). Jopling described their visit to Paris in 1963 in a foreword of the last book published by Møller-Christensen (1978:8-9) with the words, 'I remember marvelling at the unconcerned way in which Vilhelm walked through the streets carrying skulls in a string bag, oblivious of the sidelong glances of passers-by' (Figure 13).

In 1939, he had finally settled down as a physician (family doctor) in Roskilde less than 100 km from Copenhagen, the capital of Denmark. In 1956, he changed to a less demanding position as medical consultant for the local schools. This change allowed him to devote more time to his most favourite work in medical history and palaeopathology. In 1964, at the age of 61, Møller-Christensen was appointed Professor of Medical History at the Museum of Medical History, then part of the University of Copenhagen. It was the first time Møller-Christensen was offered a full-time position at the university where students benefited from the field of research he had contributed to so much. However, in 1956, Møller-Christensen had already applied for a position as head of the Laboratory of Physical Anthropology at the University of Copenhagen. At that time the laboratory was part of the Institute of Anatomy where almost all skeletal material from Denmark and Greenland was stored (and still is). Sadly, a number of serious personal and professional conflicts got in the way of his appointment to the position where he would certainly have initiated many scientific projects and would probably have raised the quality of research to international standards.

After a leprosy conference in Rio in 1962, Møller-Christensen went to San Angola in Brasil, a small "leprosy" town, and to Manaus, the capital of the Amazon where 5% of the population had leprosy at that time. He studied the patients together with Prof. Georg Klingemüller from Germany. Not unexpectedly, they found the same facial bone changes in the patients as in the leprous skeletons. Several years later, Møller-Christensen published a comparative study on other scientists' observations of leprosy patients, his own in Bangkok, and on his skeletal studies. He concluded that the disease developed in an almost identical patterns in all groups (Møller-Christensen, 1974 and Figure 14).

In 1966, at the age of 63, Møller-Christensen remarried. His second wife, Martha, was a nurse and a cousin of one of his former "students", the well known Johannes G. Andersen (Andersen 1969). How they met, I have never found out, but finding Møller-Christensen's house was never a problem, for who else would name their home "Spina"?

As Professor of Medical History at the University of Copenhagen, Møller-Christensen also had to take on many administrative duties. His ability to raise money enabled him to open the Museum of Medical History in Copenhagen to the public in 1969. The Danish Royal Family attended the opening and reception, a celebration Vilhelm was very proud of indeed. Many of his later publications dealt with aspects of medical history at the Faculty of Medicine.

I first met Møller-Christensen the day I defended my Ph.D. thesis on palaeopathology at the University of Copenhagen (Bennike 1985). Unfortunately, he never came to the Laboratory of Biological Anthropology where I had been a student. I surely would have benefited enormously had he been my supervisor. I certainly feared his critique of and comments on my thesis. Great was my relief when he commented on my work positively and welcomed me to "his" field of research by bringing me one of the best gifts he could think of, his own dissertation on The History of the Forceps (1938).

Vilhelm had to retire from the university in 1973 when he turned 70, but he continued to publish and to work in the field of palaeopathology (Møller-Christensen, 1974, 1976). In 1978, he published his last monograph, but sadly without help from the Museum of Medical History which he had done so much for. In the book he concludes that his examination of 300 leprous skeletons with *facies leprosa nasalis* and *facies leprosa maxillaris* from the Middle Ages proved that 1) atrophy of the anterior nasal spine was always combined with inflammatory changes of the nasal cavity, and 2) that atrophy of the alveolar process of the maxilla was always combined with the inflammatory changes of the oral surface of the hard palate (Møller-Christensen, 1978). He felt that a study of the correlation between leprous changes in the nasal cavity and atrophy of the anterior nasal spine, and on the correlation between leprosy changes of the soft palate and atrophy of the alveolar process of the maxilla, should be carried out as he impatiently points out in his last book, 'We are still waiting for a clinical report on.....' (Møller-Christensen, 1978:135).

Having spent a large part of his later life on the history of leprosy, Vilhelm Møller-Christensen retained his respect and concern for the many people he had met who suffered from the disease. He writes, 'The skeletal material, representing 650 persons, is so far unique. They are the remains from the surroundings of Næstved. They were housed in the hospital and cared for. No return was possible. Just as the River Jordan runs into the Dead Sea, so that all wreckage carried by the river will come to rest in the Dead Sea, so it has been with the lepers –the outcasts- who were admitted to a St. Jørgen's hospital. Rich or poor, squire or beggar, when they first came here, journey's end sooner or later was the leprosy graveyard' (Møller-Christensen, 1978: 35).

References

Andersen, J.G. 1969 *Studies in the Medieval diagnosis in Denmark; an osteological, historical and clinical study.* Copenhagen:Cisters Bogtrykkeri.

Andersen, J. G. and Manchester, K. 1992 The rhinomaxillary syndrome in leprosy: a clinical, radiological and palaeopathological study. *Int. J. of Osteoarchaeology* 2:121-129.

Bennike, P. 1985 *Palaeopathological Studies of Danish Skeletons. A study of demography, disease and injury.* København:Akademisk Forlag.

Brinch, O. and Møller-Christensen, V. 1949 On comparative investigations into the occurrence of dental caries in archaeological skulls with a preliminary report on investigations of caries frequency in medieval times in Denmark. *Odontologisk Tidsskrift* 4:357-80.

Danielsen, K. 1968 Leprogenic odontodysplasia. *Int. J. Leprosy* 36:552-666.

Danielsen, K. 1970 Odontodysplasia leprosa in Danish Mediaeval Skeletons. *Tandlægebladet* 74:603-625.

Greene, G. 1960 *A Burnt-Out Case.* London: Penguin Books.

Ingelmark, B.-E. 1956 De funktionellt anatomiske förhållandena i ryggraden med särskild hänsyn till dess småleder. *Göteborg Universitets Årsskrift* 62: 1-48.

Ingelmark, B.-E., Møller-Christensen, V. and Brinch, O. 1958 Spinal joint changes and dental infection. *Acta Anatomica Supplement* 36: 17-19.

Isager, K. 1936 *Skeletfundene ved Øm Kloster. Til belysning af middelalderlig Patologi og Klosteret som Hospital.* København: Lewin & Munksgaard.

Jørgensen, D. 1956 The deciduous dentition. *Lægeforening* 4:258-262.

Larsen, Ø. 1973 Vilhelm Møller-Christensen (Interview). *Tidsskrift for Den norske lægeforening* 4: 258-262.

Lechat. M.F. and Chardome, J. 1955 Altérations radiologiques des os de la face chez les lépreux Congolais. *Ann. Soc. Belge. Med. Trop.* 35: 603-611

Lundström, A.and Lysell, L. 1963 An anthropological examination of a group of Medieval Danish skulls with particular regard to the jaws and occlusal conditions. *Acta Odont. Scand.* 11:111-128.

Lunt, D.A. 1969 An Odontological Study of Medieval Danes. *Acta Odont Scand.* 55 (Supplement 27).

Lunt, D.A. 1978 Molar attrition in Medieval Danes. In: P.M. Butler and K.A. Jowsey (eds): *Development, function and evolution of teeth.* London: Academic Press, pp. 465-82.

Melsom, R.S. 1953 *Changes in the maxillary bone in Leprosy. A clinical and roentgenological examination.* Sixth Congress Internacinal de Leprologia, Madrid, pp. 747-750.

Michman, J. and Sager, P. 1957 Changes in the anterior nasal spine and alveolar process of the maxillary bone in leprosy. *Int. J. Leprosy* 25: 217-222.

Møller, P. and Møller-Christensen 1943 Skeletfund fra Æbelholt Kloster. *Aarbøger for Nordisk Oldkyndighed og Historie* (no volume cited): 183-200.

Møller, P. and Møller-Christensen, V. 1951 Et kvindekranium med spor efter kræftmetastaser. *Aarbøger for Nordisk Oldkyndighed og Historie* 1950: 373-380.

Møller, P. and Møller-Christensen, V. 1952 A Medieval female skull showing evidence of metastases from a malignant growth. *Acta. Path. Microbiol. Scand.* 30: 336-342.

Møller-Christensen, V. 1938 *The History of the Forceps.* København: Lewin & Munksgaard.

Møller-Christensen, V. 1951a Træk af spedalskhedens historie i Danmarks middelalder. *Med. Forum* 4:278-88.

Møller-Christensen, V. 1951b Om fundet og udgravningen af den middelalderlige leprakirkegård i Næstved. *Ciba Tidsskrift* 3: 975-76.

Møller-Christensen, V. 1952 A case of leprosy from the Middle Ages in Denmark. *Acta. Med. Scand. Suppl.* 266:101-108.

Møller-Christensen, V. 1953a Location and excavation of the first Danish leper graveyard from the Middle Ages. *Bull. Med. Hist.* 17:112-123.

Møller-Christensen, V. 1953b *Ten Lepers from Næstved in Denmark. A study of skeletons from a Medieval Danish Leper Hospital.* Copenhagen:Danish

Science Press.
Møller-Christensen, V. 1953c Changes in the maxillary bone in leprosy. *Int. J. Leprosy* 21: 6161-617.
Møller-Christensen, V. 1958 *Bogen om Æbelholt Kloster.* København: Dansk Videnskabs Forlag.
Møller-Christensen, V. 1961 *Bone changes in leprosy.* Copenhagen: Munksgaard.
Møller-Christensen, V. 1961/62 Lepra og dens historie. *Medicinsk Årbog* (No volume stated): 17-81.
Møller-Christensen, V. 1963 Skeletfundene fra Skt. Jørgens Kirke i Svendborg. *Fynske Minder* (No volume stated): 35-49.
Møller-Christensen, V. 1965 New knowledge of leprosy through palaeopathology. *Int. J. of Leprosy* 33: 603-10.
Møller-Christensen, V. 1967 Evidence of leprosy in earlier peoples. In: Brothwell, D.R. and Sandison, A.T. (eds.) *Diseases in Antiquity.* Springfield, Illinois: C.C. Thomas, pp. 295-307.
Møller-Christensen, V, 1969a Provisional results of the examination of the whole Næstved leprosy hospital churchyard –ab. 1250-1550 A.D. *Nordisk Medicinhistorisk Årsbok* Part 4: 29-36.
Møller-Christensen, V. 1969b Tandpleje i middelalderen. *Nordisk Medicinhistorisk Årsbok* Part 3: 16-24.
Møller-Christensen, V. 1969c The history of syphilis and leprosy –an osteoarchaeological approach. *Abbottempo*, Book 1: 20-25.
Møller-Christensen, V. 1973 Osteo-archaeology as a medico-historical auxillary science. *Medical History* 17: 411-18.
Møller-Christensen, V. 1974 Changes in the anterior nasal spine and the alveolar process of the maxilla in leprosy. A Clinical Examination. *Int. J. Leprosy* 42: 431-35.
Møller-Christensen, V. 1976 Leprosy and its way from the Old to the New World. *Dansk Medicinsk Historisk Årbog* (no volume stated): 107-115.
Møller-Christensen, V. 1978 *Leprosy changes in the skull.* Odense: Odense Univ. Press.
Møller-Christensen, V., Bakke, S. N., Melsom, R.S. and Waaler, E. 1952 Changes in the anterior nasal spine and the alveolar process of the maxillary bone in leprosy. *Int. J. Leprosy* 20: 335- 340.
Møller-Christensen, V. and Brinch, O. 1948 Tooth infection and spinal joint diseases in Medieval Denmark. *Paradentologia,* 1: 12-17.
Møller-Christensen, V. and Faber, B. 1952 Leprous changes in a material of Mediaeval skeletons from the St. George's Court, Næstved. *Acta. Radiol.* 37: 308-317.
Møller-Christensen, V. and Hughes, D.R. 1962 Two early cases of leprosy in Great Britain. *Man* 62: 177-179.
Møller-Christensen, V. and Hughes, D.R.1966 An early case of leprosy from Nubia. *Man* 66: 242-245.
Møller-Christensen, V. and Inkster, R.G. 1965 Cases of leprosy and syphilis in the osteological collection of the Department of Anatomy, University of Edinburgh. *Danish Med. Bulletin* 12:11-18
Møller-Christensen, V. and Jopling, W.H. 1964 An examination of the skulls in the Catacombs of Paris. *Med. Hist.* 8: 187-188.
Møller-Christensen, V. and Møller, P. 1946 Et tilfælde af formodet spedalskhed fra Danmarks middelalder. *Aarbøger for Nordisk Oldkyndighed og Historie* (no volume stated): 232-252.
Møller-Christensen, V. and Sandison, A.T. 1963 *Usura orbitae (cribra orbitalia)* in the collection of crania in the Anatomy Department of the University of Glasgow. *Path. Microbiol.* 26:175-183.
Møller-Christensen, V. and Weiss, D.L. 1971 One of the oldest datable skeletons with leprous bone-changes from the Næstved leprosy hospital churchyard in Denmark. *Int. J. Leprosy* 39: 172-82.
Sager, P. 1969 *Spondylosis cervicalis –A pathological and osteoarchaeological study.* Copenhagen: Munksgaard.
Weiss, D.L. and Møller-Christensen, V. 1971a Leprosy, echinococcosis and amulets. A study of a Medieval Danish inhumation. *Medical History* 15 (3): 260-67.
Weiss, D.L. and Møller-Christensen, V. 1971b An unusual case of tuberculosis in a medieval leper. *Danish Med. Bull.* 18: 11-14.

Figure 1: The white building in the background is the former Royal Surgical Academy in Copenhagen, built in 1787. It is now the Museum of Medical History to which the small yellow house, also called "The Leprosy Museum", belongs. All 750 skeletons from the Næstved leprosy cemetery are in boxes on the first floor, under the roof. Below, there is an interesting exhibition on medieval leprosy and a large pathological bone collection from the 19th century.

Figure 2: In 1964 Vilhelm Møller-Christensen set up the permanent leprosy exhibition in the leprosy museum

Figure 3: Møller-Christensen developed a special excavation technique of skeletons which was used at the excavation of the Næstved cemetery. He dug a ditch around each grave in order to excavate the skeletons 'as if they were laying on an operating table'. The archaeologists always wondered why the skeletons excavated by Møller-Christensen had no traces of a coffin, whereas those excavated by the archaeologists had. Perhaps this was a question of priority!

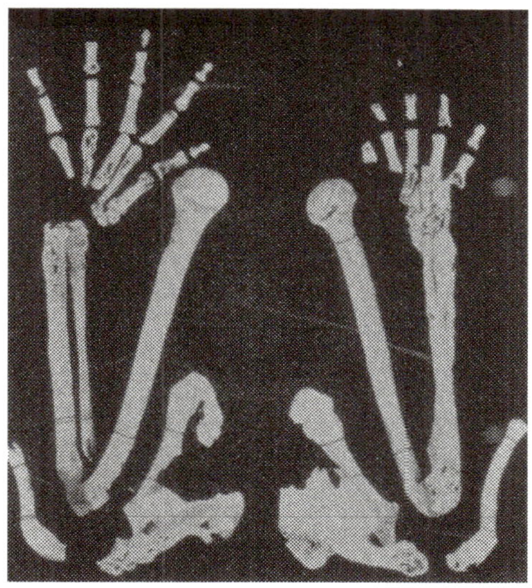

Figure 4: Pathological changes of the upper parts of the female skeleton from Æbelholt which Møller-Christensen thought indicated leprosy. After his extensive studies of leprosy skeletons, it became clear that the changes were due to another disease, perhaps ergotism.

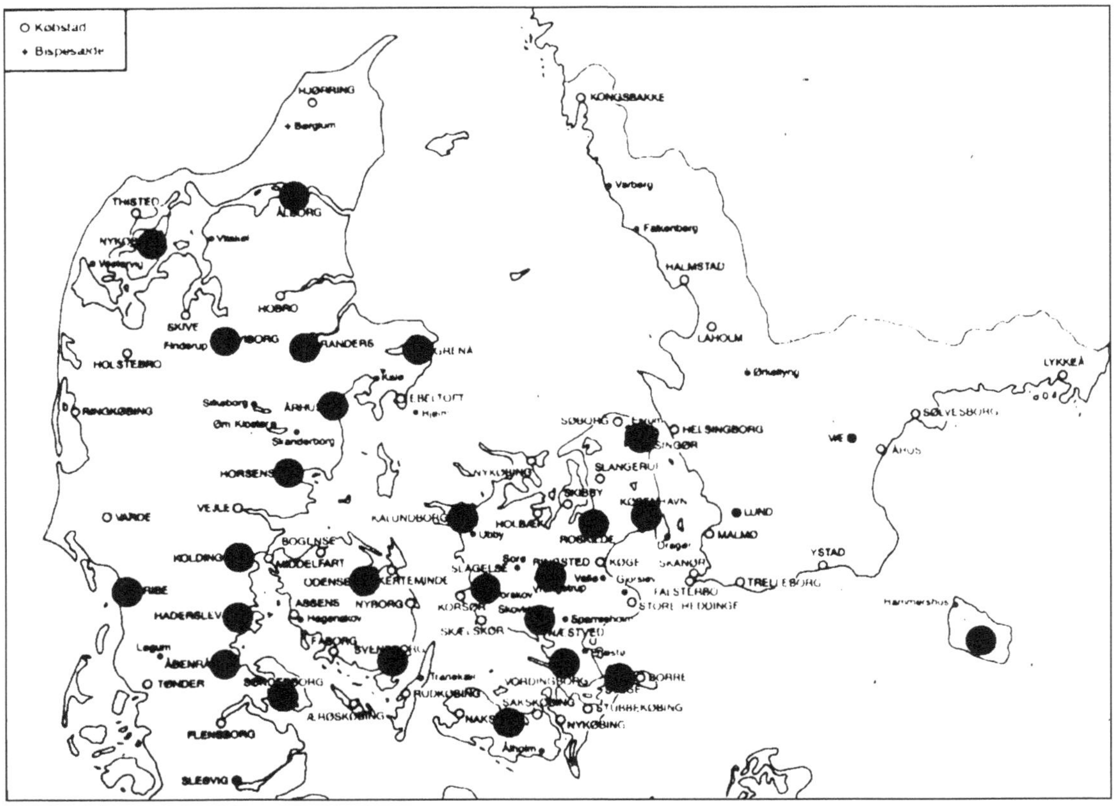

Figure 5: Denmark had 31 leprosy hospitals between 1250-1550 AD. Næstved is in the southern part of Zealand, but leprous skeletal material has also been excavated from four other leprosy cemeteries: around Odense and Svendborg on the island Funen and on the islands Møn and Bornholm, south and southeast of Zealand respectively.

Figure 6: Vilhelm Møller-Christensen always wore a beret, and was therefore easy to pick out at excavations. He was especially attentive to the smallest finds in a grave and discovered a number of remains that are seldomly described pleura plaque (Weiss & Møller-Christensen, 1971a), walls around cysts caused by echinococcus (Weiss & Møller-Christensen, 1971b), a bead in a carious tooth (Møller-Christensen 1969b, etc). Even tiny ear bones were found and kept separately for future studies!

Figure 7: At the 6th International Congress on Leprosy in Madrid 1953. Møller-Christensen brought 10 skeletons in his car to demonstrate the bone changes of leprosy. This was probably a surprising event for the participating leprologists.

Figure 8: During a meeting at the Armed Forces Institute of Pathology in 1965, the paticipants were photographed. Møller-Christensen is the tall man in the middle of the front row.

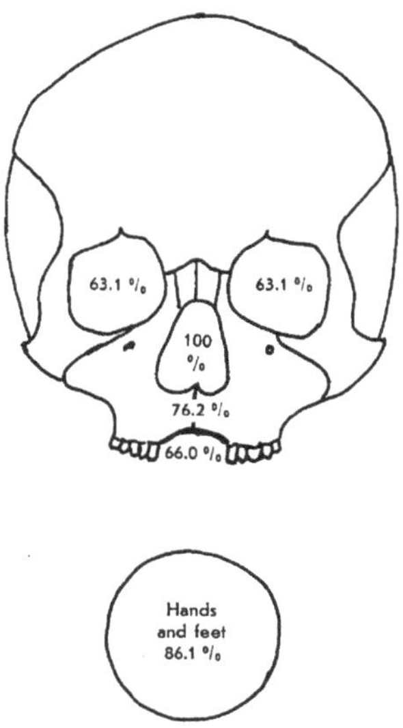

Figure 9: Møller-Christensen (1961: 39) found *cribra orbitalia* which he named *usura orbitae* in 63.1% of 247 adult leprous skulls with changes due to inflammation of the nasal cavity. Atrophy of the anterior nasal spine was seen in 76.2%, and atrophy of the maxillary alveolar process in 66% of the skulls. In 115 of the skeletons with the changes of the nasal cavity, and with preserved hands and feet, typical deformations of the hands and feet were seen in 86.1%.

Figure 10: Møller-Christensen on the roof of Museé de l'Homme studying bones in 1963.

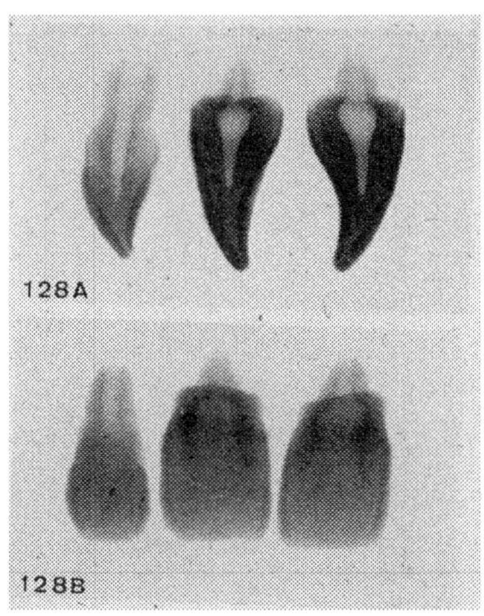

Figure 11: Stunted dental roots in subadults from the leprosy churchyard in Næstved. Similar finds have not been reported in leprosy patients or leprous skeletons from other sites. A new study of a comparison of the mechanisms leading to stunted roots in various cases are in preparation by I. Kjær, K. Danielsen and P. Bennike, n.d.

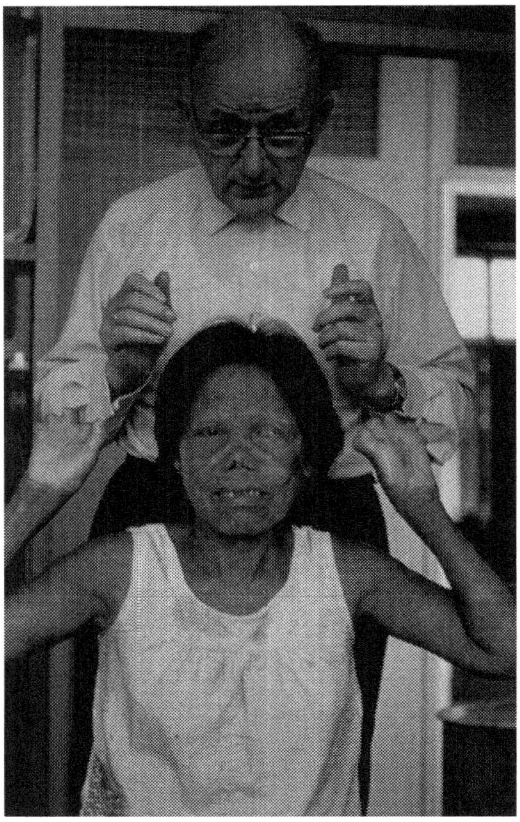

Figure 13: 'I remember marvelling at the unconcerned way in which Vilhelm walked through the streets carrying skulls in a string bag, oblivious of the sidelong glances of passers-by' (Jopling in (Møller-Christensen, 1978: 8-9).

Figure 12: Møller-Christensen went to Thailand twice, in 1959 and in 1972, to study patients suffering from leprosy. In 1972 he met one of the patients again and offered her dentures in exchange for her two loose lateral incisors.

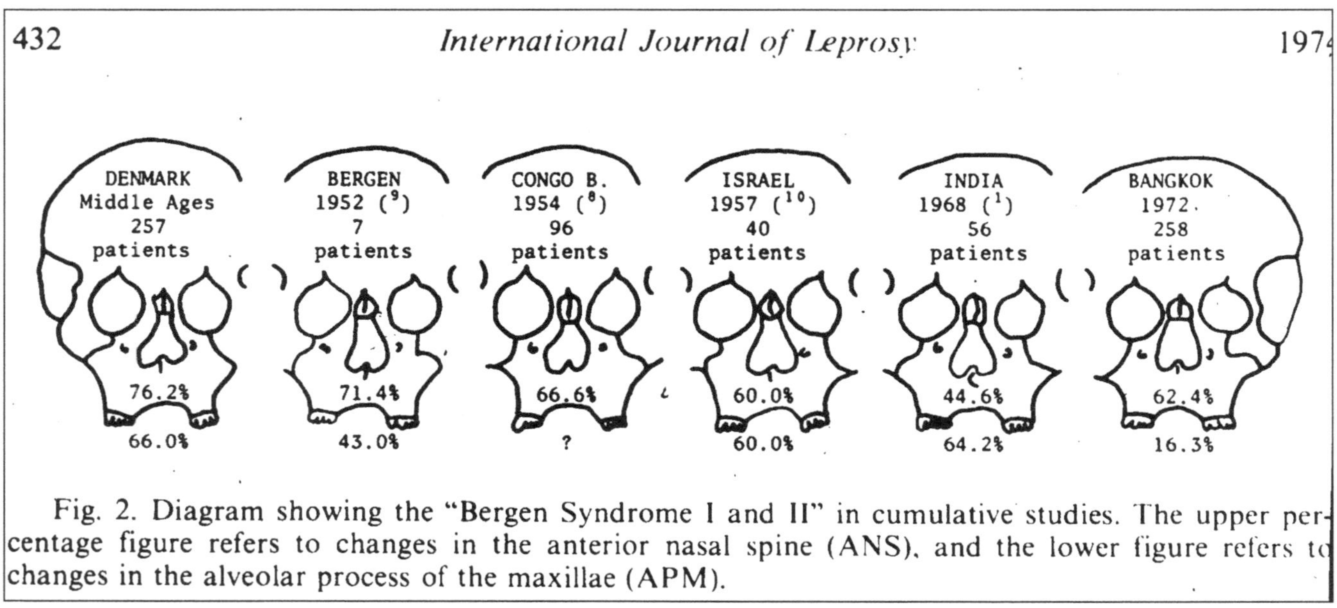

Figure 14: The diagram published in (Møller-Christensen (1974) shows changes in the anterior nasal spine (upper percentage), and changes in the alveolar process of the maxillae (lower percentage). The numbers refer to: 9: Melsom (1953), 8: Lechat and Chardome (1955), 10: Michmann and Sager (1957), 1: Andersen (1969). The low percentage of changes in the alveolar process of the maxilla in the Bangkok material may be due to the lack of radiography facilities. The numerous studies published after Møller-Christensen's first presentation of maxillary changes at the 6th International Leprosy Congress in Madrid 1953 clearly illustrate the increased interest in this phenomeon among leprologists.

A possible leprosy hospital in Stubbekøbing, Denmark

Thomas Brander and Niels Lynnerup
Laboratory of Biological Anthropology
University of Copenhagen
The Panum Institute
Blegdamsvej 3
DK-2200 Copenhagen
Denmark

E-mail: n.lynnerup@antrolab.ku.dk
t.brander@wanadoo.dk

Abstract

According to historical sources a leprosy hospital was situated in Stubbekøbing, Denmark, between 1481-1505 AD. Partial archaeological excavations were undertaken in 1994, during which 60 skeletons were located and retrieved from a burial ground. The skeletons were analysed with special attention paid to palaeopathological indicators of leprosy. However, only four skeletons displayed osteological changes associated with leprosy, and none of the changes were of a specific, diagnostic nature (i.e. *facies leprosa*). Conversely, 22 skeletons displayed more non-specific pathological changes. Thus, while not positively identifying a leprosy hospital in Stubbekøbing, the osteological data may indicate that the site had a hospital function.

Keywords

Palaeopathology, epidemiology, fractures, anaemia, male/female ratio.

1. Introduction

According to historical sources a leprosy hospital was situated in Stubbekøbing, a town on the island of Falster, south of Zealand, Denmark, between AD 1481 and 1505 (Jensen, 1994). Excavations of the supposed cemetery were undertaken in 1994. The excavations were only partial, but revealed a chapel building as well as parts of a cemetery. Sixty skeletons were retrieved, comprising probably less than half the potential burials. The objective of this study was to evaluate whether the retrieved skeletons support the hypothesis of the site being the cemetery of the leprosarium described in the historical sources.

2. Materials and Methods

The skeletons were aged and sexed following the usual standards in physical anthropology (e.g. long bone growth, auricular surfaces, sternal rib ends, dental wear and dental development, discussed in Buikstra and Ubelaker, 1994). Subadults made up 25% (N = 15) of the 60 skeletons, and 16 of the adult individuals were female, with 20 male. Nine adult skeletons could not be sexed. The total age and sex distribution is shown in Table 1. The skeletons were analysed with special attention to palaeopathological indicators of leprosy.

3. Results

Among the 31 partial or complete skulls there were no cases of *facies leprosa* as described by Møller-Christensen (1974). However, four skeletons showed changes of the hands or feet, which could be associated with leprosy (Møller-Christensen, 1961):

Case 1: Destructive changes of the distal ends of the right radius and ulna of a 20-30 year old woman

Case 2: Destructive changes of one distal phalanx and three proximal phalanges, and changes of the left tibia, fibula and calcaneum of a 20-30 year old woman.

Case 3: Destructive changes of the right 1st metatarsal of a 40-50 year old man.

Case 4: Destructive changes of the left 1st metatarsal of a 30-40 year old woman.

Numerous other pathological changes were observed on the skeletons: nine of the skeletons showed periosteal reaction, and four of these were from a group of individuals with changes indicative of leprosy. Cribra orbitalia was observed in a seven- and a nine-year old subadults. Two cases of ankylosis of the sacroiliac joint were found in males, both above 50 years of age. One case of chronic otitis media and one case of reactive arthritis were also found (Table 2).

Six skeletons showed 20 fractures, of which seven were of the long bones, and thus 1.8% of the total number of long bones showed fractures (see Table 3). In addition to the long bone fractures, there was one depressed fracture of the skull, and one individual presented 12 rib fractures. None of the skeletons with changes associated with leprosy had fractures.

4. Discussion

Since 75% of the adult skulls from the material from the well-known leprosarium at Næstved showed *facies leprosa* (Møller-Christensen, 1961), the lack of any cases of *facies leprosa* at this site could indicate that the cemetery was not used as a leprosarium in Stubbekøbing. Periosteal reactions are characteristic of, but not pathognomonic, leprosy (Andersen, 1969). However, in our skeletal material 20% of the adult sample showed periosteal reactions, and 100% of the skeletons with signs indicative of leprosy showed periosteal reactions. These figures correspond well with the results presented by Lewis *et al.* (1995) from the leprosarium of Chichester, England.

Fractures of long bones were seen in 8.9% of the adults, a pattern normally seen in cemeteries not associated with hospitals (Judd and Roberts, 1998), and this may indicate a possible hospital function for the cemetery in Stubbekøbing. On the other hand, the total proportion of fractures of the long bones was only 1.8% which is less than the 2.6% reported in the material from the Chichester leprosarium (Judd and Roberts, 1998). Male leprous patients have a higher risk of getting fractures. Examination of male leprous patients have shown that patients have lower concentrations of free testosterone and oestradiol, and that these concentrations are correlated with low BMD (bone mineral density) and to how long patients have been affected by leprosy (Ishikawa, 1997, 1999). This is perhaps due to the fact that a very high proportion of male leprosy patients have involvement of the testes (Grabstald and Swan, 1952). Indeed, all the six individuals with fractures at this site were men.

How do the results correspond with present knowledge on the epidemiology of leprosy, and the hypothesis that the material originates from leprosy sufferers? The general trend in modern data is that the detection rate of leprosy is higher in males than among females, and that the male/female ratio approaches 1:0, the higher the prevalence. There is a slight surplus of men in the skeletal assemblage from Stubbekøbing giving a male/female ratio of 1,25, which is lower than most ratios seen in modern data (Irgens and Skjaeven, 1985). This is to be expected since the prevalence of leprosy in medieval Scandinavia was high, although the 1481-1505 AD period probably is towards the end of the epidemic in Denmark. It has been shown that the age at onset of leprosy rises when the incidence of the disease decreases (Irgens and Skjaeven, 1985). In theory, a high prevalence of an infectious disease will give a low age at onset. This corresponds with the data from the Næstved leprosarium where it has been shown that most of the patients had a childhood onset (Ell, 1988). This was shown by using the degree of loss of the maxillary bone, and comparing these data with the rate of maxillary bone loss in modern leprosy patients from Malaysia (Nah *et al.*, 1985). There are, however, no cases of *facies leprosa* in the Stubbekøbing material. The high rate of cribra orbitalia (55% of the skulls) at Næstved (Møller-Christensen, 1961) also could indicate a childhood onset in medieval Scandinavia, since cribra orbitalia is a probable sign of childhood anaemia (Mittler and Van Gerven, 1994; Stuart-Macadam, 1985). A study of untreated leprosy patients has shown that 19% were anaemic and 85.7% of the lepromatous group were anaemic (Lapinsky, 1992). At Stubbekøbing only two cases of cribra orbitalia were found, which could indicate a later onset of leprosy than that seen at Næstved (if, of course, it is accepted that the cribra orbitalia is due to leprosy at all). It may well represent a general indicator of "stress" during the growth period. Stuart-Macadam (1992) has indicated that increased pathogen load may lead to iron deficiency anaemia.

Below are listed four possible explanatory models in attempting to consider the original hypothesis that a leprosarium functioned at the site of Stubbekøbing:

1. None of the 60 skeletons derive from patients with leprosy. Thus, the osteological changes found indicative of leprosy (four cases) must have other causes e.g. rheumatoid arthritis, psoriatic arthritis, osteomyelitis etc.

2. The leprosarium at Stubbekøbing had only a few patients with leprosy, and this would explain why only four cases were found with changes indicative of leprosy.

3. The patients of the supposed leprosarium were in a very poor general state of health. Therefore, the patients with leprosy would die before the disease could manifest itself in terms of osteological changes. This theory, however, does not fit with our present knowledge of the disease, since the patients with a high bacillary load (the ones most likely to develop osteological changes) have this because of a low cellular immunity.

4. If the material reflects the late phase of the epidemic in Medieval Denmark, the low number of osteological changes indicative of leprosy and the absence of pathognomonic changes may reflect a much later age at onset of the disease.

5. Conclusions

Due to the lack of pathognomonic osteological changes of leprosy (i.e. *facies leprosa*) the skeletal material from the Stubbekøbing cemetery cannot be positively identified as being associated with a leprosy hospital. It is not possible, however, to rule out that at least some leprous patients were buried at the site, since four skeletons showed changes indicative of leprosy. Indeed, it has been suggested that, for instance, in Britain, leprosy sufferers were not always strictly segregated from the non-leprous in institutions which admitted sufferers of leprosy (Roberts, this volume), or that people with leprosy may have been sometimes admitted to Poor Houses. Four explanatory models are offered to explain the osteological findings in view of our present knowledge of the epidemiology of leprosy.

Acknowledgements

Anna-Elisabeth Jensen, archaeologist and curator at the Falsters Minder Museum in Falster, Denmark, is kindly thanked for her help in providing the cultural historical and archaeological data related to the Stubbekøbing cemetery excavation.

References

Andersen, J.G. 1969 *Studies in the Mediaeval Diagnosis of Leprosy in Denmark.* Copenhagen: Costers Bogtrykkeri.

Buikstra, J.E. and Ubelaker, D. (eds) 1994 *Standards for data collection from human skeletal remains.* Arkansas: Archeological Survey, Research Seminar Series 44.

Ell, S.R. 1988 Reconstructing the epidemiology of Medieval leprosy: Preliminary efforts with regard to Scandinavia. *Perspectives in Biology and Medicine* 31 (4): 496-506.

Grabstald H. and Swan L.L. 1952 Genitourinary lesions with special reference to problems of atrophy of testes. *J. American Medical Assoc.*149: 1287-1291.

Irgens L.M. and Skjærven R. 1985 Secular trends in age at onset, sex ratio, and type index in leprosy observed during declining incidence rates. *American J. of Epidemiology* 122:695-795.

Iskikawa, S. 1997 Osteoporosis due to testicular atrophy in male leprosy patients. *Acta Med. Okayama* 51(5): 279-283.

Iskikawa, S. 1999 Osteoporosis in male and female leprosy patients. *Calcified Tissue International* 64: 144-147.

Jensen, A-E. 1994 *Excavation report.* Faalsters, Minder Museum (In Danish). Unpublished.

Judd M.A. and Roberts, C.A.1998 Fracture patterns at the Medieval leper hospital in Chichester. *Amer. J. Phys. Anthropology* 105: 43-55.

Lapinsky, S.E. 1992 Anaemia, iron-related measurements and erythropoietin levels in untreated patients with active leprosy. *J. Internal Medicine* 232: 273-278.

Lewis, M.E., Roberts, CA. and Manchester, K. 1995 Inflammatory bone changes in leprous skeletons from the Medieval Hospital of St. James and St. Mary Magdalene, Chichester, England. *International J. of Leprosy* 63 (1): 77-85.

Mittler D.M. and Van Gerven D.P. 1994 Developmental, diachronic, and demographic analysis of cribra orbitalia in the Medieval Christian populations of Kulubnarti. *Amer. J. Phys. Anthropology* 93: 287-297.

Møller-Christensen, V. 1961 *Bone changes in leprosy.* Copenhagen: Munksgaard.

Møller-Christensen, V. 1974 Changes in the anterior nasal spine and the alveolar process of the maxillae in leprosy. A clinical examination. *Int. J. Leprosy* 42(4): 431-435.

Nah, S.H., Marks, S.C. and Subramaniam, K. 1985 Relationship between the loss of maxillary anterior alveolar bone and the duration of untreated lepromatous leprosy in Malaysia. *Leprosy Review* 56: 51-55.

Stuart-Macadam, P. 1985 Porotic hyperostosis: representative of a childhood condition. *Amer. J Phys. Anthropology* 66: 391-398.

Stuart-Macadam, P. 1992 Porotic hyperostosis: a new perspective. *Amer. J. Phys. Anthropology* 87:39-47

Table 1 Age and sex distribution of the skeletal material

Age (years)	Subadults	Female	Male	Sex unknown
0-1	2			
1-6	6			
7-12	5			
13-19	2			
20-30		5	2	
31-40		8	3	
41-50		3	8	
>50			3	
>18			4	9
Total	15	16	20	9

Table 2 Pathological changes encountered

	N	Individs.	% of the adults
Periosteal reaction	9	9	20.0
Fractures	20	6	13.3
Arthrosis	15	15	33.3
Osteochondritis dissecans	4	3	6.7
Caries	10	10	22.2
Cribra orbitalis	2	2	0
Ankylosis of sacroiliac jt.	2	2	4.4
Reactive arthritis	1	1	2.2
Otitis media	1	1	2.2

N = number of cases
Individs. = number of individuals

Table 3 Fractures

Bone	N	Fractures	%	R+L bones	R+L fracures	%
Clavicle R	26	1	3.8			
L	23	0	0	49	1	2.0
Humerus R	28	1	3.6			
L	25	0	0	53	1	1.9
Ulna R	30	0	0			
L	30	1	3.3	60	1	1.7
Radius R	26	0	0			
L	27	1	3.7	53	1	1.9
Femur R	26	0	0			
L	29	0	0	55	0	0
Tibia R	28	1	3.6			
L	26	0	0	54	1	1.9
Fibula R	30	2	3.3			
L	25	0	0	55	2	3.6
Total				**379**	**7**	**1.8**

N = Number of bones
R = Right
L = Left

The stigma of leprosy: its historical origins and consequences with particular reference to the laws of Wales

John Cule
Abereinon
Capel Dewi
Llandysul
Ceredigion
SA 44 4PP
Wales

'Ni bydd atglaf o glafwr' (A "leper" has no convalescence). Welsh adage.

1. The Leprosy Colony

In the year 1873, the year that Armauer Hansen identified the causative organism, a Belgian priest, Father Damien arrived in Hawaii to work amongst the 800 leprosy patients there. He was revolted by the conditions "of filth and moral degradation", a significantly pejorative association. Nearly all "were prostrate on their beds, in damp grass huts, and their constitutions badly broken down. The smell of their filth, mixed with the exhalation of their sores, was simply disgusting, and unbearable for a newcomer". Rather unusually, for few missionaries did, McDougall records that Damien himself developed lepromatous leprosy, on that Hawaian island of Molokai three years later, and died there in 1889 (McDougall, 1996). There was yet no specific treatment for a disease known for more than two thousand years, when "lepers" still, as the proverb says, had no convalescence.

Although this was a mass picture presented by 800 sufferers, it had long since needed only the sight of a single sufferer to produce fear and loathing. The mutilated victim and thoughts of the continuing misery to which disabilities would reduce a person with leprosy, reminded the onlooker that proximity was said, by all who "knew" about these things, to be dangerous. An idea of its transmissibility had long been appreciated, and could become their own fate.

2. Classical descriptions of leprosy

Aretaeus of Cappadox (AD 150-200) gave words in the latter half of the second century AD to enhance the stigma when he wrote, "And since [those infected with this disease] are like this, who would not shun them, or who would not turn away in disgust, even if he were a father, or a son, or a brother? Since there is a fear lest the disease should be passed on by contact. Hence, many have banished men that were dearest to them into mountains and deserts." (Adams, 1856).

This response was triggered by the immediate physical revulsion felt toward the disease known to Aretaeus as *elephantiasis*, because the potential appearance of somebody with leprosy was thought to resemble the beast of the same name. This term was used by later Greek physicians, becoming lengthened to *elephantiasis aretaeus* or *elephantiasis graecorum*. The clinical identity of this *elephantiasis* to modern leprosy was affirmed by Simpson (1872 - of chloroform fame). In his book "On the causes and symptoms of chronic diseases", after describing leprosy's points of resemblance to the elephant, Aretaeus continues, 'The disease is also called Leo, on account of the swelling and toughening of the face.....; and Satyriasis from the redness of the cheeks, and the irresistible and shameless impulse *ad coitum*. Moreover, it is also called the Heracleian affection, insomuch as there is none greater and stronger than it' (Adams, 1856:368-373).

The profligate sexual behaviour, again referred to by Aretaeus as 'libidinous desires of a rabid nature" was also attributed in more modern times to sufferers from tuberculosis, but the licentiousness of a "leper" became a byword in the Middle Ages. Dols (1979) discusses the mystery why Aretaeus, a contemporary of Galen, does not refer to the humoral theory or mention Galen's views in his description. The Greeks left a basic pathogenesis for the Arabs to learn, describing a polymorphous disease but somewhat weak clinically on the skin lesions, failing to note the characteristic loss of cutaneous sensation (*ibid*).

3. The Arabs

The Arabs recognised lepromatous leprosy as *judham*, the Arabic root of which means to cut off or mutilate. Dols (*ibid*) states that 'the neurological signs and symptoms of the disease were noted by Albucasis' (Abu' l-Qásim -az-Zahráwi, b 936 near Cordoba). There is some confusion about the Arab ideas of its contagiousness. A number of pious traditions of the Prophet assert the spread by contagion, which are completely denied by other versions that allege that the disease comes directly from God. This matter, like many others concerning the history of leprosy, is still not resolved. The general medical opinion of the time appears to have believed in its mild contagious nature, while preserving ideas of its hereditary transmission, but not necessarily accepting its cause as resulting from illicit sexual intercourse as was the belief of many in Europe (*ibid*). Contagion was the major cause for separating the "leper" from society in Hellenistic medicine.

To the fear of contagion was added a dread of social and moral defilement, as noted by Aretaeus. The development of these ideas in Christian Europe is instructive when compared with the perception of leprosy in contemporary Muslim Society. The Mediaeval Church greatly accentuated the moral aspect of the disease; leprosy became a divine punishment for the damned sinner (Dols, *ibid*). This brought added misery.

In modern terminology, the term *'elephantiasis'* is reserved for a disease caused by *Filariasis bancrofti*, to which disease the Arabs had already given the name of *da'l fil*. Haly Abbas ('Al-Majusi Ali ibn al-Abbas d. 994 AD) in his famous work *Kamil al Sinah* or *Liber Regius* mentions *'judham'* but not *da'l fil*, attributing it to a collection of black bile (Volume 1, p.310).

Avicenna (980-1037 AD) referring to the nomenclature of diseases in his *Canon* [*l-Quanun*, Volume I, p.78, Bulaq, 1877] refers to two diseases which resemble animals, namely *d'al fil* (elephant disease) or "elephantiasis" and *d'al Asad* (lion disease) or "leontiasis". He goes on to describe the disease *al-Judham* [Volume III, p140] as *d'al Asad* or "leontiasis", '... for it gives a glowering expression to the face of the sufferer and makes him look like a lion (Isaacs, pers. comm.). To make the confusion worse *elephantiasis graecorum* needs to be distinguished from the term *lepra Arabum*, which is synonymous with the *elephantiasis* of modern terminology.

4. Biblical leprosy

The above had nothing to do with Biblical leprosy, an unfortunate misnomer for *tsara'ath*, a spiritual uncleanness remediable, if at all, by spiritual means administered by a priest or rabbi (Browne, 1975, Cule, 1970). This persisting confusion was introduced into the nomenclature 400 years earlier (c. 200 BC) by the biblical translators of the *Septuagint*, perpetuated in the *Vulgate* some 600 years later. This misnomer has been frequently discussed and most scholars now accept that the leprosy of the Bible (or *tsara'ath*) was a spiritual uncleanness requiring ritual religious remedies. However, the name and the stigma stuck fast to a condition sufficiently distressing on its own.

The first mention of leprosy in the *Bible* is in *Exodus* (4:6) when one of the signs Moses was to give to Pharaoh was to cause *tsara'ath* to appear on his own hand and then disappear. Miriam was punished for her unjustified criticism of her brother by a brief attack of the condition (*Numbers* 12:10). King Uzziah had the effrontery to offer incense on the inner altar, whereupon *tsara'ath* broke out on his forehead for 'the Lord had smitten him'. Hastily thrust out of the Temple he had to live out the rest of his days in isolation (II *Chronicles* 26:16). These are the blemishes, or the stigmata of the disfavour of the Lord. It was an uncleanness, and a punishment to be regarded with horror. In these circumstances the translators regrettably favoured the translations "leper" and "leprosy", regarding it as a result of *nega*, or smiting, or the manifestation of extreme divine displeasure. If further proof of this incorrect association with its modern meaning is required, it is in the absence of any palaeopathological evidence of the disease, that later became clinically known as leprosy, in Israel in Biblical times (see Crane-Kramer, this volume).

Nevertheless, its association has caused a lasting religious stigma persisting throughout the Middle Ages and not entirely lost today. To add to the misery of the deformed sufferer, the strong biblical condemnation was heeded in the Mediaeval period. It continued as a feature of the much later Victorian era when illness was often regarded as a punishment for sin and wrong-doing, moreover a deserved punishment meriting censure rather than sympathy. This strongly held belief had given Job a problem in the aetiology of his affliction of boils. The matter is even more complicated as the distinction between religious veneration and abhorrence is not always a clear one.

Frazer (1961:620) somewhat surprisingly records that 'the Greeks could not decide whether the Jews worshipped swine or abominated them. On the one hand they might not eat swine, but on the other they might not kill them.' The supposition was that the pig was sacred, and that originally 'the pig was revered rather than abhorred by the Israelites'. Frazer supports this with his claim that 'down to the time of Isaiah some of the Jews used to meet secretly in gardens to eat the flesh of swine and mice as a religious rite'. This led to his conclusion that 'perhaps...all so-called unclean animals were originally sacred'. A similar view has been expressed toward leprosy. Lascaratos (1996) argues that leprosy was designated a sacred disease in Byzantine times under 'the spirit of Christianity and for social reasons'. It was so termed 'because it was considered a test of man by God, while the sacred disease of the ancients (epilepsy) was no longer sacred in a man possessed by demons'.

5. Stigmatisation

The medical historiography of leprosy seems designed to accord with the literary definitions of stigmatisation. Amongst the earliest clinical signs of leprosy are changes in skin pigmentation. Historical accounts are full of descriptions of the sores evident in the more advanced stages of the disease. The *Shorter Oxford English Dictionary* (1933) opens its definitions of "stigma" with 1596, 'a mark made upon the skin by burning with a hot iron...as a token of infamy or subjection'. The second definition betokens, 'a mark of disgrace or infamy; a sign of severe censure or condemnation, regarded as impressed on a person or thing' as in a "brand", the term first used in English in this sense in 1619. Interestingly, it was not until 1632 that it is recorded as representing the crucifixion wounds of Christ, again a term of sanctity, not of anathema. Its direct pathological association appears to have begun in 1661 referring to 'a morbid spot, dot or point on the skin', and it continues so to be used in medical parlance today, without any intention of a slur on the sufferer whatsoever.

6. Differential diagnosis of leprosy

Modern early clinical recognition of leprosy may be alerted by the discoloured skin, the patches of anaesthesia and changes in the nervous system. However, the use of laboratory aids confirms the diagnosis. Before these were available, early signs of leprosy needed to be distinguished from a mulitiplicity of similar skin lesions resulting from a variety of causes, including neurofibromatosis, erythema nodosum, pellagra, *lupus vulgaris*, filariasis, leishmaniasis, yaws and syphilis. In addition, in the 16th century, leprosy was distinguished in *The Welsh Leech Book* from morphea (*Wellcome Manuscript* 417 f 5 r). This is an extensive differential diagnosis and it would not be surprising if the earlier physicians of history mistakenly included at least some of the diseases with similar features under the clinical title "*lepra*".

Leprosy, the name by which this disease came to be accepted in the Middle Ages, doubtless mistakenly included some of those diseases. Francois Boissier de Sauvages, writing half a millenium later, thought that Gilbertus Anglicus's chapter on leprosy, *De lepra*, in his *Compendium Medicinae* (circa 1270), was the best extant on Greek *elephantiasis* (Sauvages, 1795:301). He described four varieties related to the four humours, which he called *elephantia*, *leonina*, *tyria* and *alopecia*, and modes of infection which included venereal transmission (Gilbertus Anglicus, 1510). Bartolomeus Anglicus (c. 1230-50) voiced the prevalent fear of contagion, amidst his other descriptive theories of its spread (1601:354). Even today, the mode of transmission is not clear, although McDougall (1996) regards inhalation as the most likely route. Talbot (1967:75-76) has pointed out that much of this highly praised account by Bartolomeus can be found in Ricardus Anglicus at the end of 12th century and the even earlier author Haly Abbas (d. 994) in the tenth. There is, therefore, historical as well as palaeopathological evidence for the existence of Hansen's disease in Europe by the Early Middle Ages.

However, it is comforting to consider that the leprosy sufferer was not wholly one of persecution. There is evidence for an apparent alleviation of what was hitherto popularly believed to have been the inevitably miserable social fate accompanying the forced isolation of the afflicted.

7. English Law

The writ in English Law, *De Leproso Amovendo*, already in existence before 1100 AD, stated '*propter contagium morbi predicti*', to which others added '*et propter corporis deformitatem*', before the "leper" was to be removed from society (Coke, 1794), and there was a need to show physical disfigurement. The execution of the writ involved a religious rite, seeming to us nowadays to have a macabre and horrifying unreality, but which was, in fact, similar to the service read for the preparation of a hermit. The priest led the sick person to the church, "as a dead man", covered with a black cloth, and chanting the *Responsorium*, "*Libera me Domine*". After Mass, which was heard on the knees beneath a black cloth covering two trestles, and after the priest had cast earth upon the sufferer as in a burial service, who was 'dead to the world, but alive again unto God'. The person was then taken forth into the open field and had to listen to a list of recited prohibitions (Translation in Clay, 1966, Appendix A). This service, being found in the *Sarum Manual*, was likely to have been in general use in the southern half of England and Wales.

Although the fear of leprosy was great, efforts were made to mitigate the hardships caused by statutory separation. Alms and hospitals were provided, but the plight of the "leper" was still a miserable one, despite the accepted Christian duty of charity, or *caritas*, toward him. This duty was clearly and practically recognised by the physicians, who enjoined care in making a diagnosis, which carried with it such a deprivation of civil liberties. John of Gaddesden (1280?-1361), in his *Rosa Anglica*, asked that 'no one is to be adjudged a "leper", and separated from intercourse of mankind, until the figure and form of the face is actually changed'. The advice seems caring and thoughtful.

It is of course possible that the great fear of infectivity may have been determined by a type of leprosy that was then more infectious than appears to be the case today. However, despite the ever present fear of infection, the practice of the rule for the segregation of a "leper" was not always in accord with the precept. The writ *De Leproso Amovendo* was enforced, but this legislation was often local and not consistent. Exeter allowed the sufferers to roam freely in the town, a custom which was already old in 1163. The law did not attempt to interfere with those who did not flagrantly annoy their neighbours, and kept discreetly to themselves. In many places, despite the avowed purpose of the ("leper") hospital, there were infirmaries and almshouses where afflicted and other inmates were mixed together.

8. Leprosy in Wales and the Border

In my own country of Wales the *Inspexus and Confirmation* of the 1400 deed of foundation of the Cardiff leper hospital was a mixed one for 'leprous, poor and feeble persons'. One has to be a little circumspect about a possible general meaning for "leprous", but the dedication of its chapel to St Mary Magdelen gives added credence (*Patent Rolls*, 1400). The site of the hospital can be seen on Speed's map of Cardiff in the top right corner, outside the East Gate at the end of 'Cokkerton Stret' – 'to spite' (hospital). This reminds one that there are further corruptions in Wales of Magdelen to Merlin. Merlin's Bridge in Haverfordwest was originally Maudlins Bridge near the "leper" hospital. Furthermore, the frequent use in place names of *claf*, usually a general term for "a sick person", or "diseased" as an adjective to signify "leper" or "leprous" can be confusing. The 16th century dictionary of Syr Thomas Wiliems, himself a practising physician, gives *gwahan-glwyf* (separation sickness) and *brêch yr Iddewon* (Jew's Pox) amongst the terms meaning leprosy (Figure 1). The last indigenous sufferer known as a "leper" in Wales was a Tenby man called Alson Walter. In 1576, Thomas Brown gave evidence at a hearing of the Causes in the Star Chamber that for over forty years he had known of this man as the only residential "leper" at the Tenby Hospital. He had been there for twelve years and was paid 'a groat a week, and some weeks two groats, and some weeks none'. Interestingly, this hospital had been transferred three years earlier with a rental value of 65 shillings to one Robert Recorde, a nephew of Dr Robert Recorde, the mathematical physician of that town (RCAHMC, 1925:397-400).

8.1 A "leper" hospital seal

The seal of Saint Giles, patron saint of the "leper" hospital at Salop, in the Welsh Borders, shows two figures: the hind, by whose milk Saint Giles was nurtured in the desert and a squarish object representing the clock dish or alms basket with clapper, which "lepers" were obliged to employ in begging charity. The other Salopians paid for their admission. The poverty of the thirteenth and early fourteenth centuries reduced many "leper" hospitals and other charities to such penurious conditions, that they were unable to maintain their functions.

8.2 Punishment of the "leper"

Rigid regulations were applied to all receivers of charity, whether leprous or pauper, who were provided with places in almshouses. None other was perhaps quite so retributive as the punishment for defaulting "lepers" of the Greenside Hospital, Edinburgh. There the penalty for infringement of the rules confining the leprous within its walls was death. That this might not be seen as an empty threat, a gallows was erected for the immediate execution of offenders (Arnot, 1799:258). At the other end of the scale, the penalty exacted from recalcitrant "lepers" at Shireburn, was somewhat counterproductive, being expulsion from the hospital (Surtees, 1816:284-286).

In Wales, there was an almshouse for the "lepers" at Ystrad Marchell, founded by Owen Cyfeiliog, 'that was burnt because of their fornication' (Brooke o Vowdwy, 1902:355). This exciting reason may well have been an excuse for appropriating the funds. Early in the reign of Edward III, the Welsh monks of the neighbouring abbey of Strata Marcella had been removed because they had lived *vitam enormiter dissolutam*. It is perhaps not without significance that they were replaced by English monks (Phillimore, 1920:245).

It is likely that the restrictions placed on the "lepers" in these institutions, although built for their care and protection, were not always adequately compensated for by the extent of the maintenance offered. The "leper" sometimes was forced to beg, to supplement the meagre charity of the foundation. If the main support was found to be in the bowl, the forbidden tavern might seem jollier for a person with better prospects, than praying and cleaning-up at the hospital, with its score of other irksome duties. Expulsion for indiscretions might then become a desirable end by enjoyable means. As an unknown verminous beggar, expelled "lepers", away from their home village, would need overt gross stigmata to be recognised for what they were, particularly if they chose to abandon their distinctive dress. Perhaps if employment was sought, there was a risk that marks would be recognised by an employer, but the penalties of discovery were no worse than the their original state. Guy de Chauliac (1300-1368) recognised the problems caused by a positive diagnosis of the disease and called for great circumspection in the examination and judgement of "lepers" (Ogden, 1971:377-389).

8.3 The Laws of Wales

In the ancient Laws of Wales, known as the *Laws of Hywel Dda*, illustrated in the beautiful *Peniarth Ms 28*, a "leper" was considered not a "whole man", and because of this precept consequently was not allowed to take the office of a judge. 'There are three persons no one of whom by law can be a qualified judge. One of them is a person with a defect, such as being deaf, or blind or leprous, or an insane person....' (Richards, 1954:99). However, the promulgation of the law was not wholly inimical to the interests of the leprous. Although suffering the deprivation of certain privileges as a consequence of the loss of civil rights, a person with leprosy was exempted from the payment of certain penalties. The "leper" was exempted from the penalties of being a relative of a murderer, in an equal degree to exclusion from the privileges attached to being a relative of the murdered.

The Laws meticulously controlled revenge, a troublesome Welsh custom, which was killing off or maiming the king's subjects, by setting out in detail the payment for a killing. This payment, known as *galanas*, was made by the family of the killer to that of the killed, within specified relationships on both the male and female sides. The "leper" was exempted from the obligations of galanas in *Llyfr Blegywryd*: 'If there be a relative of the murderer or of the murdered who is a cleric in holy orders, or a leper, or dumb, or an idiot, he neither pays nor receives any of the *galanas*. And they are not to take vengeance for a person murdered, and vengeance is not to be taken on any one of them for *galanas*, and they cannot in any way be compelled to pay or receive anything for galanas' (Richards, 1954:46). There are also interesting references concerning the right of a wife in the Welsh *Laws of Hywel Dda* to leave her leprous husband and retain her dowry. As usual, there were three defensible grounds in such cases. Separation was legally permitted for leprosy, for lack of marital relations or for foul breath.

8.3.1 Canon Law

Hywel Dda was said to have journeyed to Rome to seek the Pope's *nihil obstat* on his legal draft. Papal attitudes toward the leprous and marriage were important to Welsh catholics. Any sanction of Rome for the wife to leave her leprous husband, by the end of the twelfth century, seems to have become in direct contravention of Canon Law. Pope Alexander III commanded the Archbishop of Canterbury in 1180 to use his power to conserve the marriage of the leprous, *Lepra superveniens non dissolvit matrimonium*. 'Our command is that you should not delay in strongly persuading wives to follow husbands and husbands to follow wives who have contracted leprosy, and to care for them with conjugal love. But if they cannot be persuaded to do this you should strictly order both of them to observe continence whilst either of them is alive. And if they refuse to accept your order, you should excommunicate them' (Friedberg, 1955).

Alexander III also decreed that "lepers" should not be excluded from marriage and that they may marry and have intercourse (Baionensi Episcopo). 'But if lepers do not wish to remain continent, and they find someone willing to marry them, they are allowed to enter into matrimony' and also, 'If it happens that a man or wife should by divine judgement become a leper, and the sick spouse demands marital rights from the healthy partner, what is demanded must by general Apostolick precept be paid; to which precept no exception in this case may be found' (Friedberg, 1955). However, in 1186, Pope Urban III, in a decretal to the Bishop of Florence, found that leprosy was sufficient cause for betrothed persons not to marry, 'Betrothal cannot compel people to get married, if leprosy occurs between the time of betrothal and the proposed marriage' (*ibid*).

9. The Stigma

For hundreds of years the stigma attached to patients known as "lepers" was associated with the blemish of *tsara ath*, a Biblical mistranslation. The shame became as great as the fear of the disease itself, until it became part of the syndrome. The Jews of Jerusalem changed the name of the Leper Hospital from the Hospital for the Tsara'atish to the Hospital for Hansen's Disease, although the reason for this is disputed. However, there is a more recently revealed stigma, adding somewhat ironically to the new title of the Jerusalem Hospital. Armauer Hansen was suspended by his fellow countrymen from practising medicine, when he was stigmatised following his instillation of a culture of *Mycobacterium leprae* into a patient's eye to study its effect. The patient already had tuberculoid leprosy, but did not develop lepromatous leprosy, nor was her eye affected (L.M. Irgens, personal communication).

Authors note

My interest in leprosy was stimulated by my earlier research into the history of leprosy in Wales and the Marches. Much of the present paper is directly derived from work at that time. A larger reference list will be found in Cule (1970).

References

Adams, F. (editor and translator) 1856 *The extant works of Aretaeus, the Cappadocian*. London: Sydenham Society (Note: Wigan's edition gives the Roman form "Herculean" in place of "Heracleian" in Aretaeus' *Cappadox De Causis et Signis Acutorum et Diuturnorum Morborum Libri Quatuor*. Liber Secundus, p. 91, ed.Wigan, J., Oxford, 1723).

Anglicus, Bartolomaeus 1601 *De rerum proprietatibus* (c.1240). Minerva reprint. Frankfurt edition.

Anglicus, Gilbertus 1510 *Compendium medicinae*. Fo. 337-346, fo. 344. Lyons.

Arnot, H. 1799 *History of Edinburgh*.

Brooke o Vowdwy, *dlyvyr John Brooke o Wowdwy* 1902 *Wrexham MS 1. Hist. MSS Committee Report on MSS in the Welsh Language*. Volume II, Part I ('a losgwyd am odineb y kleivion').

Browne, S.G. 1975 Some aspects of the history of leprosy: The Leprosie of Yesterday. *Proc. Royal Society of Med.* 68: 485-493.

Clay, R. M. 1966 *Mediaeval Hospitals of England.* London: Frank Cass (Appendix A (translation): *Manuale ad Usum Insignis Ecclesiae Sarum* in the York Manual. The Publications of the Surtees Society, 1874, Volume 63, pp. 105-7).

Coke, E. 1794 *The Institutes of the Laws of England. The First Part*. 15th edition, Lib.2, Cap. II, Section 201. 135b. London.

Cule, J 1970 The diagnosis, care and treatment of leprosy in Wales and the Border in the Middle Ages. *Transactions of the British Society for the History of Pharmacy* 1(1): 29-58.

De Sauvages, F.B. 1795 *Nosologia Methodica*. Ord. V, Volume 2. Venice.

Dols, M.W. 1979 Leprosy in Mediaeval Arabic medicine'. *Journal of the History of Medicine* 34(3):314-333.

Frazer, J.G. 1961 *The Golden Bough*. Volume 2. London: MacMillan and Co. Ltd. (St Martin's Library).

Friedberg, E.A. (ed) 1955 *Corpus Juris Canonici, Leipzig: 1876-82*. Reprint, Graz. Volume 2, col. 690. 691. (Baionensi Episcopo).

Lascaratos, J. 1996 The second Sacred Disease: Earlier euphemistic equivalent of Hansen's Disease. *International Journal of Dermatology* 35 (5): 376-378.

McDougall, A.C. 1996 Leprosy. In F.E.G. Cox (ed): *Illustrated history of tropical diseases*. London, The Wellcome Trust, pp. 60-71.

Ogden, M.S. (ed) 1971 *Guy de Chauliac,The Cyrurgie of Guy de Chauliac*. London, New York and Toronto: Oxford University Press (The Early English Text Society).

Patent Rolls 1400 *2 Henry IV. Part II. m.32.*

Phillimore, E. 1920 Mont.Coll., V, 138-141. *Archaeologia Cambrensis* 20 (6th series): 245.

Richards, M. (translation) 1954 *The Laws of Hywel Dda (The Book of Blegywryd)*. Liverpool. Royal Commission on the Ancient and Historical Monuments and Constructions in Wales and Monmouthshire 1925 *An Inventory of the Ancient Monuments in Wales and Monmouthshire. VII - County of Pembroke, London 5.* (Citing P.R.O. Chantry Certificates no.75 South Wales. P.R.O. Court of Augmentations: Particulars for Leases. Co. Pembroke. File 225. m. 88).

Simpson, J.Y. 1872 *Archaeological essays.* Volume 2. Edinburgh.

Surtees, R. 1816 *History and Antiquities of the County Palatine of Durham.* Volume1: Constitutiones Domus de Shireburn. London.

Talbot, C.H. 1967 *Medicine in Mediaeval England.* London: Oldbourne History of Science Library.

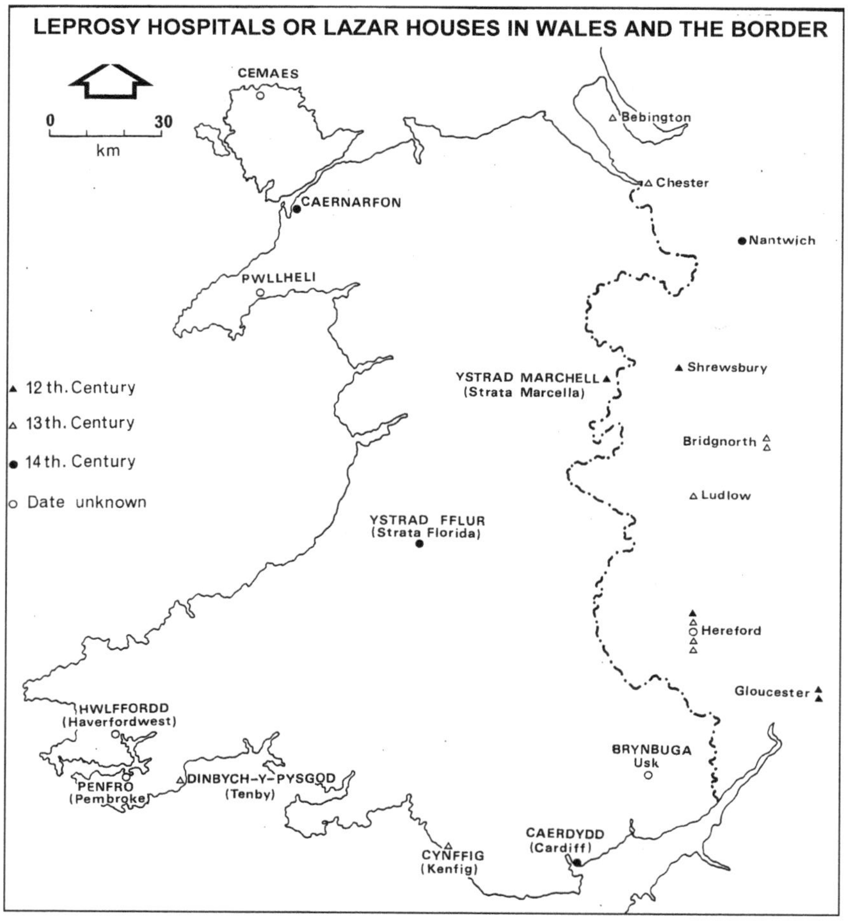

Figure 1: Location map of Leprosy Hospitals or Lazar Houses in Wales and the Border.

The history of leprosy in the territory of the Czech Republic

Milan Dokládal
Department of Anatomy
Division of Physical Anthropology
Faculty of Medicine
Masaryk University
CS - 62500 BRNO
Kamenice 3
Czech Republic

Abstract

Although the greatest number of people with leprosy were found in Western Europe, leprosy epidemics in the 12th to 14th centuries did not avoid Central Europe. The author assesses the number and distribution of leprosy hospitals in the territory of the Czech Republic. More detailed information is given on the most important leprosy hospitals in the country (Prague and in Brno) founded by the Holy Order of St. Lazarus. Finds of skeletons with leprous changes are, in the Czech Republic for the present time, very rare and not reliably verified.

Keywords

Leprosy, history, leprosy hospitals, Czech Republic.

In the Middle Ages, leprosy was a common disease in almost all parts of Europe. Its frequency and dissemination oscillated above all during the Crusades in the 11th to 13th centuries, and leprosy epidemics culminated in the 14th century. At this time several thousand leprosy hospitals existed in Europe. Although the highest number of leprosy patients were registered in Western and Southern Europe (Germany, France, Italy, Spain etc.) leprosy epidemics did not avoid Central and Eastern Europe. In today's territory of the Czech Republic, the historical lands of Bohemia and Moravia, before the year 1992, in the western parts of Czechoslovakia, also reported many thousands of leprosy sufferers, but the exact number is not known and neither is the number of leprosy hospitals. As written documents on the epidemics of leprosy are very rare and quite incomplete, it was very difficult to compile the following data on the palaeoepidemiology of this disease from the numbers reported in various Czech historical publications, and in selected papers on-the history of medicine.

The oldest, largest and most important leprosy hospitals in Czechoslovakia were in Prague and Brno, which were, in the Middle Ages, the two most important centres of Bohemia and Moravia. In Prague leprosy sufferers were concentrated in the area of the originally Romanesque church of St. Linhart, in Prague Old Town. The original site of this small church, built in the 12th and 13th centuries, is now only indicated by the naming of Linhart's street today. The church was called "*Capella pauperum aegrotorum*", and was the first "asylum" for leprous patients in Prague from the end of the 12th century, and it existed also into the 13th century. When St. Linhart's church was redeveloped into a gothic church, leprous patients were moved out from the inner town to outside the city walls close to St. Lazarus Church where a new leprosy hospital was founded by an unknown benefactor in around AD 1281. At the beginning of the 14th century, at the end of the Přemyslides Dynasty, two deep pits were excavated close to this church, where it is believed leprosy cadavers were buried together with victims of a famine (Nový, 1987, Rozsívalová, 1991, Secký, 1928, Štech *et al.*, 1946).

In Brno, the historical capital of Moravia, there were two leprosy hospitals. The earlier information on the leprosy hospital on Krenová (Kröna) Street is from the year 1293. It was officially called "*Capella leprosorum*". The founding of this leprosy hospital was the result of a contribution by an unknown owner of the rich St. Peter's estate and farm, situated in the same part of the town. The second leprosy hospital in Brno was part of the hospital of St. Stephen's Church, founded in the first half of the 14th century. It was situated outside the centre of the town walls close to the Svratka River. These leprosy hospitals were, at the end of the 14th century, combined to form the first town hospital (Zapletal, 1952, Obermajer, 1994). The collective administration of both leprosy hospitals and the town hospitals is noted in the historical town archives of Brno and Moravia from 1344. St. Stephen's hospital was for a long time called "*Hospitale leprosorum*" but the majority of the hospital patients were gradually replaced by patients suffering from other infectious diseases. The largest hospital of the Middle Ages in Brno was situated in today's Pekarská street; St. Ann's Hospital was built at the site of an old leprosy hospital in the 13th century. The Brno leprosy hospitals had very close contacts with the three biggest leprosy hospitals in Slovakia (Bratislava, Nitra and Levoča), and with the leprosy hospitals in Austria (Spital Klagbaum Leprahouse in Vienna, Landstrasse, Mülk, Selzbach, Salzburg etc.).

Other leprosy hospitals and centres were founded and existed for a relatively shorter time in the 13th and 14th centuries in several smaller historic towns of Bohemia and Moravia: Jindrichuv Hradec in Southern Bohemia,

Litomyšl, Chrudim and Kutná Hora in Eastern Bohemia, Olomouc, and Znojmo in Central and Southern Moravia, etc. However, more detailed information about these leprosy hospitals is not available. In the monastery of Zderaz, in the countryside not far from Prague, monks carried out spiritual services for the patients and poor in the St Lazarus leprosy hospital.

Most of the leprosy hospitals in the territory of the Czech Republic were founded and supported by the French Knighthood Order of St. Lazarus (Ordre Militaire et Hospitaliaire de Saint Lazare). This Order was founded in Palestine in the 10th century and, in 1154, it was transmitted to Paris via King Louis VII, and later to many countries of Europe. The Order's main aim was to found leprosy hospitals and to treat the leprosy patients. Leprosy epidemics ended in Central Europe in the 14th and 15th centuries. In the 16th century there were no more leprosy reported in the territory of the Czech Republic. The only patient treated for leprosy in the Czech. Republic in the 20th century died in 1937 (he caught the infection during his stay in South America).

Finds of skeletons with leprous changes in the Czech Republic are, for the present, very rare (and many of them are not reliably verified). There has not yet been the opportunity to open medieval charnel-houses situated close to the largest leprosy hospitals and their cemeteries. Although there are no people suffering from leprosy in the Czech Republic today, in Prague an important charity organization supports the fight against leprosy in Asia and Africa, and Czech missionaries and doctors work in different countries throughout the world to help to combat leprosy.

References

Nový, R. 1987 The hospitals in the pre-husite period in Prague (In Czech). *Documenta Pragensia* 7: 63.

Obermajer, J. 1994 The leper-houses sometimes and today. (In Czech). *Zdravi* 14: 20 - 21.

Rozsívalová, E. 1991 Some remarks to the history of leprosy. (In Czech). *Casopis lékaru ceských* 120: 701-702.

Secký, R. 1928 *The Hospitals in Old Prague*. Prague (In Czech).

Štech, V.V., Wirth, Z. and Vojtíšek, V. 1946 *Past Prague*. Prague (In Czech).

Zapletal, V. 1952 The old hospitals in Brno (In Czech). *Lékařské Listy* 7:347-351.

The palaeoepidemiology of leprosy: an overview

Michel F. Lechat
109 Rue des Trois Tilleuls
B-1170 Brussels
Belgium

Email : michel-francois.lechat@epid.ucl.ac.be

Abstract

This paper reviews the distribution and spread of leprosy in populations from the dawn of history until the advent of effective therapy in the mid-nineteenth century. Special attention is given to the expansion and subsequent decline of the disease in Europe, its introduction in the New World, and its diffusion in secluded areas, islands, genetic isolates or migrants. Epidemiological patterns as recorded by history suggest hypotheses regarding the dissemination of leprosy in prehistory, as well as inviting speculations on its mode of transmission.

Keywords

Leprosy, epidemiology, elephantiasis, Mycobacteria, lazaret

NB: This paper has subsequently been published in a slightly different form in *International J. Leprosy* 67(4):460-470 (1999)

1. Introduction: palaeoepidemiology and the ecology of diseases

The study of the palaeoepidemiology of an infectious disease should ideally start by tracing its origin, that is the prime encounter of humans with the germ, its reservoirs and its vectors, under the conditions of a long vanished ecology. Next, the subsequent interplay of the host and the parasite should be followed as they travel in concert through the ages and places, overtaking new populations.

For diseases of recent emergence, the matter is often hard enough. However, through observation and experimentation, some hypotheses on their origin and mechanism of transmission can at least be tested. For ancient diseases, however, the approach can be purely speculative. Let us take cutaneous leishmaniasis as an example. In a recent book on the history of tropical diseases (Miles, 1996:206-229), the following account is given of its origin: 'Sandflies have been in existence for 30 million years and, feeding on the juice of plants in which flagellate protozoa live, they presumably became infected with the ancestors of Leishmania parasites, ... lizards, and later, small rodents became the sandflies' chosen prey ... (then) man as a hunter-gatherer, became infected when he came into contact with these rodents...'

The more complex the natural history of a disease, the easier it becomes to propose some sort of sensible scenario for its origin. This may seem a paradox. The reason however is simple. The epidemiological factors to be taken into account, such as the biology of an obligatory vector, the dispersion of an intermediate host, or the biotope of an animal reservoir, raise many constraints which curtail our freedom to make hypotheses. Concerning leprosy, for this ill-famed centuries-old-epitome-of-all-scourges, there is no clue to lead us to reasonable guesswork, no mosquito and no extra-human reservoir but for an occasional monkey or an exotic mammal such as the armadillo. Unless further evidence proves otherwise, humans are the only host, as well as the only reservoir, for the etiological agent *Mycobacterium leprae*. Since the bacillus has never been cultivated *in vitro*, it looks as though it has lost the genetic machinery required to grow outside the human body. It may thus be safely assumed that it is an old mycobacterium that has diverged and became an obligatory parasite of humans early in the course of evolution. The nearly completed deciphering of its genome and the comparison of it with that of *M. tuberculosis*, and possibly of other mycobacteria, should help to elucidate its origin.

Transmission of leprosy is rather unimaginative: people to people, from an infected individual to a susceptible person. Leprosy is a cosmopolitan disease. Notwithstanding differences in climates and environments, at one moment or another it has occurred in all types of habitats occupied by humans. While there is no suggestive palaeontological evidence regarding its transmission in proto-historic populations, its spread and geographic distribution over the centuries show a number of singularities. Yet, historical observations may possibly provide some clues for broadening the debate.

2. Early records of leprosy at the dawn of history

Written records of diseases with signs and symptoms resembling those of leprosy suggest that it could have existed in ancient times in Egypt, in India, and in China. In Egypt, Huspati (Horus-Dên), a quasi-mythical Thinite king of the First Dynasty (c.3500 BC), is reported in a papyrus dating from ± 1500 - 1200 BC as having suffered from a disease whose signs are evocative of leprosy (Scott, 1943). In India, a precise clinical description was given in a compilation of texts dating from about 600 BC (Sushraka Samhita) – Dharmendra (1978). In China, the Nei Ching, a collection of medical texts assembled between the 8th and the 3rd century BC, but attributed to Huang-Ti, a legendary emperor who lived around 2600 BC (Skinses, 1985), lists among the signs of leprosy the loss of eyebrows together with nodules and ulcerations of the skin.

That the records attesting to the existence of leprosy in earlier times were actually compiled much later does not necessarily preclude their veracity, for they were generally meant to be a recollection of traditional knowledge. The congruence of the given descriptions with the clinical signs of the disease is a much more delicate issue. For example, according to Biblical experts, the condition described in Leviticus under the term of "Tsara'ath" does not refer specifically to leprosy but is said to encompass a large variety of skin diseases or blemishes of all kinds (Jeanselme, 1934). Archaeological finds suggestive of leprosy dating from these early periods are few. In Egypt, indisputable evidence of the existence of leprosy is relatively recent. Four skulls from the 2nd century BC present lesions of the nasal bones specific for leprosy (Dzierzykray-Rogalski, 1980). There is therefore no conclusive evidence to settle the discussion about of the first appearance of leprosy.

3. Expansion of leprosy in history and the invasion of Europe

While leprosy presumably already existed in Egypt and further East in early history, it made a belated entry into Greece and Rome. Literary sources of Greek and Latin origins raise peculiar problems of semantics. The group of skin conditions referred to as "lepra" in the Hippocratic texts does not apparently include the disease we have come to know in modern parlance as leprosy. Conversely, the disease entity most precisely and accurately described by the Latin authors Celsus (25 BC - 37 AD) and especially Areteius of Cappadocia (c. 200 AD) that we may recognise as leprosy, is designated by the term "elephantiasis" (Andersen, 1969). According to Pliny the Elder (23 - 79 AD), elephantiasis, by now designating leprosy, was introduced in Rome around 62 BC by the armies of Pompey returning from the campaign against Mithridates, King of Pont (*ibid*). That the ailment was not common in Rome may be surmised from the fact that it was not included in the defects rescinding the sale of slaves, as was the case for phthisis (tuberculosis), fevers, eyesores and mental disorders (Jeanselme, 1934).

Following the Roman conquests, leprosy spread to Gaul as well as to Germany, where Galen (± 150 AD) signals its presence, and thereafter to the whole of Europe during the Middle Ages (Chaussinand, 1955), although the possibility of a prior contamination of Spain by the Phoenicians as early as the 10th Century BC has been theorized by some authors (Contreras and Miquel, 1973). The Vikings brought it to Scandinavia. In Iceland, a bishop was to be deposed in 1413 because his deformities prevented him from celebrating Mass (Jeanselme, 1934). At the time of the Crusades, the bacillus shuttled back and forth between Europe and the Near East.

The fear of contagion was great. Gloomy rituals were set up to exclude patients from social life. Leprosy hospitals were reportedly built by the thousands from Spain to Scotland, which at least attest to a wide dissemination of the disease. However, figures of that sort can be misleading. In all likelihood, judging from their ruins, most of these institutions were housing a small number of patients at a time, and most probably not all were suffering from leprosy. That at least some of the people consigned to these leprosy hospitals were actually affected with leprosy is, however, confirmed by the excavations conducted in Denmark by Møller-Christensen in the 1950s which demonstrated characteristic bone changes in a number of skeletons (e.g. see Møller-Christensen *et al.*, 1952). In addition, this author identified specific alterations in the maxillary bones which, after having been validated radiologically in living patients (Melsom, 1953, Lechat and Chardome, 1955), can now be used as a standard to verify the diagnosis of leprosy in osteoarchaeological material.

There are two main conclusions to this story. Firstly, it is most likely that leprosy had not been previously introduced to Europe by Indo-Europeans during the successive waves of migrations, be it the Celts, the Germans, the Greeks or the Italians. Either the disease did not, or did not yet, exist at the site where these migrations originated, somewhere in the steppes or the plateaus of Central Asia 7000 or 8000 years ago, or the conditions for its transmission were not met during the long westward trek and *M. leprae* "got lost" *en route*. Secondly, there is no climatic or other geographic factor determining the distribution of leprosy. From the sun-scorched banks of the Nile to the frost of Iceland, the determining factor responsible for the movement of the bacillus are humans, with their distribution, mode of life and social structure.

4. Leprosy in the western hemisphere

After having reached its peak in Europe, leprosy worked its way into fresh areas in the Western Hemisphere. There is no record of the disease being prevalent in the First Nations people before the arrival of Europeans. Since the conquistadores and the first colonisers came from regions of Europe, mainly Spain and Portugal, where leprosy at the time was not uncommon, and were in addition accompanied by physicians and priests supposedly well acquainted with its signs and symptoms, it is unlikely that cases of the disease in native populations would have escaped attention. Anthropomorphic earthenware from the pre-Colombian era once believed to represent leprosy lesions are now said to suggest rather leishmaniasis, ("*espundi*") or blastomycosis (Krumbach, 1986).

In conclusion, it is accepted that "mongoloid" people who

migrated across the Bering Strait landbridge some 20,000 years ago did not carry the leprosy bacillus with them. It came to the Americas with European invaders first, via slaves from Africa second, and finally with labourers imported from China. Among the Conquistadores and the first colonisers, Jimenez de Quesada, the founder of Bogota, was affected and contaminated some persons of his entourage (Contreras and Miquel, 1973). It is, however, the slave trade which apparently played the major role in the introduction of leprosy into the New World. The number of slaves deported from trading stations in Africa between 1511 and 1787 oscillates between 10 million and, based on figures from the British Admiralty, 50 million (Gonzalez Prendes, 1963). Most of the slaves were imported from the Western coasts of Africa extending from Senegambia (Senegal) to Benguella (Angola), and from as far away as Mozambique on the Indian Ocean. It is not known how common leprosy was at that time in that part of the world. Judging from the high prevalence recorded a few centuries later during the colonial period, it can be presumed that the disease had been widely disseminated in Africa for a long time.

The number of these so-called involuntary migrants who ultimately reached the American shores was, however, much less than those captured in the African hinterland. "Sanitary control" was carried out before shipment to exclude people in poor health or with visible defects, although it is reported that in later years, as the contraband trade developed, the prevalence of diseases among smuggled slaves increased sharply (Scott, 1943). Of those embarked, it is said that a large proportion, at times up to more than one half of the load, died or were otherwise disposed of during the ocean crossing or the "Middle Passage" as it came to be called, insurance policies being an additional incentive not to keep sick patients on board until termination of the journey (*ibid.*). Despite the above checks, the number of leprosy patients among the slaves delivered at their destination would not have been insignificant. In addition, some of them still in the latent stage, having been infected in their homeland, were bound to develop the clinical disease after arrival.

That the disease was present among the African slaves is attested by the fact that in Cuba the bills of sale stipulated that for leprosy, and likewise for mental disorders, contrary to other ailments, the buyer had the license of returning the object of the transaction within two months if found affected by these conditions (Gonzalez Prendez, 1963).

The immigration of Chinese labourers to Cuba, at the time a colony of the Spanish Crown began in 1847, the very same year the slave trade was officially abolished. In the 57 years that followed, 202 patients native of China were hospitalized in the San Lazaro Hospital in Havana (*ibid.*).

5. Migrations and secluded, remote or otherwise singular areas

5.1 Slow diffusion in contiguous areas and mosaic patterns

The spread of leprosy in areas adjacent to each other was observed as recently as the present century among the aborigines of Wandammen Bay, Irian Jaya and in the Northern territory of Australia (Leiker and Sloane, 1954, Hargrave, 1980). In Western Australia, the disease was reportedly imported by crews of pearl-fishing boats and was propagated inland from the coast with successive intervals setting up small foci along the rivers (Lechat, 1986).

This pattern of diffusion may explain the slow dispersion of leprosy in prehistory, the disease remaining sequestered in secluded populations for untold centuries until some social break-up set it free to flare elsewhere. The more compartmentalized the group, the slower will be the diffusion.

Different opportunities for contact could also explain the fragmented distribution of the disease observed in some parts of Africa, where strongly contrasted prevalences ranging from high to low may coexist in adjacent areas. The same differences are described for tribes of Papua New Guinea living in adjacent valleys. Although the suggestion is admittedly quite speculative, this mosaic pattern or patchwork could reflect, up to the present, some vestige of the respective isolation of tribes until the recent past. The more exchanges between contiguous areas, through trade or otherwise, the faster will be the spread of leprosy and the more homogenous will become the distribution of prevalence.

5.2 Transmigration

There are historical records of the transmigration of populations with a high prevalence of leprosy into an environment already occupied by other groups with low or zero prevalence. Such transplantations may be regarded as "natural experiments" where the spread of the disease may be observed as the two populations come into contact.

The transmigration of leprosy patients does not necessarily result in the dissemination of the disease in the new location. Transmission may remain largely limited to and persist for generations within immigrant groups if they do not mingle with the host population. It is then cultural isolation, reducing the opportunities for infective social contacts, rather than geographic confinement, which will restrain transmission to the outside. An example of such migrations is given by the French colonists who settled in Tracadie, New Brunswick and Louisiana, following their expulsion from Nova Scotia at the time of the "Grand Dérangement" in 1755. The Acadian focus, which is scattered among the parishes of Louisiana, has been active up to this day, with many cases still occurring in the Cajun population (Feldman and Sturdivant, 1973).

An outbreak of high prevalence was also observed in the penal institutions of French Guiana (Floch, 1951). From 1852 until 1939, 71,667 male convicts were transported to this French territory. The number of cases which have occurred since then has not been recorded. A calculation of the incidence for the years 1939 to 1948 yields a figure of 4.15 cases per 1,000 person-years. All ethnic groups, European, Arabs, Asians and Africans, were indiscriminately affected militates due to social factors such as living in overcrowded and unhygienic conditions which were significant determinants in the transmission of the disease.

According to a different pattern, the arrival of patients in a previously unaffected area does not result in the production of new foci. It could be that the newcomers will adopt the mode of life of the host population, as ill-defined as the term might be, which in some way will curb the transmission of the bacillus. Illustrative in this respect is the story of Scandinavian immigrants who settled in the Upper Mississippi Valley in the 19th Century (Washburn, 1950, Doull, 1962). The disease died out after the second generation and social and economic factors have been invoked to explain this observation.

Although not sufficiently documented, these observations seem to confirm that social contact is a determining factor in governing the transmission of the disease. Both of these patterns would be difficult to reconcile with a supposed intervention of an extra-human reservoir.

5.3 Virgin communities: the Pacific Islands

The Pacific Islands constitute a sort of laboratory in which one can watch the introduction of leprosy in hitherto virgin populations. The diffusion of leprosy in a number of those islands reveals two principal epidemiological patterns, i.e. outbreaks of leprosy, and transmission limited to a few families.

Outbreaks are characterized by a rapidly increasing incidence following the occurrence of the first cases. After a few years, a large part of the population may be affected. These outbreaks are often subsequent to the arrival of a particular individual said or known to have had leprosy. The legend of the introduction of leprosy into previously unaffected communities by Chinese labourers and crews of whaling ships must be accepted with caution. More interesting examples are those in which the first case is clearly defined, sometimes with the name and the medical history clearly recorded.

The unparalleled story of the Nauru epidemics is truly remarkable (Wade and Ledowsky, 1952). The introduction of leprosy in this, at that time, German colony is ascribed to a woman native of the Gilbert archipelago (now the Republic of Kiribati), who arrived in Nauru in 1911 or 1912. Recognized on arrival as having leprosy and barred from entry for that reason by the medical officer in charge, she was however allowed to stay by the Governor. She took a 13-year-old girl as a servant and died two years later. In 1920, the girl, by then 22, in turn developed leprosy. Soon thereafter three additional cases were detected. All were isolated. A few months later, the pandemic of Spanish influenza reached Nauru, killing 30 per cent of the population as well as all the leprosy patients but one. The next year, 1922, 25 new cases appeared. By the end of that year, after a survey had been conducted among the entire Nauruan population of 1,113 persons, the number of cases diagnosed had increased to 139, yielding a prevalence of 125 per 1,000. Soon, almost every family had at least one member affected. By the end of 1929, no fewer than 438 cases had occurred among the autochthonous population, which corresponds to a 7.5-year period, and a prevalence of 36.5% (1922 to mid 1929). The epidemic thereafter subsided, partly due to the forced emigration of the population to the neighbouring island of Truk as well as to the elimination of the patients during the Second World War occupation by the Japanese military. It is however not yet completely extinct. The latest information reports an average of two cases detected per year between 1982 and 1994 (1994 population: 9,900), in spite of early detection and effective multiple chemotherapy (WHO, 1995).

In other islands the spread of leprosy is reported to have been limited to a few families, as in Niue Island (Dempster, 1949) and Eastern (American) Samoa (Sloan, 1954). These island outbreaks raise interesting points. To what extent may genetic factors or lack of immunity resulting from the absence of exposure to some other mycobacteria explain these different patterns? This remains an unsolved question.

5. Colonia Tovar: a genetic isolate

Colonia Tovar, a small and until recently isolated community in Venezuela, presents a strange illustration of high prevalence coupled with high inbreeding. Between 1843 and 1856, 146 immigrants from Germany settled in this isolated valley in the State of Aragua (Convit *et al.*, 1952, Lechat *et al.*, 1967). In 1950, about a century later, a survey of 1,126 inhabitants (almost the entire resident population) disclosed a prevalence of 100 per 1,000 (113 cases), in sharp contrast with the rest of the State of Aragua where the prevalence did not exceed 1.35 per 1,000.

The original site of contamination remains unknown. Did the disease stem from a contact with an occasional case in the creole population upon arrival in America, or was it present in the immigrant population at their place of origin in the Black Forest? Did the high prevalence result from the existence of some environmental peculiarity specific to the Valley. Could inbreeding have played a role, which would suggest the intervention of genetic factors in the susceptibility to leprosy (Convit *et al.,* 1952)? Was the containment of the disease in the population due to geographic as well as cultural isolation, verging on spontaneous segregation? These questions remain unanswered.

6. Leprosy on the decline

In Western Europe, after having reached a peak in the 12th and 13th centuries, leprosy then started to decline until it had all but disappeared by the turn of the 20th century. Residual foci persisted in the Baltic states, in Greece, in the Iberian Peninsula and in the Balkans. The very term of "decline" is however misleading, since it may apply to two quite distinct epidemiological indicators, that is the prevalence depending on the number of patients affected with the disease at a given moment, or the incidence corresponding to the number of new cases occurring during an interval of time.

An exhaustive study of the annual incidence rates (and cohort-based incidence rates recording the year of birth) for the period 1855 - 1920 has been carried out in Norway by Irgens (1981). In the coastal part of the country where 97.8% of the 6,652 patients with year of onset of leprosy within the observation period were living, the prevalence fell from 35.2 per 10,000 in 1855 to 1.2 in 1920. During the same period, the incidence declined 100 times from 33.4 to 0.3 per 100,000.

What was the cause of the decline? Several reasons have been put forward to explain the downturn in Europe in the late Middle Ages, such as higher fatality rates in leprosy patients during plague epidemics in the 14th century (the Black Death), the modification of eating patterns, or the development of immunity associated with the spread of tuberculosis (Sansaricq, 1965). This last hypothesis deserves more than a cursory mention. Numerous studies have demonstrated some degree of association in the cell-mediated immunity against *M. leprae* (Mitsuda test) and *M. tuberculosis* (PPD). After much debate, there is mounting evidence that vaccination by a combination of BCG with killed *M. leprae* confers a significant degree of protection against leprosy (Karonga Prevention Trial, 1996, Gupte *et al.*, 1998). Competition between the two bacilli could therefore account at least in part not only for the recession of leprosy, but also for the variations observed in its spread and its distribution.

One may indulge in some speculation. A number of observations in animals have led to the general acceptance that exposure to one or another Mycobacterium influences subsequent responses to other mycobacterial species (Stanford, 1978). For example, the observed protection against tuberculosis imparted by BCG is influenced by previous exposure to environmental mycobacteria. Such exposure could either enhance or reduce the susceptibility to *M. leprae.* In this respect, it should be noted that vaccination by a preparation of *M. avium intracellulare* (ICRC vaccine) has been found to confer a degree of protection against leprosy similar to that produced by the combination of BCG with killed *M. leprae* (Gupte *et al.*, 1998). Mycobacteria are ubiquitous in the environment. They could possibly modulate transmission not only by increasing or lowering resistance to infection by *M. leprae*, but also by acting on the poorly elucidated immunological mechanisms which govern the transformation of the subclinical infection into an overt disease, either by shortening or lengthening the duration of the latent stage, and/or by altering the relative proportion of individuals who respectively develop the multibacillary (with high infectivity) or the paucibacillary (with low infectivity) clinical types.

The distribution of a large range of mycobacteria could, therefore, have played a determining role in the susceptibility of individuals and populations to infection by *M. leprae* and, as a consequence, could have exerted an effect on the dissemination of the disease and on its time trends. The palaeoepidemiology of leprosy could be only a chapter in the immunogenicity of mycobacteria, past and present.

7. Conclusions

In the context of the 1991 World Health Assembly (WHA) Resolution on the "Elimination of Leprosy as a Public Health Problem" (World Health Assembly, 1993:117-118), a world-wide programme of treatment with a combination of highly effective drugs (Multiple Drug Therapy, or MDT) was launched.

Over the last fifteen years or so, some 10.7 million patients have indeed been discharged (WHO, 1998a). The number of registered patients has fallen from 5.4 million worldwide in 1985 to about 800,000 at the start of 1998 (WHO, 1998b). Due to the large scale implementation of highly effective drugs for the treatment and cure of leprosy, it is therefore becoming almost impossible to study the epidemiological trends of leprosy under natural conditions.

Nonetheless, there are still perhaps some enclaves, so-to-speak "fossils" of primeval leprosy, where the natural epidemiology of the disease could be observed. In the context of the WHO programme, a most innovative initiative has been launched under the title of SAPEL, or Special Action Projects (WHO, 1998a). It consists of the identification of populations where imaginative and unusual approaches should be designed in order to reach and treat the patients. SAPEL's projects include, among others, pygmies in equatorial forests, nomads in subsaharan deserts, settlers in virgin territories, riverine populations of the Amazon basin and tribal communities in the hills of Southeast Asia. Bridging the present with the past, epidemiological information collected during these projects could improve our understanding of the dynamics of leprosy in remote and little known human groups. Living under conditions which at times still bear some resemblance to the ecology of the past, a study of these communities could contribute both to the study of palaeoepidemiology and to the management of leprosy control in the future. The study of the epidemiology of the past should not only bring an intellectual gratification, but also stimulate us to be better equipped to face new challenges in the future.

References

Anderson, J.G. 1969 *Studies in the Medieval diagnosis of leprosy*. Thesis for the Doctor of Medicine. Danish Medical Bulletin 16 (Suppl.16).

Chaussinand, R. 1955 *La Lèpre*. Paris : Expansion Scientique Française.

Contreras, F. and Miquel, R. 1973 *Historia de la lepra en España*. Madrid : Grafica Hergon

Convit, J., Gonzales, C.L. and Rasi, E. 1952 Estudios sobre la lepra en el grupo etnico alemán de la Colonia Tovar. *Int. J. Leprosy* 20 : 185-193.

Dempster, G.O.L. 1949 Leprosy in Niue Island. A note on the history of the disease. *Int. J. Leprosy* 17 : 411-414

Dharmendra, R. (ed) 1978 *History of spread and decline of leprosy*. Bombay : Kothari Medical Publishing.

Doull, J.A. 1962 The epidemiology of leprosy. Present status and problems. *Int. J. Leprosy* 30: 48-66.

Dzierzykray-Rogalski, T. 1980 Paleopathology of the ptolemaic inhabitants of the Dakhleh Oasis (Egypt). *J. Human. Evol*. 9 : 71-74.

Feldman, R.A. and Sturdivant, M. 1973 *100 years of leprosy in Louisiana: An epidemiological analysis*. Tenth International Leprosy Congress, Bergen, 1973, Abstracts, pp. 116-117.

Floch, H. 1951 La lèpre au bagne Guyanais. Son évolution durant un siècle (1852-1950). Ses particularités. *Int. J. Leprosy* 19 : 283-295.

Gonzalez Prendes, M.A. 1963 *Historia de la lepra en Cuba*. La Habana : Academia de Ciencias de la Rep. de Cuba.

Gupte, M.D. and 13 other authors 1998 Comparative leprosy vaccine trial in South India. *Indian J. Leprosy* 70 : 369-388.

Hargrave, J.C. 1980 *Leprosy in the Northern Territory of Australia. With particular reference to the aborigines of Arnhem Land and the arid regions of the Northern Territory*. Australia : Government Printer of the Northern Territory.

Irgens, L.M. 1981 Leprosy in Norway: An epidemiological study based on a national patient registry. *Leprosy Review* 51 (Suppl.1) : 1-130.

Jeanselme, E. 1934 *La lèpre*. Paris : G. Doin Ed.

Karonga Prevention Trial Group 1996 Randomised controlled trial of single BCG, repeated BCG, or combined BCG and killed *Mycobacterium leprae* vaccine for prevention of leprosy and tuberculosis in Malawi. *Lancet* 34 : 17-21.

Krumbach, H. 1986 Zur Frage der Lepra im prä-und post-kolumbischen Amerika. In J.H. Wolf (ed) : *Aussatz; Lepra; Hansen-Krankheit. Ein mensheitsproblem in Wandel*. Volume 2. Wurzburg: DAHW, pp. 201-209.

Lechat, M.F. 1986 *Prospect for eradication of leprosy*. Conference for the closing of the last leprosy hospital in Australia, Perth, Western Australia, 1986.

Lechat, M.F. and Chardome, J. 1955 Altérations radiologiques des os de la face chez les Lépreux Congolais. *Ann. Soc. Belge. Med. Trop*. 35 : 603-611.

Lechat, M.F., Bile, T. and Rasi, E. 1967 A study of Blood Groups and Leprosy in the Population of Colonia Tovar, Venezuela. *Int. J. Leprosy* 35 : 488-493.

Leiker, D.L. and Sloan, N.R. 1954 Leprosy in Netherlands New Guinea. *Int. J. Leprosy* 22: 431-439.

Melsom, R.S. 1953 Changes in the maxillary bone in leprosy. A clinical and roentgenological examination. Sixth Congress Internacional de Leprologia, Madrid, pp. 747-750.

Miles, M.A. 1996 New World leishmaniasis. In F.E.G. Cox(ed): *The Wellcome Trust Illustrated History of Tropical Diseases*. London : The Wellcome Trust, pp. 206-229.

Møller-Christensen V., Baake, S.N., Melsom, R.S. and Waaler, E. 1952 Changes in the anterior nasal spine and the alveolar process of the maxillary bone in leprosy. *Int. J. Leprosy* 20 : 335-340.

Sansarricq, H. 1965 *La lèpre*. Paris: Ellipses.

Scott, H.H. 1943 The influence of the slave trade in the spread of tropical disease. *Trans.Roy.Soc.Med.Hyg*. 37 : 169-188

Skinsnes, O.K. 1985 Understanding of leprosy in ancient China. *Int. J. Leprosy* 53 : 289-307.

Sloan, N.R. 1954 *Leprosy in American Samoa*. Technical Paper Number 62, Mimeographed. Noumea : South Pacific Commission.

Stanford, J.L. 1978 The importance of environmental mycobacteria. In B.R. Chatterjee B.R. (ed): *Leprosy. Etiobiology of Manifestations, Treatment and Control*. Calcutta : Statesman Commercial Press, pp. 224-244.

Wade, H.W. and Ledowsky, Y. 1952 The leprosy epidemic at Nauru: A review with data on the status since 1937. *Int. J. Leprosy* 20 : 1-29.

Washburn, W.L. 1950 Leprosy among Scandinavian settlers in the Upper Mississippi Valley, 1864-1932. *Bull. Hist. Med*. 24 : 123-148.

World Health Assembly 1993 *Handbook of resolutions of the WHA and the Executive Board. Volume 3, 1985-1992.* 3rd edition. Geneva: World Health Organisation.

World Health Organisation (Regional Office for the Western Pacific) 1995 *Epidemiological review of leprosy in the Western Pacific Region 1982-1994*. Manilla, World Health Organisation.

World Health Organization 1998a *Action programme for the elimination of leprosy. Status Report*. Geneva: W.H.O./LEP/98.2.

World Health Organization 1998b Progress towards leprosy elimination. *Weekly Epid. Rec*. 73 : 188-190.

Infant and childhood leprosy: present and past

Mary E. Lewis
School of Conservation Sciences
Bournemouth University
Talbot Campus
Fern Barrow
Poole
Dorset
BH12 5BB

Email: mlewis@bournemouth.ac.uk

Abstract

Leprosy is positively diagnosed on the skeleton when pathological changes to the maxilla, known collectively as the "rhinomaxillary syndrome" are evident. Such lesions denote the most debilitating form of the disease, or lepromatous leprosy. A more tentative diagnosis is made when characteristic lesions on the lower legs, hands and feet are observed as a result of secondary neuropathy and subsequent trauma. Typically, palaeopathological evidence for leprosy has been limited to adult skeletal remains even though today leprosy is seen to develop in later childhood, and lepromatous leprosy has been identified in infants and young children. This paper describes the pictorial and documentary evidence for the disease in children from past European populations and explores the pathogenesis of leprosy in children before the introduction of chemotherapy, in an attempt to explain why it is rarely diagnosed in archaeological samples.

Keywords

Non-adult skeletons, archaeology, epidemiology, transmission, leprosaria, Medieval Europe

1. Modern clinical evidence for childhood leprosy

Today, miscarriages are common in women suffering from lepromatous leprosy, and they usually result from bacteraemia or transplacental infection. The children that do survive have a low foeto-placental weight and suffer from fetal distress (Duncan, 1980). Once born, children of mothers with lepromatous leprosy are described as small and feeble, and after the first six months of life grow more slowly than their healthy peers (Duncan, 1985). Although they suffer from the same ailments as other children, those exposed to leprosy at this early age are particularly susceptible to infantile convulsions, marasmus, skin diseases, gastric and respiratory problems, and up to 42% of these children may die (Gomez *et al.*, 1922).

Of the children born into lepromatous families, 25-50% develop clinical signs of the disease before the age of five years (Melsom *et al.*, 1982). In many cases, leprosy heals spontaneously but may re-occur when the child is older. In 1956, Lara and Nolasco surveyed 287 children presenting single leprous lesions in early childhood: 78% of these cases healed spontaneously, 12% of the children had incomplete healing and may have relapsed in later life, and 10% had developed lepromatous leprosy by their early teens. In their detailed study of early leprosy lesions in children from Culion Island in the Philippines, Gomez and colleagues (1922) found that young children with the disease commonly develop single lesions with some loss of pigmentation, or "macules", and that active bacilli are rarely found in the nasal region. By the age of 13 years nasal lesions become more apparent and infiltrations, anaesthesia and nodules are common symptoms, with a loss of thermal and tactile discrimination (Gomez *et al.*, 1922, Noussitou *et al.*, 1978).

The most common forms of leprosy in children are classified as tuberculoid, borderline or "indeterminate" where an immune status cannot be determined and, if left untreated, 25% of these children will go on to develop "spectrum" leprosy (Sundharam, 1990, Waters, 1988). Children rarely present with lepromatous leprosy, or develop paralysis and the gross deformities characteristic of that spectrum of the disease. However, when lepromatous leprosy does occur, nodules are common on the face and ears, and ulcerations on the elbows and legs are persistent (Noussitou *et al.*, 1978). In fact, "spectrum leprosy" has been identified in children between the ages of 6 to 12 years, and one of the earliest accounts of lepromatous leprosy was in a six month old child, who had developed the signs and symptoms of the disease by the age of four months (Ramu, 1959).

Periods of physiological stress, for example during a growth spurt or puberty, are often associated with an increase in leprosy in areas where it is endemic (Badger, 1964). Rao *et al.* (1972) found that in India, there was an increase in leprosy in males between the ages of 15 and 19 years, and that elevated levels were common in females between 10-14 years. However, there were no significant differences in the sex ratios of the disease until adulthood. Furthermore, levels of infant and childhood leprosy increase as the disease becomes more endemic, with up to 20% of children infected in hyperendemic areas (Christian,

1981; Noussitou *et al.*, 1978). Here, the sex ratios are equal and cases of lepromatous leprosy more common (Badger, 1964). In areas where the disease is less endemic, lepromatous leprosy usually only occurs during adulthood and males are more susceptible to the disease, presumably because they are more mobile and more likely to become exposed (Ell, 1989).

2. Modes of transmission

Discovering the mode of transmission of a disease is vital in order to control its spread and in most endemic countries people are exposed to *M. leprae* during childhood. In India, one study suggested that 50% of women and 30% of men who develop leprosy have done so before the age of twenty (Duncan, 1985). For this reason, the exposure time and transmission of leprosy in children has been widely studied and, before vaccinations, led to babies being removed from their families within 14 days of their birth to avoid contagion (Rogers and Muir, 1946). However, some evidence has suggested that such measures are ineffective and that children are being exposed to the disease much earlier in their lives.

2.1 Congenital leprosy

Studies into congenital leprosy are hindered by the nature of the disease in infants who usually display ill-defined, single leprous lesions that may be overlooked, ignored or actively concealed by the mother in attempt to protect the child from any social stigma. Nadkarni *et al.* (1988) found single leprous lesions in children to occur more commonly in areas that could be shielded by clothing. These lesions may then heal spontaneously, leaving the child to develop the disease in later life, thus disguising any evidence of infantile or even congenital leprosy (Figure 1). Congenital leprosy is a controversial subject in the clinical literature but evidence for this form of the disease has been argued since the 1890s (Navarro, 1890). Before the mid 1980s, infantile leprosy was thought to be unlikely due to the slow generation time of *M. leprae* and a long incubation period of two to five years (Jopling and McDougall, 1988). However, in 1985, Brubaker *et al.* reported 91 cases of leprosy in children under one year of age and suggested that the infants had become exposed to the infection *in utero*. Such findings led researchers to argue that favourable conditions, in a highly susceptible host, may reduce the generation time of the bacillus from 13 days to just 26 hours (Brubaker *et al.*, 1985; Girdhar *et al.*, 1989).

The placenta is highly vascular and a breach of its integrity may lead to large amounts of the bacilli reaching the fetus (Duncan *et al.*, 1983). In addition, pathogens may be spread through the umbilical vein or through an ingestion of amniotic fluid (Jordan and Spencer, 1949). In order to prove congenital transmission, researchers needed to prove that the child was infected before birth, rather than by the mother immediately after birth (Melsom *et al.*, 1980). In 1913, Rabinowitsch identified acid-fast bacilli in the heart blood of a fetus during an autopsy of a woman with lepromatous leprosy, and argued for a haematogenous spread of the infection. Crozier and Cochrane (1929) later reported on a child born with an ulcer on its heel which may have been leprous in origin. Although the mother was not infected with leprosy the father had the lepromatous form of the disease. It was suggested that *M. leprae* may have been transmitted to the foetus via the sperm. Since these early discoveries, Melsom *et al.* (1980) have reported IgA antibodies of *M. leprae* in the cord sera of children with leprous mothers. As IgA cannot be passed from the mother to the child across the placenta, the presence of these anti-*M. leprae* antibodies suggest the live passage of bacilli across the placenta and their active proliferation (Melsom *et al.*, 1982).

2.2 Gastrointestinal infection

Some researchers have suggested a gastrointestinal route of infection (Pedley and Geater, 1976). In 1967, Pedley isolated *M. leprae* in the breastmilk of a woman with lepromatous leprosy, as well as in a non-lactating woman (1968a), and in the breast secretions of a male hypertrophied breast (1968b). Girdhar *et al.* (1981) located bacilli in 8% (1 of 13) of women with tuberculoid leprosy, 12.5% (1 of 8) of the borderline group and 64% (9 of 14) of untreated lepromatous cases.

2.3 Respiratory infection

The respiratory tract is one of the most common portals of entry for leprosy, as bacilli are transmitted via droplets in the breath, or from the highly infectious nasal discharge. Close contact of an infant or child with its parents and other relatives makes them particularly susceptible to this mode of infection (Jopling and McDougall, 1988).

2.4. Skin-to-Skin Contact

Cochrane (1959) suggests that prolonged intimate contact between children and their parents leads to a skin-to-skin transmission of the disease. For example, he cites that in India, where it is often the practice for the father to sleep on the veranda and the children to sleep with their mother, in these cases the husband usually escapes infection whereas the children contract the disease.

3. Historical evidence for childhood leprosy

3.1 *Documentary*

'..*what an awful infallible certainty, as surely as we inherited it from our fathers do we transmit it* [leprosy] *to our children.*' Thompson, 1882 (cited in Lewis, 1987: 296).

References to the congenital form of leprosy were common in medieval Europe but may have resulted from confusion with congenital syphilis, and a lack of knowledge about how leprosy was transmitted. For example, many considered leprosy to be a venereal disease, perhaps due to the downgrading of some women from a less obvious tuberculous pole, to the more evident lepromatous form of the disease during pregnancy (Duncan *et al.*, 1981). In addition, some medieval medical examiners may have confused leprosy with venereal syphilis, due to the superficial similarity of the lesions on the face and legs (but see Crane, this volume).

One of the earliest written accounts of childhood leprosy comes from a Chinese bamboo book dating from around 200 BC, and tells of a villager called Bing, who had developed leprosy by the age of three years (Skinsnes, 1980). King Baldwin IV of Jerusalem was said to have suffered from anaesthesia of the limbs as a child, and to have become blind at 23, and it has been suggested that these disabilities were caused by leprosy (Lowe, 1947). Medieval literature also reveals that between the 16th and 18th centuries, new-born children were fed purges to clear mucous from their mouths and to protect them from leprosy (Fildes, 1988).

Much of the evidence for childhood leprosy in the past comes from references to children entering leprosaria. For example, Richards (1977:69) cites an entry in the medieval literature from Denmark which states that 'even two small leper children reported to a meeting of the parish...in 1658...were admitted to the hospital...' (*ibid.*: also see pages 33 and 44). In Finland, records from the leprosarium in Sjählö indicate that 12 children with signs of the disease were living there during the seventeenth century (Fagerlund, 1886). However, not all the references to children prove that they were suffering from the disease, and they may have been allowed to enter the leprosarium with their sick parents to prevent them from becoming orphaned and destitute. Richards (1977: 60) cites the case of a young woman from Sweden who had entered the hospital with her parents as a child and, although she never developed the disease, remained there for 16 years. Some children may have been born to women with lepromatous leprosy inside the leprosarium, or to the healthy wives of the inmates. However, husbands and wives were often segregated or advised to divorce when they entered a leprosarium (Brody, 1974), and male infertility, often caused by the disease (Jopling and McDougall, 1988), would have decreased the number of children born in the hospital. That children were considered particularly vulnerable to the disease, either morally or physically, is indicated from edicts dating from the fifteenth century, which state that people with leprosy were not to touch children, not even their own (Mercier, 1915).

3.2 *Artistic*

Artistic representations of children with leprosy are rare. However, a painting from the thirteenth century in Europe, the '*Wonder of Dominicis*' by Gaddi (died 1366), depicts a child with a claw hand and paralysis of the feet indicating that the child suffered from lepromatous leprosy (Gron, 1973). A second painting from the seventeenth century shows a child petitioning to enter a leprosarium (by Bol 1616-1690); the child appears to have the nodular form of the disease, but it has been suggested the boy had a scalp infection rather than leprosy (Goldman, 1968). Perhaps the most famous portrait of a child with lepromatous leprosy was illustrated by Danielsson and Boeck in 1848 and is of a 13-year-old boy in a Scandinavian hospital (Figure 2).

4. Archaeological evidence for childhood leprosy

Bone changes recognisable in skeletal material usually occur at the lepromatous pole of the disease and can take decades to develop. The diagnosis of leprosy from skeletal remains relies on the quality of bone preservation, the severity of the disease at the time of the individual's death, and their immune status. Skeletal lesions of leprosy are both primary, and secondary to the neuropathy that follows infection (Table 1) but in isolation each lesion may represent non-specific infection or neuropathy. Therefore, a combination of infective and neuropathic lesions in the skeleton is needed before a positive diagnosis can be made. Children may experience the same range of leprosy signs as adults. However, their early death will reduce the number of chronic cases and prevent the more debilitating bone changes developing in their skeletons.

In 1922, Gomez *et al.* reported that, in children with leprosy, bacilli in the nasal region were rare. However, an accumulation of bacilli in this area is thought to result in facial changes seen in the more advanced cases of lepromatous leprosy, and recognised skeletally as rhinomaxillary syndrome (RMS) (Andersen and Manchester, 1992). In fact, Mirander (1970) suggested that clinical cases of rhinomaxillary syndrome in children were unusual and bone absorption only occurred at the incisors. Therefore, early changes associated with RMS may be overlooked in non-adult skeletons. It may be for this reason that archaeological evidence for leprosy in children is limited.

Leprogenic odontodysplasia (LOD) refers to the development of concentrically stunted roots in the developing permanent incisors, as a direct result of *M. leprae* bacilli, and is thought to be indicative of childhood exposure to the disease (Figure 3). Danielsen (1968) first discussed these dental defects in four children with rhinomaxillary changes from Næstved in Denmark, and found them in individuals dying before the age of 20 years. Danielsen (1970) suggests that the resorption of the alveolar bone and tooth deformities appeared to be related, to and may have been caused by an infiltration of bacilli into the pulp cavity. Although the dental defects varied in their severity, they were only ever identified with the rhinomaxillary syndrome. However, no clinical cases of these dental deformities have been reported (Roberts, 1986) and to date, no new archaeological examples have been published.

Leprogenic periostitis may occur secondary to neuropathy

in the lower legs and extremities. Neuropathy results in anaesthesia, and neglected ulcerations allow toxins from pyogenic bacteria to spread up the muscle planes of the legs, inflaming the periosteum and resulting in new bone formation on the tibiae and fibulae (Lewis *et al*., 1995). These lesions are normally bilateral in the lepromatous form of the disease and are more easily recognised than the unilateral periostitis associated with tuberculoid or high resistance leprosy (Manchester, pers. comm.). Although ulcerations of the legs and elbows have been reported in clinical cases of childhood leprosy (Noussitou *et al*., 1978), those entering the archaeological record may die before the chronic deformities of the legs and feet, recognised in adults develop.

4.1 Archaeological cases of infant and childhood leprosy

Due to the superficial and non-specific lesions associated with congenital leprosy today, it is unlikely that, if children born to lepromatous mothers survived, any skeletal lesions would be evident. However, a few cases of young children with skeletal evidence of lepromatous leprosy in England and Denmark have been discovered. In 1958, Brothwell reported on the skull of a child aged between 14 and 16 years from Scarborough Castle (skull 4.3832). The skull had evidence of nasal aperture absorption, a pitted palate, and loss of the nasal spine. The child was located in an ordinary burial ground suggesting that they were not segregated, despite displaying the saddle nose associated with the condition. The skull of an 8 to 9 year old child of unknown provenance is held in the skeletal collections at the University of Bradford and, again, shows the features characteristic of RMS but, despite its young age, it does not have any evidence of leprogenic odontodysplasia (Figures 4 and 5). In both cases, the combination of pitted lesions and smooth remodelling in the rhinomaxillary area are characteristic of leprosy. However, in isolation these lesions would not be pathognomonic of the disease. For example, pitting of the nasal surface could be indicative of a non-specific inflammatory response such as rhinitis. Rhinomaxillary changes associated with treponemal disease are characteristically more aggressive and irregular (see Cook, this volume).

Møller-Christensen's famous study of the skeletons excavated from St. Jorgen's Hospital in Næstved revealed that 13% of the skeletons displayed lesions characteristic of lepromatous leprosy before the age of 17 years (1961). Of particular interest are three young children with rhinomaxillary changes and leprogenic odontodysplasia, aged between 8 and 10 years old. Despite having evidence for the lepromatous pole of the disease, these children did not have postcranial changes associated with neuropathy. Møller-Christensen (1978) also published evidence for rhinomaxillary syndrome in 12 juveniles (aged 14-20 years), nine of whom had evidence for postcranial changes. In 1996, Anderson published 11 cases of leprosy in non-adults from the later Medieval site of St. John Timberhill in Norwich. The children were aged between 12 and 20 years, and five had evidence of rhinomaxillary syndrome with associated postcranial changes. However, the older age of these individuals, compared to the younger non-adults with the disease at Næstved, suggests that children need to survive longer with the disease to develop the more debilitating changes on the hand, foot and lower leg bones.

Both the documentary and archaeological evidence for childhood leprosy suggests that in Scandinavia, and in particular Denmark, when children were diagnosed with leprosy they entered leprosaria. However, a review of the skeletons excavated from Sts. James and Mary Magdalene's Hospital in Chichester, England did not reveal individuals under the age of 17 years with skeletal lesions associated with leprosy. Nevertheless, 64% (7) of the individuals with rhinomaxillary syndrome were young adults aged between 17 and 25 years. Reichart (1976) reported that, in untreated cases of leprosy, facial lesions can take up to 17 years to develop, indicating that the young adults with RMS at Chichester may have had a childhood exposure to the disease (Lewis, 1993). Nevertheless, as with many of the adult skeletons that are discovered in leprosy cemeteries with no lesions indicative of the disease, we cannot know how many children entered the hospital with soft tissue lesions did not survive long enough for them to manifest on the skeleton.

5. Discussion and conclusions

The diagnosis of children with leprosy in the past and present is hindered by the subtle nature of the disease in young individuals and its potential to spontaneously heal. However, documentary and artistic evidence suggest schildhood leprosy did occur so how do we explain the scarcity of the disease in non-adults from the archaeological sites? Children born to mothers with lepromatous leprosy today are small and feeble and often fail to thrive. Therefore, in the past, it seems unlikely that many of these children would have survived beyond infancy. Children who develop the disease usually suffer from the indeterminate, borderline or tuberculoid forms of leprosy which are harder to identify palaeopathologically due to the non-specific nature of the lesions in the skeleton and lack of rhinomaxillary changes. Even if RMS does develop, the rhinomaxilla needs to be complete in order to identify the earliest changes. Unfortunately, the maxilla is very fragile and is often damaged during burial, excavation and/or curation. In order to identify leprogenic odontodysplasia, the roots of the maxillary permanent incisors need to be fully formed and so defects in individuals under the age of seven or eight years can go undetected. In addition, these defects are associated with RMS, but as this condition advances the maxillary incisors are often lost. Without the changes of the skull, we rely on the characteristic lesions of the hands, feet, tibiae and fibulae. These changes occur in adults and older children as a result of trauma to affected hands and feet, and the neglect of ulcerations on their feet that allow environmental bacteria to enter the wounds. The high susceptibility of children with leprosy to other infections means that they may have died before they became mobile and gross deformities developed.

Although few leprosy cemeteries have been excavated in medieval Europe, there appears to be a difference in the way children with the disease were treated in England and in Scandinavia. Evidence of children with lepromatous leprosy from lay cemeteries in England is not conclusive proof that they were not recognised or segregated. However, the lack of evidence for children with the disease at Chichester, seems to suggest that they did not enter leprosaria. In Scandinavia, evidence from Finland, Sweden and Denmark indicates that children were allowed to enter these hospitals, and this is substantiated by the evidence of children as young as eight years with leprosy at Næstved. Modern epidemiological evidence suggests that in areas where leprosy is hyperendemic, males and females are equally exposed to the disease and signs and symptoms usually begin in childhood; where leprosy is less endemic, lepromatous leprosy is less common and usually occurs in early adulthood. Until more evidence of infant and childhood leprosy is found we cannot start to address the more fundamental questions about the disease in the past. Fewer cases of lepromatous leprosy in the children from Britain, as opposed to Scandinavia, may help us to determine whether leprosy in Britain was less endemic than elsewhere.

This review has suggested that although there are very few cases of children with possible leprosy in the archaeological record, other evidence for the lepromatous form of the disease in young adults suggests that there was childhood exposure to the disease in the past, just as there is today. The paucity of evidence may relate to the difficulty in diagnosing the less severe forms of the disease in the skeletal material. However, it is likely that the small number of children with the disease in the archaeological record may indicate that leprosy was not hyperendemic in medieval Europe.

Acknowledgements

I am grateful to Malin Holst for her translation of the Rabinowitsch (1913) article from German, and Heikki Vuorinen for providing a copy and translation of Fagerlund's *Finlands Leprosorier*. I would also like to thank Charlotte Roberts and an anonymous reviewer for their comments on earlier drafts of this paper.

References

Andersen, J.G. and Manchester, K. 1987 Grooving of the proximal phalanx in leprosy: a palaeopathological and radiological study. *Journal of Archaeological Science* 14:77-82.

Andersen, J.G. and Manchester, K. 1988 Dorsal tarsal exostoses in leprosy: a palaeopathological and radiological study. *Journal of Archaeological Science* 15:51-56.

Andersen, J. and Manchester, K. 1992 The rhinomaxillary syndrome in leprosy: a clinical, radiological and palaeopathological study. *International Journal of Osteoarchaeology* 2:1-9.

Andersen, J.G., Manchester, K. and Roberts, C.A. 1994 Septic bone changes in leprosy: a clinical, radiological and palaeopathological review. *International Journal of Osteoarchaeology* 4:24-30.

Andersen, J.G., Manchester, K. and Rozina Shahzady, A. 1992 Diaphyseal remodelling in leprosy: a radiological and palaeopathological study. *International Journal of Osteoarchaeology* 2:211-219.

Anderson, S 1996 Leprosy in a Medieval Churchyard in Norwich. In S. Anderson (ed): *Proceedings from the First Annual Meeting of the Osteoarchaeology Research Group*. Oxford: Oxbow Books, pp. 31-37.

Badger, L.F. 1964 Epidemiology. In R.G. Cochrane and T.F. Davey (eds.): *Leprosy in Theory and Practice*. Bristol: John Wright and Sons Ltd., pp. 69-97.

Barton, R.P.E. 1979 Radiological changes in the paranasal sinuses in lepromatous leprosy. *The Journal of Laryngology and Otology* 93:597-600.

Brody, N.S. 1974 *The disease of the soul: leprosy in Medieval literature*. London: Cornell University Press.

Brothwell, D.R. 1958 Evidence of leprosy in British archaeological material. *Medical History* 2:287-291.

Brubaker, M.L., Meyers, W.M. and Bourland, J. 1985 Leprosy in children one year of age and under. *International Journal of Leprosy* 53:517-522.

Christian, M. 1981 The epidemiological situation of leprosy in India. *Leprosy Review* 52:35-42.

Cochrane, R.G. 1959 *Leprosy in Theory and Practice*. Bristol: John Wright and Sons Ltd.

Cooney, J.P., and Crosby, E.H. 1944 Absorptive bone changes in leprosy. *Radiology* 42:14-19.

Crozier, C.G. and Cochrane, R.G. 1929 Leprosy in an infant. *British Medical Journal* 1 501-502.

Danielsen, K. 1968 Leprogenic odontodysplasia (*dens leprosus*). Transactions of the Ninth International Leprosy Congress, 16-21 September, London. *International Journal of Leprosy* 36: 1-7.

Danielsen, K. 1970 Odontodysplasia leprosa in Danish medieval skeletons. *Saertryk af Tandlaegebladet* 74:603-625.

Danielsson, D. and Boeck, W. 1973 Traite de la spedalsked (Regulations concerning lepers). (Translation of 1848). *Bulletin of the New York Academy of Medicine* 49:567-568.

Duncan, M.E. 1980 Babies of mothers with leprosy have small placentae, low birth weights and grow slowly. *British Journal of Obstetrics and Gynaecology* 87:471-479.

Duncan, M.E. 1985 Leprosy in young children - past, present and future. *International Journal of Leprosy* 53:468-473.

Duncan, E.M., Melsom, R., Pearson, J.M.H., Menzel, S.

and Barnetson, R.S.C. 1983 A clinical and immunological study of four babies of mothers with lepromatous leprosy two of whom developed leprosy in infancy. *International Journal of Leprosy* 51:7-17.

Duncan, M.E., Melsom, R., Pearson, J.M.H. and Ridley, D.S. 1981 The association of pregnancy and leprosy. *Leprosy Review* 52:245-262.

Ell, S.R. 1989 Three times, three places, three authors and one perspective on leprosy in Medieval and early modern Europe. *International Journal of Leprosy* 57:825-833.

Fagerlund, L.W. 1886 *Finlands Leprosorier I.* Helsingfors: Finska Litterature-sällskapets tryckeri.

Fildes, V.A. 1988 *Wet Nursing: A History from Antiquity to the Present.* New York: Basil Blackwell.

Girdhar, A., Girdhar, B.K., Ramu, G. and Desikan, K.V. 1981 Discharge of *M. Leprae* in milk of leprosy patients. *Leprosy in India* 53:390-394.

Girdhar, A., Mishra, B., Lavania, R.K., Bagga, A.K., Malaviya, G.N. and Girdhar, B.K. 1989 Leprosy in infants - report of two cases. *International Journal of Leprosy* 57:472-475.

Goldman, L. 1968 Favus mistaken for leprosy by artists of the Renaissance. *Arch. Dermatol.* 98:660-661.

Gomez, L., Basa, J.A. and Nicolas, C. 1922 Early lesions and the development and incidence of leprosy in the children of lepers. *The Philippine Journal of Science* 21:233-256.

Gron, K. 1973 Leprosy in art and literature. *International Journal of Leprosy* 41:249-283.

Hauhnar, C.Z., Mann, S.B.S., Sharma, V.K., Kaur, S., Mehta, S. and Radotra, B.D. 1992a Maxillary antrum involvement in multibacillary leprosy: a radiologic, sinuscopic and histologic assessment. *International Journal of Leprosy* 60:390-395.

Hauhnar, C.Z., Kaur, S., Sharma, V.K. and Mann, S.B.S. 1992b A clinical and radiological study of the maxillary antrum in lepromatous leprosy. *Indian Journal of Leprosy* 64:487-494.

Hjorting-Hansen, D.D.S., Kloft, B. and Henning Schmidt, M.D. 1965 Leprotic granuloma in the maxilla. *International Journal of Leprosy* 33:83-88.

Job, C.K. 1963 Pathology of leprous osteomyelitis. *International Journal of Leprosy* 31:26-33.

Jopling, W.H. and McDougall, A.C. 1988 *Handbook of Leprosy.* Oxford: William Heinemann Medical Books.

Jordan, J.W. and Spencer, H. 1949 A case of congenital tuberculosis. *British Medical Journal* 5:217-219.

Kulkarni, V.N. and Mehta, J.M. 1983 Tarsal disintegration (T.D.) in leprosy. *Leprosy in India* 55:338-370.

Lara, C.B. and Nolesco, J.O. 1956 Self-healing or abortive, and residual forms of childhood leprosy and their probable significance. *International Journal of Leprosy* 24:245-263.

Lewis, G. 1987 A lesson from Leviticus: leprosy. *Man* 22:593-612.

Lewis, M.E. 1993 *Diagnostic features of leprosy in the Chichester skeletons.* University of Bradford, MSc thesis, Unpublished.

Lewis, M.E., Roberts, C.A. and Manchester, K. 1995 Inflammatory bone changes in leprous skeletons from the medieval hospital of St. James and St. Mary Magdalene, Chichester, England. *International Journal of Leprosy* 63:77-85.

Lowe, J. 1947 Comments on the history of leprosy. *Leprosy Review* 18:54-64.

Melsom, R., Duncan, M.E. and Bjune, G. 1980 Immunoglobulin concentration in mothers with leprosy and in healthy controls and their babies at the time of birth. *Leprosy Review* 51:19-28.

Melsom, R., Harboe M. and Duncan, M.E. 1982 IgA, IgM and IgA anti-*M-leprae* antibodies in babies of leprosy mothers during the first 2 years of life. *Clinical Experimental Immunology* 49:532-542.

Mercier, C. 1915 *Leper houses and Medieval hospitals.* London: H.K. Lewis

Mirander, R.P.G. 1970 Childhood leprosy. *International Journal of Leprosy.* 38(1): 96.

Møller-Christensen, V. 1961 *Bone changes in leprosy.* Copenhagen: Munksgaard.

Møller-Christensen, V. 1978 *Leprosy changes of the skull.* Odense: University Press.

Møller-Christensen, V., Bakke, S.N., Melsom, R.S. and Waaler, E. 1952 Changes in the anterior nasal spine and the alveolar process of the maxillary bone in leprosy. *International Journal of Leprosy* 20:335-343.

Nadkarni, N.J., Grugni, A., Kini, M.S. and Balakrishnan, M. 1988 Childhood leprosy in Bombay: a clinico-epidemiological study. *Indian Journal of Leprosy* 60:173-188.

Navarro, R. 1890 Congenital leprosy. *British Medical Journal* 2:766.

Noussitou, F.M., Sansarricq, H. and Walter, J. 1978 *Leprosy in Children.* Geneva: World Health Organisation.

Pedley, J.C. 1967 The presence of *M. leprae* in human milk. *Leprosy Review* 38:239-242.

Pedley, J.C. 1968a Presence of *M. leprae* in the breast secretions of a non-lactating woman with lepromatous leprosy. *Leprosy Review* 39:111.

Pedley, J.C. 1968b Presence of *M. leprae* in the nipple secretion and lumina of hypertrophied mammary gland. *Leprosy Review* 39:67.

Pedley, J.C. and Geater, J.G. 1976 Does droplet infection play a role in the transmission of leprosy? *Leprosy Review* 47:97-102.

Rabinowitsch, M. 1913 Leprabacillen im kreisenden blute der lepra - kranken und im herzblute eines leprafötus (Leprosy bacilli in the circulating blood of patients and the heart-blood of a leprous foetus). *Berliner Klinische Wochenscrift* 50:252-253.

Ramu, G. 1959 Adult type of lepromatous leprosy in a child of 6 months. *Indian Journal of Child Health* (No volume stated): 231-314.

Rao, P.S.S., Karat, A.B.A., Kaliaperumal, V.G. and Karat, S. 1972 Prevalence of leprosy in Gudiyatham Taluk; South India. Part 1: Specific rates with

reference to age, sex and type. *International Journal of Leprosy* 40:157-163.
Reichart, P. 1976 Facial and oral manifestations in leprosy. *Oral Surgery* 41:385-399.
Richards, P. 1977 *The Medieval leper and his Northern heirs*. New Jersey: D.S. Brewer, Rowman & Littlefield.
Roberts, C.A. 1986 Leprogenic odontodysplasia. In E. Cruwys and R.A. Foley (eds.): *Teeth and Anthropology*. Oxford: British Archaeological Reports, International Series 291, pp. 137-147.
Rogers, L. and Muir, E. 1946 *Leprosy*. Bristol: John Wright and Sons Ltd.
Singh, I., Kaur, S., Khandelwal, N., Kaur, I. and Deodhar, S.D. 1994 Arthritis in leprosy: clinical, laboratory and radiological assessments. *International Journal of Leprosy* 62:428-433.
Skinsnes, O.K. 1980 Leprosy in the archaeologically recovered Bamboo Book in China. *International Journal of Leprosy* 48:333.
Sundharam, J.A. 1990 Leprosy in childhood. *Indian Pediatrics* 27:1126-1128.
Thompson, W.M. 1882 *The Land and the Book*. London: Nelson.
Waters, M.F.R. 1988 Leprosy. *Medicine International* 55:2294-2297.

Table 1. Primary and secondary lesions associated with leprosy in the skeleton.

AREA	PRIMARY BONE LESIONS	SECONDARY BONE LESIONS
Rhinomaxillary	* Localised inflammatory granulomata on the nasal and oral surfaces of the palatine process [1]	* Atrophy of the nasal aperture, anterior nasal spine and alveolar process [7,8] * Diffuse inflammatory pitting of nasal and oral palatine processes [8] * Maxillary sinusitis [9,10,11]
Dentition	* Leprogenic odontodysplasia[2,3] (concentric constriction of the permanent dental roots)	* Loss of maxillary incisors as a result of alveolar recession [8,12]
Hands and Feet	* Lepromatous osteomyelitis [4] * Sensory, motor and autonomic neuropathy [5] * Bone cysts	*Sensory neuropathy:* * Pyogenic infections [6] * Septic arthritis and ankylosis [13,14] * Tarsal and carpal disintegration [15] * Loss of proprioception [5] * Fractures, subluxation and dislocation [13, 16] *Motor neuropathy*: * Claw-hand and foot deformities [17, 19] * Drop-foot and wrist deformities [13, 15] * Navicular squeezing and dorsal barring [15,18] * Palmer grooves [17] Autonomic neuropathy * Concentric remodelling of the phalanges [12, 13, 16] * Achroosteolysis [13] * Knife-edge remodelling of metatarsals and metacarpals [12, 13] * Osteoporosis [5]
Upper and lower limbs	* Lepromatous osteomyelitis [4,6]	* Diffuse periostitis [6] * Septic arthritis and ankylosis [13] * Ossification of the interosseous membrane [19]

[1]Hjorting-Hansen, *et al.*, 1965; [2]Danielsen, 1968; [3]Danielsen, 1970; [4]Job, 1963; [5]Jopling and McDougall, 1988; [6]Andersen, *et al.*, 1994; [7]Møller-Christensen, *et al.*, 1952; [8]Andersen and Manchester, 1992; [9]Barton, 1979; [10]Hauhnar, *et al.*, 1992a; [11]Hauhnar, *et al.*, 1992b; [12]Møller-Christensen, 1961; [13]Andersen, *et al.*, 1992; [14]Singh, *et al.*, 1994; [15]Kulkarni and Mehta, 1983 ; [16] Cooney and Crosby, 1944; [17] Andersen and Manchester, 1987; [18]Andersen and Manchester, 1988; [19]Lewis, *et al.*, 1995.

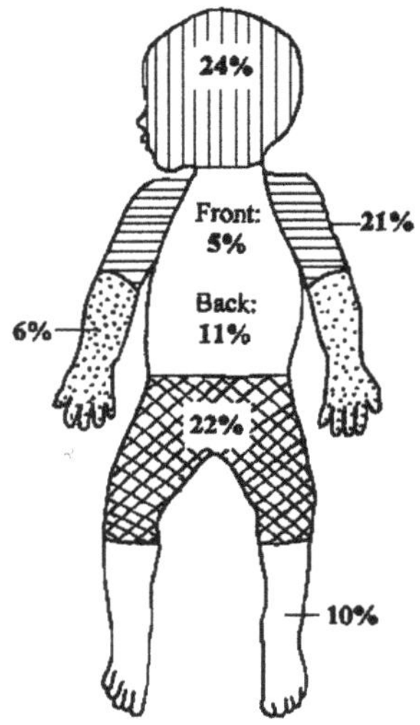

Figure 1: Diagram showing the areas most commonly affected by single leprous lesions in the child (after Nadkarni *et al.*, 1988)

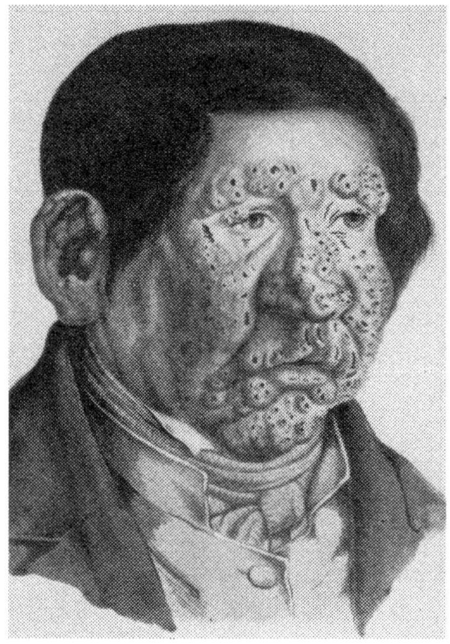

Figure 2: Thirteen year old child with leprosy from Scandinavia, illustrated by Danielsson and Boeck in 1848.

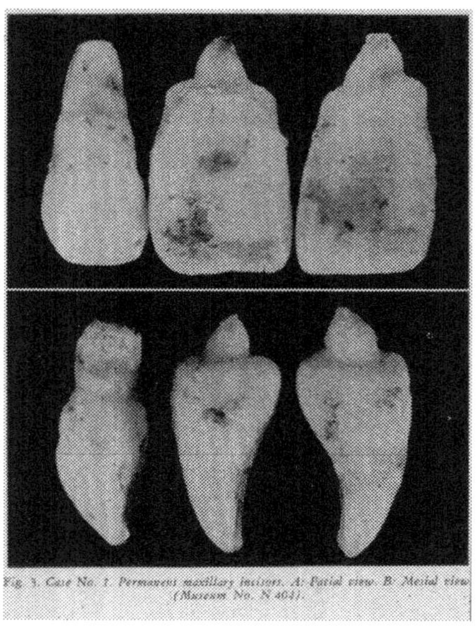

Figure 3: Central and lateral maxillary incisors from Næstved 404, showing concentric remodelling of the roots characteristic of leprogenic odontodysplasia (from Danielsen, 1970).

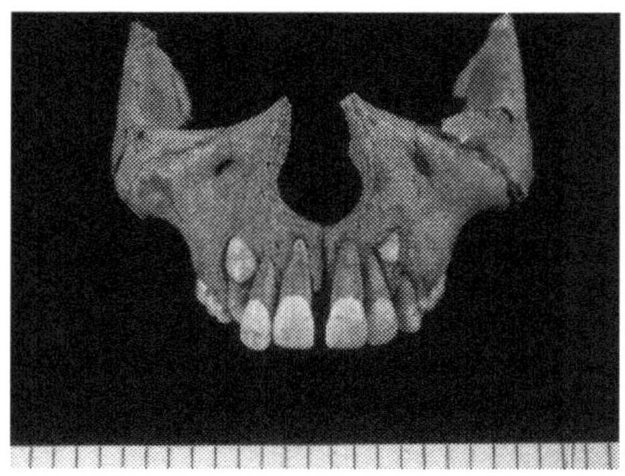

Figure 4: Anterior view of the Bradford skull showing remodelling of the nasal aperture and absorption of the alveolar bone between the central incisors. The tooth roots are exposed as a result of postmortem damage.

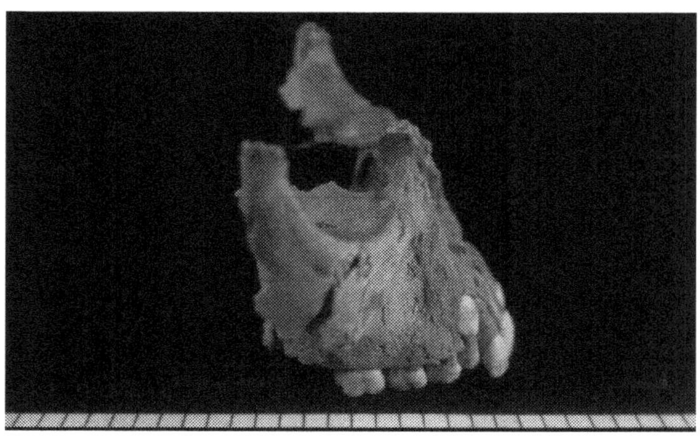

Figure 5: Lateral view of the Bradford skull showing loss of the anterior nasal spine which could have resulted in saddle nose in life.

The myth of the spread of leprosy with the crusades

Piers D. Mitchell
Wellcome Trust Centre for the History of Medicine
University College
University of London

Home (postal) address
2 Milton Mansions
Queen's Club Gardens
London W14 9RP
England

E-mail: p.mitchell@clara.co.uk

Abstract

Many authors have proposed that returning crusaders in the 12th and 13th centuries were responsible for an epidemic of leprosy in mediaeval Europe. However, critical assessment of the theory raises serious doubts about the interpretation of this evidence.

While it is true that the number of leprosaria greatly increased at the time of the crusades, the number of general hospitals did too. This suggests a change in social attitudes to caring for the sick, rather than more leprosy patients. The larger sections on the disease in medical texts may have been a consequence of the extra information available from the translation of Arabic manuscripts at that time. The epidemiology is not suggestive of an epidemic. The proportion of cases of lepromatous and tuberculoid leprosy excavated from leprosaria cemeteries are the opposite of that expected at the time of an epidemic. Furthermore, it is known that leprosy was present in Europe over 500 years before the crusades and any epidemic should have happened then. There is little evidence to blame the crusades for a leprosy epidemic. Any true increase in prevalence in the mediaeval period might be better ascribed to increasing population density and urbanisation.

Keywords

Crusades, epidemiology, leprosaria, mediaeval, medical texts.

1. Leprosy in the mediaeval period

Leprosy was one of the most important diseases in the mediaeval period. This is despite the fact that it does not have a high mortality rate nor is it highly infectious. When considering the most important diseases at this time, other conditions spring to mind such as bubonic plague (Amundsen, 1977) or tuberculosis (Mitchell, 1999, Stirland and Waldron, 1990) rather than leprosy. However, leprosy was of such importance because of the huge social consequences which resulted from a diagnosis of this disease. No other disease had hospitals set up specifically for people with that illness. Few other diseases had so many laws passed in an attempt to control the problem. No other disease was so associated with social stigma that people lost their families, livelihood and social position, and were cast out of their community.

In consequence, much work has been published on both the historical and archaeological evidence for leprosy in the mediaeval period. Areas of particular interest include the criteria for a diagnosis of leprosy (Andersen, 1969, Demaitre. 1985, Dols. 1979), confusion between true leprosy and other diseases with similar signs and symptoms (Hudson, 1963, Huls, 1975), the medical treatment of the disease (Roberts, 1987), the role of the leprosarium (Roberts, 1986), the social coping mechanisms introduced by communities (Mitchell, 2000a), the prevalence of the different forms of the disease over the centuries and the possible immunological interaction with other infections (Manchester, 1984, Ell, 1987), diet (Ell, 1985) and social status (Dols, 1983, Ell, 1986). One further area of discussion has been the manner in which leprosy spread from one region of the world to another and many theories have been proposed as to when and how this infection managed to reach almost all areas of the globe. An assessment of just one aspect of this question is outlined here, namely whether the crusades can be blamed for causing an epidemic of leprosy in Europe during the mediaeval period.

2. The nature of leprosy

For those not familiar with this complex disease a short summary will allow better understanding of certain arguments outlined in the study. Leprosy is an infectious disease caused by a bacterium, *Mycobacterium leprae* (Ji and Grosset, 1994, WHO, 1985). It is found principally in nerves, the skin and around the nose and is spread mainly by coughing and sneezing, but also by contact with ulcerated skin. It is surprisingly difficult to transmit from person to person as it is not very infectious. The Mycobacterium is so slow growing that it often takes many years before any signs or symptoms are noticed. Few of these signs and symptoms are actually caused by the bacteria themselves and most of the problems result from the body's attempts to kill off the organisms, as nearby tissues tend to be damaged in the process.

Different people react in different ways to the disease and the response can be displayed as a spectrum, with tuberculoid leprosy at one extreme and lepromatous leprosy at the other. These forms are not caused by different strains or subspecies of *M. leprae*, but rather it is the response of the infected person to the disease which governs the signs and symptoms they will develop. In tuberculoid leprosy white blood cells attack and destroy the bacteria but the nerves may become damaged in the process. The result may be an area of skin with loss of sensation and altered colour, weakness of muscles, and dry skin prone to deep cracks as the ability to sweat is lost. These changes predispose the skin to ulceration and loss of function or deformity of the affected area due to muscle weakness. As only part of a limb is typically involved, it may be possible to conceal the disease with the appropriate clothing and, in any case, the disease may heal spontaneously given time. If an excavated skeleton has classic changes of leprosy on the bones of just one foot or hand then the tuberculoid form is most likely. Lepromatous leprosy is placed at the other end of the spectrum and in this forms the body attempts to destroy the bacteria with antibodies, which normally break holes in bacterial cell walls. However, this is ineffective with this type of bacteria as they have very thick cell walls and also hide from the immune system by getting inside human cells. In consequence, the bacteria can slowly multiply to cause nodules and plaques in the skin and destruction of the nose. Other signs and symptoms include loss of eyebrow hair, hoarse voice, blindness, muscle weakness and loss of sensation in the skin of the hands and feet, which can progress to ulceration and deformity. This form is extremely hard for a person to conceal as it involves the face. Unlike the tuberculoid form, it does not heal spontaneously but continues to progress insidiously. If an excavated skeleton has changes suggestive of leprosy around the nose, hands and feet then the lepromatous form is most likely.

3. Leprosy in the crusades

If it is argued that a disease was transferred to Europe by crusaders then it must first be shown that the disease was present in the Latin East. These states were established in AD 1099 when Jerusalem was captured at the end of the first crusade. The states which made up the Latin East covered the coast of the eastern Mediterranean from what is now Turkey to Egypt, as well as Cyprus. While the dominant culture was French, the Frankish states were populated by individuals from almost every country in Europe and lasted for two hundred years, until the last castles on the mainland were lost in 1302.

It is known that leprosy was present in the Frankish states as archaeological cases from the region have been found dating from before the time of the crusades (Zias, 1991). None of the excavation of crusader sites so far undertaken have been of leprosaria cemeteries (Mitchell, 1994, Mitchell, 1997, Rose, 1998) so it is difficult to get an objective estimate of the prevalence of disease there. Perhaps the most famous individual known to have had the disease was King Baldwin IV of Jerusalem, who died in 1185 aged about twenty-three (Mitchell, 2000b). He developed borderline tuberculoid leprosy as a child, first noticed as an area of skin on his right arm which had lost sensation (Figure 1). When he reached and passed through puberty his immune system stopped fighting the infection with white blood cells and used antibodies instead, termed "downgrading". This resulted in his developing the lepromatous form of the disease so that he went blind, developed a hoarse voice, his nose was destroyed, he developed ulcers on his deformed hands and feet, and had to be carried about on a litter. The need to keep the kingdom's army up to strength meant that those who developed leprosy were not just segregated, as was the case in Europe, but joined a specially formed military order called the Order of St. Lazarus (Shahar, 1982). This enabled the affected individuals to fight in the king's army and defend stretches of city walls while preventing them from mixing freely with the rest of the population. The accuracy of the diagnosis of leprosy in the mediaeval period has long been a point for discussion. While there is evidence that in the same region during the Byzantine period other diseases with similar symptoms and signs, such as psoriatic arthritis, may have been confused with leprosy (Zias and Mitchell, 1996), doctors at the time of the crusades appear to have had a reasonable knowledge of the disease. Medical texts certainly included all the major signs and symptoms of leprosy (Singer, 1949, Campbell and Colton, 1955-60) but it is not known for sure how many of these were required to be present before a diagnosis could be made. Some doctors tried to improve diagnostic accuracy by using certain tests, such as pricking the skin to see the colour of the fluid which oozed out (Figure 2). A number of leprosaria have been identified in the Latin East (Mitchell, 2000a) and contemporary chronicles tell of how some of the more eccentric and devoutly religious crusaders and pilgrims gave up their wealth and status to care for the people in these institutions (Kedar, 1983). One, named Alberic, wore a rough goat-hair shirt, cut his hair and beard asymmetrically and shouted rude remarks at pilgrims who passed his leprosarium while on their tour of the holy places. However, he also looked after the patients, carrying the weakest about on his shoulders and washing their ulcerated feet. While doctors had no cure for leprosy at that time, they did take the standard approach to the treatment of illness used in the mediaeval period (Campbell and Colton, 1955-60, Singer, 1949). This involved the modification of diet, oral drugs and ointments, bloodletting and bathing in sulphurous hot springs. At that time it was believed that most diseases were the result of an imbalance of various "humours" of the body and that these could be restored to balance using these techniques. They had no concept of infection by a microorganism as causing diseases such as leprosy and so were completely baffled by it.

4. The epidemic theory

Of all the theories concerning the spread of leprosy from one region to another, perhaps the most frequently quoted is that soldiers in the crusading armies contracted leprosy during their stay in the eastern Mediterranean and brought it back to Europe with them (Castiglioni, 1947:353, Clay 1966:36, Foucoult, 1967:6, Harrington, 1920:7, Zambaco Pasha, 1914:97). The theory goes on to suggest that this resulted in an epidemic of leprosy in Europe during the mediaeval period. In support of this, it is known that hundreds of thousands of people did go on crusade between 1095 and 1291 AD and a proportion managed to returned home alive. Leprosy is known to have been present in the Latin East, so it is quite plausible that some of these crusaders may have contracted the disease there. If soldiers and camp followers did return with leprosy then they may have passed on the disease to others in their towns and villages on their return. At the time of the crusades the numbers of leprosaria built in Europe are known to have increased dramatically. At the same time many new laws were passed concerning the diagnosis and segregation of people with the disease. One interpretation of this evidence is that there was a dramatic rise in the prevalence of leprosy in Europe during the 12th and 13th centuries, which was concurrent with the return of crusaders from the eastern Mediterranean. In other words, the crusades were responsible for an epidemic of leprosy in the mediaeval Europe.

4.1 The historical evidence

There are two different ways that historical texts can be used to assess the "crusades" theory. The first is to investigate the significance of the rise in the number of leprosaria established in Europe in each century during the mediaeval period. It is important to determine if the number of leprosaria in a region is actually representative of the prevalence of the disease in that area, or more dependent on other factors. Study of the establishment of leprosaria in Britain has shown that there were few leprosaria until the eleventh century. Their expansion was then rapid and one study of over two hundred well documented leprosaria (Roberts, 1986) identified over fifty foundations dating from the twelfth century, and around eighty in the thirteenth century. From then on the number of new foundations declined with about sixty in the fourteenth century and hardly any new leprosaria being founded in the fifteenth and sixteenth centuries. However, it is known from archaeological evidence that leprosy was present in Britain for over seven hundred years before the foundation of leprosaria became at all common (Manchester and Roberts, 1989). An assessment of the role of the hospital in mediaeval Europe may give the answer as to why leprosaria were mainly founded from the eleventh century onwards. Using a similar approach to the study of leprosaria over time, a breakdown of the formation of general hospitals in Britain by Roberts shows an almost identical pattern. While a small number of hospitals were established in the eleventh century, there was then a dramatic rise peaking in the fourteenth century, declining to relatively small numbers again in the sixteenth century. If it is argued that the number of leprosaria foundations reflect the prevalence of leprosy then a rise in the number of hospitals should mean that other diseases were becoming much more common too. Clearly this is not likely and suggests that another reason was behind the foundation of both leprosaria and hospitals at that time. It seems likely that it was a change in social attitudes to the way in which the poor and sick were cared for. The statutes of these establishments also showed that they were often paid for by rich benefactors who provided funding on the condition that the people who lived there prayed daily for the soul of the benefactor. For these reasons it became popular for an individual with some spare funds to establish a leprosarium or hospital. In fact, records of some leprosaria show that by the end of the crusader period it was not uncommon for them to be half empty (Richards, 1977:83). This did not prevent the foundation of yet more leprosaria by others who wanted prayers for their soul too, despite the fact that they were not necessarily needed. Clearly the foundation of new leprosaria cannot be regarded as an accurate way to estimate changes in the prevalence of leprosy over time.

The second approach to establishing the validity of the "Crusader theory" is to look for any change in the amount of information on leprosy contained in medical texts. It could be argued that common diseases would have had more information written about them than rare diseases, as seen in modern medical books. If this was the case, then a change might crudely reflect the prevalence of the disease at the time the text was written. To study this idea the medical works of three well known authors were consulted (Celsus, 1935-8, Adams, 1844-7, Campbell and Colton, 1955-60). Aulus Cornelius Celsus wrote his text in 1st century BC Italy, Paulus Aegineta in the 6th-7th century AD eastern Mediterranean and Theodorich Borbognoni in 13th century Italy, the time of the crusades. The number of words written on the subject of leprosy was determined for each text, as were the number of words written about all skin diseases. One problem which arises is that texts written in different languages would use a different number of words to say the same thing. To allow comparison, editions of each text were chosen in which the same language was used. While it could be said that the Latin version might be the most scholarly language to use, English editions were chosen so that both archaeologists and historians would be in a position to check the data for themselves. In any case, a similar pattern would still arise if each text was in Latin. The sections counted were not necessarily those under the heading of "leprosy", "lepra", "elephas graecorum" or other terms used at certain times in the past to refer to leprosy. Each section was assessed for symptoms and signs which are compatible with the disease leprosy as we understand it today. The main criteria were destruction of the nose, loss of eyebrow hair, hoarseness of the voice, blindness, loss of skin sensation, ulceration and deformity in the fingers and toes.

The findings are summarised in Table 1 and show that there was a clear and gradual increase in the amount of text written about leprosy from the time of Celsus to Theodorich. It could be argued that as medical knowledge progressed (and plagiarism became more common), books became larger so that the information written about all diseases increased over the centuries. However, the amount of text devoted purely to leprosy as a percentage of the amount of text on all skin diseases also shows a gradual rise over time. Celsus wrote little about leprosy, making up a very small proportion of his section on skin disease. Seven hundred years later, Paulus Aegineta wrote more on leprosy, and this contribution also made up a larger proportion of his section on skin disease. By the time Theodorich wrote about leprosy another seven hundred years later still, there was an even larger section on leprosy, making up about forty percent of his entire section on skin disease. This suggests that the importance of leprosy to medical authors working around the Mediterranean increased gradually from classical times to the mediaeval period. If it was more important to medical authors it might be that it was becoming more common over that time too. If this was the case, then the significant proportion of the population with leprosy in mediaeval Europe was not a consequence of an epidemic, but rather a gradual increase in number of people with the disease.

4.2 Archaeological evidence

While the historical evidence has failed to provide any support for the epidemic theory, it is possible that information from archaeological excavations may. The first question has to be when leprosy was introduced into Europe. If the first skeletal evidence for the disease dates from the time of the crusades, then they could be a plausible mode of disease transfer. However, human skeletal remains with the classical pathological changes of leprosy have been found in northern Europe as early as the 4th and 5th centuries AD (Manchester, 1984, Manchester and Roberts, 1989). Therefore, it seems that the crusades cannot be blamed for introducing the disease.

Another question might be whether the epidemiology of the disease changed as a result of the crusades. It is theoretically possible that leprosy was present in Europe with a low prevalence for hundreds of years, limited by the low population density and the fact that most people lived their entire lives within a few miles of their village. It is conceivable that an influx of infected crusaders may have altered the epidemiological balance and allowed the disease to become much more common, perhaps to epidemic proportions. One way to determine if leprosy is epidemic in a population, rather than endemic, is to look at the proportion of people with the lepromatous and tuberculoid forms of the disease (WHO, 1985). Study of the natural history of a leprosy epidemic was possible in the early part of this century when the disease reached new populations in the Pacific Islands (Bray, 1934, Leiker, 1960, Wade and Ledowsky, 1952). At that time, doctors had sufficient understanding of the disease to distinguish tuberculoid from lepromatous leprosy but the antibiotics used to cure the disease today had not yet been discovered. When leprosy is first introduced into a population the vast majority of cases are tuberculoid leprosy and only a small percentage are lepromatous. A significant proportion of the population have the disease, often up to a third, and all ages are affected. This epidemic phase lasts a number of decades as leprosy has a long symptomless incubation period and then takes many years to slowly cause the disfiguring changes which allow its diagnosis. After the epidemic phase comes the endemic phase. Leprosy is said to be endemic in a community when the disease has been there for some time and the incidence of new cases is stable. The majority of new cases are seen in children or young adults, with the peak age at presentation between ten and twenty years old. Only 1 to 5 percent of a susceptible community develop the disease once it has become endemic, and a much higher proportion of these will have the lepromatous form. Using this information we can look at the evidence for tuberculoid and lepromatous leprosy in the excavated cemeteries of leprosaria in Europe and see if the proportion of each type fits the epidemic or endemic pattern. One limiting factor to this approach is that those people segregated into leprosaria were likely to be the more disfigured cases, while those with less obvious changes may well have been able to conceal their disease. The result is that those individuals buried in a leprosarium cemetery may not have been representative of all those in the population who had leprosy. However, if the majority of cases excavated are tuberculoid then this would suggest that the disease was in the epidemic phase in that population, although tuberculoid leprosy is difficult to diagnose skeletally.

A number of leprosaria in use before and during the time of the crusades have been excavated. The leprosarium of St. John Timberhill in Norwich, England identified a number of cases of leprosy (Anderson, 1998) which have been radiocarbon dated to the preconquest period (before 1066 AD). The vast majority of cases were lepromatous, rather than tuberculoid, which is compatible with endemic leprosy, as would be expected at that time. The leprosarium of St. James and St. Mary Magdalene in Chichester, UK was in use from the twelfth to the seventeenth centuries (Lee and Magilton, 1989). That part of the cemetery in use first (area 1) would have been contemporary with the crusades and again shows a striking predominance of lepromatous over tuberculoid cases (Table 2). The results of excavations such as these give no evidence for an epidemic of leprosy at the time of the crusades.

6. Conclusion

Having reviewed all the available evidence regarding the epidemic theory the conclusions are strikingly clear. The increase in the number of leprosaria in mediaeval Europe does not suggest an epidemic of leprosy at that time, rather a change in the way people with disease were treated by society. The increase in the amount of information on leprosy in medical texts suggest a gradual increase in the importance of the disease, and so perhaps its prevalence, over the centuries but does not suggest an epidemic at the time of the crusades. The archaeological evidence is harder to interpret but certainly does not support the concept of an epidemic at that time.

It is quite possible that as many soldiers with concealable signs and symptoms took leprosy with them on crusade, as returned with the disease, having contracted it in the Latin East. There is not much convincing evidence yet for blaming the crusades for dramatically changing the epidemiology of leprosy in mediaeval Europe. Until such time as new evidence comes to light then this widely held concept should really be discarded.

References

Adams, F. (ed) 1844-7 *The Seven Books of Paulus Aegineta*. Three volumes. London: Sydenham Society, Volume 2, pp.1-5.

Amundsen, D. W. 1977 Medical deontology and pestilential disease in the late Middle Ages. *Journal of the History of Medicine and Allied Sciences* 32: 403-21.

Andersen, J. G. 1969 Studies in the medieval diagnosis of leprosy in Denmark. *Danish Medical Bulletin* 16 (Supplement): 1-142.

Anderson, S. 1998 Leprosy in a medieval churchyard in Norwich. In S. Anderson (ed): *Current and Recent Research in Osteoarchaeology*. Oxford: Oxbow Books: pp.31-7.

Bray, G. W. 1934 The story of leprosy at Nauru. *International Journal of Leprosy* 2: 319-23.

Campbell, E. and Colton, J. (eds) 1955-60 *The Surgery of Theodorich, ca. AD 1267*. Two volumes. New York: Appleton-Century-Crofts, Volume 2, pp.167-82.

Castiglioni, A. 1947 *A History of Medicine*. Translated by E.B. Krumbhaar. London: Routledge and Kegan Paul.

Celsus, Aulus Cornelius 1935-8 *De Medicina*. Translated by W.G. Spencer. Three volumes. London: Heinemann, Volume 1 p.343.

Clay, R. M. 1966 *The Mediaeval Hospitals of England*. London: Frank Cass.

Demaitre, L. 1985 The description and diagnosis of leprosy by fourteenth century physicians. *Bulletin of the History of Medicine* 59: 327-44.

Dols, M.W. 1979 Leprosy in medieval Arabic medicine. *Journal of the History of Medicine and Allied Sciences* 36: 314-33.

Dols, M.W. 1983 The leper in medieval Islamic society. *Speculum* 58(4): 891-916.

Ell, S.R. 1985 Diet and leprosy in the medieval West: the noble leper. *Janus* 2: 113-29.

Ell, S.R. 1986 Leprosy and social class in the Middle Ages. *International Journal of Leprosy and Other Mycobacterial Diseases* 54(2): 300-5.

Ell, S.R. 1987 Plague and leprosy in the Middle Ages: a paradoxical cross-immunity? *International Journal of Leprosy and other Mycobacterial Diseases* 55(2): 345-50.

Foucoult, M. 1967 *Madness and Civilization. A History of Insanity in the Age of Reason*. Translated by R. Howard. London: Tavistock.

Harrington, J. 1920 *The School of Salernum*. London: Oxford University Press.

Hudson, E.H. 1963 Treponematosis and pilgrimage. *American Journal of the Medical Sciences* 246: 645-56.

Hulse, E.V. 1975 The nature of biblical leprosy and the use of alternative medical terms in modern translations of the bible. *Palestine Exploration Quarterly* 107: 87-105.

Ji, B. and Grosset, J. 1994 Leprosy. In P. D. Hoeprich, M. C. Jordan and A. R. Ronald (eds): *Infectious Disease*. Philadelphia: J.B. Lippincott, pp.1008-17.

Kedar, B.Z. 1983 Gerard of Nazareth. A neglected twelfth-century writer in the Latin east. *Dumbarton Oaks Papers* 37: 55-78.

Lee, F. and Magilton, J. 1989 The cemetery of the hospital of St. James and St. Mary Magdalene, Chichester - a case study. *World Archaeology* 21(2): 273-82.

Leiker, D. L. 1960 Epidemiological and immunological surveys in Netherlands New Guinea. *Leprosy Review* 31(4): 241-59.

Manchester, K. 1984 Tuberculosis and leprosy in antiquity: an interpretation. *Medical History* 28: 162-73.

Manchester, K. and Roberts, C. 1989 The palaeopathology of leprosy in Britain: a review. *World Archaeology* 21(2): 265-72.

Mitchell, P.D. 1994 Pathology in the Crusader Period: human skeletal remains from Tel Jezreel. *Levant* 26: 67-71.

Mitchell, P.D. 1997 Further evidence of disease in the Crusader Period population of Le Petit Gerin (Tel Jezreel, Israel). *Tel Aviv* 24(1): 169-79.

Mitchell, P.D. 1999 Tuberculosis in the crusades. In G. Pálfi, O. Dutour, J. Deák and I. Hutás (eds): *Tuberculosis. Past and Present*. Szeged, Hungary: TB Foundation and Budapest, Hungary: Golden Book Publishers Ltd., pp.43-9.

Mitchell, P.D. 2000a The evolution of social attitudes to the medical care of those with leprosy within the Crusader States. *Cahiers du GRHIS (Rouen)* 11: 21-8.

Mitchell, P.D. 2000b An evaluation of the leprosy of King Baldwin IV of Jerusalem in the context of the medieval world. Appendix. In B. Hamilton (ed): *The Leper King and his Heirs. Baldwin IV and the Crusader Kingdom of Jerusalem*. Cambridge: University Press, pp.245-58.

Richards, P. 1977 *The Medieval Leper and his Northern Heirs*. Cambridge: D.S. Brewer.

Roberts, C. 1986 Leprosy and leprosaria in medieval Britain. *M.A.S.C.A. Journal* 4(1): 15-21.

Roberts, C. 1987 Leprosy and tuberculosis in Britain: diagnosis and treatment in antiquity. *M.A.S.C.A. Journal* 4(4): 166-71.

Rose, J.C. 1998 Crusader Period disease in Jordan. *Abstracts of the 12th European Meeting of the Paleopathology Association, August 26-29th, 1998, Prague-Pilsen, Czech Republic.* Prague: Charles University p.77.

Shahar, S. 1982 Des lepreux pas comme les autres. L'Ordre de Saint-Lazare dans le royaume latin de Jerusalem. *Revue Historique* 267: 19-41.

Singer, C. 1949 A thirteenth century clinical description of leprosy. *Journal of the History of Medicine and Allied Sciences* 4: 237-9.

Stirland, A. and Waldron, T. 1990 The earliest cases of tuberculosis in Britain. *Journal of Archaeological Science* 17: 221-30.

Wade, H.W. and Ledowsky, V. 1952 The leprosy epidemic at Nauru; a review with data on the status since 1937. *International Journal of Leprosy* 20(1): 1-29.

World Health Organisation (WHO) 1985 *Epidemiology of Leprosy in Relation to coNtrol.* Technical Report Series No. 716. Geneva: World Health Organisation.

Zambaco Pasha, D.A. 1914 *La Lèpre a Tranvers les Siècles et les Contrées.* Paris: Masson.

Zias, J. 1991 Leprosy and tuberculosis in the Byzantine monasteries of the Judean Desert. In D. J. Ortner and A. C. Aufderheide (eds): *Human Paleopathology: Current Syntheses and Future Options.* Washington D.C.: Smithsonian Institution Press, pp.197-9.

Zias, J. and Mitchell, P.D. 1996 Psoriatic arthritis in a fifth-century Judean Desert monastery. *American Journal of Physical Anthropology* 101: 491-502.

Table 1: Volume of text on leprosy by representative medical authors from the classical to mediaeval periods (English language versions).

	Celsus 1st Century BC	**Paulus Aegineta** 6th-7th Century AD	**Theodorich** 13th Century AD
Number of words on leprosy	250	2,000	5,280
Text on leprosy as % on skin disease.	2.5%	12%	39%

Table 2: Tuberculoid leprosy in the mediaeval leprosarium cemetery of St. James and St. Magdalene, Chichester, England (12th to 17th centuries AD).

	Area 1 (oldest)	**Area 2** (intermediate)	**Area 3** (latest)
% total cases of leprosy in each area which were tuberculoid (total n=83)	16%	21%	35%

Adapted from Lee and Magilton (1989)

Figure 1: Prince Baldwin having his numb forearm pinched by his schoolmates and the examination by William of Tyre. British Library, MS Henry Yates Thompson 12, fol. 152r. History of Outremer by William of Tyre, this copy written 1250-60 AD. Reproduced with the kind permission of the British Library, London.

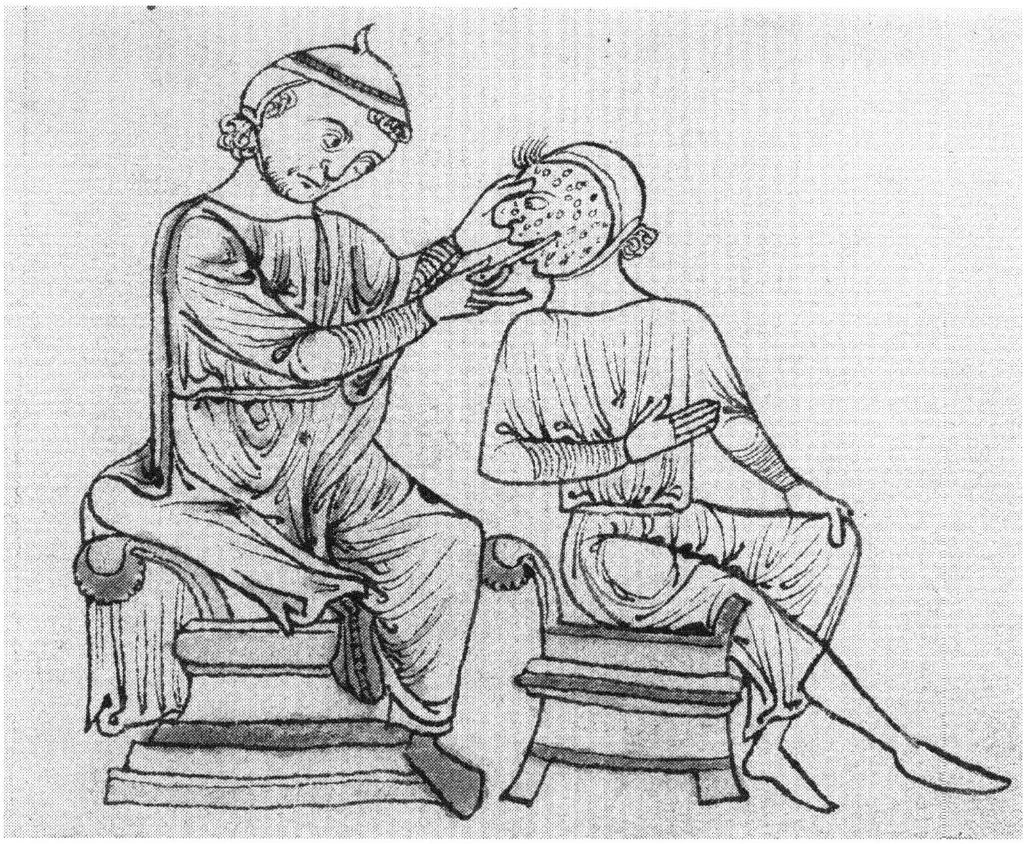

Figure 2: A doctor examining a man with skin disease. Trinity College MS 0.1.20, fol. 264r. Anglo-Norman translation from c.1230-1250 AD of the Chirurgia of Roger of Parma, written c.1180 AD. Reproduced with the kind permission of the Master and Fellows of Trinity College, Cambridge, England.

Leprosy in Roman period skeletons from Kellis 2, Dakhleh, Egypt

Joseph E. Molto
Department of Anthropology
Lakehead University
Thunder Bay
Ontario
Canada P7B 5E1

E-mail: jemolto@flash.lakeheadu.ca

Abstract

The Roman period Kellis 2 (K2) cemetery in the Dakhleh Oasis, Egypt, has yielded two young adult male burials (K2-B6 and K2-B116) with osseous lesions pathognomonic of lepromatous leprosy. The burials are radiocarbon dated to the early-mid 4th century A.D. Two other young adult males (K2 B9 and K2-B222) show skull changes suggestive of leprosy. Preliminary morphological evidence places B6 and B116 in the local population, although they are not maternally related. Moreover, δ15 N isotopic evidence suggests they lived elsewhere prior to their death. This, and their inclusion in the general cemetery population hints at a level of tolerance towards leprosy at Kellis. A crude prevalence rate of 3.7% indicates that leprosy was possibly endemic to the region. These and four Ptolemaic period skeletons from Dakhleh are the earliest corporate evidence of leprosy in antiquity and raise important questions about the ultimate origins and evolution of this mycobacterial disease.

Keywords

Leprosy, Roman period, Dakhleh Egypt

1. Introduction

This paper describes new cases of leprosy from Dakhleh, a remote Oasis in Egypt's western desert (Figure 1). The skeletons are associated with the ancient town of Kellis, which was occupied for approximately 700 hundred years from the Ptolemaic period well into Roman times (Molto, in press, 1997). Burials from the early occupation of Kellis (i.e. Ptolemaic and early Roman) were interred in rock-cut tombs in Kellis 1 (Cook, 1994), while those from the later Roman period were interred in individual mudbrick interments in Kellis 2 (Sheldrick, 1997). The Kellis 2 (K2) burials are primary interments oriented east-west with their heads to the west. At least two male burials from K2 (K2-B6 and K2-B116) have skeletal changes pathognomonic of lepromatous leprosy. Two others males (K2-B9 and K2-B222) are listed as possible sufferers of this disease. The skeletons are described below and briefly placed within a population framework.

2. Material and Methods

The site of K2 is ideal for paleoepidemiological research as bone preservation is excellent and demographically the sample is large and highly representative. Of the 378 burials excavated to date, ~ 60% are subadults, and the sex ratio is statistically balanced (Molto, 1999). Moreover, those with leprosy are included with the general cemetery population (Figure 1).The skeletal changes of leprosy are well known and have been reported by many researchers, including Andersen *et al*. (1992), Andersen and Manchester (1987, 1988), Aufderheide and Rodríguez-Martin (1998), Brothwell (1981), Crane (1993), Jopling (1971), Lewis *et al*. (1995), Michman and Sager (1957), Paterson and Job (1964), Moller-Christensen (1961, 1967, 1978), Roberts and Manchester (1995) and Steinbock (1976).

A strong light source was used to examine the bones of the endonasal cavity of the Dakhleh crania. Following photography, the inferior nasal conchae were sampled for future *Mycobacteria leprae* ancient DNA research. Radiographs were taken only of the hand and foot bones. Skeletal sex determination was based on the Phenice method (Phenice, 1969) and skeletal robusticity criteria (e.g. supraorbital ridge and mastoid processes). Adult age was estimated from the morphogenesis of the pubic symphysis (Katz and Suchey, 1986, Katz and Suchey, 1989) and the sternal rib ends (Loth and Iscan, 1989), the fusion of epiphyses (i.e. iliac crest and medial clavicle), the fusion of S1-S2, and the degree of dental attrition relative to the Kellis 2 population sample. Burials 6 and 116 have respective calibrated AMS radiocarbon dates of 435 ± 50 years A.D. (Isotrace # T0-5476) and 445 ± 50 years AD (Isotrace # T0-6103). The overlapping of the dates is of interest given the proximity of these burials within the cemetery. Archaeological dating based on coins found in the village suggests that Kellis may have been abandoned late in the fourth century (Bowen *et al.*, 1993), or in the lower end of the range noted for the radiocarbon dates. Since the village and the cemetery are only partly excavated, the exact temporal range of Kellis 2 is unknown. A conservative estimate based on archaeological and radiocarbon dating, is c. 300-450 A.D. (Molto, 1994, Sheldrick, 1997). Radiocarbon dates for burials K2-9 and K2-222 are not available so their temporal affiliations with burials 6 and 116 are unknown.

3. Results

3.1 The skeletons affected by leprosy

Burial 6 is the complete well preserved skeleton of a male with a pubic symphysis age estimate of 28.7 ± 6.5 years (Phase III) and a rib age estimate of 26-32 (Phase 4). Both the clavicular and iliac epiphyses are fused, while S1-S2 is open. These data suggest an individual in his late 20's (~29 years), which is supported by the dental attrition (Figure 2a). Figures 2a and 2b show the osseous changes to the rhinomaxillary region. Resorption of the anterior alveolus of the maxilla is advanced, resulting in the loss of the incisors, right canine and premolar. Concomitant with this tooth loss is a build-up of calculus on the lingual surface of the mandibular incisors and canines (not shown in the figures). The anterior nasal spine and the margins of the nasal aperture are resorbed, as is the inferior border of the right nasal bone. In the nasal cavity the conchae and the endonasal floor show evidence of pitting and periosteal reaction. The infection has thinned and eroded the palate, producing perforations that are skeletally patent with the oral cavity in the mid left palate and the left transverse palatine suture (Figure 2a). These changes are pathognomonic of rhinomaxillary syndrome (Andersen and Manchester, 1992). Figure 3 illustrates key osseous changes in the hands of burial 6. Periosteal new bone occurs on the palmar surfaces of all the right metacarpals and on both the palmar and dorsal surfaces of the left. The 3rd left metacarpal is particularly affected. Unfortunately, the distal phalanges were not recovered. While diaphyseal resorption of the proximal phalanges and the metacarpals is absent, palmar grooving occurs on the anterior distal margins of the proximal phalanges of both hands (Figure 4). These grooves are associated with the early development of claw-hand deformity (Andersen and Manchester, 1987). The distal end of the 5th right proximal phalanx is eroded and its dorsal surface has considerable reactive periostitis. None of the seven mid-phalanges show erosive changes. Osseous changes in the feet are less advanced than those of the hand and consist entirely of periosteal new bone formation. This occurs on the right calcaneum, the dorsal surfaces of the right metatarsals and the 1st, 4th and 5th left metatarsals. It is most pervasive on the fifth metatarsals. There are only six foot phalanges preserved with one, a left distal phalanx of the 1st metatarsal, showing arthritic changes and a healed fracture. The latter is probably unrelated to leprosy. Erosive changes are absent and the metatarso-phalangeal joints are normal. The only other pathological change that may be associated with leprosy is "active" periostitis on the right distal fibula, and the mid-to proximal medial tibial diaphyses (Lewis *et al.*, 1995).

Burial 116 is the intact skeleton of a male with respective pubic symphysis and rib age estimates of 24.3 ± 3.6 (Phase II) and 20-24 (Phase 3) years. All epiphyses except the sternal end of the clavicle are fused, while S1-S2 is open. Dental attrition agrees with a skeletal age in the early twenties (~23 years). Unfortunately, many of the middle and distal phalanges in the hands were not recovered. The pathological changes in the skull are shown in Figures 5a and 5b. The anterior nasal spine is resorbed but the margins of the pyriform aperture are normal. There is no evidence of inflammatory changes at prosthion, although the palate adjacent and posterior to the incisive foramen (not shown in the figures) is extremely pitted. Moreover, reactive osseous changes occur on the floor of the right endonasal cavity near the midline and on the inferior nasal conchae (Figure 5a). These skull changes are characteristic of rhinomaxillary syndrome (Andersen and Manchester, 1992).

The hands show significant pathological osseous changes (Figure 6). All distal phalanges present (five right and the left 1st) show various stages of acroosteolysis, while most of the proximal phalanges have enlarged nutrient foramina and palmar grooving, signalling the development of claw-hand deformity (Andersen and Manchester, 1987). This deformity has also produced "pitting" of the heads of the latter bones (Figure 7) on their dorsal margins, likely as a response to chronic hyperflexion. A lateral view of claw-hand deformity in the second right phalanx is profiled macroscopically and radiologically in Figures 8a and 8b. A narrowing of the medullary space and the cortical bone on the palmar surface of the distal end is evident in the radiograph (Figure 8b). Periostitis is present on the lateral and dorsal surfaces of both phalanges (Figure 8a). Of significance for clinical assessment is the fact that the periostitis is not clearly visible on the radiograph. Periostitis also occurs on the dorsal surfaces of the proximal and middle phalanges of the right middle digit and 1st and 2nd metcarpals. On the left hand slight periostitis occurs on the palmar and lateral surfaces of the second metacarpal, but is otherwise absent on the other metacarpals and the proximal phalanges. The osseous changes on the left hand were generally less pronounced than those on the right.

Figure 9 shows a dorsal view of the metatarsals and phalanges of both feet. The left foot shows advanced resorptive changes which have lead to the destruction of the distal phalanges, 4th and 5th metatarsals and the fracturing of the 2nd and 3rd metatarsals. Mid-diaphyseal remodelling of the right proximal phalanges is also advanced, particularly in the third digit. The articular surfaces of the middle and proximal phalanges in all but the right 5th phalanx of both feet are generally osteopenic and porotic. The changes in the right foot are less advanced but illustrate the typical development of metatarsal diaphyseal remodelling (Andersen *et al.*, 1992). This is illustrated macroscopically and radiographically in Figure 10 (a and b). The distal end of the 3rd metatarsal is completely resorbed and the 2nd shows partial remodelling. Together, these bones illustrate the developmental pattern of diaphyseal resorption in the distal part of the 3rd metatarsal which, in this case, is knife-edge rather than concentric remodelling. The other metatarsals are normal except for some slight periostitis on the distal-dorsal surface of the 5th metatarsal. Ulceration and sepsis of the left midfoot has ankylosed most of the cuboid and the medial and lateral cuneiform bones (Figure 11). The calcaneus lateral to the anterior talar facet has considerable reactive new bone formation with several osteolytic foci. Figure 12 illustrates key changes in the navicular bone which is the focal support of the longitudinal arch (Andersen and Manchester, 1988). The anterior articular facets are completely destroyed and the bone is wedge-shaped with large dorsal exostoses that are a sequel to the navicular's superior displacement. Dorsal exostoses also occur on the cuneiforms and the anterior dorsal margin of the talar neck. Exostoses near the medial talar tubercle are also interpreted as a concomitant result of ligamentous stress following collapse of the medial mid-foot. The plantar surface of all the tarsals show destructive and osteosclerotic changes from ulceration and sepsis. The changed biomechanics of the foot resulted in a large exostosis on the postero-medial margin of the calcaneum. The only evidence of osseous changes in the right tarsals are dorsal exostoses on the navicular bone, the middle and lateral cuneiforms. Based on the changes to the right 2nd and 3rd metatarsals noted above, these exostoses represent the early development of *pes planus* from loss of motor function and the start of collapse of the longitudinal arch. Slight active periosteal new bone is present on the distal right ulna and radius, and the left distal tibia and fibula. The fibula is particularly affected as diffuse and striated periostitis occurs throughout the diaphysis. A faint trace of periostitis is present on the right distal anterior tibia.

3.2 Possible cases of leprosy

Burial 9 is the complete skeleton of a male with pubic symphysis and rib age estimates of 28.8 ± 5.9 years and 24 to 28 years, respectively. The medial clavicular and iliac epiphyses are both fused, as is S1-S2. Dental attrition suggests an individual in his early 20s. The palate is osteopenic and perforated in the right palatine bone along the right transverse palatine suture (Figure 13). In addition, the inferior nasal conchae (not shown in Figure 13) show slight inflammatory (pitting) changes. Periosteal reactive changes occur on the right medio-proximal fibula. In the vertebral column, a large lytic lesion occurs on the right infero-anterior centrum of the L4. This lesion is well circumscribed and has penetrated the cartilaginous plate. This skeleton is listed as "possibly" leprous on the basis of the skull changes; the vertebral lesion could possibly be the result of tuberculosis.

Burial 222 is the complete skeleton of a young adult male with respective pubic symphysis and rib age estimates of 23.4 ± 3.6 years and 26 to 32 years. The dental pathology supports a skeletal age in the late 20s. The anterior nasal spine is resorbed, as are the margins of the nasal aperture (Figure 14). Enlarged foramina are present, and are likely related to the absorptive/inflammatory process occurring along these margins particularly on the right side. The inferior nasal conchae are pitted, suggesting an acute inflammatory response. Active bilateral periostitis occurs on the supero-medial surfaces of the tibia.

Table 1 summarizes the skeletal changes in these burials. The skeletal changes in B6 and B116 are pathognomonic of lepromatous leprosy (Møller-Christensen, 1961). The osseous rhinomaxillary changes in B9 and B222 are typical of the early stages of lepromatous leprosy, although the periostitis in the infracranial bones is unlikely to be related to this infection (Crane, 1993). The fact that these individuals are also young adult males is additional, albeit provisional, support for this diagnosis, as skeletal leprosy is most common in this cohort (Crane, *ibid.*).

3.3 Differential diagnosis

A number of diseases show similar early skeletal changes to those found in lepromatous leprosy. These include tuberculosis, treponematosis, sarcoidosis, glanders and several of the mycoses (Crane, 1993). However, none of these diseases produces the "advanced pattern" of skeletal changes in the rhinomaxillary region and hands and feet observed in burials K2-6 and K2-116. The diagnosis of leprosy in burials 9 and 222, although presumptive, is strengthened by the identification of lepromatous leprosy in the population. The osteolytic vertebral lesion in burial 9 could, however, represent an infection from *M. tuberculosis*. Although tuberculosis could be a potential cause of the palatal lesions in this burial, early endonasal infection by *M. leprae* is more probable. The diagnosis of leprosy in B9 and B222 awaits molecular confirmation (Molto, 1998, 1999).

4. Discussion

Early cases of leprosy are rare in the palaeopathological literature, with the oldest evidence dating to the Ptolemaic (c. 250 BC) period in the Dakhleh Oasis (Dzierzykray-Rogalski, 1980). Although the skeletal lesions in the four male skulls were only briefly described by Dzierzykray-Rogalski (1980), the diagnosis of leprosy is generally accepted by palaeopathologists (Aufderheide and Rodríguez-Martin, 1998, Manchester, 1984, 1991, Steinbock, 1976). Moreover, the commingled nature of the burials precluded associating affected hand and foot bones with the skulls. The completeness of the Kellis 2 skeletal remains, and the variation in the osseous impact of this disease in these four 4th century A.D. burials, adds to the corpus of information on the pathogenesis and antiquity of leprosy. A combination of factors influences the pathogenesis of leprosy. These include, the age of initial infection, the length of the incubation period and the immune status of the individuals (Aufderheide and Rodríguez-Martin, 1998). Leprosy is usually contracted by immuno-compromised children, with clinical disease developing after a long incubation period of at least 3 years (Manchester, 1984). Although estimates vary, it is acknowledged that in populations with endemic leprosy few individuals (~1%) will develop clinical disease. Of this limited group only a small percentage (~15%) will show diagnostic skeletal involvement (Chamberlain *et al.*, 1931). The high rate of bone involvement (>50%) found in leprosaria patients (Møller-Christensen, 1961, 1967) is clearly a sampling bias, as expected when individuals are culled from a population because of a disease (Crane, 1993).

In Kellis 2, only 108 adolescent and adult burials were examined for leprosy. The prevalence of 3.7% (4/108), though an extremely crude estimate (see Waldron, 1991), suggests that leprosy was endemic in the Oasis using Chamberlain *et al.*'s (1931) model. Of course, this assumes that the affected individuals were indigenous to ancient Dakhleh. The skeletal changes in K2-6 and K2-116 would be unequivocally associated with gross clinical evidence of leprosy. The rhinomaxillary disease in B222 may also have been clinically recognizable (i.e. "saddle nose"). Of interest is the fact that of the four burials, only B9 was disturbed postmortem in a cemetery with anthropogenic disturbance in greater than half of the graves. The presence of clinical leprosy in burials 6, 116 and 222 may have deterred potential grave robbers if the looting was an "inside job", as suggested by Sheldrick (1997). The inclusion of individuals with the disease in the general population (Figure 1) may, however, suggest a level of tolerance towards leprosy at Kellis in the Roman period. Whether this is a concomitant of Christian acceptance of individuals with leprosy is a question of considerable significance for future research (Parr *et al.*,1998).

Given the low probability of finding leprosy in ancient populations (Steinbock, 1976), the large number of cases found in ancient Dakhleh, begs the question of the origin of this mycobacterial disease. Dzierzykray-Rogalski (1980) suggested that the individuals afflicted with leprosy in his Ptolemaic Dakhleh samples were elite "Europoids" brought to Dakhleh to clauster this disease: 'The Oasis was used as a deportation place of the leprosy-ill belonging to the ruling class' (Dzierzykray-Rogalski, 1980:73). Objective evidence for the presumptive ancestry of these individuals, or the supposed "negroid" affiliations of the recipient Dakhleh population, is not provided by Dzierzykray-Rogalski.

Current research on skeletal variation, hair structure and mtDNA markers on the ancient Dakhleh population samples, clearly refutes the "negroid" affiliations of the Dakhleh population. For example, the hair (n=156 samples) of the ancient Dakhlens is generally dark brown and wavy (Molto, 2001, 1998) and its microstructure is more caucasoid than negroid (Charles Linch, pers. comm.). Moreover, the HPA I mtDNA marker, which is found in 76% of modern sub-Saharan Africans, is absent in both the late period (N=5) and Kellis 2 (N=14) remains from Dakhleh (Parr, pers. comm.). Finally, the morphogenetic data from these populations show a slowly evolving deme from the Late Period through to Roman times (Molto, 1994). The most parsimonious genetic interpretation of the Dakhleh population samples in the time period under consideration is that of an indigenous regional North African population, distinct from their sub-Saharan neighbours. This does not, however, imply that the population was a genetic isolate. Documentary evidence from the Roman period of Kellis identifies considerable contact with other peoples, particularly those in the Nile region (Worp, 1995). While the population affiliation of the Ptolemaic burials with leprosy is equivocal, those with this disease in Kellis 2 appear to fit craniometrically with the male subset of this population sample (Table 2).

Briefly, the majority of these measurements and indices fall within one standard deviation of the K2-sample. Of additional interest is the fact that the individuals with unequivocal leprosy (B6 and B116) are not maternally related. As shown in Table 3, they represent two separate mitochondrial consensus sequences (Anderson *et al.*, 1981), and are dissimilar to each other at HVI loci 16223, 16298, and 16311 (Parr *et al.*, 1998). These individuals otherwise share close spatial (see Figure 1) and temporal (see radiocarbon dates) proximity in the cemetery and, in addition, both may have been recent arrivals to Kellis prior to their deaths. The latter derives from the interpretation of the δ15N isotopic data. B6 and B116, with respective δd15N values of 16.5% and 14.5 %, are outliers relative to the rest of the Kellis 2 sample (δ15N = 17.9 ± 1.1%; N= 61). Current isotopic research has demonstrated a positive enrichment (increasing values) of 15N with increased aridity (Ambrose and DeNiro, 1986a, 1986b, Schwarcz *et al.*, 1999). Given the hyperarid conditions in Dakhleh today and in the Roman period (Fairgrieve and Molto, 2000), the most parsimonious interpretation of these data is that the leprous individuals inhabited a less arid region, likely the Nile, prior to coming to Dakhleh (Dupras, 1999). Questions about the relationships of these individuals to each other and the Kellis population are fundamental to the interpretation of leprosy at Dakhleh. The research is ongoing and will incorporate multivariate analyses of osteometric, non-metric and molecular data. Medical historians hypothesize that leprosy originated in Indo-China during the pre-Christian millennium (Aufderheide and Rodríguez-Martin, 1998, Steinbock, 1976). The supporting evidence is entirely literary and iconographical in nature (Manchester, 1984). Though valuable, palaeopathologists consider such information as limited in terms of the diagnosis of ancient infectious diseases (Steinbock, 1976). Recently, Grmek (1983) has suggested an African origin for leprosy. The accumulating evidence for leprosy in ancient Dakhleh, and its absence from early sites in Asia and intermediate regions, provide increasing support for this hypothesis, although it does not negate the possibility of an Asian origin for this notorious disease.

Acknowledgements

I thank Professor Tony Mills the director of the Dakhleh Oasis Project for providing the opportunity for the bioarchaeological research to be undertaken in the Dakhleh Oasis. The radiographs were processed in the Dakhleh Oasis by Dr. Peter Sheldrick whose technical and medical expertise in the field is always appreciated. I also thank Dr.

Scott Hamilton, Professor of Archaeology and Mr. Jerome Capricci, Research Associate, Paleo-DNA laboratory, both at Lakehead University, for their comments and editorial suggestions. This research was supported by a Lakehead University Senate Research Infrastructure Grant.

References

Ambrose, S. and DeNiro, M. 1986a Reconstruction of African human diet using bone collagen carbon and nitrogen isotopic ratios. *Nature* 319:321-324.

Ambrose, S. and DeNiro, M. 1986b The isotopic ecology of East African mammals. *Oecologia* 69:395-409.

Andersen, J.G. and Manchester, K. 1987 Grooving of the proximal phalanx in leprosy: A palaeopathological and radiological study. *J. Archaeological Science* 14: 77-82.

Andersen, J.G. and Manchester, K. 1988 Dorsal tarsal exostoses in leprosy. A palaeopathological and radiological study. *J. Archaeological Science* 15:51-56.

Andersen, J.G. and Manchester, K. 1992 Rhino-maxillary syndrome in leprosy: A clinical, radiological and paleopathological study. *International J. of Osteoarchaeology* 2:121-129.

Andersen, J., Manchester, K. and Shahzady, A. R. 1992 Diaphyseal remodelling in leprosy: a radiological and palaeopathological study. *International J. of Osteoarchaeology* 2: 211-219.

Anderson, S., Bankier, A.T., Barrell, B.G., de Bruijn, M., Coulson, A., Drouyin, J. and Eperon, I.C. 1981 Sequence and organization of the human mitochondrial genome. *Nature* 290: 457-465.

Aufderheide, A. and Rodríguez-Martin, C. 1998 *Cambridge encyclopedia of human palaeopathology*. Cambridge: University Press.

Bowen, G., Hope, C. and Kaper O.E. 1993 A brief report on the excavations at Ismant El-

Kharab in 1992-93. B.A.C.E. (no volume stated): 17-28.

Brothwell, D. 1981 *Digging up bones*. Ithaca: Cornell University.

Chamberlain, W.E., Wayson, N.E. and Garland, L. 1931 Bone and joint changes of leprosy. A roentgenological study. *Radiology* 17: 930-939.

Cook, M. 1994 The mummies of Dakhleh. In A. Herring and L. Chan (eds): *Strength and diversity:a reader in physical anthropology*. Toronto: Canadian Scholars Press Inc., pp.259-277.

Crane, G. A. 1993 *Paleopathological and paleoepidemiological analysis of skeletal remains from the mediaeval leper hospital of St. James and St. Mary Magdalene, Chichester, England.* Unpublished Masters Thesis, Department of Anthropology, University of Toronto.

Dzierzykray-Rogalski, T. 1980 Paleopathology of the Ptolemaic inhabitants of Dakhleh Oasis. *J. Human Evolution* 9:71-74.

Dupras, T. 1999 *Dining in the Dakhleh Oasis, Egypt: Determination of diet using documents and stable isotopic analysis*. Unpublished PhD Thesis, McMaster University, Hamilton, Ontario, Canada.

Fairgrieve, S. and Molto, J. 2000 Cribra orbitalia in two temporally disjunct population samples from the Dakhleh Oasis, Egypt. *Amer. J. Physical Anthropology* 111(3):319-331.

Grmek, M.D. 1983 *Diseases at the dawn of western civilization: research on the pathological reality of the Prehistoric Greek world*. Paris: Payot.

Jopling, W. 1971 *Handbook of leprosy*. London: William Heinemann Medical Books, Redwood Burn Ltd.

Katz, D. and Suchey, J.M. 1986 Age determination of the male os pubis. *Amer. J. of Physical Anthropology* 69: 427-435.

Katz, D. and Suchey, J.M. 1989 Race differences in pubic symphyseal aging patterns in the male. *Amer. J. of Physical Anthropology* 80:167-172.

Lewis, M., Roberts, C. and Manchester, K. 1995 Inflammatory bone changes in leprous skeletons from the medieval hospital of St. James and St. Mary Magdalene, Chichester, England. *Int. J. of Leprosy* 63(1):77-85.

Loth, S.R. and Iscan, M.Y. 1989 Morphological assessment of age in the adult: the thoracic region. In M.Y. Iscan (ed): *Age markers in the human skeleton*. Springfield, Illinois: Charles. C. Thomas, pp. 105-136.

Manchester, K. 1984 Tuberculosis and leprosy in antiquity: an interpretation. *Medical History* 28:162-173.

Manchester, K. 1991 Tuberculosis and leprosy: Evidence for the interaction of disease. In D. Ortner and A. Aufderheide (eds): *Human Paleopathology: Current syntheses and future options*. Washington, DC.: Smithsonian Institution Press, pp. 23-35.

Michman, J. and Sager, F. 1957 Changes in the anterior nasal spine and the alveolar process of the maxillary bone in leprosy. *International J. of Leprosy* 25: 217-222.

Møller-Christensen, V. 1961 *Bone changes in leprosy*. Copenhagen: Munksgaard.

Møller-Christensen, V. 1967 Evidence of leprosy in earlier peoples. In D. Brothwell and A.T. Sandison (eds): *Diseases in antiquity*. Springfield, Illinois: Charles C. Thomas, pp 295-306.

Møller-Christensen, V. 1978 *Leprosy changes of the skull*. Odense, Denmark: University Press.

Molto, J.E. (2001) The skeletal biology and paleoepidemiology of population samples from Tirghi and Kellis, Dakhleh, Egypt. In M. Marlow and A. Mills (eds): *The Oasis Paper: Monograph 6*. Oxford: Oxbow Books: 81-100.

Molto, J.E. 1997 *Morbidity and mortality in Kellis 2, Dakhleh, Egypt*. Paper presented at the Second Dakhleh Oasis Conference, Royal Ontario Museum, Toronto, Canada.

Molto, J.E. 1998 *Disease history from the Dakhleh oasis, Egypt*. Paper presented at the Scholars Symposium for the Society for the Study of Egyptian

Antiquities, Toronto Ontario, November 8th.

Molto, J.E. 1999 *Death! Not quite on the Nile. Paleoepidemiology of ancient populations from the Dakhleh Oasis, Egypt.* Paper presented to the Society for the Study of Egyptian Antiquities Lecture Series, March 11th.

Parr, R., Walters, S., Molto, J. and Capricci, J. 1998 *A tale of two lepers from the Dakhleh Oasis, Egypt: A Preliminary report on chemical and molecular analysis.* Poster presentation at the 25th Annual Paleopathology Association Meeting, Salt Lake City, Utah, March 31st.

Paterson, D.E. and Job, C.K. 1964 Bone changes and absorption in leprosy. A radiological, pathological and clinical study. In R.G. Cochrane and T.F. Davey (eds): *Leprosy in theory and practice.* 2nd edition. Bristol: John Wright and Sons Ltd. pp. 425-446.

Phenice, T.W. 1969 A newly developed visual method of sexing the os pubis. *Amer. J. of Physical Anthropology* 30:297-301.

Roberts, C. and Manchester, K. 1995 *The archaeology of disease.* 2nd edition. New York, Cornell University Press.

Schwarcz, H.P., Dupras, T. and Fairgrieve, S. 1999 N15 enrichment in the Sahara. In search of a global relationship. *J. Archaeological Science* 26:629-636.

Sheldrick, P.1997 *Archaeology of the Kellis 2 cemetery.* Paper presented at the 2nd Dakhleh Oasis Conference, Toronto, Ontario, Canada. Unpublished.

Steinbock, R.T. 1976 *Paleopathological diagnosis and interpretation.* Springfield, Illinois: Charles C. Thomas.

Waldron, T. 1991 Rates for the job. Measures of disease frequency in paleopathology. *International Journal of Osteoarchaeology* 1:17-25.

Webb, P.A. and Suchey, J.M. 1985 Epiphyseal union of the anterior iliac crest and medial clavicle in a modern multiracial sample of American males and females. *Amer. J. of Physical Anthropology* 68:457-466.

Worp, K.A. 1995 *Greek papyri from Kellis: 1.* Dakhleh Oasis Project Monograph 30. Oxford, Oxbow Press.

Table 1: Summary of key skeletal changes in the burials with leprosy from Kellis 2, Dakhleh Oasis, Egypt

Skull changes	**B6**	**B116**	**B9**	**B222**
Resorption of anterios nasal spine	P	P	A	P
Resorption of anterior maxilla alveolus	P	P	A	A
Remodelling of nasal aperture	P	A	A	P
Resorption of nasal bones	P	A	A	P
Inflammation oral surface of palate	P	P	P	P
Inflammation of endonasal palatal surface	P	P	P	P
Inflammation of nasal conchae	P	P	A	P
Palatal thinning	P	A	P	A
Palatal perforation	P	A	P	A
Leprogenic odontodysplasia	A	A	A	A
Hand changes				
Carpal inflammation	A	A	A	A
Metacarpal inflammation	P	P	A	A
Phalangeal inflammation	P	P	A	A
Distal phalanx erosion	P	P	A	A
Claw hand deformity	P	P	A	A
Metatarso-phalangeal joint destruction	A	P	A	A
Septic joint destruction	A	A	A	A
Ankylosis	A	A	A	A
Foot changes				
Tarsal inflammation	P	P	A	A
Metatarsal inflammation	P	P	A	A
Phalangeal inflammation	P	P	A	A
Diaphyseal concentric remodelling	A	P	A	A
Knife edge remodelling	A	P	A	A
Dorsal exostoses	A	P	A	A
Metatarso-phalangeal joint destruction	A	P	A	A
Ankylosis	A	A	A	A
Sepsis	A	P	A	A
Long bone periostitis				
Ulna	A	P	A	A
Radius	A	P	A	A
Tibia	P	P	A	P
Fibula	P	P	P	A

Table 2: Select craniometrical data and indices comparing K2 burials with leprosy to the K2 male control sub-sample

Measurements	**Kellis (with leprosy)**				**Kellis (without leprosy)**			
	B6	**B9**	**B116**	**B222**	**No.**	**Mean**	**S.D.**	**Range**
Cranial length	181	183	173	182	41	179	6.28	167-190
Cranial breadth	143	138	134	138	41	137	4.6	127-149
Cranial height	128	130	128	139	40	132	4.8	124-139
Basion-prosthion Height	X	95	88	91	36	91	4.9	74-100
Nasion-prosthion height	X	68	66	65	36	70	4.5	63-85
Bizygomatic breadth	130	133	121	124	38	128	3.9	120-133
Nasal breadth	25	25	22	24	37	24	1.7	21-27
Nasal height	50	52	49	49	38	52	2.3	43-47
<u>**Indices**</u>								
Cranial	79	75.4	77.5	75.8	40	75.8	2.7	70.3-80.2
Upper facial height	X	51.1	54.5	52.4	36	55.3	2.6	51.1-62.5
Nasal	50	48.1	44.9	49	37	46.6	3.9	41.1-52.1

Table 3: MtDNA sequences for 13 Kellis 2 individuals (left column) relative to the Anderson *et al.* (1981) standard (top). Note the different bases (bold) between the two individuals with leprosy (6 and 116) at loci 16223,16298 and 16311. These differences indicate these individuals are not maternally related.

	16217 C	16223 C	16234 C	16261 C	16264 C	16269 A	16278 C	16294 C	16296 C	16298 T	16304 T	16311 T
2		T										C
6		T										C
15		T										
16							T					
27								T	T			
31		T										
72				T								
76							T					
82		T			T	G						
87	T		T									
101								T	T		C	
116										C		
139		T										

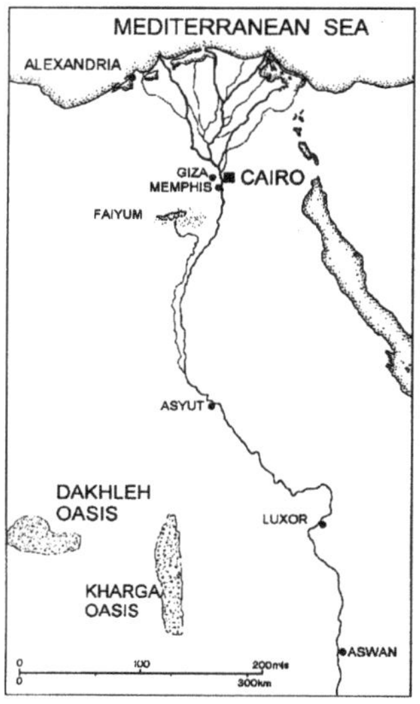

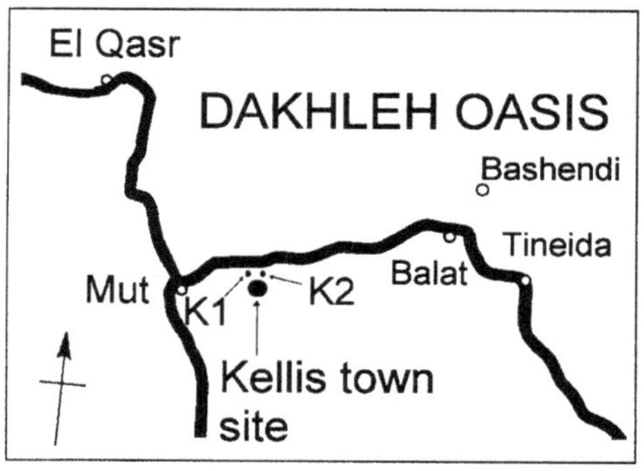

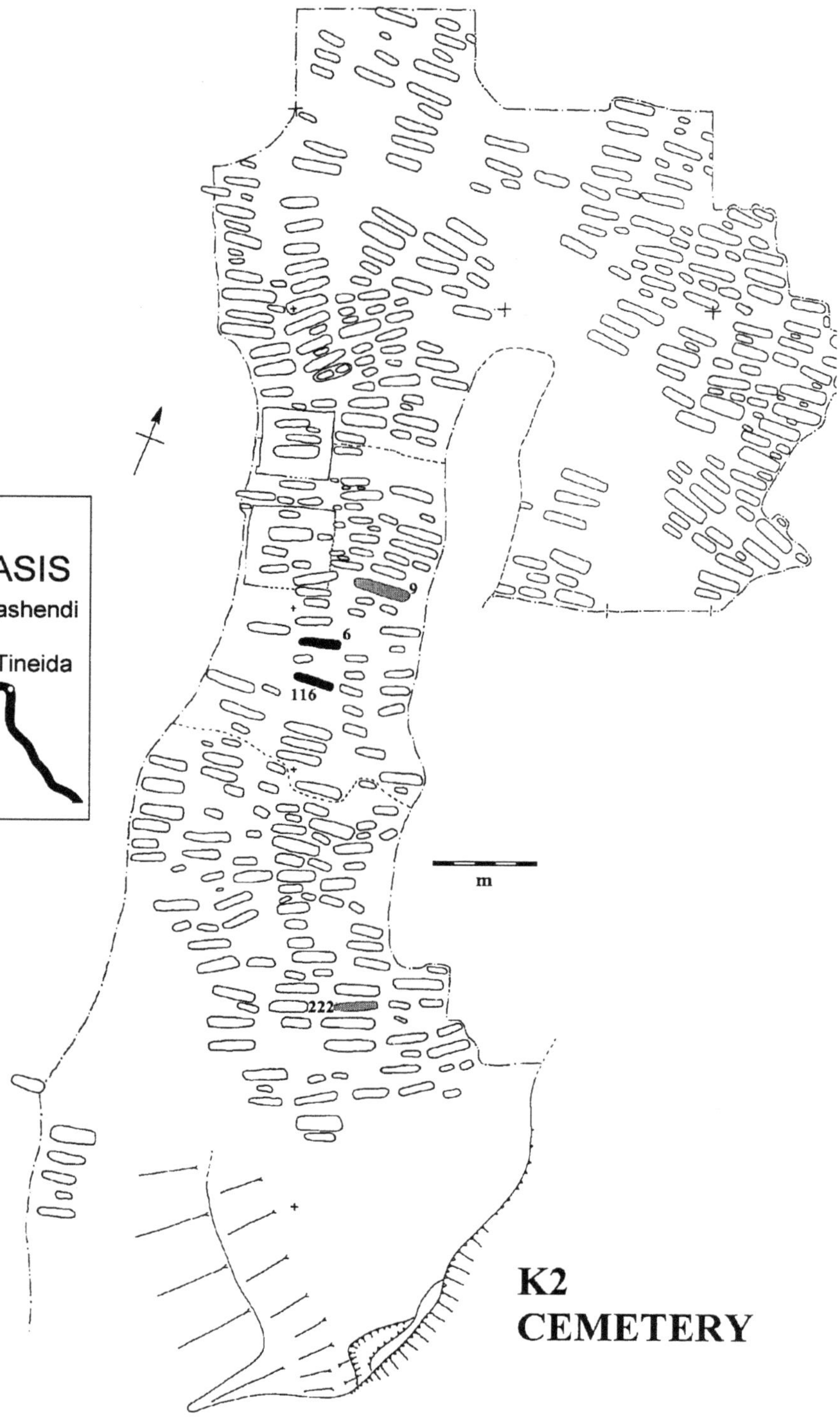

Figure 1 :

Diagrammatic representation of Kellis 2, Dakhleh, Egypt. Burials with lepromatous Leprosy shown in black. Possible lepers in grey.

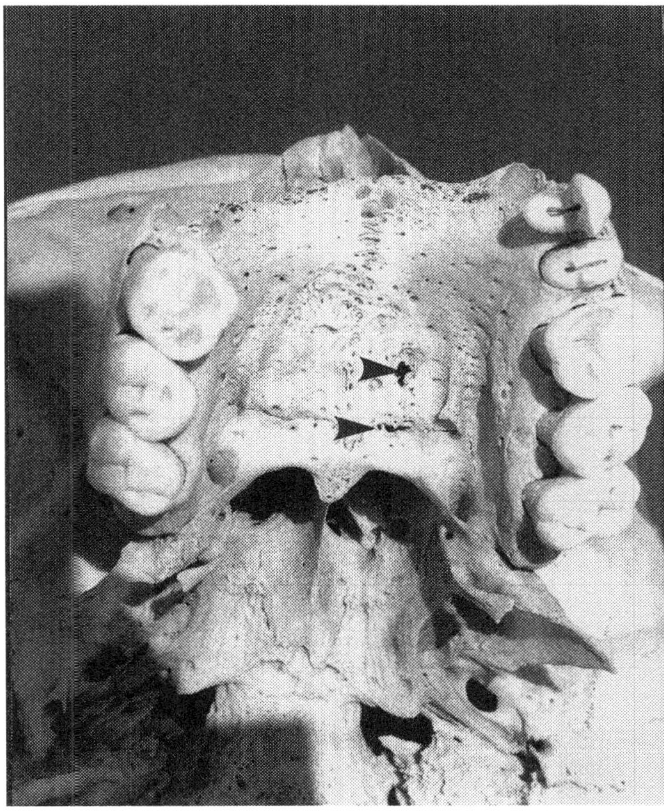

Figure 2a: The hard palate of K2-B6 showing the inflammatory changes (pitting, resorption), including perforations of the left maxilla and the left transverse palatine suture (arrows). Resorption of the alveolar bone has resulted in the antemortem loss of all the anterior teeth (except the left canine) and the right PM1. The dental attrition is characteristic of K2 individuals in their late 20s.

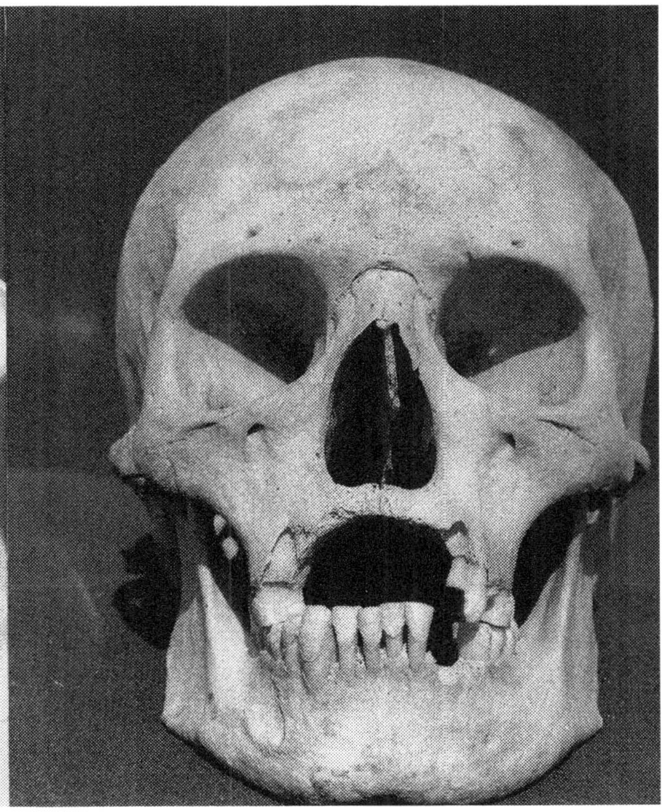

Figure 2b: Frontal view of K2-B6 showing resorptive changes of the anterior nasal spine, the anterior maxillary alveolus, the lateral margins of the nasal aperture and the right nasal bone. Not visible are the inflammatory changes of the inferior nasal conchae and the floor of the nasal cavity.

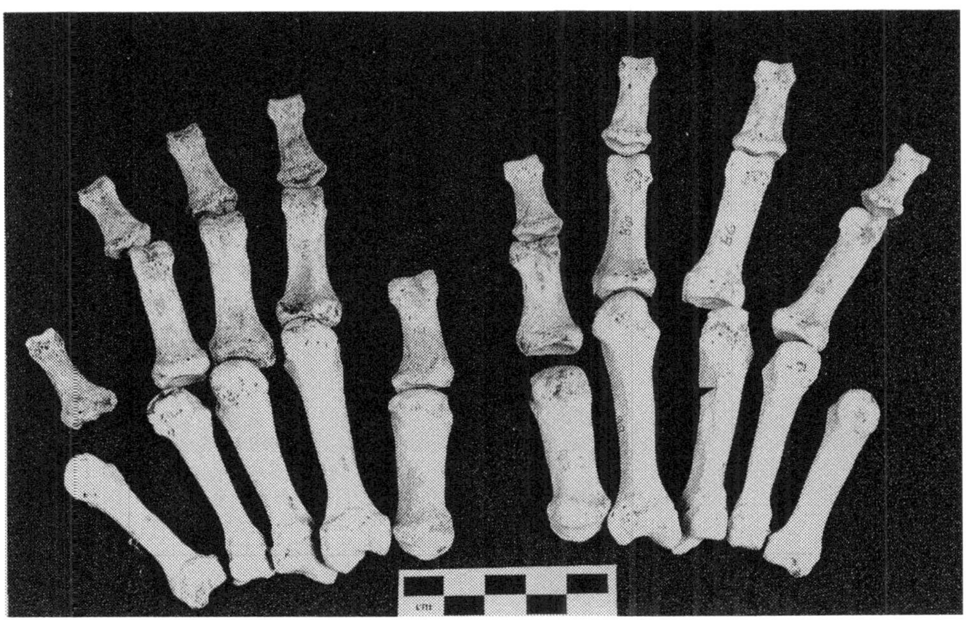

Figure 3: Palmar view of the metacarpals and phalanges of K2-B6. Palmar grooves occur at the distal ends of the juxta-articular areas of the proximal phalanges. Periostitis is present on the dorsal surface of these bones and the metacarpals, as well as the palmar surface of the left 3rd metacarpal (see arrow) and metacarpal

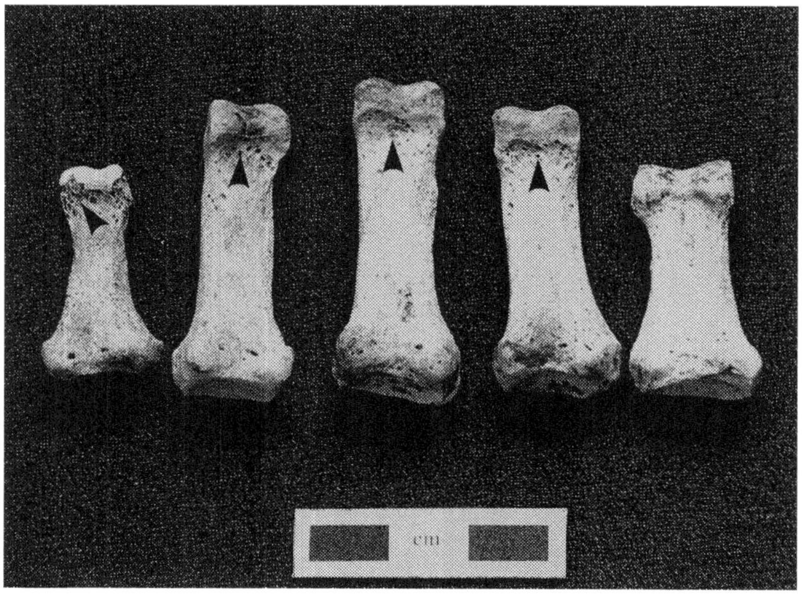

Figure 4: Close-up of the palmar surface of the right proximal phalanges of burial 6. Note the distinct palmar grooves (arrows) on the distal ends on each bone and the osteolysis on the fifth phalanx.

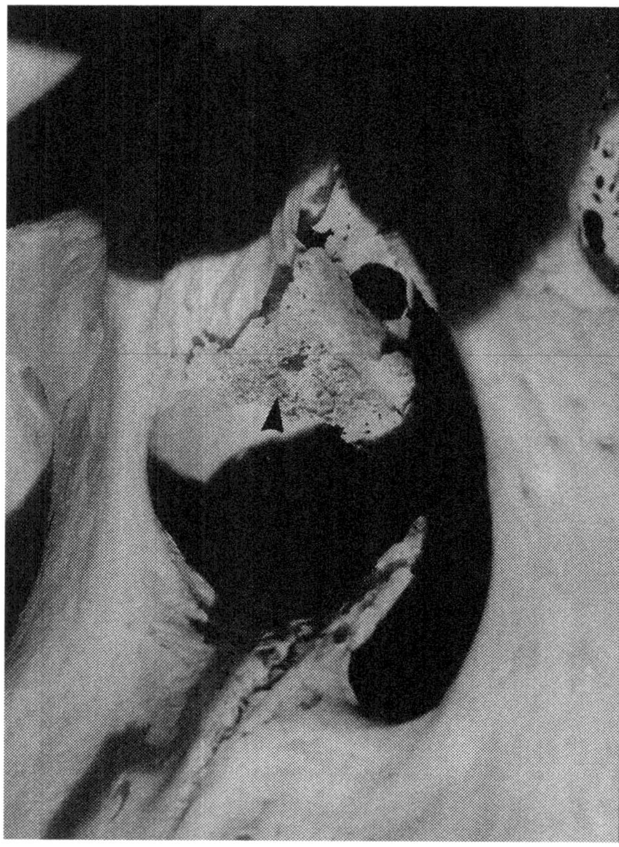

Figure 5a: Close-up of the nasal aperture of K2-116. Note the resorbed nasal spine and the osteolysis and pitting of the right inferior nasal concha (arrow)

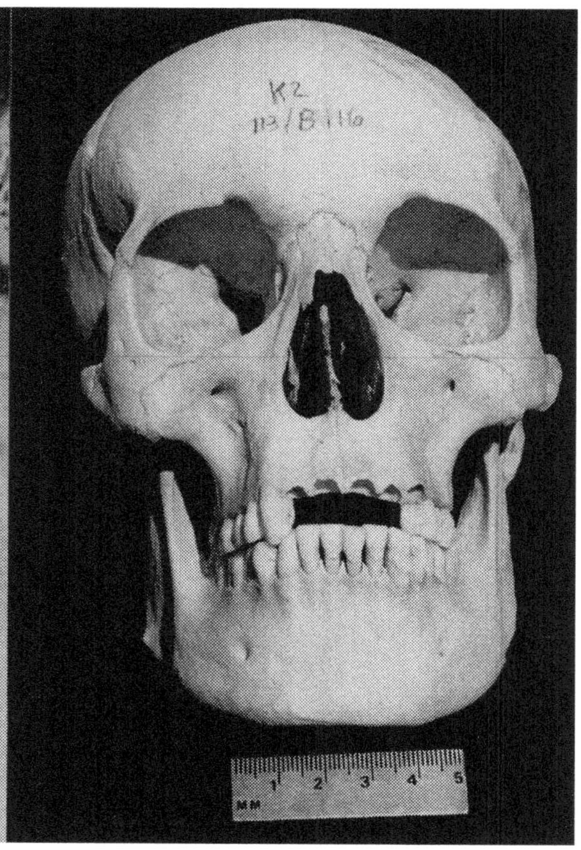

Figure 5b: Frontal view of the skull of K2-116. The anterior nasal spine is resorbed as is the anterior border of the nasal bones. The pyriform aperture is normal.

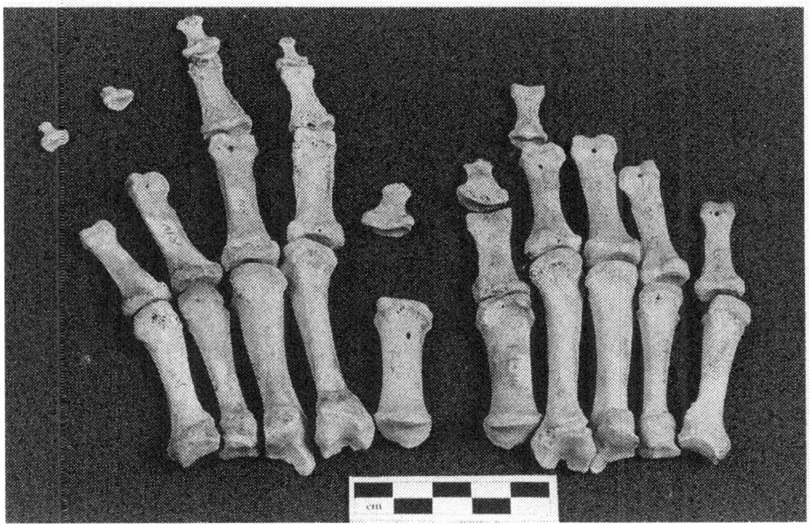

Figure 6: Palmar surface of the hands of K2-116. Note the palmar grooves in the proximal phalanges, the enlarged nutrient foramina and the resorptive changes of the distal phalanges.

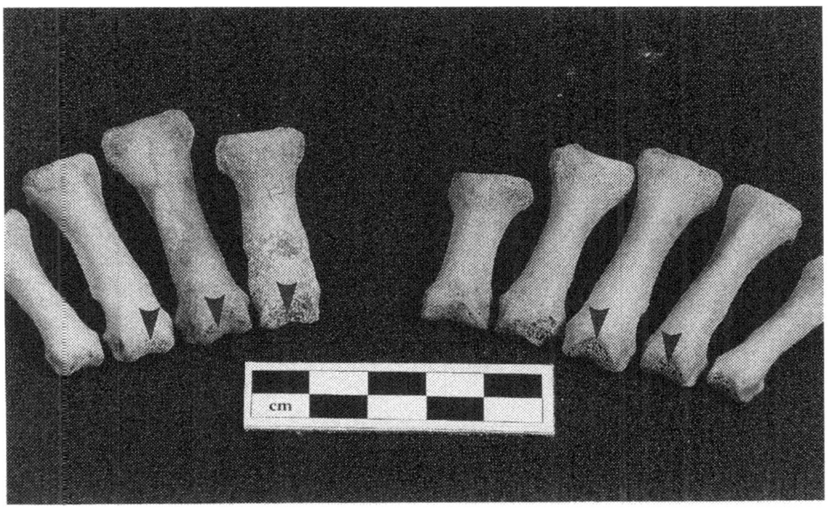

Figure 7: View of the distal-dorsal articular surfaces of the proximal hand phalanges of K2-116. Note the extensive pitting (arrows) possibly because of disuse atrophy from chronic hyperflexion.

a

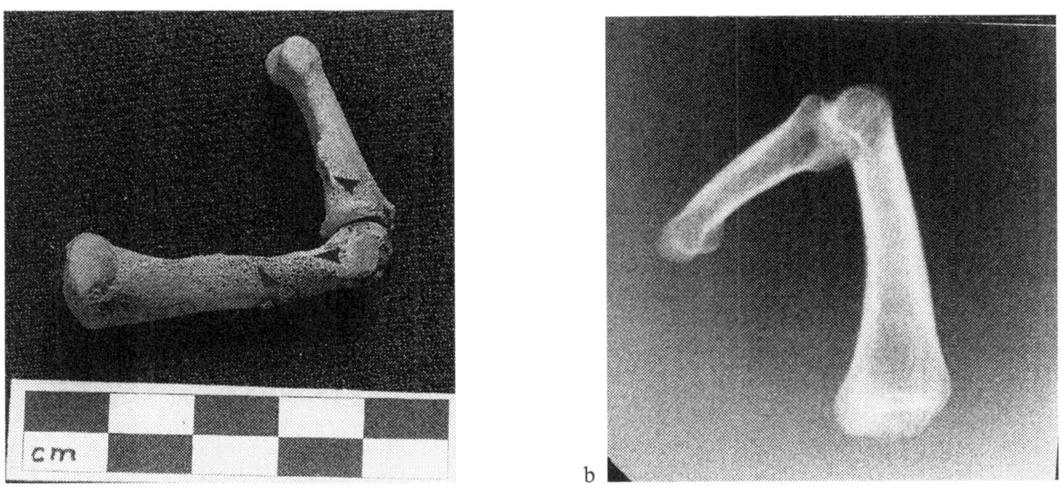

Figure 8: Lateral view of the proximal phalanges of the 2nd right digit showing the respective gross (8a) and radiological (8b) views of "claw-hand" deformity. Note the periosteal reactive changes (arrows in 8a) which are obscured in the radiograph.

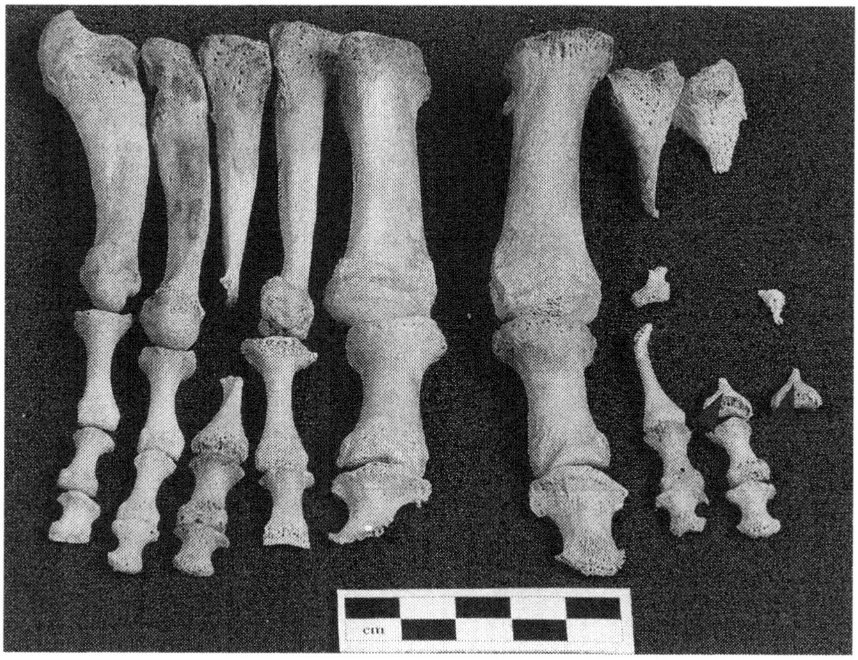

Figure 9: The foot phalanges and metatarsals of K2-116. Note the presence of shark-tooth deformity (arrows) and the mid-diaphyseal remodelling of the proximal phalanges.

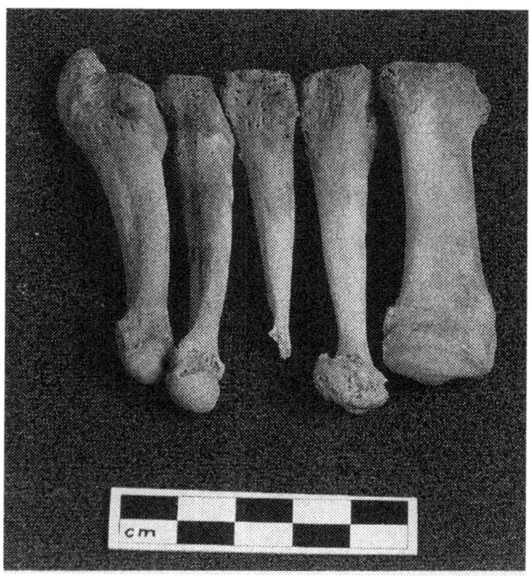

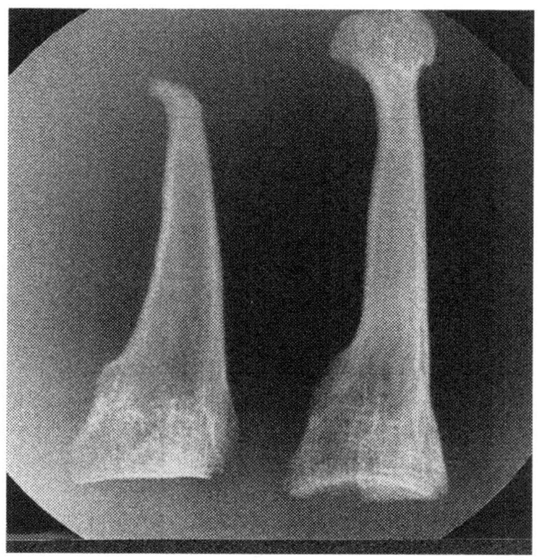

Figure 10a: Dorsal view of the right metatarsals of K2-116 showing distal third remodelling of MT II and MT III. The latter bone shows the classic "knife-edge" and "pencilling" deformity. Note the absence of metatarsophalangeal joint sepsis. The radiograph (Figure 10b) illustrates the absorptive pattern in these bones which shows the maintenance of the cortical bone despite loss in overall diameter.

Figure 11: Lateral view of the left foot of K2-116. The third and fourth metatarsals were completely resorbed and septic destruction of the cuboid, the medial and middle cuneiforms and the antero-lateral calcaneum has occurred. Note the extensive periosteal reactive changes, the pitting on the lateral surface of the calcaneus and the advanced concentric resorption of the 2nd and 3rd metatarsals (arrows).

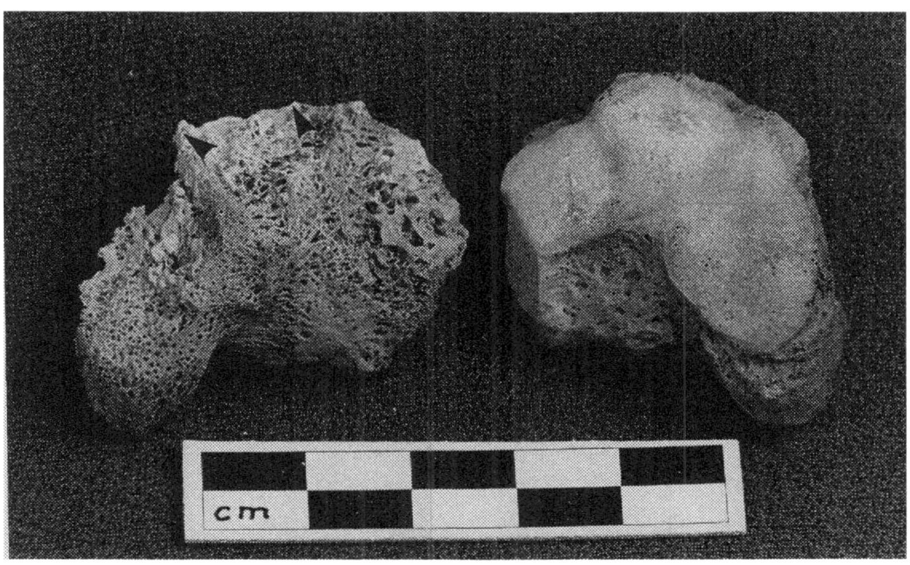

Figure 12: Anterior view of the left and right navicular bones showing osteolytic destruction of the facets on the left side. Dorsal talar exostoses (arrows), a concomitant of the navicular displacement, are present on this bone and the dorsal surfaces of the cuneiforms, the vestigial cuboid and the talus.

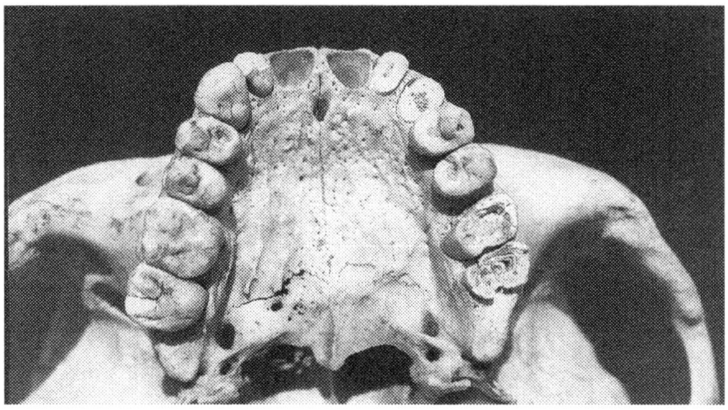

Figure 13: Palate of K2-9 showing osteolytic process of the right palatine bone and the transverse palatine suture. The pitting of the anterior region adjacent to the incisive foramen is normal. This individual is classified as a "possible" leprosy on the basis of these changes.

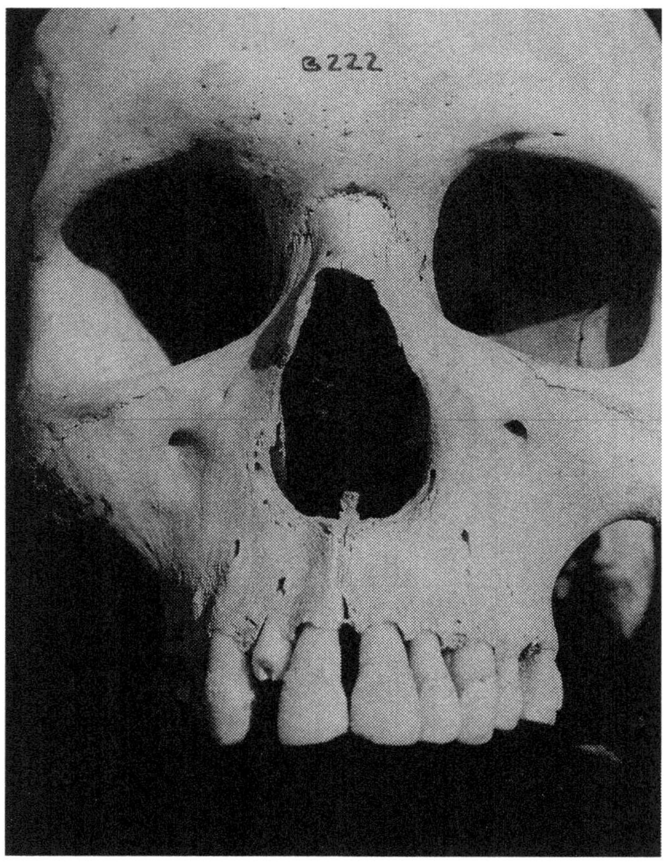

Figure 14: Frontal view of the skull of K2-222. The anterior nasal spine and the margins of the nasal aperture are resorbed. Note the enlarged foramina on the margins, particularly on the right side. This individual is classified as having "possible" leprosy on the basis of these osseous changes.

Evidence for leprosy in Medieval Ireland

Eileen Murphy [1] and Keith Manchester [2]
[1]School of Archaeology and Palaeoecology
Queen's University of Belfast
Belfast
BT7 1NN
Northern Ireland

Email: eileen.murphy@qub.ac.uk

[2]Calvin Wells Laboratory
Department of Archaeological Sciences
University of Bradford
Bradford
BD7 1DP
England

Abstract

Evidence for leprosy in Ireland has been largely restricted to historical references to the disease, and the occurrence of place-names which may have been derived from *lobhar*, the Irish language word for "leper". In addition, research has indicated that at least fifty leprosaria existed in Ireland during the Medieval period. Ireland was not isolated from social and economic contact with Britain and Continental Europe during Medieval times but, to date, only one definite palaeopathological case of leprosy has been identified. In spring 1997, excavations took place in the interior of St. Patrick's Church in Armoy, Co. Antrim, and the remains of the feet from an adult individual were recovered which are considered to display bone changes indicative of lepromatous leprosy. The palaeopathological lesions apparent in the feet of the burial from Armoy are presented and discussed within the context of a review of the evidence for the disease in Ireland.

Keywords

Leprosy, Medieval, palaeopathology, Armoy, Ireland

'The rickets are of late very rife in Ireland, where few years agoe they were unknown; so on the contrary it hath been almost quite freed from another disease, one of the very worst and miserablest in the world, namely, the leprosie which in former times used to be very common there, especially in the province of Munster; the which therefore was filled with hospitals, expressly built for to receive and keep the leprous persons. But many years since Ireland has been almost freed from this horrible and loathsome disease, and as few leprous persons are found there as in any other countrie in the world, so that the hospitals erected for their use, having stood empty a long time, are quite decayed and come to nothing' (Boate, 1652).

1. Introduction

Recent research would indicate that leprosy was not common in Europe until the Medieval period, when the growth and decline of leprosy hospitals is considered to indicate the prevalence of the disease. Studies of those interred in Danish Medieval leprosy hospital cemeteries undertaken by Møller-Christensen (1978) tend to support this view by demonstrating that at least 69% of the individuals buried displayed bone changes specific to leprosy. In Britain, there was a great increase in the number of leprosy hospitals built during the 12th and 13th centuries followed by a notable downtrend, so that by the 15th and 16th centuries very few new foundations were recorded (Roberts, 1986:16-18). The emergence of tuberculosis as a highly prevalent disease during later Medieval times has been suggested as the reason for the virtual disappearance of leprosy in Europe by the 16th century. Tuberculosis is caused by the bacterium *Mycobacterium tuberculosis* which is closely related to the leprosy bacillus. It has been suggested, therefore, that leprosy declined in prevalence as the number of cases of tuberculosis increased during the later Medieval period (Manchester, 1991). While detailed studies have been undertaken on the nature and prevalence of leprosy in Britain, very little has been noted on the history of the disease in neighbouring Ireland, with the exception of the historical approaches adopted by scholars such as Belcher (1868), MacArthur (1949, 1953) and Lee (1967, 1974-75, 1996).

2. Secondary evidence for leprosy in Medieval Ireland

2.1 Leprosy hospitals

Lee (1996) has recorded evidence for at least fifty hospitals with known associations with leprosy in Medieval Ireland, the majority being located in Munster and Leinster and relatively few recorded in either Connacht or Ulster. The geographical location for these hospitals generally correlates, therefore, with those areas of the island that were under greatest Anglo-Norman influence during the Medieval period, and it has been suggested that the hospitals were an introduction of the Anglo-Normans (Lee, 1996:16). In the Gaelic areas of Ulster and Connacht, hospitalisation may have been of a different nature. The main patron saints of leprosy sufferers are St. Stephen, St. John the Baptist, St. Mary Magdalene (Lee, 1967:74-75) and St. Giles, and many of the Irish Medieval leprosy hospitals were dedicated to the former three saints (Lee, 1974-75:225).

There are numerous references pertaining to leprosaria contained within Irish Medieval sources. One of the earliest references to a leprosy hospital may be found in the "Annals of Innishfallen" under the date 869, where it is recorded, 'Devastation of Armagh by Arlaf, so that the city was burned with its houses and hospitals (*nosocomiis* or leper houses)' (Fleetwood, 1983:13). However, it has been convincingly argued by MacArthur (1953:14-15) that this reference to early leprosaria has arisen from two mis-translations of the word *nosocomiis*. Some of the earliest probable leprosy hospitals are those which were founded during the 12th century, including the hospital which was opened by the Benedictines in Waterford in 1185, and the hospital dedicated to St. Brigid that was erected at Kilbrixy, Co. Westmeath, in 1192 (Fleetwood, 1983:13).

A leprosy hospital was erected near St. Stephen's Church in Waterford by King John during the early 13th century. Local history dictates that he founded the hospital in thanksgiving after he (or his sons) recovered from an illness related to an over indulgence of salmon and cider during a visit to the nearby town of Lismore. An inquisition relating to the hospital was held in 1661 and it is recorded that 'we find and present that Juan Murphy, sent unto Widow Bennett, was enfected with the Leprosie, and in the time of the usurped power was presented to the then Commissioner of Revenue, who denyed to give her any releefe, whereupon she miscarried and died in a miserable condition'. Supposed leprosy sufferers were admitted to the hospital up until 1723, when it is recorded that Drs. Reynet and Donogan certified that one Mary Tobin suffered from the disease. The leprosy hospital was converted into a general infirmary around 1740 (Fleetwood, 1983:14).

There is substantial evidence to indicate that a number of leprosy hospitals once existed in Dublin. A reference relating to St. Stephen's leprosarium, situated outside the walls of the city near St. Kevin's Gate, can be found in the "Dublin Corporation Records" for 1230 (Lee, 1996:46). The hospital is also included in Speed's 1610 map of Dublin where it is located on the site which was occupied by the Mercer's Hospital during the 18th century, and is now the location of the Royal College of Surgeons (Fleetwood, 1983:30). The hospital of St. James was founded in approximately 1220 near the present day Townsend Street formerly called Lazar's Hill (Lee, 1996:48-49). The oldest of the Dublin leprosy hospitals is considered to be that of St. John the Baptist, located outside Newgate in Thomas Street, which was founded in the latter years of the 12th century (Fleetwood, 1983:14).

In Ulster, a leprosarium is thought to have been located in the parish of Kilclief, Co. Down. In 1592, the townland of Glebe comprised three territories - Spittle Quarter, Carrowreagh and Fermeannes. A plot referred to as the 'Spital-field' was contained within the Spittle Quarter and during the mid-19th century the local inhabitants are said to have remembered a building located in this plot (Reeves, 1847:218). Reeves considered this building to have represented the remains of the leprosarium referred to in a document of 1387, which stated that, 'Robert de Vere, Marquis of Dublin &c., committed to Nicholas Lepyng, clerk, the custody of the Lepers' house of St. Peter, nigh Kylceth in Ultonia', while a second historical reference, dated to 1415, recorded that 'the King committed to John Fitz-Richard, chaplain, John Molyn, and Walter Sely, the custody of the hospital of lepers' houses of St. Nicholas of Down, and St. Peter of Kyicleth, with their lands and appurtenances, to be held while in the King's hands, rent free' (Reeves, 1847:218). The leprosarium of St. Nicholas is considered to have been located in Downpatrick (Reeves, 1847:232), and Purdon (1896:269) records that this hospital was dedicated by the De Laceys and the De Burgos.

2.2 "Squints"

A characteristic feature of many of Ireland's Medieval churches is the occurrence of a notably small window in one of the church walls. It is generally considered that the purpose of these "squints" was to enable people afflicted with contagious diseases to observe religious services from the exterior of a church, thereby avoiding contact with the apparently healthy members of the community. The west gable of the 15th century church at Carrigafoyle, Co. Kerry, contains a small disfigured window which was known as a "leper hole" (Toal, 1995:236), while local tradition relates that a "Lepper Window" is to be found in one of the chancel walls of the Medieval church of St. Nicholas in Carrickfergus, Co. Antrim. It is interesting to note that a leprosarium is also thought to have existed within the town during the Medieval period (McSkimin, 1909:164). MacArthur (1953:12), however, is sceptical of the entire "squint" theory and noted that no contemporary evidence exists which suggests that these windows were in fact specifically associated with leprosy; he considers the association to be nothing more than a modern interpretation.

2.3 Place-name evidence

Another form of evidence which may relate to the occurrence of leprosy in Ireland can be found in place-names. In the Irish language the word *lobhar* is translated as "leper" and many townlands with anglicised versions of this word may have had associations relating to leprosy, such as Drumalour (the ridge of the "lepers") or Ballylowre (the town of the "lepers") (Weymouth, 1938:52, Lee, 1967:73). In County Kerry, for example, there is a tradition that a holy well existed in the vicinity of Rattoo Abbey. The well was known locally as *Tobar na lour*, which translates as the "well of the leper", and people with leprosy are reported to have been cured there (Toal, 1995:230).

Townlands which indicate that they were associated with an ancient hospital, although not necessarily a leprosarium, generally include the elements "Spital", "Spittle" or "Spidal". The mid-17th century Cromwellian 'Civil Survey' for County Limerick, for example, records that, 'Spittle Land, otherwise called Leaps house of Athdate [Adare], being one Tenent and a garden together with three plantacon accres whereof two are neare the sd spittle house and th other is Eastewarde the River of Maige, and within ye Earle of Kildare his lands oweing to the said Earle 12d chiefly p ann. And notwithstanding the said Thomas, and his father have been possessed thereof yet it is conceived they were soe possessed in right of ye Leapers and not by inheritance' (Simington, 1938:145).

2.4 Historical sources

There are numerous references pertaining to leprosy and leprosy hospitals contained within the Medieval Irish documentary sources, and a comprehensive discussion of their contents has been undertaken by Belcher (1868) and Lee (1996). A bequeathal recorded in 1360 or 1370, for example, records the donation of, 'a gift from Ellena Multon to God and the Blessed Virgin, and St. Stephen, the proto-martyr, and to the poor lepers of the city of Dublin, resident therein, of two acres of ground whereon stood the chapel of St. Stephen, near Dublin, with a small meadow called ?Mary's, to the east of the said chapel, to hold in perpetual charity for the support of the said poor for ever' (Belcher, 1868:39).

In the "Calendar of Ancient Records" for Dublin under the date of '1491 Fourth Friday after 24th June' it is stated that, 'every fre man and woman of this cite enfected with lepyr be received and take into the house of Saint Stewnes within the fraunches of this cite frely, without anny fyne paying to anny person or persones. And whosoever attempt the contrary to pay xl. s' (Fleetwood, 1983:30). This may be considered as a case of the town officials attempting to segregate leprosy sufferers in Dublin from the settlement's healthy population, and may suggest that people with leprosy in Medieval Ireland were accorded the same treatment as those in the rest of Europe. There are certainly numerous references in Irish texts to land and money being donated to leprosy hospitals, which may indicate that Irish society did indeed share the same moral responsibility for the care of the unfortunate sufferers, as occurred elsewhere in Europe.

Many of the stories relating to the Early Christian Irish saints make reference to their association with "lepers". St. Brigid of County Kildare is said to have gained an enormous reputation for the healing of the sick, particularly those with leprosy (Flanagan, 1990:25). St. Mac Nissi of County Antrim is recorded as being gifted with miraculous powers and it is alleged that a blind man and a "leper" were cured by washing in a fountain of water accompanied by the saint's prayers (Flanagan, 1990:90). Indeed, one of the earliest references to St. Patrick contained in Colgan's "Acta Sanctorum" stated that the saint kept a person with leprosy in his house and washed his sores with his own hands (Fleetwood, 1983:12). These stories must be treated with some scepticism, however, and during the late 19th century a Dr. Frazer even went so far as to write in a letter that, 'as for the old saints, I believe all kinds of chronic diseases of the skin combined with dirt were conveniently called Leprosy' (Purdon, 1896:271).

Gerald of Wales, writing in the late 12th century, makes reference to the occurrence of leprosy in Ireland, and stated that, 'as they [the Irish] did not boil or roast their meat, it was full of juices, and produced the leprosy, a disease very common, and for which leper-houses were everywhere to be found. They were taught that the bad effects of this and other gross ailments were corrected by aqua vitae' (Weymouth, 1938:52). It is impossible, however, to know whether Gerald was referring to the disease leprosy as we understand it today.

3. The validity of the secondary sources

Knowledge pertaining to the history of leprosy, and indeed any disease, relies on artistic, literary and palaeopathological evidence, all of which are dependent on the clinical manifestations of the disease. Evidence derived from artistic and literary sources may be considered to be of secondary importance to that from actual skeletal or mummified remains since it is subjective in nature, and is reliant on the interpretation of the disease by contemporary observers. Within the context of specific and pathognomonic bone changes, however, palaeopathological diagnosis may be irrefutable.

The Greek word *lepra* means a scale and was used to describe skin diseases such as psoriasis. Indeed, by Medieval times the term "leprosy" had become associated with a whole range of skin conditions and diseases as diverse as smallpox and the bubonic plague, all totally unrelated to the disease leprosy as we understand it today (MacArthur, 1949:186). It is impossible for us to be certain, therefore, that when early writers refer to "leprosy" that they are referring to leprosy, as caused by *Mycobacterium leprae*, and not some other disease.

MacArthur (1949:183-184) provides a clear example of how cautious we should be when attempting to obtain information about leprosy from the early historical sources. He questions the translation of the Gaelic word *clamtrusca*, which is included in the "Annals of Ulster" reference for 950 relating to a great epidemic that affected the Viking population in Dublin. In English translations of the text the word is translated as "leprosy". The same epidemic is referred to as *clamhtrusccadh* in the "Annals of the Four Masters", which John O' Donovan also translated into English as leprosy. MacArthur gleaned further insights into the true nature of this epidemic from the English translation of the "Annals of Clonmacnois", undertaken by the Elizabethan writer Ma Geoghagan who stated that, 'the pox (which the Irishmen called the *Dolor Gentilium*) ran all over Ireland this yeare', thereby indicating that he understood the disease to involve the occurrence of pustules on the skin. More specific information pertaining to the nature of the disease can be found in his entry for the year 675 where he states, 'there raigned a kind of great leprosie in Ireland this yeare called the pox, in Irish *bolgagh*'. The word *bolgach* translates as smallpox and MacArthur considers it most likely to be the case that the epidemic of 950 was in fact smallpox, as opposed to leprosy.

4. Primary evidence for leprosy in Ireland

4.1 Palaeopathological evidence

The cemeteries associated with a number of Medieval religious foundations have been the subject of archaeological investigations in recent years, including some sites that have direct historical associations with leprosy. For example, St. Stephen's Hospital in Dublin has historical documentation that connects this site with the disease. Surprisingly, however, no palaeopathological cases were discovered during the analysis of the skeletal remains retrieved during an excavation programme in 1992 (Laureen Buckley, 1998, pers. comm.). Similarly, the historical documentation reports that a leprosy hospital was built by Nicholas FitzMaurice at Ardfert, Co. Kerry in 1312 (Lee, 1996:36). Ardfert has a cathedral, churches and a friary of Medieval date, and the hospital may have been part of this religious complex, although the analysis of skeletal remains excavated from St. Brendan's Cathedral during the period from 1989 to 1992 failed to identify a single palaeopathological case of leprosy (Barra Ó Donnabháin, 1998, pers. comm.). It is possible, however, that this was because those who died with the disease were buried in some other cemetery in the area, or that they showed no bone changes at their time of death.

Elsewhere in Ireland, two possible cases of Medieval leprosy have recently been identified during post-excavation analysis of human skeletal remains. The two possible cases originate from Cashel, Co. Tipperary (Miriam Clyne, 1998, pers. comm.) and Mount Offaly, Cabinteely, Co. Dublin (Linda Lynch, 1999, pers. comm.). Cashel has clear historical associations with leprosy, and it is recorded that in c. 1230 a leprosy hospital was built in the vicinity of the town by Sir David Latimer, the seneschal of Cashel (Lee, 1996:40). Mount Offaly, Cabinteely, however, has no known associations with leprosy. To date, only one definite case of Medieval leprosy has been recorded in the published palaeopathological record for Ireland (Murphy, 1997, Murphy and Manchester, 1998).

During spring 1997, excavations took place at St. Patrick's Church in Armoy, Co. Antrim, under the direction of Declan Hurl of the Environment and Heritage Service of DOE NI (Hurl, 1998). The upper layers revealed a large quantity of disarticulated human remains, while in the lower levels the remains of fifty-six individuals of all ages and both sexes were recovered. An Early Christian foundation, the papal taxation of 1306 valued "Ecclesia de Ethirmoy" at £4 11s. 8d (Reeves, 1847:81). The burials recovered during the excavation inside the western part of the present 19th century church are considered to have lain outside the walls of the Medieval church. This was located at the eastern end of the present church and these skeletons are thought to be of Medieval date. In the 1622 "Ulster Visitation Book", the Medieval church was described as being "unrepayred" suggesting that it had become abandoned or had fallen into ruin (O'Laverty, 1887:449). The exact date when the Medieval church had fallen into decline is not known, but the "Ordnance Survey Memoirs" of 1838 record that, 'the old church was destroyed by the accumulation of graves that had been formed about the windows and doors of it' (Day *et al.*, 1994:6), which suggests that people had been burying their dead in the church interior for a considerable period of time. A recent calibrated AMS radiocarbon date obtained for the foot bones indicated that the individual had died at some time during the period between 1444 and 1636 AD (Declan Hurl, 1999, pers. comm.).

Changes indicative of lepromatous leprosy were clearly evident in both feet (Figure 2). Dorsal tarsal exostoses were present on several of the tarsal bones, and these arise as a consequence of loss of motor function which leads to the collapse of the longitudinal arch of the foot, and the development of *pes planus* or flat foot (Andersen and Manchester, 1988:52). Surface inflammatory pitting was evident on the dorsal and plantar surfaces of most tarsals, which indicates that there was overlying soft tissue infection. Diaphyseal remodelling in which the medio-lateral diameter was diminished was apparent in all metatarsals, causing them to have a "sucked candy" appearance. In conjunction with acroosteolysis this has resulted in the proximal ends of the metatarsals displaying the "knife-edged deformity". It has been suggested that diaphyseal remodelling occurs as a result of sympathetic neuropathy and alteration of the peripheral nerve vascular bed dynamics which then selectively stimulates extracortical osteoclastic and endosteal osteoblastic reactions (Andersen *et al.*, 1992:211). All of the metatarsals displayed inflammatory pitting on their surfaces, a finding which is again indicative of overlying infection. The only phalanx present was the right first proximal phalanx which displayed the "cup and peg" deformity of the first metatarsophalangeal joint. Since the destruction of the first proximal phalanx was so advanced it is highly probable that all of the other phalanges had been destroyed as a consequence of the disease process, rather than not having been recovered during the excavation (Figure 3).

As the bone changes were so advanced in both feet it is possible that the individual would have been suffering from the lepromatous form of leprosy. In addition, the severity of the bone changes in the feet would suggest that it was quite possible that he or she would have displayed the characteristic changes of *facies leprosa*, with a collapsed nose, missing upper front teeth, loss of eyebrows, and nodules. It is possible that the disease would have progressed to a stage were the person was also hoarse and blind. It is likely that their hands would have had mere stumps for fingers, as was certainly the case with regard to the toes. The feet and lower legs would probably have had ulcers, which the person could not feel due to their damaged nerves. As a consequence of this, as they went about their everyday existence they would have been susceptible to a variety of infections from these open sores.

5. Conclusion

While there are three known Medieval leprosy hospitals in Ulster, it is worth noting that the Armoy skeleton was recovered from a site with no known associations with the disease. The presence of an isolated case of late Medieval/early Post-Medieval leprosy at Armoy is compatible with the history of the disease, since it is unlikely that there would have been any thriving leprosaria and associated cemeteries in existence at this time. It is therefore probable that the few individuals with the disease would have been buried in the same cemeteries as everyone else.

Ireland was not isolated from social and economic contact with Britain and Continental Europe during the Medieval and early Post-Medieval periods (O'Neill, 1987), and the existence of large numbers of Irish hospitals associated with leprosy would suggest that the disease did exist in Ireland throughout the Medieval period on a similar scale to that evident in the rest of Europe. It is probable that the development and decline of leprosy in Ireland followed the same course as it did throughout the rest of Europe, and that it had largely disappeared by the end of the 17th century.

It is possible to gain an insight into the decline of leprosy from "Smith's County History of Waterford", which was written before 1745. In an account of the Waterford leprosaria it is stated that, 'the Leper House of Hospital of St. Stephen, in this city, situated in St. Stephen-street, was first endowed by the family of the Powers, with the lands called Leperstown, in the parish of Killea, about five miles from this town.....Formerly about fifty poor used to receive a yearly allowance by the master's hands. But as it was thought that a publick infirmary would best answer the intent of the pious benefactors, since the leprosy is not a disease now much complained of, it has been thought proper to endow an infirmary for the reception of such sick and wounded poor as shall offer themselves to the attending physician or surgeon to be cured....' (Belcher, 1868:41).

Acknowledgements

We would like to thank Declan Hurl, Environment and Heritage Service, DOE NI, who directed the excavation at St. Patrick's Church, Armoy, for permitting us to publish this palaeopathological case study in advance of the main excavation report. We are very grateful to Dr. Don Ortner of the Department of Anthropology, Smithsonian Institution, Washington D.C. for verifying the diagnosis of leprosy. We would also like to thank Miriam Clyne and Linda Lynch for their information on the two other possible cases of Irish leprosy, and Colán MacArthur for his communications regarding his father's (Lt. Gen. Sir William MacArthur) research on early leprosy. We are also grateful to Laureen Buckley and to Barra Ó Donnabháin, Department of Archaeology, University College Cork, for their respective information on the excavations at St. Stephen's Hospital, Dublin, and St. Brendan's Cathedral, Ardfert, Co. Kerry. Our thanks must also be expressed to Tony Corey of the Environment and Heritage Service, Department of the Environment for Northern Ireland, for preparing the photographs and Dr. Colm Donnelly, School of Archaeology and Palaeoecology, the Queen's University of Belfast, for his comments on the text.

References

Andersen, J. G. and Manchester, K. 1988 Dorsal tarsal exostoses in leprosy: A palaeopathological and radiological study. *Journal of Archaeological Science* 15:51-56.

Andersen, J. G., Manchester, K. and Shahzady Ali, R. 1992 Diaphyseal remodelling in leprosy: A radiological and palaeopathological study. *International Journal of Osteoarchaeology* 2:211-219.

Belcher, T. W. 1868 Notes on the Mediaeval leper hospitals of Ireland. *The Dublin Quarterly Journal of Medical Science* 46:36-45.

Boate, G. 1652 *Ireland's naturall history 1604-1650: Being a true and ample description of its situation.* London: John Wright.

Day, A., McWilliams, P. and Dobson, N. (eds.) 1994 *Ordnance Survey Memoirs of Ireland Vol. 24. Parishes of County Antrim IX, 1830-2, 1835, 1838-9.* Belfast: The Institute of Irish Studies.

Flanagan, L. 1990 *A chronicle of Irish saints.* Belfast: Blackstaff Press.

Fleetwood, J. F. 1983 *The history of medicine in Ireland.* 2nd edition. Dublin: Skellig Press.

Hurl, D. P. 1998 Adjusting your sites: The ecclesiastical enclosure at Armoy, Co. Antrim. *Archaeology Ireland* 12:10-11.

Lee, G. A. 1967 Leprosy and certain Irish place names. *Dinnseanchas* 2:71-75.

Lee, G. A. 1974-75 The leper hospitals of the Upper Shannon area. *Journal of the Old Athlone Society* 1:222-229.

Lee, G. A. 1996 *Leper hospitals in Medieval Ireland.* Blackrock: Four Courts Press.

MacArthur, W. P. 1949 The identification of some pestilences recorded in the Irish Annals. *Irish Historical Studies* 6:169-188.

MacArthur, W. 1953 Medieval "leprosy" in the British Isles. *Leprosy Review* 24:8-19.

Manchester, K. 1991 Tuberculosis and leprosy: evidence for interaction of disease. In D. J. Ortner and A. C. Aufderheide (eds): *Human paleopathology. Current syntheses and future options.* Washington DC: Smithsonian Institution Press, pp. 23-35.

McSkimin, S. 1909 *The history and antiquities of the county of the town of Carrickfergus.* Belfast: Mullan and Son, James Cleeland, Davidson and M'Cormack (revised by E. J. M'Crum).

Møller-Christensen, V. 1978 *Leprosy changes of the skull.* Odense: University Press.

Murphy, E. 1997 *Archive report on the human remains from St. Patrick's Church, Armoy, Co. Antrim.* Unpublished report prepared for the Environment and Heritage Service, Department of the Environment of Northern Ireland.

Murphy, E. and Manchester, K. 1998 "Be Thou Dead to The World": Leprosy in Ireland, evidence from Armoy, Co. Antrim. *Archaeology Ireland* 12:12-14.

O'Laverty, J. 1887 *An Historical Account of the Diocese of Down and Connor, Ancient and Modern, Vol. IV.* Ballynahinch: Davidson Books (1982 reprint).

O'Neill, T. 1987 *Merchants and mariners in Medieval Ireland.* Dublin: Irish Academic Press.

Purdon, H. S. 1896 Mediaeval hospitals for lepers near Belfast. *Ulster Journal of Archaeology* 2: 268-271.

Reeves, W. 1847 *Ecclesiastical antiquities of Down, Connor and Dromore.* Ballymena: Braid Books (1992 reprint).

Roberts, C. 1986 Leprosy and leprosaria in Medieval Britain. *M.A.S.C.A. Journal* 4:15-21.

Simington, R. C. 1938 *The Civil Survey AD 1654-56, Volume 4: County of Limerick.* Dublin: Irish Manuscripts Commission, The Stationary Office.

Toal, C. 1995 *North Kerry Archaeological Survey.* Kerry: Brandon Book Publishers Ltd., FAS and the NKAS Committee.

Weymouth, A. 1938 *Through the leper-squint.* London: Selwyn and Blount.

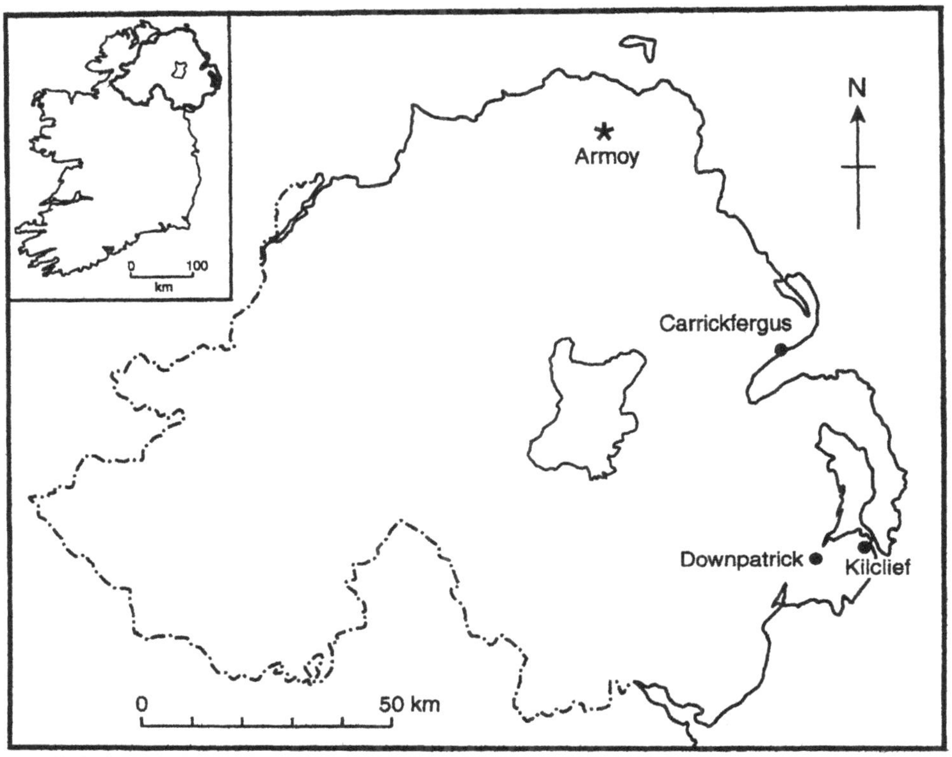

Figure 1: Map showing the location of Armoy, Co. Antrim, and the sites of hospitals in Ulster with detailed historical associations with leprosy.

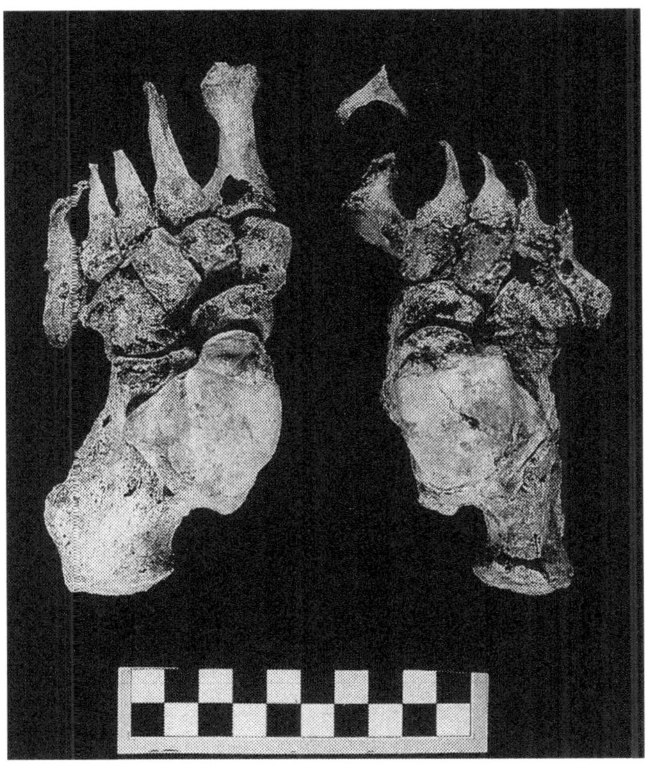

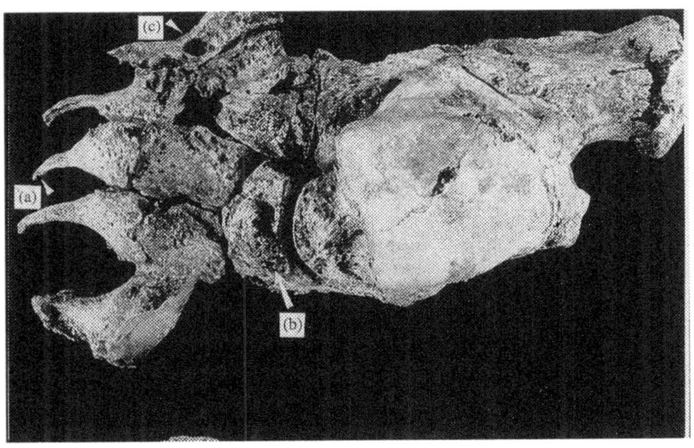

Figure 3: Detail of the right foot of the leprosy sufferer from Armoy, Co. Antrim. (a) knife-edged remodelling, (b) dorsal tarsal exostosis, (c) osteomyelitic cloaca.

Figure 2: General view of the feet of the leprosy sufferer from Armoy, Co. Antrim.

The history of the treatment of leprosy and the use of hydnocarpus oil

Shanthakumar Thomas Oommen
Colaco Hospital
Bendoor Well
Mangalore 575 002
India

E-mail: tomoommen@yahoo.com

Abstract

The major challenge in the management of leprosy is to avoid resistance, relapses or re-infection. The future of leprosy management is directed towards prevention of deformities. History is our guide to the future, for there can be no vision without a sense of history. In India, leprosy was considered to be a punishment for sin. In China, explanations regarding causes and treatment of leprosy followed preconceived notions. For centuries the only drug found to be effective against leprosy was the extract of the hydnocarpus fruit. The healing of experimental wounds in animals was significantly better when they were administered the oil of hydnocarpus orally or topically. This oil has both anti-leprosy and wound-healing properties, and it could be considered as a useful adjunct in the management of leprosy.

Keywords

Chaulmoogra, hydnocarpus, leprosy chemotherapy.

1. Introduction and the history of leprosy

A study of the remote past of the management of leprosy reminds us that for several centuries the earliest method of treatment was to contain the disease by isolating the patients. Those who had the disease were treated as social outcasts and were not permitted to be involved with the festivities and religious celebrations of their communities. They were only allowed to observe the proceedings through a crack in the wall, which was called the hagioscope (hagios = holy; scope = view); and wherever they went they had to sound bells or clappers to warn those around them that a victim of leprosy was passing by.

Although there is very little evidence to show that true leprosy (as we know it today) was actually present in Mesopotamia, perhaps the earliest evidence yet to be found for the segregation of patients suffering from disfiguring skin diseases is in Mesopotamia (Sigerist, 1955). Located between the east and the west, this land was on a major trade route and must have been visited from remote antiquity by a variety of plagues that also took their course from the east to the west. Various codes protected owners against losses. Since slaves in Mesopotamia were objects of property, Article 278 of the Code of Hammurabi (circa 1800 BC) enforced that the sale of a slave was considered invalid if he suffered from a contagious disease called *sibtu*, possibly leprosy.

Out of this awareness developed a form of public health action such as the "isolation of persons with communicable diseases". The Council of Lyons, in 583 AD, restricted free association with leprosy patients. In 644 AD, the Lombard King, Rothari, issued an edict that all people who had leprosy should be isolated. After the 10th century the number of leprosaria grew enormously, and by the beginning of the 13th century there were about 2000 such houses in France alone, and about 19,000 throughout Europe. By the 17th century AD, the disease had almost completely disappeared from Europe (Rosen, 1958:62-65).

The earliest reference to leprosy in the Chinese literature was in the Chou dynasty of the sixth century BC. If the Nei Chang of Huang Ti is correct, then leprosy was known in China for more than 5000 years. In this medical classic it was described that the first symptoms of the disease was the numbness of the skin, and a sensation of "worms creeping". The eyesight became blurred and on the skin there were dark scaly patches. Four quotations are found in the classic description which have some bearing on leprosy:

- those suffering from *ta feng* have stiff joints, the eyebrows and beard fall off

- the wind scatters through the muscles and comes in conflict with the *wei chi* or defensive forces. The channels, being clogged, lead to the flesh becoming nodular and ulcerated and, because of the stagnant movement of the defensive forces, numbness results

- The vital spirits degenerate and turn cloudy, causing the bridge of the nose to change colour and rot, and the skin to ulcerate. The winds and chills lodge in the blood vessels and cannot be rid of. This is called *lei feng*

- For the treatment of *lei feng* prick the swollen parts with a sharp needle and let the foul air out until the swelling subsides (Wong and Wu, 1936).

The Sung and Yuan dynasties also had very similar views

of the disease. Treatment included diaphoretics, purgatives, and strong poisons such as arsenic, scorpions and snakes. The earliest mention of the use of chaulmoogra oil for the treatment of leprosy in China was by Chu Tan-chi of the 14th century. He disapproved of its use because of its side effects (Wong and Wu, 1936).

Leprosy in India has been mentioned in the Vedic writings dated around BC 1400. In the Sushruta Samhida of 600 BC, one finds a reasonably good account of the clinical features of leprosy as it was known then, and the various measures adopted for its treatment (Filliozat, 1964, Haldar, 1977). In India, leprosy was grouped among the diseases that were considered to be a punishment for a sin in a previous life. Someone with leprosy was regarded as the most sinful person and their condition was attributed to the fact that they must have committed a most heinous crime in their previous life.

According to the Code of Manu, leprosy resulted in this life if one stole a Brahmin's gold. However, to kill a hundred cows was equal to killing one Brahmin, to kill a hundred Brahmins was equal to killing one woman, to kill a hundred women was equal to killing one child, and to kill a hundred children was equal to telling one lie. Only those whose sins equalled that of telling one hundred lies would develop leprosy. The prescription for the treatment of leprosy included worshipping the snake or the sun, or getting bitten by a cobra (Juggi, 1973a:181). Lowe (1942) felt that the Laws of Manu should not be regarded as conclusive proof of the prevalence of leprosy. However, when studied in relation to Hindu medical writings of a similar period, they provide strong evidence that leprosy was common.

2. Treatment of leprosy

In the 15th century an Indian physician named Gopalkrishna produced a comprehensive work on the materia medica of India, which he called The *Rasendrasara Sangraha*. One substance that was included in the book was called *galita kushtari rasa*, which was used for the treatment of leprosy. This was a formulation containing mercury, sulphur, copper, iron and vegetable oils. The patient was instructed to rub the medication between the palms of his hands with the extract of the plant *Sida cordifolia* until the globules of the mercury disappeared, and then to warm his palms before the fire. This was quite a novel and noble method of drug administration for a disease that was considered to be a divine curse (Juggi, 1973b:101 and 117). In the Middle East, since the 8th century, Islamic medicine has been using mercury and other substances dissolved in oil and applied to the skin with the patient by a fire (Campbell and Calton, 1960). This idea may have spread to India during the medieval period with Islam. In 1954, a British surgeon of the Indian medical Service, Mouat, formally introduced the use of chaulmoogra oil as a therapy for leprosy. In 1903, Engel Bay introduced a preparation which he named *Antileprol*, which contained ethyl esters of hydnocarpus oil and was claimed to be a definite cure for leprosy (Cochrane, 1959).

It is said that, in the 19th century, the daughter of a foreign officer in India had contracted leprosy. Being the daughter of a dignitary it was considered unacceptable to have her isolated or sent to a leprosy hospital. The Commissioner of the District was introduced to a leading Indian physician named Baba Vaidya who treated the girl with the oil of hydnocarpus orally, embedded in a banana to mask the taste and to reduce the abdominal side effects. The girl was cured of her leprosy. Since the physician would not accept any monetary remuneration for his services, the commissioner ordered that a statue of the physician be erected in the centre of the town. The statue of Baba Vaidya stands in the heart of the town of Ponda in Goa to this day.

At the International Leprosy Congress held in 1893, the only point on which all leprologists agreed was that the disease was incurable. Before the advent of any specific chemotherapy for leprosy, the only substance that had proven anti-leprosy activity was the extract of the *Hydnocarpus wightiana*, which grew in many parts of the Indian subcontinent. According to an Asian legend, a Burmese prince who was banished from his kingdom because he had contracted leprosy was healed when he ate the fruit of the hydnocarpus. Rogers (1916) first reported its efficiency when injected into the lesions, but the treatment was extremely painful as the injections were oil-based. Oral formulations were unacceptable due to the high incidence of gastro-intestinal side effects. He even administered it intravenously (Rogers, 1916). Although Paul Ehrlich conceived the idea of the magic bullet for chemotherapy, it was only in 1908 that Fromm and Whittman synthesised dapsone for the treatment of leprosy. However, since adequate studies had not been done to determine the margin of safety for the drug, it was considered to be highly toxic and was returned to the shelf for nearly 35 years.

Meanwhile, in the early 1900s, Dr Fredrick B Powers, the Scientific Director of the Wellcome Research Laboratories, published his epoch-making series of publications on phytochemicals, including his study on chaulmoogric acid and hydnocarpic acid, which contributed most valuably to the fight against leprosy. Powers had started his research in Flueckiger's Pharmaceutical Institute at the University of Strasbourg, and continued his study at the Philadelphia College of Pharmacy at the University of Wisconsin in the U.S.A., after which, he had taken over as Director of the Fritsche Brothers in Passaic, New Jersey (Rosen, 1958).

The fight against leprosy took a pivotal turn on December 9th 1940. On that day, four months after his arrival at Carville, Louisiana, and as Medical Officer-in-Charge, Dr Guy Faget wrote a letter to his friend, which led to sulfones being tried against leprosy. By February 1941, the first group of leprosy patients were started on injections of Promin, supplied *gratis* by Parke-Davis medical company.

By 1943, Faget and his colleagues were able to make a heartening preliminary report; they described their clinical trial with Promin as 'the most encouraging experiment ever undertaken'. Later the same year another sulfone, Diasone, was tried at Carville, Louisiana, U.S.A. with success.

The discovery of penicillin in 1922 changed the history of chemotherapy. However, in 1957, two years after the death of Sir Fleming, and commenting on the future of penicillins, George Bankoff (1957) said, 'Once produced synthetically, the use of penicillin will increase a thousand-fold. The incurable diseases like tuberculosis, leprosy and cancer will receive a fair trial. There are already signs that leprosy – the scourge of mankind from the beginning of ages – is responding. Soon perhaps, leprosy will remain only a name in the medical dictionary, a name that future generations will be tormented by only in nightmares. There will be no need to cast suffers into horrible enclosures and leave them helpless and desperate.'

A year before the death of Dr Alexander Fleming in 1954, Dr Vincent Barry of Dublin synthesised Clofazimine, a rhiminophenazine dye that was found to be effective against isoniazod-resistant tubercle bacilli. The following year it was also found to be effective against leprosy, and in 1962 it was used clinically by Browne and Hogerzeil to treat leprosy patients. Meanwhile, several other remedies had been tried, none of them with any significant success. These included potassium iodide, arsenic, antimony, copper, sera vaccines (including the diphtheria toxin), and aniline dyes (Cochrane, 1959).

The 1970s and 1980s saw an increase in the number of drugs that were effective against leprosy, with rifampicin, ethionamide, prothionamide, aminoglycosides, tetracyclines and quinololones being added in quick succession, and in various combinations into what is today called the multi-drug regimen (Hastings and Franzblau, 1988). However, none of them were without adverse effects, which limited their acceptability and use. By 1998, at the Leprosy Congress in Beijing, specialists from various disciplines were not just listing the various drugs that were currently available for the treatment of leprosy, they were also saying that the disease could possibly be eliminated by the end of AD 2000. The major challenge in the management of leprosy is to avoid drug resistance, relapses, re-infection or deformities. Furthermore, the prevention of deformities has to be the major thrust in the management of leprosy today, since these are what make the sufferer appear grotesque and disfigured, leading to social ostracism.

The "Law of signatures" states quite simply that if a disease is endemic in any particular geographic location, nature itself provides a cure in the same geographical location. This law was found to be true for diseases like malaria. It is not surprising, therefore, that the oil of hydnocarpus should be taken up again as a useful remedy for leprosy. In an experimental study undertaken at the Department of Pharmacology in Manipal, South India, it was found that the rate of wound healing of experimentally inflicted animals was significantly better in animals fed on the oil of hydnocarpus or, where the oil was applied topically over the wound. The wounds healed by primary intention, the granulation tissue was stronger, and the amount of hydroxyproline formed in the collagen tissue was significantly higher than in the control animals. This showed that the oil of hydnocarpus certainly possessed wound-healing properties (Oommen *et al.*, 1999, Oommen *et al.*, 2000). The chemical structure of the hydnocarpic and chaulmoogric acids of the oil show that they contain an unsaturated double bond in a cyclopentene ring structure, thereby suggesting that they have a free-radical scavenging property and can, thus, promote wound healing. The oil of the hydnocarpus has both wound-healing and anti-leprosy properties and could be considered as a useful adjunct (not as an alternative) in the management of leprosy.

Acknowledgements

The author wishes to acknowledge the assistance received from Dr Derek Browne which made the oral presentation of the paper at Bradford possible, and the Wellcome Trust, U.K. for the grant received to carry out the research needed to prepare and revise this paper.

References

Bankoff, G. 1957 *The conquest of disease: the story of Penicillin.*

Campbell, E. and Calton, J. 1960 *The surgery of Theodoric c.1267.* Volume 2. Appleton Cantary Crofts Inc.

Cochrane, R.G. 1959 *Leprosy in theory and practice.* Bristol: John Wright and Sons Ltd.

Filliozat, J. 1964 *The Classic Doctrine of Indian medicine: its origin and Greek parallels.* Delhi: Oriental Booksellers and Publishers.

Haldar, J.R. 1977 *Medical science in Pali literature.* Calcutta: Indian Museum.

Hastings, R. and Franzblau, S. 1988 *Chemotherapy of leprosy.* Annual Review of Pharmacology.

Juggi, O.P. 1973a Causation of disease. In: *History of science and technology in India.* Volume 3. Second Edition. Calcutta: Atma Ram & Sons.

Juggi, O.P. 1973b Medicine in medieval India. In: *History of science and technology in India.* Volume 3. Second Edition. Calcutta: Atma Ram & Sons.

Lowe J. 1942 Comments on the History of Leprosy. *Indian Medical Gazette*, November 1942, 77: 680-681.

Oommen, S.T., Rao C.M. and Raju, C.V.N. 1999 The effect of oil of hydnocarpus on wound healing. *International Journal of Leprosy* 67 (2): 154-158.

Oommen, S.T., Rao C.M. and Raju, C.V.N. 2000 The effect of oil of hydnocarpus on excision wounds. *International Journal of Leprosy* 68 (1) : 69-70.

Rogers, L. 1916 Preliminary note on the intravenous injection of gynocardate of soda in leprosy. *British Medical Journal* 2: 1-7.

Rosen, G. 1958 *A history of public health.* New York: M.D. Publishers Inc.

Sigerist, H.E. 1955 Preventive and archaic medicine. In: *History of medicine.* Volume 1. Oxford: University Press, pp. 381-398.

Wong, K.C. and Wu, L. 1936 *Chinese medicine.* Second Edition . National Quarantine Service.

Historical and palaeopathological evidence of leprosy in Hungary

György Pálfi[1], Albert Zink[2], Christian J. Haas[23], Antonia Marcsik[4], Olivier Dutour[5] and Andreas G. Nerlich[2,6]

[1]Department of Anthropology
National Museum of Natural History
Ludovika tér 2.
H-1083 Budapest
Hungary
Email: palfigy@hotmail.com

[2]Department of Pathology
Ludwig-Maximilians-University
Thalirchner Str. 36
D80337 Munich
Germany
Email: Albert.Zink@lrz.uni-muenchen.de

[3]Department of Pathology
Friedrich-Alexander University
Nurnberg-Erlangen
Germany
Email: molekular.pathologie@klinikum-augsburg.de

[4]Department of Anthropology
University of Szeged
Szeged
Hungary
Email: marcsik@bio.u-szeged.hu

[5]Department of Biological Anthropology
UMR 6578 CNRS
Faculté de Médecine
Université de la Méditerranée
Marseille
France
Email: Olivier.Dutour@medecine.univ-mrs.fr

[6]Department of Pathology
München-Bogenhausen Hospital
Munich
Germany
Email: Andreas.Nerlich@lrz.uni-muenchen.de
Correspondence to György Pálfi

Abstract

According to medical historical data, the first authentic written evidence of leprosy in Hungary dates from 1082; the prevalence of the disease increased around the 12th century, it may have reached its peak in the 13-14th centuries and declined around the 16th century. The skeletal evidence of the disease is sporadic, as leprosy hospital cemeteries have not been excavated until recently, and most of the palaeopathological cases come from non-leprosy hospital cemetery osteoarchaeological samples. The first recognised evidence of leprosy (advanced-stage rhinomaxillary changes and leprous alterations of the feet) comes from the 10th century cemetery of Sarretudvari (S-202) (Pálfi, 1991). The same series contained a second possible case (S-237), of a skeleton showing bony ankylosis in the tarsal bones of the right foot, with new bone formation and erosive changes, and slight periosteal reaction of the hard palate and the nasal region of the fragmentary skull (Pálfi, 1997). A similar case to S-202, showing leprous pathological alterations of the feet, has recently been discovered in the 14th to 15th century skeletal sample of Opusztaszer (Marcsik, 1998). Although in the cases of Opusztaszer and S-202 the morphological changes are typical, the results of palaeomicrobiological studies are negative. Thus, in areas related to damage of peripheral nerves there is no (or insufficient) mycobacterial DNA preserved to permit its amplification by PCR. However, the presence of *M. leprae* DNA in the hard palate sample of S-237 clearly attributes this case to leprosy (see Haas *et al.*, this volume, Palfi and Marcsik, 1999). This pattern has also been recorded in five skeletons with less specific maxillary alterations from the 10th to 11th-century cemetery of Puspokladany, where Spigelman *et al.* (this volume) identified *M. leprae* DNA in two specimens.

Keywords

Medical history, palaeopathology, skeletal leprosy, palaeomicrobiology, mycobacterial DNA, Hungary.

1. Historical data

According to medical historical data, the first authentic written proof of leprosy in Hungary dates back to 1082 AD (Fekete, 1874, Palla, 1956). It is a document signed by King Ladislas I about the territorial limits of the episcopate of Veszprem, Western Hungary, which refers to a *"puteus leprosorum"* near the city of Fehervar. This expression can be translated as "the well of lepers", but also as "the pit of lepers". Documents dating from further back in time show that in the early medieval period people with leprosy lived in primitive dwellings built around wells, outside the cities. However, "the pit of lepers" may also be explained as a reference to their primitive pit-dwellings, which probably existed in the 11th century. What is important is that some isolation of leprosy sufferers already existed in Hungary around the end of the 11th century (Fekete, 1874, Magyary-Kossa, 1929, Palla, 1956).

This isolation became more controlled and organized during subsequent centuries when several leprosy hospitals, or leprosaria, were established in the Carpathian Basin. As in other European countries, in Hungary the disease appeared more frequently between the 13th and 15th centuries, becoming an epidemic that peaked around the 14th century (Zubriczky, 1924, Magyary-Kossa, 1929). However, several leprosaria were also established in the 15th century in the northern and eastern parts of the country. There are also some data on the foundation of leprosy hospitals in Transylvania even in the second half of the 16th century (Vida, 1994). The leprosy church of Segesvar in Transylvania (Figure 1) is one example of these late-founded leprosaria. In this case, the external pulpit of the church permitted people with leprosy, housed in external buildings, to attend the mass outside the church (Vida, 1994).

As for the tending or treatment of people with leprosy in the Medieval period, we have some data concerning the life of Saint Elisabeth of the Arpad dynasty, from the first half of 13th century. St. Elisabeth, daughter of King Andrew II, is known as the Hungarian patron saint of the leprous. She established many leprosy hospitals and, according to the chronicles, she also treated the patients herself (Magyary-Kossa, 1929, Vida, 1994). In a 15thcentury illustration from the St Elisabeth Church of Kassa, St Elisabeth is shown tending a leprous patient (Figure 2). She is frequently portrayed with the leprous, for example on another painting from the 15th century (Figure 3), where she is bathing some leprosy sufferers in her leprosy hospital.

A contradictory story in Hungarian medieval history is the suspected leprosy of King Louis I, or Louis the Great (Figure 4). A Medieval Italian chronicle revealed that Louis I, King of Hungary, died with long-standing leprosy in 1382 AD. This picture, which was painted around 1360 at least 20 years before the death of the King, portrays him in white gauntlets. Since there is no other Hungarian king represented in gloves and with slightly deformed facial features in the chronicles, the account of the Italian "Chronicon Regiense" might be considered accurate (Vida, 1994). However, the life history of King Louis the Great seems to contradict the possibility that he suffered from leprosy. If he really had been ill, he would certainly have been deposed from the throne in Medieval Hungary (he was the most successful conqueror of the Hungarian Kingdom and also became the king of Poland and Naples). The country had reached its maximal extension during his 40 years reign. The fact that he often attended campaigns even in old age may be thought to be rather incompatible with a disease like leprosy. Unfortunately, palaeopathology cannot help to clarify this issue, since the bony remains of the king can be found in a mixed ossuary, containing the skeletal remains of approximately 400 individuals.

2. Palaeopathological data

As for the palaeopathology of leprosy in Hungary, despite the thousands of skeletons having been already examined, the skeletal evidence of the disease is sporadic (Marcsik and Palfi, 1999). It is likely that the disease did not exist in the so-called Avar Age (6th to 9th centuries AD), which is perhaps the best known period from a palaeopathological point of view, thanks to a study of more than 4000 skeletons (Marcsik and Palfi, 1999, Palfi and Marcsik, 1999). However, there are a few sporadic cases from the Hungarian Conquest Period (10th century). Despite the clear epidemiological context of the disease in the Medieval period, the lack of leprous cases is not so surprising. As has already been suggested, the leprous were isolated from the non-infected population relatively early in medieval Hungary, and leprosy hospital cemeteries have never been excavated. Therefore, our medieval palaeopathological cases come from non-leprosy hospital cemetery osteoarchaeological series.

2.1 Two cases of leprosy from the series of Sarretudvari (10th century A.D.)

2.1.1 Morphological alterations

The first recognized palaeopathological evidence of leprosy comes from the 10th century cemetery of Sarretudvari, Eastern Hungary. The skull of a well-preserved but incomplete skeleton of a 50-60-year-old female shows the signs of possible *facies leprosa* (Figure 5). The resorption of the anterior nasal spine seems to be an antemortem process. The lateral margins of the pyriform aperture are enlarged and remodelled (Figure 6). The alterations of the feet are more characteristic. The metatarsals, particularly those of the left foot, show marked concentric diaphyseal remodelling. There are signs of osteomyelitis and ankylosis of the third right metatarsal and loss of the distal part of the first metatarsal. The progressive development of acro-osteolysis led to a peg deformity and the possibility of loss of some phalanges during life (Figures 7 and 8). The alterations described above are very similar to those seen in some leprous skeletons described by Møller-Christensen (1961). The observed cranial and post-cranial alterations suggest leprosy as the most probable cause of these changes (e.g. Ortner and Putschar, 1985, Andersen and Manchester, 1992, Andersen *et al.*, 1992, Roberts and Manchester, 1995). Although the surfaces of the bone are slightly damaged by post-mortem erosion, signs of dorsal exostoses can be seen on some of the right tarsal bones. This may be a consequence of abnormal mechanical stress on the tarsals due to collapse of the foot arch (Andersen and Manchester, 1988). There is also bilateral periosteal apposition of bone on both tibiae and fibulae. These changes can indicate a leprous infection, but could also be the consequence of a secondary pyogenic infection, related to osteomyelitis in the metatarsals in this case. Following a macroscopic and a radiological differential diagnosis, we published this case in 1991 as the first osteoarcheological evidence of leprosy in Hungary (Palfi, 1991, Olah, 1990).

The 10th century series of Sarretudvari provided a second possible case of leprosy, but in this second case the morphological charges are much less conclusive (Palfi, 1997). This middle-aged adult male skeleton (S-237) showed pathological changes in the small bones of the feet. Bony ankylosis in the tarsals of the right foot with extensive asymmetrical periarticular new bone formation and erosive changes in the right tarsal joints was noted (Figures 9 and 10). The left side was significantly less affected, with only minor erosions of the talo-calcaneal joint. The skull, which only consisted of several fragments, showed slight periosteal reactions of the hard palate and the nasal region.

2.1.2 Molecular biological analysis

The molecular biological analysis of these two cases was carried out at the Department of Pathology, University of Munich. In the case of S-202, for technical reasons, only a sample of bone from the left distal part of the metatarsal was available for analysis, but nothing from the rhinomaxillary area. As for the second case (S-237), a bone sample was obtained from the hard palate (S-237a) and another from the right metatarsal (S-237b).

Using our DNA extraction and purification protocol (Haas *et al.*, this volume), it was possible to isolate *M. leprae* DNA from both specimens. This was confirmed by the specific amplification of a 202 bp fragment of the human beta-actin gene, which was detected in all the analysed samples. This merely excludes the inhibition of the PCR reaction or the absence of amplifiable ancient DNA. Using primers directed against the repetitive element 1 (RLEP1) of the *M. leprae* genome, the expected amplification product of 372 bp was detectable only in the sample S-237a (Figure 11). No specific PCR products with this primer pair were obtained in samples removed from the foot bones of both Hungarian cases (S-237b and S-202) (Figure 11). Finally, the specificity of the amplified RLEP1 segment in the skull bone sample (S-237a) was confirmed by direct sequencing of these PCR products. The obtained nucleotide sequences were identical to the respective published sequences of recent *M. leprae.* Thus, the presence of *M. leprae* in the hard palate sample of S-237 also clearly attributes this case to leprosy. However, despite the macro-morphological evidence, no *M. leprae* DNA was detected in the two foot bone samples, not even in the specimen where the hard palate sample was positive for *M. leprae*. This observation suggests that peripheral bones affected in leprosy may contain either significantly fewer bacteria or, more likely, they are caused by secondary processes, such as secondary osteomyelitis due to damage to the surrounding soft tissues and infection with non-specific bacteria. While the typical morphology of leprous bone alterations in the case of S-202 suggests that the amount of bacterial DNA might have been too small, the more "atypical" lesions in the foot bones of S-237 are more in favour of secondary osteomyelitic changes of the feet (Haas *et al.,* 2000).

3. Other palaeopathological data of leprosy from Hungary

Another skeleton with leprous pathological changes of the feet (diaphyseal remodelling, acro-osteolysis), very similar to the case of S-202, has recently been discovered in the 14th to 15th century cemetery of Opusztaszer, Southern Hungary (Marcsik, 1998). It is interesting that the molecular biological analysis of this case, which was carried out by Spigelman and colleagues (Donoghue *et al.*, this volume), also gave a negative result, similar to S-202.

Five cases of less specific maxillary alterations from the 10th to 11th century cemetery of Puspokladany, Eastern Hungary should be discussed. The palaeopathological study of this large cemetery (629 graves) did not reveal typical changes of leprosy. However, five skeletons showed slight, early-stage pathological alterations of the maxilla. Unfortunately, the bones of hands and feet are missing in all of these cases. As for the morphological changes, slight periosteal new bone formation can be observed in the inferior meatus of the left maxilla and on the inner surface of the nasal cavity of the case from grave 222. In another skeleton (grave 503), the border of the nasal aperture is slightly atrophied, but the only really remarkable pathological sign, similar to the previous case, is the presence of periosteal apposition of new bone on the inner surfaces of the maxilla, in the nasal cavity area. Although these last cases do not present macroscopic evidence of leprosy, the molecular biological study carried out by Donoghue *et al.* (this volume) identified *Mycobacterium leprae* DNA in these two specimens.

4. Conclusions

The history of the Hungarian Kingdom is not complete without discussing the history of leprosy. It is sufficient to mention St. Elisabeth, the daughter of King Andrew II, and the Hungarian patron saint of the leprous, or Louis I, King of Hungary, who died from probable long-standing leprosy. Although the first historical written proof of leprosy in Hungary dates from the end of the 11th century, palaeopathological evidence attests to the presence of the disease from the middle of the 10th century ("Hungarian Conquest Period"). From this period, seven probable leprous cases (representing two skeletal series: Sarretudvari (2), and Puspokladany (5)) have been reported, and in three of them the infection by *M. leprae* was confirmed by molecular biological analysis in two different laboratories. It is interesting that, until now, no evidence of leprosy has been found despite the study of several thousands of skeleton, from the previous historic period, the so-called "Avar Age" (6th to 9th centuries AD). These data may suggest that the disease arrived in the Carpathian Basin with ancient Hungarians from the East, around the end of the 9th century. However, with this level of knowledge, we cannot really formulate such precise palaeoepidemiological theories. Further palaeopathological studies of samples from earlier periods in the Carpathian Basin (e.g. Bronze Age, Roman Period) may give some new information concerning the first appearance of the disease in this area. Until now only two medieval cases of leprosy has been reported (Marcsik, 1998, Pálfi *et al.*, 2001), despite the medical historical data suggesting a high prevalence of the disease in the 13th to 15th centuries, the skeletal evidence of the disease is sporadic, as leprosy hospital cemeteries have not been excavated until recently.

Based on our results, we can conclude that in the so-called "osteologically typical" cases, when the bone alterations are related to damage of the peripheral nerves, there is probably insufficient usable mycobacterial DNA preserved (or none at all), which would allow amplification by standard or nested PCR methods. These advanced-stage cases are important data for paleoepidemiological studies. However, during morphological analysis of skeletons, we also have to pay special attention to the probable early manifestations of the rhinomaxillary changes (Ortner, this volume), which may provide more adequate material for subsequent palaeomicrobiological analyses. In the case of fragmentary skulls, and especially when the cortex is eroded by taphonomic processes and therefore the external face of the pyriform aperture cannot be examined, the study of the inner surfaces of the maxillary bones can be particularly important. The active-stage periosteal new bone formation on the inner nasal cavity area may also be included among the so-called "early-stage rhinomaxillary changes".

References

Andersen, J.G. and Manchester, K., 1988 Dorsal tarsal exostoses in leprosy: a palaeopathological and radiological study. *Journal of Archaeological Science* 15: 51-56.

Andersen, J.G. and Manchester, K., 1992 The rhinomaxillary syndrome in leprosy: a clinical, radiological and palaeopathological study. *International Journal of Osteoarchaeology* 2: 121-129.

Andersen, J.G., Manchester, K. and Shahzady, A.R., 1992 Diaphyseal remodelling in leprosy: a radiological and palaeopathological study. *International Journal of Osteoarchaeology* 2: 211-219.

Fekete, L. 1874 *A magyarországi ragályos és járványos kórok rövid történelme.* Debrecen: No publisher cited.

Haas, C.J., Zink, A., Pálfi, Gy., Szeimes, U., Nerlich, A.G. 2000 Detection of leprosy in ancient human skeletal remains by molecular identification of *Mycobacterium leprae. American Journal of Clinical Pathology* 114: 428-436.

Magyari-Kossa, Gy. 1929 *Magyar orvosi emlékek.* Eggensberger-Rényi Kiadó. Budapest: I. Kötet.

Marcsik, A., 1998. Az Opusztaszeri csontvazanyag paleopatologias elvaltozasai. In Gy.L. Farkas (ed): *Opusztaszer-Monostor lelohely antropologiai leletei.* Szeged: JATE Embertani Tanszék, pp. 97-155.

Marcsik, A. and Pálfi, Gy. 1999 Presence of infectious diseases in ancient populations in Hungary. In M. Henneberg and C. Oxnard (eds.): Perspectives in Human Biology, Volume 4: Issue 1 : *Is human evolution a closed chapter*?. Nedlands : Centre for Human Biology, The University of Western Australia, pp. 159-166.

Møller-Christensen, V. 1961 *Bone changes in leprosy.* Copenhagen: Munksgaard.

Olah, S. 1990 *Sárrétudvari-Hízóföld honfoglalás kori temetôjének történeti embertani értékelése.* Doctoral Thesis, József Attila University, Szeged.

Ortner, D.J. and Putschar, W.G.J. 1985 *Identification of pathological conditions in human skeletal remains.* Washington D.C.: Smithsonian Institution Press.

Pálfi, Gy. 1991 The first osteoarchaeological evidence of leprosy in Hungary. *International Journal of Osteoarchaeology* 1 : 99 - 102.

Pálfi, Gy. 1997 Maladies dans l'antiquité et au Moyen-Âge. Paléopathologie comparée des anciens Gallo-Romains et Hongrois. *Bull. et Mém. de la Soc. d'Anthrop. de Paris.* 9 (1-2): 1- 206.

Pálfi, Gy. and Marcsik, A. 1999 Paleoepidemiological data of tuberculosis in Hungary. In: G. Pálfi, O. Dutour, J. Deák and I. Hutás (eds.): *Tuberculosis: Past and Present.* Szeged, Hungary: TB Foundation and Budapest, Hungary: Golden Book Publisher Ltd., pp. 531-539.

Pálfy, Gy., Pap, I. and Fóthi, E. 2001 Mycobacteriális fertőzések új paleopatológiai esetei. In I. Isépy, Z.

Korsós and I. Pap (eds): *II. Kárpát-medencei Biológiai Szimpózium.* Budapest: Magyar Biológiai Társaság és Magyar Természettudományi Múzeum, pp. 325-331.
Palla, A. 1956 *A veszprémi "puteus leprosorum".* Budapest: Communicationes ex Bibliotheca Medicae Hungarica.
Roberts, C. and Manchester, K. 1995 *The archaeology of disease.* Ithaca, New York: Cornell University Press.
Vida, M. 1994 *Muveszet es orvostudomany a tortenelmi Magyarorszagon. Medicina in artibus in Hungaria.* Budapest: Magyar Kepek Kiado and Semmelweis Orvostorteneti Muzeum, Konyvtar es Leveltar.
Zubriczky, A. 1924 *Az európai nagy leprajárvány története.* Budapest: Magyar. Tud, Társ, Sajtóvállalata, RT.

Figure 1: The leprosy church of Segesvar (15th century) in Transylvania (Sigisoara, Romania). Courtesy of M. Vida, Semmelweis Orvostorteneti Muzeum, Konyvtar es Leveltar, Budapest.

Figure 2: St. Elisabeth bathes a patient in her leprosy hospital (15th century, St Elisabeth Church of Kassa, Kosice, Slovakia). Courtesy of M. Vida, Semmelweis Orvostorteneti Muzeum, Konyvtar es Leveltar, Budapest

Figure 3: St. Elisabeth bathes a patient in her leprosy hospital (by Severinus (1484), St Egyed Church, Bartfa). Courtesy of M. Vida, Semmelweis Orvostorteneti Muzeum, Konyvtar es Leveltar, Budapest.

Figure 4: King Louis I, or Louis the Great, King of Hungary, who died from probable long-standing leprosy in 1382 AD (by Miklos Medgyesi, around 1360). Courtesy of M. Vida, Semmelweis Orvostorteneti Muzeum, Konyvtar es Leveltar, Budapest.

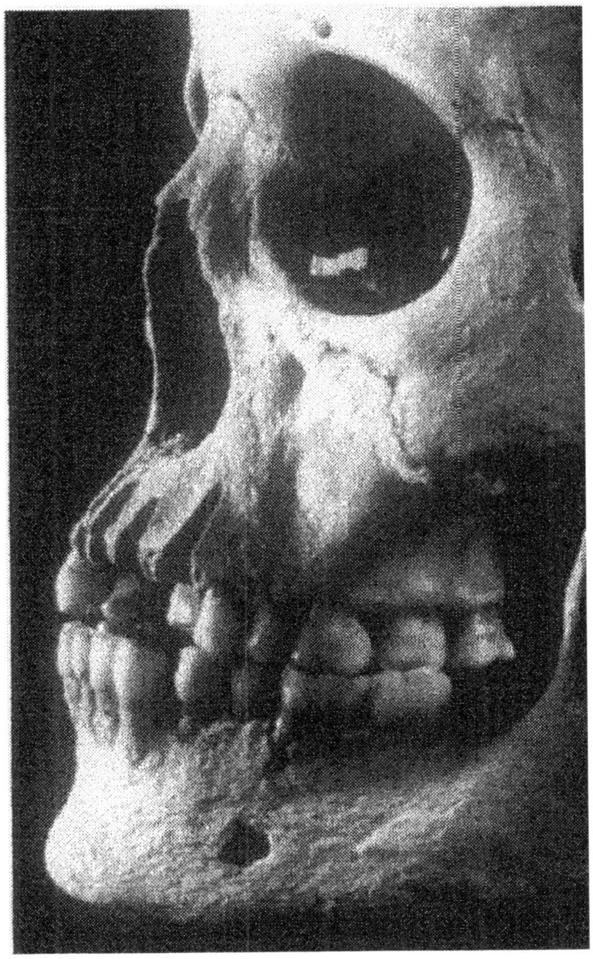

Figure 5: Erosion of nasal margins and loss of anterior nasal spine (S-202, 10th century)

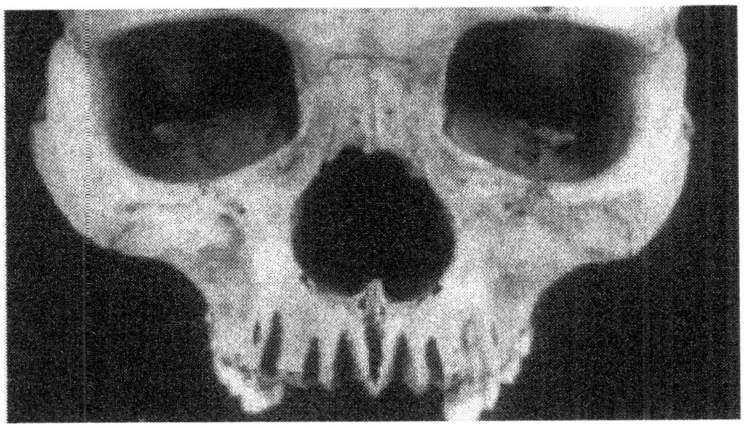

Figure 6: *Facies leprosa* (S-202, 10th century)

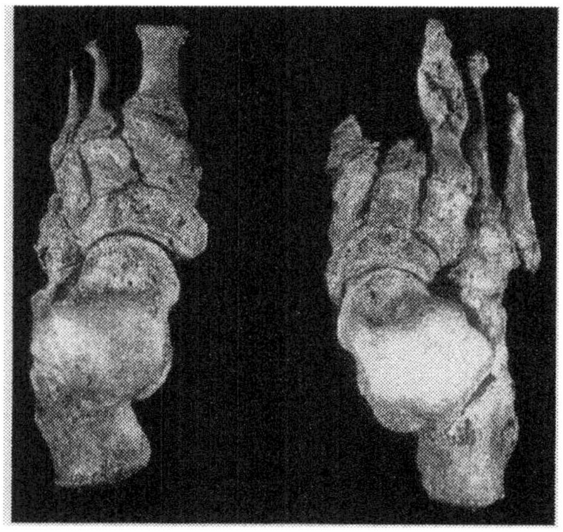

Figure 7: Severe changes of the feet: acro-osteolysis and diaphyseal remodelling of metatarsals, and dorsal exostoses on tarsals (S-202, 10th century)

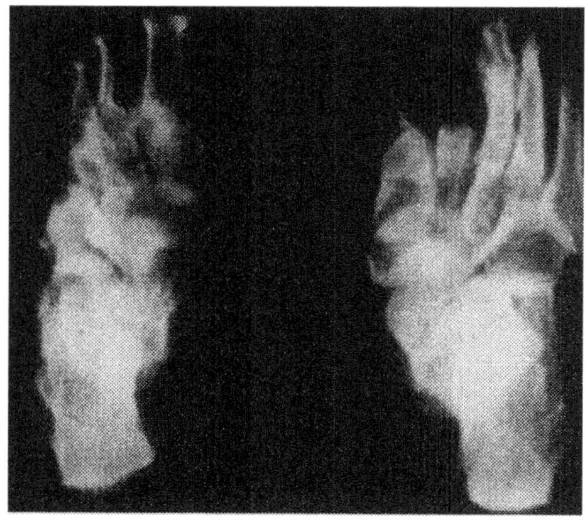

Figure 8: Radiographs of the feet with marked concentric diaphyseal remodelling of the metatarsals, and osteomyelitis and ankylosis of the third metatarsal (S-202, 10th century). Radiograph by J. Kovacs.

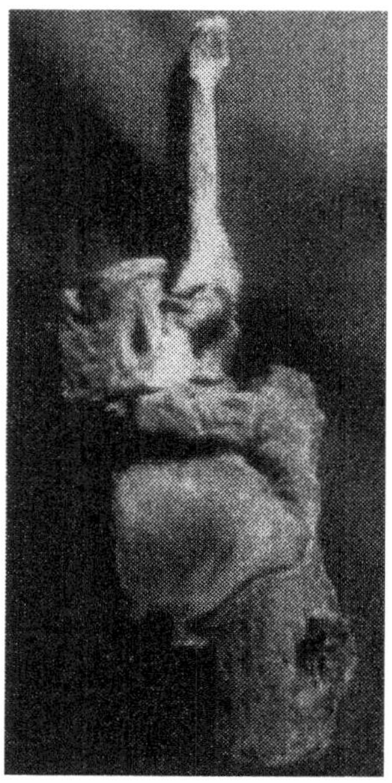

Figure 9: Bony ankylosis in the tarsals of the right foot with extensive asymmetrical periarticular new bone formation (S-237, 10th century).

Figure 10: Periarticular new bone formation and erosive changes in the right tarsal joints (S-237, 10th century).

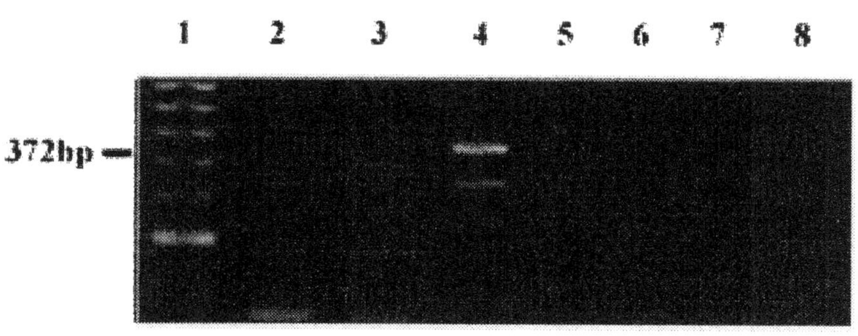

Figure 11: Amplification of a 372 bp fragment of RLEP1 specific for *M. leprae*. Lane 1: 50 bp ladder; Lane 2: case S-202; Lane 3: case S-237b; Lane 4: case S-237a; Lane 5-8: blank controls.

The antiquity of leprosy in Britain: the skeletal evidence

Charlotte Roberts
Department of Archaeology
University of Durham
South Road
Durham DH1 3LE
England

Email: C.A.Roberts@durham.ac.uk

Abstract

Leprosy has had a long history in Britain and yet most of the evidence comprises historical documentation and iconographic evidence. The primary evidence for the disease, in skeletal remains, has been reported sporadically over the years. This paper brings together the extant data so far from published and unpublished work, considering age and sex distribution, distribution pattern and funerary context. A total of 128 individuals were affected in 41 archaeological sites from a total of 8253 burials revealing skeletons (1.55%). Most examples came from the later Medieval period, most were male, and the majority of sites revealing leprous individuals were not leprosy hospital cemeteries. The results are discussed with reference to the limitations of the data, with suggestions for future work.

Keywords

Leprosy, Britain, frequency, Historic period, skeletal evidence

1. Introduction

'Few diseases have a richer cultural heritage, and none so rich a mythology' (Fine, 1982:1).

Leprosy is a disease with a long history whose occurrence and character today in some parts of the world indicate that, socially, it is still considered a stigmatising condition that leads to ostracism and isolation (Joseph and Rao, 1999). Of course, the stigma attached to any disease is often the result of inappropriate conceptualisation of what the disease is and how it is transmitted; leprosy is not the only disease today and in the past that has attracted stigma. The stigma associated with leprosy today (as in the past) often leads to the isolation and segregation of people from their communities and a lack of social integration with their peers, so essential for their well being. While many people working with human remains derived from archaeological sites aim to identify health problems seen in individual skeletons and in populations as a whole, often there is a lack of attention paid to the effect of disease on these people's lives. Likewise, today in hospital situations the emphasis tends to be on the disease, rather than on the person with the disease ("clinical labelling"). There is a move away from considering the character, personality and lifestyle of the person affected to looking at the clinical manifestations of the disease itself and its treatment. Leprosy provides an opportunity to look at a disease that has a complex and interesting epidemiology, that has affected people and populations from past to present, and has also had an associated stigma.

Leprosy is rarely seen in Britain today, except in immigrant populations (Van Buynder *et al.*, 1999). For example, there have been 1358 cases notified since 1951 with a peak of 467 between 1960 and 1969, then a decline. Between 1990 and 1997 40% of cases had sufficient anaesthesia to cause functional impairment. Currently there are 128 patients on the central registry (70% male), and 71% have significant complications such as loss of vision, deformity and anaesthesia. Thus, even today in Britain, leprosy is causing misery to some people. Leprosy was apparently a common infectious disease during the later Medieval period in Britain from about the 12th century AD, although the first skeletal evidence comes from the 4th century (Reader, 1974). Most of the evidence for its prevalence comes from documentary sources (Richards, 1977), with reports in the palaeopathological literature indicating skeletal evidence for the disease. Sources suggest that the infection was feared and that people with it were diagnosed and segregrated into leprosy hospitals founded mainly from the 12th to 16th centuries AD (Roberts, 1986). However, this may not necessarily have been the case for all parts of Britain at all periods of time. The skeletal evidence for leprosy has been sporadically reported over the years but until now there has been no real synthesis of data. This paper aims to bring together the extant skeletal evidence for leprosy in Britain and consider its limitations, discussing it with reference to the historical data for its presence.

2. Material and Methods

The skeletal data collated for this study comprises unpublished work by Manchester and Roberts (1986) which surveyed museum and archaeological unit curated skeletal collections, and the consideration of all published (and some unpublished) data on the skeletal evidence for leprosy (references provided in Table 1). All sites date from some time during the Romano-British period (4th century AD) to the post-Medieval periods (18th-19th centuries AD), and all populations considered are from settled communities living in permanent housing, domesticating plants and animals, and relying on agricultural produce whether in a rural or urban environment. The diagnostic criteria taken as being indicative of leprosy followed the recommendations of Andersen and Manchester (1987, 1988, 1992), Andersen *et al.* (1992, 1994), Jopling and McDougall (1988), Møller-Christensen (1961). The following scenarios were accepted as being indicative of leprosy, and Figures 1a and b, 2 and 3 illustrate some of the cases:

1. *Facies leprosa* (or rhinomaxillary syndrome)
2. *Facies leprosa* and foot changes
3. *Facies leprosa* and hand changes
4. *Facies leprosa* with tibia, fibula and foot changes
5. Tibia, fibula and foot changes
6. Hand, tibia, fibula and foot changes

Periosteal new bone formation of the tibiae and fibulae alone were not accepted as a criterion for leprosy diagnosis because many conditions can lead to this change. For example, it may appear alone as a result of trauma, as part of one of the treponemal syndromes, or as a result of non-specific infection. Patterning of bone changes associated with leprosy is key to diagnosis as many of the changes in isolation may be the result of other pathological conditions. Diagnosis was conducted using macroscopic analysis, although biomolecular techniques of diagnosis for leprosy are developing (see this volume) and problematic cases in the future will be aided by these developments.

3. Results

The results are presented for sites that revealed evidence for leprosy (Table 1). There are many more sites from British contexts where leprosy has not been identified. Unfortunately it is not possible to determine absolute frequencies for leprosy in Britain because of problems with the data. Data produced has been presented as the percentage of individuals affected from all sites examined and, for most sites, it is not known whether all bones of all skeletons were available for study. Furthermore, it is not possible to assess exactly how many skeletons from the Romano-British period through to the late and post-Medieval periods have been studied, and therefore absolute frequencies of people affected cannot be given. This is a problem in palaeopathological study generally in the U.K., a problem that is being addressed currently.

A total burial population of 8253 skeletons from 41 archaeological sites were considered, revealing 128 affected individuals. This comprises a frequency of 1.55%. If we consider that around 50,000 burials have been analysed in Britain to date this would mean a frequency of leprosy of approximately 0.26% (individuals affected), but this must be seen only as a very general estimate; it also covers about 1500 years. Two sites were Romano-British in date (4th century AD), twelve sites were from the Anglo-Saxon period (5th-11th centuries), and 27 sites were later (12th –16th centuries) or post-Medieval (post-16th century). The sites revealing leprous individuals derive mainly from the south and east of England with very few in Wales, Scotland or Ireland (Figure 4). Taking the data by period, the first evidence (dated to the 4th century AD) comes from the Romano-British period from sites in Dorset and Gloucestershire (south and south-west England respectively). Two individuals are affected, comprising 0.14% of the 1480 Romano-British burials. More examples of leprosy come later in the early Medieval (or Anglo-Saxon/post-Roman) period, around the 5th-11th centuries AD. From a total of 2031 individuals, 18 (0.89%) were affected, but numbers increase in the later and post-Medieval periods (12th century to the 19th century). One hundred and eight individuals were identified with leprosy from a total of 4742 burials (2.28%). Because the majority of cemeteries of late and post-Medieval date are not stratified to the extent that it is possible to correlate specific skeletons with a more closely defined date, it was not feasible to ascertain whether leprosy fluctuated through that time period.

Thirteen sites produced multiple individuals with leprosy, which might be expected if a cemetery was attached to a leprosy hospital or even non-leprosy hospital, but for only five of the twelve sites was this the case. However, in at least one case (Timberhill, Norfolk), despite there being a large number of individuals with skeletal changes of leprosy (24 of 181 individuals), there was no such known hospital association (Anderson, 1998).

Of the 110 individuals with leprosy where a definite sex could be attributed, two thirds were male. Unfortunately for many of the sites considered, the ratio of males to females was not available, and therefore it is difficult to determine whether the sex frequency of leprosy is a true reflection of the disease's predilection for affecting males more than females. Of the 98 aged individuals with leprosy, most were in the age bracket 30-50 years, with only nine non-adults affected. Again, the age profile of many of the sites utilised for this survey was inadequate and, therefore, whether this reflects the true age predisposition is unclear.

When funerary context is considered a wide range of different types is seen. The two Romano-British cemeteries of Poundbury and Cirencester were both rural in derivation, while the twelve Anglo-Saxon contexts consisted of a churchyard cemetery (1), rural cemeteries with no associated churchyard (8), an early phase of a cathedral cemetery (1), a large pit (1) and one documented leprosy hospital. For the later and post-Medieval periods the range of contexts include, for the 27 sites, general Medieval cemeteries (4), a cist cemetery (1), cemeteries associated with abbeys (2), friaries (4), churches (3), a chapel (2), a priory (1), cathedrals (2), non-leprosy hospitals (3), leprosy hospitals (4) and a monastery (1).

4. Discussion

While absolute frequency rates for leprosy cannot be determined, results suggest that there was an increase in leprosy through time, which correlates with the historical data. This is expected since people in the later and post-Medieval periods in particular were living in close contact with each other in mostly urban environments, where droplet transmission of leprosy would have been possible. The number of cases, however, is low. Perhaps this supports suggestions by MacArthur (1953:18) that there were few leprous individuals in Britain during the Medieval period, and (1925) that the term probably represented many conditions which were interpreted incorrectly. Most cases come from the south and east of England with few in the north and west, although the late and post-Medieval cases tend to be situated further north into northern England and Scotland (Figure 4). However, we know from historical data that the last indigenous case of leprosy in the British Isles was admitted to Edinburgh Hospital in Scotland in 1798 from the Shetland Islands (Browne, 1975), and that leprosy hospitals founded in the 15th and 16th centuries tended to be further north into Scotland (Roberts, 1986). The lack of evidence in Wales, Scotland and Ireland may be explained in part because of the acidic soils in those areas which does not preserve bone well. Furthermore, many cemetery sites will probably lie unexcavated in the rural areas of these places where new building development is not as much of a threat to archaeological evidence; perhaps the concentration of sites revealing leprous individuals in the south and east reflect the amount of archaeological fieldwork undertaken. However, even more relevant is that many skeletons from archaeological sites lack hand, foot and facial bones for examination, and it is unclear how many of the individuals in this study suffered postmortem loss of bone. Although the 200 plus leprosy hospital foundations from the 11th to the 16th centuries AD in Britain cannot be taken as indicative of the frequency of the disease, their distribution matches the skeletal evidence (Figure 5). This may suggest population density at the time or that people were more charitable during those time periods. However, it is suggested that these hospitals were opened also for the non-leprous, and certainly in the later periods of their use they became general hospitals when leprosy declined from about the 14th century onwards.

In many clinical studies males are reported to be more affected than females (e.g. Van Buynder *et al.*, 1999) but there may be many reasons for this pattern. For example, in some parts of the world females do not have the privileges of social freedom that many western countries have and, therefore, may not have access to diagnosis. Women often have to stay at home for domestic responsibilities and, furthermore, they may not be comfortable in exposing their bodies to male physicians for diagnosis of early skin lesions. It may also be possible that females could have a stronger immune response to leprosy than males; Ortner (1998) suggests that females may, in fact, possess this attribute. However, if females were stronger in this sense then we may expect to see more extensive bone damage on female skeletons rather than males (see Wood *et al.*, 1992), or alternatively they may be less likely to show bone change even though they had the disease. Interestingly, a recent study suggests that women in India with leprosy actually have a better quality of life than men. This perhaps reflects their acceptance of the disease because of their secondary role in society (Joseph and Rao, 1999). However, the effect on males, e.g. the ability to maintain employment, may be more devastating.

With respect to age and leprosy, although a correlation was made, it should be remembered that a person with leprous bone change could have developed that change any number of years before their death; thus, is age relevant in the discussion? Leprosy can also have a very long incubation period and this will affect when bone changes develop, so that people need to survive well into adulthood to develop bone changes (for example, skeletons of children with rhinomaxillary syndrome from Naestved, Denmark showed few postcranial changes, Mary Lewis, pers. comm.). The many extrinsic and intrinsic factors that affect whether infection develops or not in a person further complicates the picture.

The funerary contexts from which the leprous individuals in this study derive clearly cover a wide range and are not all from leprosy hospital cemeteries which might be expected if the historical data for diagnosis and segregation is to be believed. However, only five leprosy hospitals were identified from a total of 38 sites considered. This raises a number of questions and comments. If leprosy was a stigmatised disease and people were ostracised in the later Medieval period, then why do we find most of the leprous individuals in British contexts in non-leprosy funerary situations? Some suggestions can be made: the person with leprosy may have been accepted within their social group despite being diagnosed as leprous. The generic statement that the leprous in the later Medieval period were segregated may not hold true. Perhaps diagnosis was not accurate? Simpson (1842) indicates that this may have been the case, and it is suggested that diagnostic methods were often unconventional. Although some diagnosticians may have known how to diagnose leprosy, others may not have had the necessary knowledge. Finally, the person may not have, externally, appeared leprous and, therefore, may not have been diagnosed or segregated. Furthermore, if a person had a low resistant form of the disease, the manifestation of leprosy in their bodies may not have been particularly obvious. It may have been only those with lepromatous leprosy who were diagnosed, while those with less severe forms were left to function relatively normally within their social group. Nevertheless, people could have evaded detection, knowing that their lives would be over once they entered a leprosy hospital, or they could have made themselves look leprous to gain charity. It therefore cannot be predicted exactly where leprous individuals lived, died and were buried in Britain because of many confounding factors.

Clearly, the frequency of leprosy, as seen in skeletal material from archaeological sites in Britain, cannot be taken as the absolute frequency of this disease during the Historic period for a variety of reasons. The sites considered here are only those which have revealed leprosy and, therefore, the "individual" frequency refers only to the total number of individuals for all the sites examined. Many more cemetery sites have been excavated and examined but no evidence of leprosy has been found. Thus, the skeletal material considered here is only a small sample of the original living population and is therefore biased (see Waldron, 1994 for more discussion on sample representivity). Consequently, age and sex distribution will also be biased, and therefore any comments on the demographic profile of the leprous individuals identified must subsequently be flawed. Furthermore, leprosy is known to affect the skeleton in up to only 15% of people (Steinbock, 1976:198). Therefore, only the very tip of the iceberg of the leprosy problem in Britain is being seen in the skeletal evidence. Nevertheless, higher frequencies may be expected in leprosy hospital cemeteries; for example, 77% of 202 individuals in the Medieval leprosy hospital at Naestved, Denmark had bone change (Andersen, 1969) which is equivalent to modern leprosy hospital rates.

There will also inevitably have been loss or damage to facial, hand and foot bones for many skeletons from these sites and, therefore, true frequencies for leprous bone change cannot be suggested. Nevertheless, the rate of leprous involvement of the skeleton is low in untreated leprosy today, and therefore a low percentage frequency rate may be expected. Furthermore, whether a person develops bone change will depend on their immune status, age at contraction and the length of the incubation period. There are, in addition, a number of social and environmental factors that are necessary for leprosy to be contracted. For example, a well balanced nutritious diet will help build a strong immune system and prevent a person being infected, and crowded living conditions will help leprosy to be transmitted. A question that has not yet been answered is whether those individuals buried in leprosy hospital cemeteries with no bone damage actually had leprosy. For example, only 36 of the 351 individuals from the Chichester leprosy hospital have bone change accepted as indicative of leprosy (Ortner, pers. comm.), although around 70% of the skeletons from the Naestved leprosy hospital cemetery (Danish) had leprous bone change. Biomolecular techniques and analysis should help in the future and could provide absolute frequency rates.

The data presented here tells us something about leprosy in Britain but the data are biased in a number of ways. There is much more work that needs to be done in Britain in order to ascertain the real frequency of leprosy. This is a pre-requisite for comparative work in Europe and to be able to comment on how data from Britain fits in with the overall origin, evolution and palaeoepidemiology of leprosy throughout the world. Future meticulous and careful excavation, recording and analysis should provide the data needed to contribute to a world history of leprosy.

Acknowledgements

The Science and Engineering Research Council and The Wellcome Trust partially funded some of this research, Don Ortner provided data on burials from the Chichester Medieval Leprosy hospital and Figure 3, and Jean Brown (formerly of the Archaeological Sciences Department, University of Bradford), and Trevor Woods (Department of Archaeology, University of Durham) produced and reproduced, respectively, the figures. Too many colleagues to mention helped in various ways with the data on skeletal evidence from specific sites, and curators who allowed access to those skeletal collections analysed by the author are thanked for their help. Finally, this paper is dedicated personally to Keith Manchester; if it wasn't for Keith I would never have "met" leprosy.

References

Andersen, J. 1969 Studies in the Medieval diagnosis of leprosy in Denmark. *Danish Medical Bulletin* 16(Supplement): 1-142.

Andersen, J. and Manchester, K. 1987 Grooving of the proximal phalanx in leprosy: a palaeopathological study. *J. Archaeological Science* 14(1):77-82.

Andersen, J. and Manchester, K. 1988 Dorsal tarsal exostosis in leprosy: a palaeopathological and radiological study. *J. Archaeological Science* 15:51-56.

Andersen, J. and Manchester, K. 1992 The rhinomaxillary syndrome in leprosy: a clinical, radiological and palaeopathological study. *International J. Osteoarchaeology* 2:121-129.

Andersen, J., Manchester, K. and Ali, R. 1992 Diaphyseal remodelling in leprosy: a radiological and palaeopathological study. *International J. Osteoarchaeology* 2:211-219.

Andersen, J., Manchester, K. and Roberts, C.A. 1994 Septic bone changes in leprosy: a clinical, radiological and palaeopathological review. *International J. Osteoarchaeology* 4:21-30.

Anderson, S. 1990 *The human skeletal remains from Staunch Meadow, Brandon, Suffolk.* Ancient Monuments Laboratory Report 99/90. Unpublished.

Anderson, S. 1998 Leprosy in a Medieval churchyard in Norwich. In S. Anderson (ed): *Current and recent research in osteoarchaeology. Proceedings of the 3rd Meeting of the Osteoarchaeology Research Group.* Oxford: Oxbow Books, pp. 31-37.

Bishop, M. W. 1983 Burials from the cemetery of the hospital of St Leonard, Newark, Nottinghamshire. *Transactions of the Thoroton Society of Nottinghamshire* 87:23-35.

Boulter, S. 1992 *Death and disease in Medieval Grantham.* Undergraduate Dissertation, University of Sheffield, Department of Archaeology and Prehistory. Unpublished.

Brothwell, D.R. 1958 Evidence of leprosy in British archaeological material. *Medical History* 11: 287-291.

Brothwell, D.R. 1977 On a mycoform structure in Orkney, and its relevance to possible further interpretations of so-called souterrains. *Bulletin of the Institute of Archaeology, London* 14:179-190.

Brothwell, D.R. and Krzanowski, W. 1974 Evidence of biological difference between early British populations from Neolithic to Medieval times, as revealed by eleven commonly available cranial vault measurements. *J. Archaeological Science* 1:249-260.

Brothwell, D.R., Powers, R. and Hirst, S.M. 2000 The pathology. In P. Rahtz, S.Hirst and S.M. Wright : *Cannington cemetery. Excavations 1962-3 of prehistoric, Roman, post-Roman and later features at Cannington Park Quarry, Near Bridgwater, Somerset.* Britannia Monograph Series No. 17. London: Society for the Promotion of Roman Studies, pp. 195-256.

Browne, S.G. 1975 Some aspects of the history of leprosy: the leprosy of yesteryear. *Proc. Royal Society of Medicine* 68:485

Cardy, A. 1989 Carmelite Friary on the Green, Aberdeen. Supplement to Stones (ed), 1989.

Cardy, A. 1997 The human bones. In P. Hill: *Whithorn and St Ninian. The excavation of a monastic town*, Gloucester: The Whithorn Trust and Sutton Publishing, pp. 519-592.

Chundun, Z. and Roberts, C.A. 1995 Human skeletal remains (pp. 214-218). In P. Cardwell: Excavation of the hospital of St Giles by Brompton Bridge, North Yorkshire. *Archaeological Journal* 152:109-245.

Duhig, C. 1998 The human skeletal material. In T.Malim and J. Hines: *The Anglo-Saxon cemetery at Edix Hill (Barrington A), Cambridgeshire.* York: Council for British Archaeology Research Report 112, pp. 154-199.

Farley, M. and Manchester, K. 1989 The cemetery of the leper hospital of St Margaret, High Wycombe, Buckinghamshire. *Medieval Archaeology* 33:82-90.

Fine, P.E.M. 1982 Leprosy: the epidemiology of a slow bacterium. *Epidemiologic Review* 4:161-188

Hacking, P. and Wakely, J. n.d. *The human skeletal remains from Abingdon Abbey*. Unpublished report.

Jopling, W. and McDougall, A.C. 1988 *Handbook of leprosy*. London: Heinemann Books.

Joseph, G.A. and Rao, P.S.S. 1999 Impact of leprosy on the quality of life. *Bulletin of the World Health Organisation* 77(6):515-517.

King, S. E. 1994 *The human skeletal remains from Glasgow Cathedral*. Unpublished report.

Knüsel, C.J. and Göggel, S. 1993 A cripple from the Medieval hospital of Sts James and Mary Magdalen, Chichester. *Int. J. Osteoarchaeology* 3:155-165

Lee, F. and Magilton, J. 1989 The cemetery of the hospital of St James and St Mary Magdalene, Chichester – a case study. *World Archaeology* 21(2):273-282

Lunt, D.A. 1996 Dental and anatomical reports: The dentitions. In E. Proudfoot (ed): Excavations at the long cist cemetery on the Hallow Hill, St Andrew's, Fife 1975-7. *Proceedings of the Society of Antiquaries of Scotland* 126:387-454.

Lunt, D.A. 1997 The dentitions. In M.J. Rains and D.W Hall: *St Mary's Church, Kirk Hill, St Andrews*. Glenrothes, Tayside and Fife Archaeological Committee, pp. 131-137.

MacArthur, W.P. 1925 Some notes on old time leprosy in England and Ireland. *J. Royal Army Medical Corps* 45(6):410-422.

MacArthur, W.P.1953 Mediaeval leprosy in the British Isles. *Leprosy Review* 24:8-19.

Magilton, J. and Lee, F. 1989 Leper hospital of St James and St Mary Magdalene, Chichester. In Roberts, C.A., Lee, F. and Bintliff, J. (eds): *Burial archaeology. Current research, methods and developments*. British Archaeological Reports British Series 211, pp.249-265.

Manchester, K. 1981 A leprous skeleton of the 7th century from Eccles, Kent, and the present evidence of leprosy in early Britain. *J. Archaeological Science* 8:205-209

Manchester, K. and Roberts, C.A. 1986 *Palaeopathological evidence of leprosy and tuberculosis in Britain*. S.E.R.C. Report for Grant Number 337.367

Mays, S. 1989 *The Anglo-Saxon human bone from School Street, Ipswich, Suffolk*. English Heritage, Ancient Monuments Laboratory Report 115/89, Unpublished.

Mays, S. 1991 *The Medieval burials from the Blackfriars Friary, School Street, Ipswich, Suffolk.* English Heritage, Ancient Monuments Laboratory Report 16/91. Unpublished.

Møller-Christensen, V. 1961 *Bone changes in leprosy*. Copenhagen: Munksgaard.

Møller-Christensen, V. and Hughes, D.R. 1962 Two early cases of leprosy in Great Britain. *Man* 287:177-179.

Murphy, E. and Manchester, K. 1998 'Be thou dead to the world': leprosy in Ireland, evidence from Armoy, Co. Antrim. *Archaeology Ireland* 12(1):12-14.

Ortner, D.J. 1998 Male/female immune reactivity and its implications for interpreting evidence in human skeletal paleopathology. In A.L. Grauer and P. Stuart-Macadam (eds): *Sex and gender in paleopathological perspective*. Cambridge: University Press, 79-82.

Ortner, D.J., Manchester, K. and Lee, F. 1991 Metastatic carcinoma in a leper skeleton from a Medieval cemetery in Chichester, England. *Int. J. Osteoarchaeology* 1: pp.91-98

Powell, F. 1996 The human remains. In A. Boddington: *Raunds Furnells. The Anglo-Saxon church and churchyard.* London, English Heritage Archaeological Report 7, pp. 113-124.

Reader, R. 1974 New evidence for the antiquity of leprosy in early Britain. *J. Archaeological Science* 1:205-207.

Richards, P. 1977 *The Medieval leper and his northern heirs.* London: D.S. Brewer.

Roberts, C.A. 1986 Leprosy and leprosaria in Medieval Britain. *Museum Applied Science Center for Archaeology Journal* 4(1):15-21

Roberts, C.A. 1993 A note on the skeletons C and E from burial 62/63/64. In A.G. Kinsley: *Broughton Lodge. Excavations on the Romano-British settlement and Anglo-Saxon cemetery at Broughton Lodge, Willoughby-on-the-Wolds, Nottinghamshire 1964-8.* Nottingham Archaeological Monographs 4, p. 58.

Roberts, C.A. and Manchester, K. 1995 *The archaeology of disease.* Ithaca, New York: Cornell University Press.

Simpson, J.Y. 1842 Antiquarian notices of leprosy and leper hospitals in Scotland and England. *Edinburgh Medical and Surgical* J. 57:121-156.

Steinbock, R.T. 1976 *Paleopathological diagnosis and interpretation.* Springfield, Illinois: Charles C. Thomas.

Stones, J.A. (ed) 1989 *Three Scottish Carmelite friaries. Excavations at Aberdeen, Linlithgow and Perth 1980-1986.* Edinburgh: Society of Antiquaries of Scotland, Monograph Series No. 6.

Taylor, G.M., Widdison, S., Brown, I.N. and Young, D. 2000 A Mediaeval case of lepromatous leprosy from 13th-14th century Orkney, Scotland. *J. Archaeological Sci.* 27:1133-1138.

Van Buynder, P., Eccleston, J., Leese, J. and Lockwood, D. 1999 Leprosy in England and Wales. *Communicable Disease and Public Health* 2(2):119-121.

Waldron, T. 1994 *Counting the dead. The epidemiology of skeletal populations.* Chichester: Wiley.

Wells, C. 1962 A possible case of leprosy from a Saxon cemetery at Beckford. *Medical History* 6(4):383-387.

Wells, C. 1967 A leper cemetery at South Acre, Norfolk. *Medieval Archaeology* 9:242-250.

Wiggins, R., Boylston, A. and Roberts, C.A. 1993 *Report on human skeletal remains from Blackfriars, Gloucester (19/91).* Unpublished report.

Wood, J.W., Milner, G.R., Harpending, H.C. and Weiss, K.M.1992 The osteological paradox. Problems of inferring health from skeletal samples. *Current Anthropology* 33(4):343-370

Table 1 Sites revealing leprous skeletons

Site	Numbers Affected	Male	Female	Unsexed adult	Juvenile	Reference
Romano-British						
Cirencester Gloucestershire	1		1			Manchester and Roberts 1986
Poundbury, Dorset	1	1				Reader, 1974
Post-Roman/ Anglo-Saxon						
Beckford, Hereford and Worcester	1	1				Wells, 1962
Bedhampton, Hampshire	1		1			Manchester and Roberts, 1986
Burwell, Cambridgeshire	1	1				Møller-Christensen and Hughes, 1962
Cannington, Somerset	3	1	2			Brothwell *et al.*, 2000
Eccles, Kent	1	1				Manchester, 1981
Edix Hill, Barrington, Cambridgeshire	1		1			Duhig, 1998
Raunds, Northamptonshire	3	3				Powell, 1996; Manchester and Roberts, 1986
School Street, Ipswich, Suffolk	1		1			Mays, 1989
Staunch Meadow, Brandon, Suffolk	1	1				Anderson, 1990
St Catherine's, Norfolk	1			1		Wells, 1967
Tean, Isles of Scilly	1			1		Møller-Christensen and Hughes, 1962
Broughton Lodge, Willoughby-on-the-Wolds, Nottinghamshire	1	1				Roberts, 1993
York Minster, York, Yorkshire	2		2			Manchester and Roberts, 1986

Later and Post-Medieval						
Carmelite Friary, Aberdeen, Aberdeenshire	1	1				Stones 1989, Cardy 1989
Abingdon, Oxfordshire	1	1				Hacking and Wakely, n.d.
Blackfriars, Gloucester, Gloucestershire	2		2			Wiggins *et al.*, 1993
Blackfriars, Suffolk	4	3	1			Mays, 1991
Castle Hill, Scarborough, Yorkshire	1	1				Brothwell, 1958
Chichester, Sussex	36	28	7	1		Knusel and Goggel, 1993, Lee and Magilton, 1989, Magilton and lee, 1989, Ortner pers. comm., Ortner *et al.*, 1991
Exeter, Devon	2	1	1			Manchester and Roberts, 1986
Glasgow Cathedral, Glasgow, Scotland	1	1				King, 1994
Grantham, Lincolnshire	3	2	1			Boulter, 1992
Greyfriars, Chester, Cheshire	1		1			Manchester and Roberts, 1986
Guildford Friary, Surrey	1	1				Manchester and Roberts, 1986
Hallow Hill, St Andrew's, Fife Scotland	1		1			Lunt, 1996
Ilford, Essex	4	2	2			Manchester and Roberts, 1986
Kirk Hill, St Andrew's, Fife Scotland	1			1		Lunt, 1997
Newark Bay, Deerness, Orkney, Scotland	1	1				Taylor *et al.*, 2000, Brothwell, 1977, Brothwell and Krzanowski, 1977
Nonesuch Palace, Ewell, Surrey	1	1				Møller-Christensen and Hughes, 1962
Norton Priory, Cheshire	1	1				Manchester and Roberts, 1986
South Acre, Norfolk	5	1	4			Wells, 1967
Stonar, Kent	3		2	1		Manchester and Roberts, 1986
St Giles, Edinburgh, Scotland	1		1			Manchester and Roberts, 1986
St Giles, North Yorkshire	3	2	1			Chundun and Roberts, 1995
St John, Timberhill, Norfolk	28	7	11	1	9	Anderson, 1998
St Leonard's, Nottinghamshire	1		1			Bishop, 1983
St Margaret's, High Wycombe, Buckinghamshire	2		1	1		Farley and Manchester, 1989
St Patrick's, Armoy, County Antrim, Ireland	1			1		Murphy and Manchester, 1998
Stratford Langthorne Abbey, Essex	1	1				Manchester and Roberts, 1986
Whithorn, Scotland	1		1			Cardy, 1997

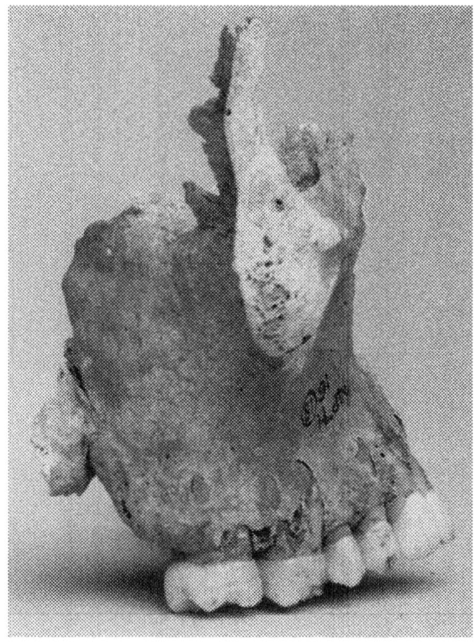

Figure 1(a) Superior view of nasal surface of palate from an individual from the later Medieval site of Norton Priory, Cheshire; pitting and new bone formation with loss of anterior nasal spine

Figure 1(b) Lateral view of 1(a) showing loss of anterior nasal spine

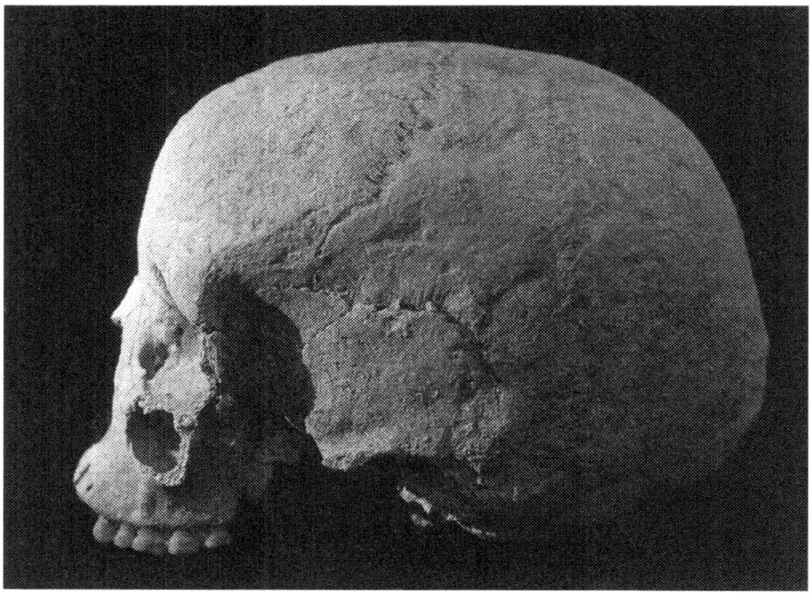

Figure 2. Lateral view of Anglo-Saxon individual from Bedhampton, Hampshire, showing loss of anterior nasal spine

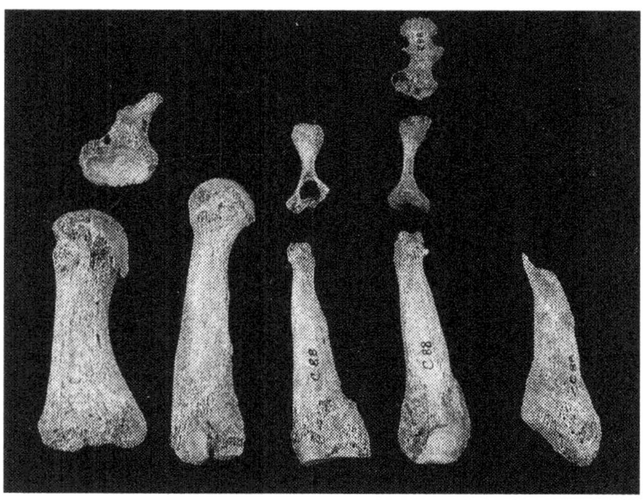

Figure 3. Right metatarsals and some phalanges from an individual from the later Medieval leprosy hospital cemetery of Chichester, Sussex showing resorption and pencilling of the distal metatarsals, and concentric atrophy of proximal phalanges with joint surface destruction

Figure 4. Distribution map of leprous individuals from Britain

Figure 5. Distribution map of leprosy hospitals founded between the 11th and 16th centuries AD in Britain

Traces of leprosy from the Czech Kingdom

Eugen Strouhal[1], Ladislava Horáčková[2], Jakub Likovský[1], Lenka Vargová[2], Jan Daneš[3]

[1] Institute for the History of Medicine and Foreign Languages
First Medical Faculty
Charles University
Kateřinská 32
121 08 Prague 2
Czechia
E-mail: eugen.strouhal @ lf1.cuni.cz

[2] Institute of Anatomy
Medical Faculty
Masaryk University
Kamenice 3
662 43 Brno
Czechia
E-mail: lhorac @ med.muni.cz

[3] Radiodiagnostic Clinics
First Medical Faculty
Charles University
Prague
Czech Republic
E-mail: jan.danes @ lf1.cuni.cz

Abstract

Leprosy in the Medieval Czech Kingdom has yet to be thoroughly studied. Traces of the disease are, however, found in three independent data sources. Historical texts mention "leper" houses in several towns, including the one adjacent to the St. Lazarus chapel outside the Old Town of Prague, operating from the mid-13th to the end of the 15th century AD. Iconographic evidence of *facies leprosa* and thickening of the toes have been recently recognized in one of the "Three Apostles" from an anonymous painting dated AD 1510 in the National Gallery in Prague. In addition, a male skull from an ossuary sample (n=554) at Křtiny near Brno, displays osseous changes suggestive of the rhinomaxillary syndrome of leprosy. The diagnosis was confirmed by the isolation of *Mycobacterium leprae* DNA in a bone sample. This is the first osteoarchaeological evidence of leprosy published from the territory of the former Czech Kingdom.

Keywords

Leprosy, history, iconography, palaeopathology, *M.leprae* DNA.

1. Introduction

Investigative research into the history of leprosy in the Medieval Czech Kingdom has been limited to date. However, given the high prevalence of this notorious disease in Medieval Europe it is logical to assume that the Czech countries were not spared its scourge. In this study three independent sources, namely historical texts, iconography and palaeopathology, are used to support this supposition. We argue for future studies of these types of data in order to fully appreciate the impact of leprosy on this Medieval population.

2. Historical evidence

Historical sources on the early history of hospitals in the Czech Kingdom are scarce. Their probable existence within the first monasteries of the Benedictine order, which were founded in AD 993 for monks in Břevnov west of Prague and for nuns in St. Georges basilica in the castle of Prague, can be only assumed (Figure 1). A duty to create hospitals and care for the poor and sick was part of the prescription of the order of St. John, which was established in Prague around AD 1169. At a later date in Brno (AD 1243) the Order took over the older hospital of St. Spirit. The oldest attested hospital of Virgin Mary in Thyn, a quarter of the Old City of Prague, was in operation between 1178 and 1251 AD when it was provided with royal donations (Nový, 1987), but the exact founding date is unknown (Bláhová, 1987). It fell into disuse sometime between 1267 (Hledíková, 1988) and 1274 AD (Nový, 1987).

Meanwhile, around AD 1234, a new hospital was founded by St. Agnes of Bohemia, the daughter of King Přemysl Otakar I, and it was endowed by Queen Konstanz in AD 1235. After two transitory locations it was finally tranferred to the Old Town at the Prague Stone Bridge, between AD 1247-52 (Hledíková, 1988). It was run by the brotherhood of St. Franciscus of the Czech Crusaders Order with the Red Star (Nový, 1987). However, no patients with leprosy have been mentioned in connection with any of these hospitals.

Leprosy, although an ancient disease (Brothwell and Sandison, 1967, Dzierzykray-Rogalski, 1980), became a prominent disease in European Medieval cities in the 12th and 13th centuries AD. This was connected with crowding of their increasing populations and perhaps also with the return of crusaders, some infected with leprosy, to their homelands (although see Mitchell this volume). However, in the Medieval Czech Kingdom leprosy seems to have been less prevalent than in other European countries (Štěpán, 1963).

The first mention of leprosy is connected with the Romanesque church of St. Linhartus (12th century), called *capella pauperum aegrotantium*. This denomination reflects gathering, accommodation and care of poor sick people to secure alms from patricians (Nový, 1987). The church was demolished around 1790 and its memory survived only as the name of Linhartus Street, connecting St. Maria Square with Small Square in the Old City of Prague.

Some of these St. Linhartus people were transferred into a special leprosy hospital (*domus leprosorum*) beside the Chapel of St. Lazarus around the mid-13th century. It was located in the then countryside about 400 m to the south of the fortifications of Prague Old Town, and today only the name of the Lazarská Street marks its location. The establishment was founded by an unknown benefactor (Tomek, 1893:181-183, Secký, 1928:14-15). The earliest evidence of its existence comes from the nearby excavation of two pits for dead victims of a famine dated, according to a Chronicler to AD 1281 (Rozsívalová, 1991), or 1282 (Svobodný and Hlaváčková, 1999:13). The chapel is dated to slightly after AD 1250 based on its Early Gothic style. In 1886, prior to its demolition, it was investigated and measured by the architect J. Mocker. Its appearance is known from a few preserved depictions, among which is a drawing by V. Jansa dated 1897 (Figure 2). The sick-house proper survived in a rather unattractive state immediately northwest of the chapel until its demolition in 1900 (Figure 3).

After the foundation of the Prague New Town by the Emperor Charles IV on the 2nd February 1348 AD, the hospital and chapel were gradually encircled by newly erected houses and the idea of isolation of leprous patients was thus impaired. Nevertheless, the hospital and chapel survived through the time of the Hussitic wars (1420-1436), when the chapel became the property of the Butchers' Guild. The latest evidence for its original function is the bequest of an Old City burgher Kateřina of Kozlů, dated AD 1482. At the beginning of the 18th century, the chapel was rebuilt in Baroque style. It was dessicrated and closed by the order of Emperor Joseph II in 1788 and demolished to provide space for new buildings in AD 1900-1901 (Baťková *et al.*, 1998: 124-5).

The reason for the creation of special leprosaria was probably not due to fear of contagion, which was not recognised in the Middle Ages, but due to the fear of the appearance of those afflicted with leprosy. Owing to inappropriate diagnostic knowledge, patients with other disfiguring diseases, such as elephantiasis, psoriasis, eczema etc., may have been included in the leprosaria. Despite the possibility that this isolation of leprous patients may have, however, contributed to reduce the prevalence of leprosy, the chapel and leprosarium were located on a busy road connecting Prague Old City with the settlement under the important stronghold of Vyšehrad, about 1.5 km to the south. This location may have enabled the leprous and the poor to beg for food and money from people travelling on the road.

A certain amount of care was provided for the patients and the poor, and spiritual services were provided by the monks from the monastery of the Czech Crusaders Order at Zderaz, about half a kilometre south of the St. Lazarus chapel (Nový, 1987, Hledíková, 1988). The maintenance expenses of the leprosarium were covered by gifts and bequests of the Prague patriciate from the rents in neighbouring streets of the New Town of Prague, as well as from a few villages near Prague and the Middle Bohemian town of Kouřim (Rozsívalová, 1991).

Few additional leprosaria are known in Bohemia (Figure 4). There was a house for people with leprosy in the town of Jindřichův Hradec, which started its activity in the 13th century by the order of the Templars (founded in Jerusalem). It served originally as an infirmary for male members of the Order sick with leprosy. A separate house was established nearby for sick females (Teplý, 1927). There are also reports of a military hospital reserved for victims of leprosy in the town of Litomyšl (Winter, 1892:213) and of leprosy hospitals at Kutná Hora and Chrudim (Dokládal, this volume). In Moravia, leprosaria were operating in Olomouc, Znojmo and Brno. From the latter town, two leprosy hospitals are known, one in Křenová (Kröna) Street (mentioned in AD 1293), and the other as part of the hospital of St. Stephen's Church, outside the city walls near the Svratka river, and founded in the first half of the 14th century (Dokládal, this volume).

Leprosy has also been mentioned by Medieval Czech medical authors, such as Havel of Strahov and Sigismund Albik of Uničov, in connection with *the regimenta sanitatis*; however, they do not describe the disease or its therapy (Rozsívalová, 1991).

After the 15th century leprosy gradually ceased to be a health problem in the Czech Kingdom. During the period of 1631-1714 AD no cases of leprosy were listed among the diagnoses preserved from the hospital of St. Francis in the Old Town of Prague which was run by the monks of the order of Charitable Bretheren (Firlová, 1987).

Iconographic Evidence

Possible iconographic evidence of leprosy has been recognized in a painting of the "Three Apostles" on a board of linden wood held in the National Gallery in Prague (Inv. No. O 532) and dated to around AD 1510. The artist and the picture's provenance are unknown. The board was originally painted on both sides, perhaps being a part of an altar wing or a whole wing of a dismantled altar. The painting was bought by the Gallery in 1888 from an art expert and collector, the lawyer Dr. Hugo Toman (Pešina, 1950).

The figures are of three apostles, from left to right: Saint Thadeus, Matthew and Simon. The middle figure of Saint Matthew differs from the other two by a fleshy, grossly enlarged nose (Figure 5a). The hands of the three apostles are depicted in similar fashion and their fingers do not show any deformaties. The toes of the middle apostle, however, are very broad, resembling in breadth the right big toe (Figure 5b).

It appears probable that the depicted changes in the nose and toes reflect the manifestation of leprosy in the face and feet. As differential diagnoses, post-traumatic or senile changes have to be taken into consideration.

4. Palaeopathological evidence

Recently a calvarium (labelled K 12) suspected of bearing changes caused by leprosy was discovered in a collection of 554 adult calvaria from the ossuary at Křtiny near Brno, Moravia, which is dated from the 13th to the beginning of the 18th century A.D. The preservation of the skull was good, with the exception of some postmortem surface erosion, and postmortem damage to the alveolar process (see Figures 6, 7, 9 and 10). The age at death of the individual was estimated from the cranial sutures and dentition to be 20-25 years (see Figure 8).

Concerning sex, the calvarium revealed small dimensions (length 168 mm, breadth 137 mm) and brachycrany (81.6). There was, however, a prominent glabella with a postglabellar depression, and an arched and medium high forehead. The frontal eminences were only slightly developed, and the parietal eminences were pronounced. There was no prominent external occipital protuberance (Figure 7), a slightly developed nuchal crest, medium long (28 mm) and thin (left 11 mm, right 12 mm) mastoid processes, a medium to large foramen magnum and occipital condyles, and a broad upper dental arch (Figure 8). While several features are intermediary between the two sexes, a few of them favour male sex. They gain weight in view of the young age of the individual. Ancient DNA survived in the specimen and was recovered during extraction. Using amelogenin, two bands, one of 290 bp and another of 105 bp, were successfully amplified using polymerase chain reaction (PCR), indicating that the sex of the individual was indeed male (G.M. Taylor, pers. comm.)

Pathological changes, clearly antemortem, are apparent on the edges and inside the nasal cavity, on the hard palate and on the alveolar process of the maxilla (Figures 8 to 10). The lower edges of the pyriform aperture, and the anterior nasal spine have been completely resorbed (Figure 9). Inside the nasal cavity, the perpendicular plate of the ethmoid was almost completely detached and the cribriform plate of the ethmoid bone distorted by newly formed bone. The vomer was completely absorbed, including its basal attachment, and only tiny remains of the medial nasal conchae survived. The inner walls of the nasal cavity have also been remodelled by irregular newly formed bone. Irregular new bone covers the external surface of the alveolar process of the maxilla in the region between the alveolar juga of both canines (Figure 9).

The whole extent of the oral surface of the hard palate has been completely changed by irregular deposits of new bone, some orientated radially and some forming protruding thin edges, combined with depressions and a semilunar perforation (9 x 3 mm) into the floor of the nasal cavity, oriented transversely and 20 mm posterior to the tip of the alveolar arch (Figure 11).

The pathological process continued on the external and internal aspects of the alveolar process towards the alveoli of all the incisors, which were shallow and lined with spongy bone. All incisors were lost antemortem. The left canine and 1st premolar were lost postmortem. The right canine, both premolars and the 1st molar could have been also affected by the same pathological process, as evidenced by their alveoli filled with spongy bone. They are, however, partly damaged postmortem. The right 2nd and 3rd molars were lost postmortem, their alveoli being of normal shape except for a lytic cavity at the antero-medial edge of the palatal alveolus of the second molar, which possibly developed with pathological changes on the hard palate (Figure 11).

Cranial computed tomography examination shows, on coronal sections (4 mm thick), sclerotic thickening starting at the base of otherwise unchanged nasal bones, absence of intranasal structures and perforation into the right maxillary sinus (Figures 12a, b and c). On axial sections, sclerotic changes on the palate include the perforation and closing of the alveoli (Figures 13a, b and c).

The general character of the process, combining lytic changes with new bone formation, is clearly inflammatory (periostitis and osteitis). Its localisation in the rhinomaxillary region and symmetrical distribution suggests a probable leprous origin. It corresponds exactly to *facies leprosa* (Møller-Christensen and Faber, 1952 in Steinbock, 1976: 201). Despite the lack of availability of hand and foot bones to examine, the diagnosis appears probable. Concerning differential diagnosis, treponemal disease was taken into consideration, but the pathognomonic changes on the cranial vault (i.e. caries sicca) or a saddle-shaped deformity of the nasal bones were not present in the present case.

However, in a sample of bone taken from the remnant of the dorsal edge of one of the concha nasalis media, PCR (polymerase chain reaction) for *Mycobacterium leprae* using a very sensitive method which amplifies a repetitive element in the *M. leprae* genome known as RLEP (28 copies) yielded a negative result. Primer-dimer formation was noticed in the K 12 extract and this indicates, as does the amelogenin amplification, that it was not due to the co-purification of the PCR inhibitors in the extract. However, evidence for *Mycobacterium tuberculosis* ancient DNA was positive (G. M. Taylor, *pers. comm.*).

However, in a different sample, taken from the inferior edge of the nasal bones, another DNA extraction and a nested PCR, based on primers for repetitive element, gave a final product of 99 bp and a weak but definite positive result for *Mycobacterim leprae* (Figure 14). After two months repetition, the product from the original nested PCR reaction continued to give a weak positive result after the second stage reaction, with satisfactory negative controls. The original fresh extract was negative for *Mycobacterium tuberculosis* (H. Donoghue, pers. comm.).

To summarise, based on morphological changes to the skull and the result of a DNA analysis of one of the two submitted samples, we can consider the skull of K12 the first osteoarchaeological evidence for leprosy published from the territory of the former Czech Kingdom.

5. Conclusion

Owing to the still preliminary state of the study of leprosy in the Czech Republic, the reported historic, iconographic and palaeopathological evidence of leprosy from the Medieval Czech Kingdom, in spite of their paucity, should inspire future research. The search for cemeteries attached to leprosaria should be initiated. Human skeletal remains from some of the preserved medieval charnel houses and ossuaries also warrant detailed palaeopathological examination.

Acknowledgements

We appreciate the co-operation of Dr Helen Donoghue of the Medical Microbiology Department, Royal Free and University College Medical School, London, England for providing the results of aDNA analysis. We would like to thank Michael Taylor, from Imperial College, London, England, for DNA testing of a sample of bone to identify sex and pathogene using ancient DNA analysis. Many thanks to J.Likovský for Figures 5-5a, to E.Vozárová for Figures 6-11, to J. Daneš for Figures 12-13 and to F. Donohue for Figure 14.

References

Baťková, R. and 32 other authors 1998 *Umělecké památky Prahy: Nové Město, Vyšehrad, Vinohrady (Praha 1)*. New Town, Vyšehrad, Praha: Art monuments of Prague: New Town, Vyšehrad, Vinohrady (Prague 1). Academia, Praha, pp. 124-125

Bláhová, M., 1987 Staropražské špitály. I. Vznik špitálů v Evropě a počátky špitálů v Praze. (Hospitals of Old Prague. I. Origin of hospitals in Europe and beginnings of hospitals in Prague). *Časopis Lékařů Českých*. 126: 1083 - 1085.

Brothwell D.R. and Sandison A.T. (eds) 1967 *Diseases in Antiquity*. Springfield, Illinois: Charles C. Thomas.

Dzierzykray-Rogalski, T. 1980 Leprosy in the Dakhleh Oasis (Egypt). *Africana Bulletin* 29: 135 - 140.

Firlová, I. 1987 Příspěvek k problematice sociální struktury, hospitalizace a mortality nemocných v nemocnici Na Františku v letech 1631-1714. (Contribution to problems of social structure, hospitalization and mortality of patients in the hospital of St. Francis in the years 1631-1714). *Documenta Pragensia* 7: 356 - 383.

Hledíková, Z. 1988 Staropražské špitály. III. Pražské klášterní špitály od 13.do počátku 15.století. (Hospitals of Old Prague. III. Prague monasterial hospitals from 13th to beginning of 15th century). *Časopis Lékařů českých*. 127: 57 - 59.

Møller-Christensen, V. and Faber, B. 1952 Leprous changes in a material of mediaeval skeletons from the St. George's Court, Naevsted. *Acta Radiologica* 37: 308 - 317.

Nový R., 1987 Staropražské špitály. II. Pražské předhusitské špitály. (Hospitals of Old Prague. II. Prague pre-Hussitic hospitals). *Časopis Lékařů českých* 126: 1127 - 1129.

Pešina J., 1950 *Česká malba pozdní gotiky a renesance. Deskové malířství 1450-1550.* (Czech Late Gothic and Rennaisance Painting. Board Painting 1450-1550). Praha: Orbis.

Rozsívalová, E. 1991 Několik poznámek k historii lepry. (Some notes on the history of leprosy). *Časopis Lékařů českých* 130 (24-25): 701-702.

Secký, R. 1928 *Staropražské špitály*. (Old Prague hospitals). Praha: Reprinted from *Památky* 2-9:1-55.

Steinbock, R.T. 1976 *Paleopathological diagnosis and interpretation*. Springfield, Illinois: Charles C. Thomas.

Štěpán M., 1963 Epidemie v českých zemích v době předhusitské. (Epidemics in Czech Countries before the time of Hussite Wars). *Čs. Zdravotnictví* 13: 303.

Svobodný, P. and Hlaváčková, L. 1999 *Pražské špitály a nemocnice*. (Prague Sick-houses and Hospitals). Praha: Knižnice dějin a současnosti.

Teplý, F. 1927 Dějiny města Jindřichova Hradce. (History of the Town of Jindřichův Hradec).

Jindřichův Hradec: Volume 1 (1): 60 - 63, Volume 2 (1): 213.

Tomek, V.V. 1893 *Dějepis města Prahy*. (History of City of Prague), Part 9. Praha: F. Øivnáè.

Winter, Z. 1892 *Kulturní obraz českých měst*. (Cultural Picture of Czech Towns). Praha: Matice Česká, Part 2.

Zilynskyj, B. 1987 *Špitály Nového Města pražského v pohusitské době do r.1518*. (Hospitals of the New Town of Prague after the Hussite Wars up to 1518). *Documenta Pragensis* 7(1):102.

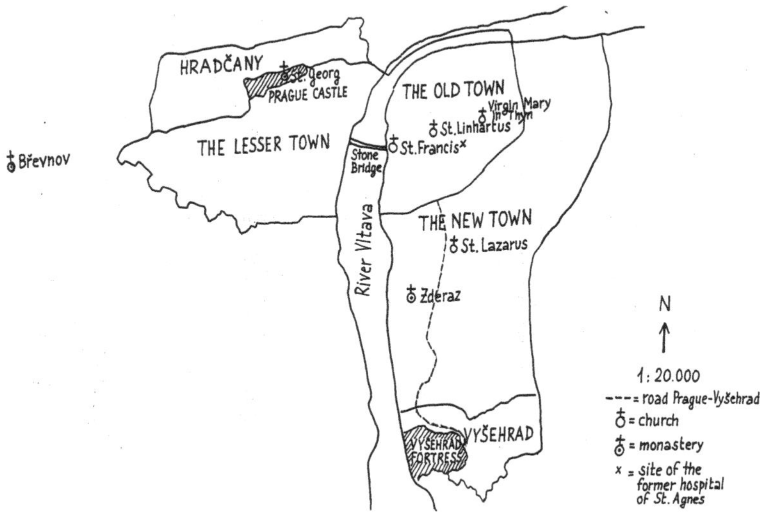

Figure 1: Map of Ancient Prague with sites mentioned in the text.

Figure 2: Ancient hospital chapel of St.Lazarus in Prague (drawing by Václav Jansa, 1897).

Figure 3: The ancient hospital (in the foreground), the chapel of St.Lazarus (behind it) and the tower of the New Town City Hall (in the background) photographed before the demolition of the St. Lazarus complex around AD 1900-01.

Figure 4: Map of the Czech Republic showing towns with leprosaria and the location of the ossuary at Křtiny.

Figure 5a: Painting of the "Three Apostles" by an unknown artist dated around 1510 (The National Gallery, Prague: reproduced from Pešina, 1950) shows details of the faces.
Figure 5b: Painting of the "Three Apostles" by an unknown artist dated around 1510 (The National Gallery, Prague: reproduced from Pešina, 1950); shows details of feet

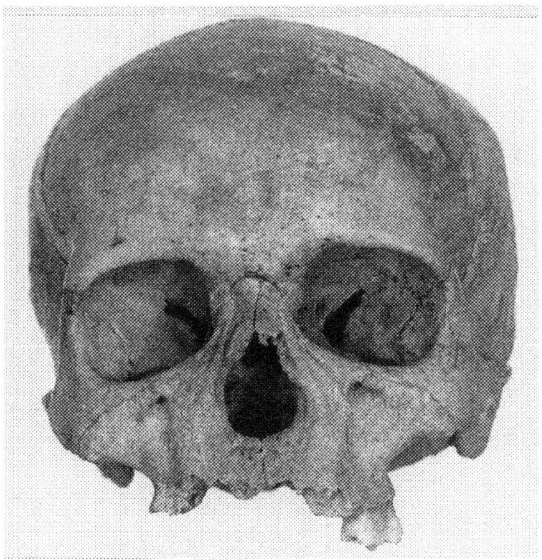

6

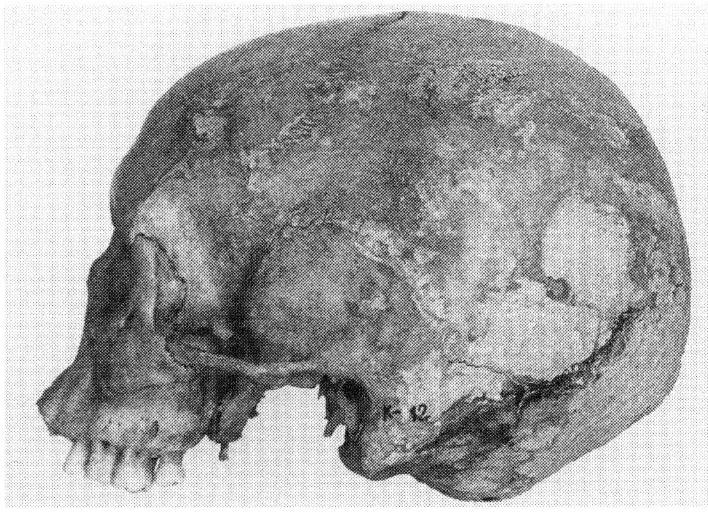

7

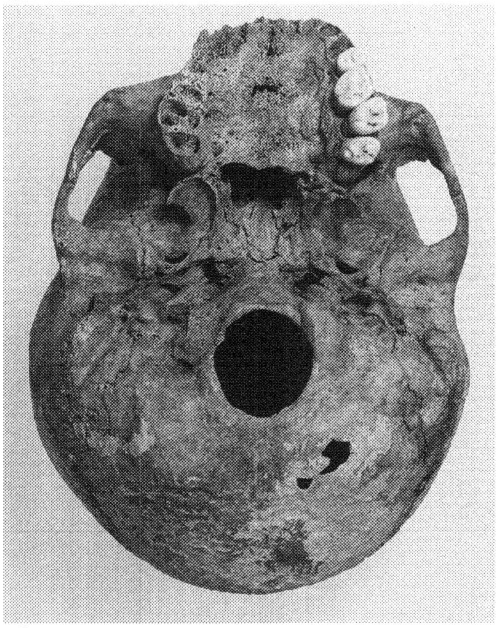

8

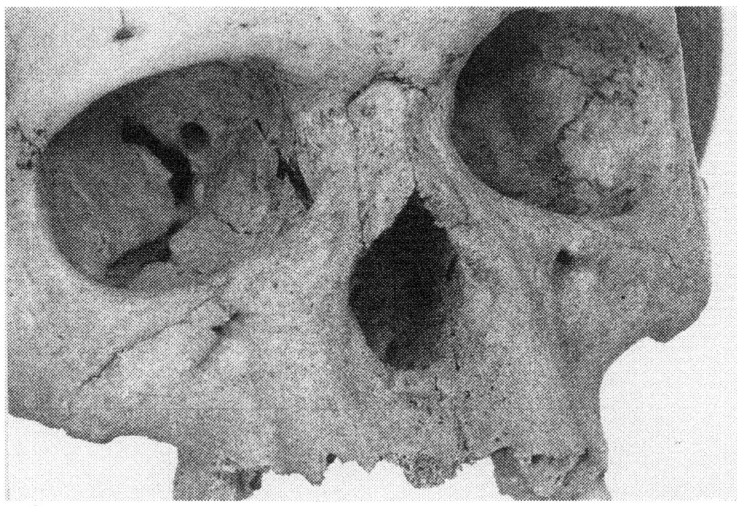

9

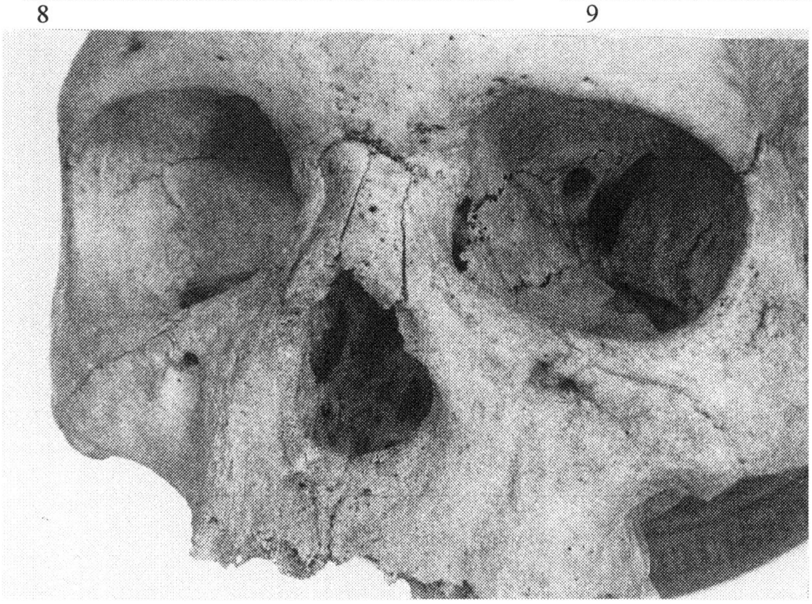

Figures 6 to 10: Skull K 12 from the ossuary at Křtiny near Brno showing details of the pathological changes of leprosy

10

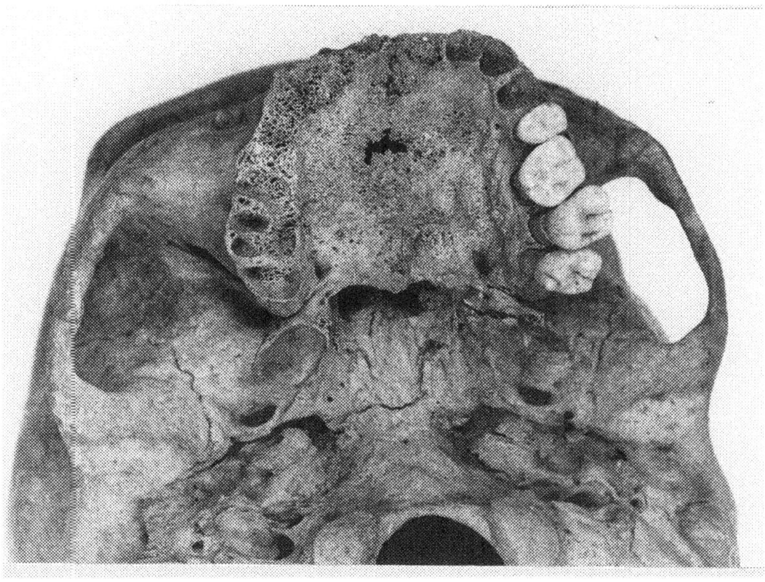

Figure 11: Detail showing the pathological changes in the hard palate and the antemortem loss of the upper incisors on both sides, and of the upper left canine and premolar, the rigth 2nd and 3rd molars.

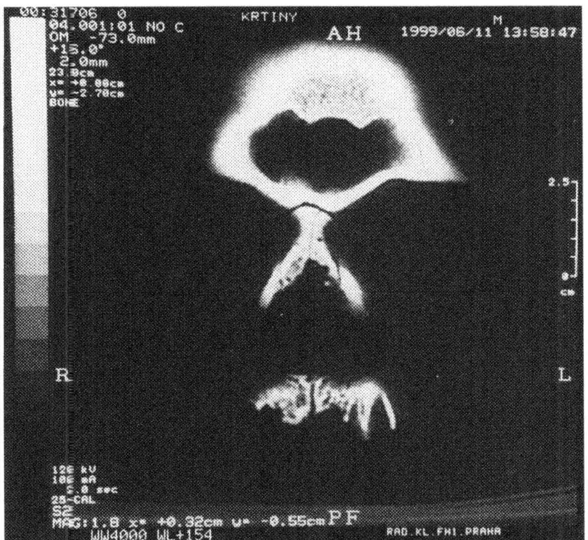

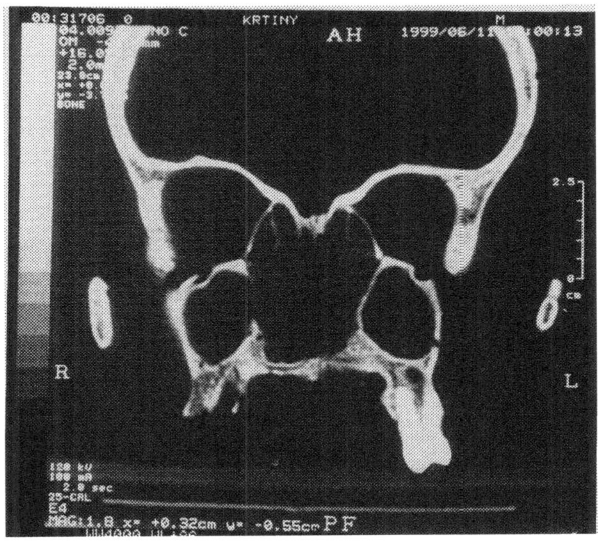

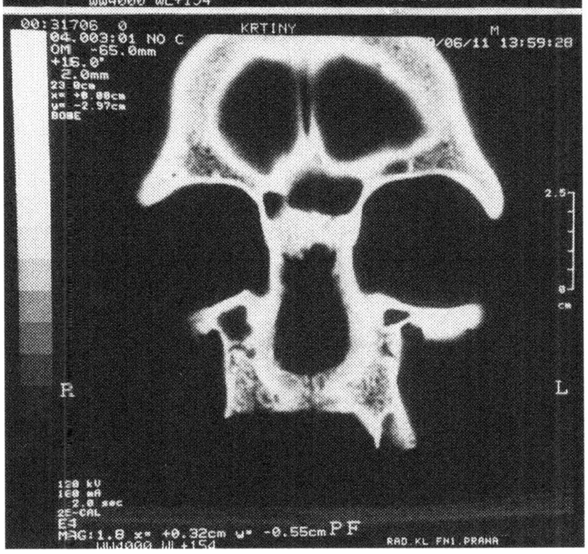

Figure 12 (a-c): CT coronal sections displaying sclerotic thickening starting at the base of otherwise unchanged nasal bones, practical absence of intranasal structures and perforation into the right maxillary sinus.

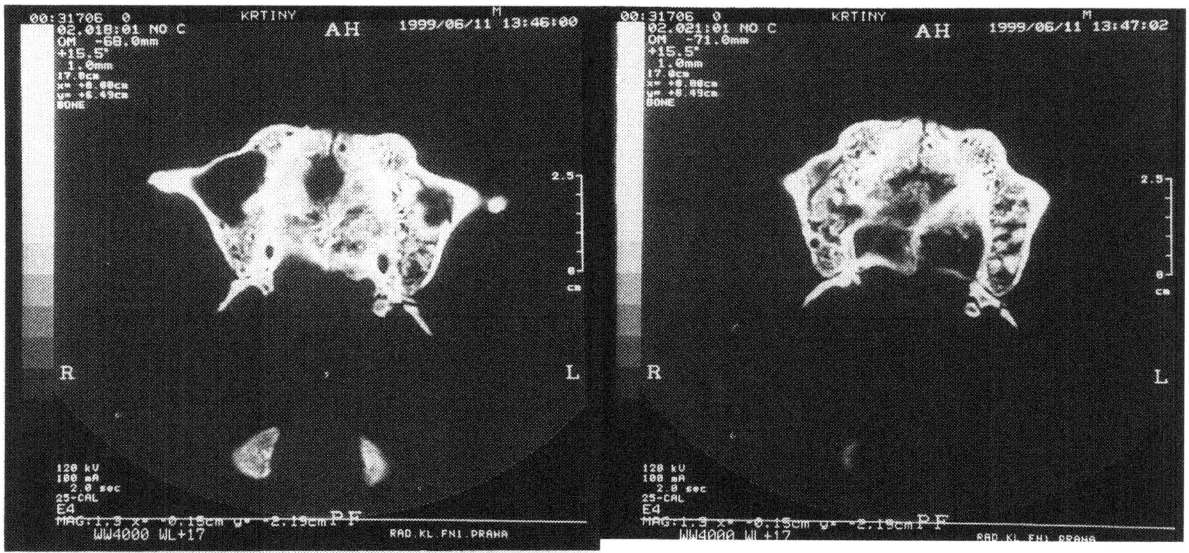

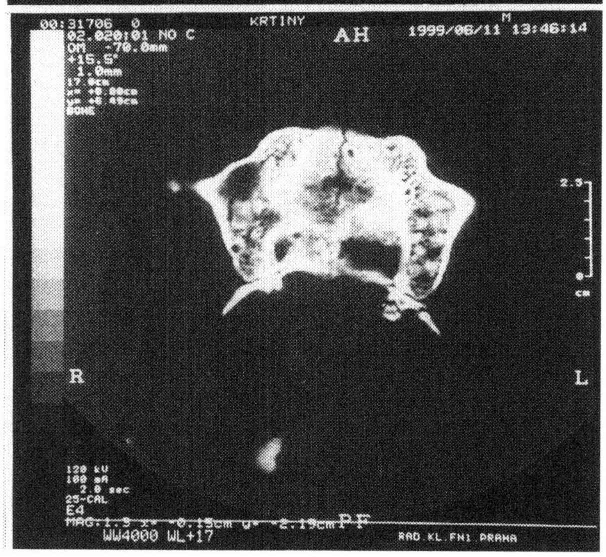

Figure 13 (a-c): CT axial sections depicting sclerotic changes on the palate including a perforation and closing of the alveoli

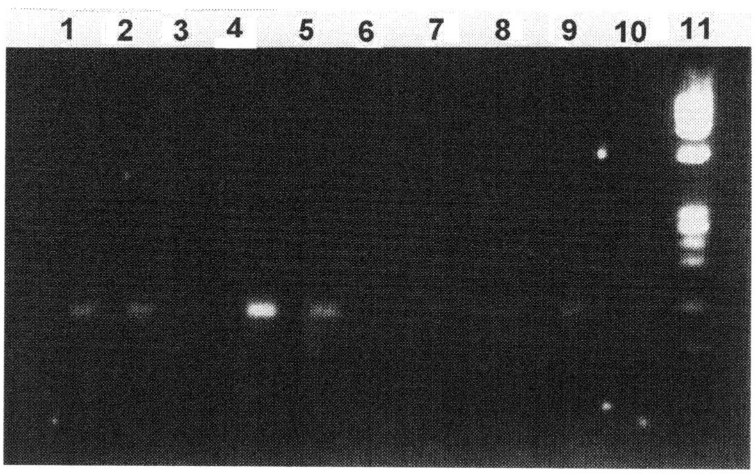

Figure 14: *M. leprae* nested PCR after 25 cycles of amplification with primers LP1/LP2 followed by 25 cycles of amplification with primers LP3/LP4 for 99 bp product from RLEP sequence. Lanes (left to right): 1: *M. leprae* 1 x 10^{-5} dilution; lanes 2-7: other archaeological samples; lane 8: K-12 silica supernate (0.5 µl stage 1 product used as template in nested PCR); lane 9: K-12 silica supernate (5.0 µl stage 1 product used as template in nested PCR); lane 10: DNA negative extraction control; lane 11: ØX174*Hae*III digest molecular markers.

Perspectives on the history of leprosy in the Pacific

Diane L. Trembly
HCR 1 Box 5285
Keaau
HI 96749
U.S.A.

Email: phrilab@interpac.net

Abstract

It has long been believed that leprosy entered the Pacific as a result of trans-Pacific voyaging, and Chinese emigration in the 19th century AD. The author's report of bones exhibiting possible leprous change dating to pre-Spanish times on the island of Guam, has refuted this assumption, at least for Micronesia. Through the study of linguistics, local legends, and early nineteeth century records, this paper explores the possibility of prehistoric leprosy elsewhere in the Pacific. The second section of the paper summarizes the epidemiology and treatment of leprosy in the Pacific since the last half of the nineteenth century.

Keywords

Leprosy, Pacific, Guam, Philippines, Kalaupapa

1. Pacific history and prehistory

Most of the islands of Indonesia were part of the Asian mainland during the ice ages, and habitation goes back into Pleistocene times. Despite a watergap that always separated New Guinea and Australia from Southeast Asia, these places were occupied as early as 40,000 - 60,000 BP and possibly even earlier. One of the islands of the Philippines was also connected to the mainland and early dates come from the Pleistocene. However, habitation in other islands of the Philippines go back to at least the third or fourth millennium BC.

Some time in the latter half of the 2nd millennium BC, some people in the islands north of New Guinea began to make a pottery called Lapita. Whether these were people who had been there for thousands of years, or were new immigrants, is not known. In the first millennium BC, the makers of this pottery migrated eastwards, through the islands of Melanesia and into the eastern Pacific, becoming Polynesian people, and reaching as far east as the Marquesas Islands early in the first millenium AD. From there they travelled north to Hawai'i, east to Rapanui (Easter Island), and southwest to New Zealand, possibly as late as 900 AD. These people were all speakers of an Austronesian language, the homeland of which is in Malaysia and Indonesia. In contrast, back in Melanesia, where the Lapita culture started, some villages speak an Austronesian language, while others speak a Papuan New Guinea language (Jennings, 1979, Bellwood, 1979, Kirch and Hunt, 1988).

Meanwhile, the islands of Micronesia, to the northwest, were also being settled. The western group, consisting of the Marianas, and Yap and Palau, were probably settled by people coming from the Philippines and/or Indonesia in the first millennium BC, which is not to say they might not have continued migrating long after that time. Other islands of Micronesia, to the east of the Marianas, were probably settled by people coming from the south, particularly from the eastern end of Melanesia and the western end of Polynesia, probably during the first millennium AD. Micronesians are linguistically and physically distantly related to Polynesians (Bellwood, 1979).

Spanish exploration of the Pacific began with Magellan in 1520 AD. He crossed the entire Pacific without making landfall until reaching Guam, in the Marianas, and he then went on to the Philippines. In the 1560s, the Spanish took control of the Philippines and the Marianas landed and did some exploring in the Solomons, and began the galleon trade between Asia and the Americas. However, with the exception of a 1595 landing in the Marquesas, they did not visit other islands. It was another 200 years before anyone else became interested in the Pacific Islands, and few islanders had never seen a European (Beaglehole, 1966, Howe, 1984).

In the late 18th century all that changed, as both Britain and France mounted major explorations. By the 19th century there was a second peopling of the Pacific. Whaling and trans-Pacific trade in furs, beche-de-mer, sandalwood, tea, silks, etc. became big business both with Europeans and Americans. Islanders signed on as crew, and Europeans and Americans sometimes jumped ship. Soon missionaries were arriving and then came the establishment of plantations and importation of plantation labourers from China, Japan and India. No place remained isolated (Howe, 1984, Daws, 1968).

2. Evidence of Leprosy in the Pacific before Euro-American Contact

Traditionally, historians have attributed the first appearance of leprosy in the Pacific to 19th century trans-Pacific voyaging, and emigration of Chinese people to Pacific islands (Browne, 1985, Roberts, 1928). Although some were phenomenal navigators, and some had developed highly complex socio-political systems, none of the Pacific peoples ever developed writing. Thus, to attempt to verify that leprosy did not exist in the Pacific prior to the 19th century, we must rely on oral history, or written accounts by the first Europeans in the islands and, where possible, examination of skeletal remains. In Fiji, the words for leprosy are not of European derivation. Furthermore, the disease is entrenched in legend (Miles, 1997, Montgomerie, 1988). However, one cannot tell if the word or words used for leprosy in the legends describe the same disease we now call leprosy, although they are the words used for leprosy today. A mission doctor in 1837 reported treating leprosy, but he did not provide a description of the disease (Miles, 1997, Montgomerie, 1988, Austin, 1949).

If there was pre-contact leprosy in Fiji, by what route did it arrive there? Bougainville, in 1768, named Vanuatu, in Melanesia, the "Leper Isles", and claimed "most" of the natives were infected with leprosy. He gives descriptions of wounds, ulcers, and dermatitis that sound like fungal infection and impetigo. As for the Solomon Islands, the next step west, nobody reported anything about leprosy. However, there are local names for it, and some associated folklore on some of the islands. To the south of Vanuatu, on New Caledonia, there is no indication of leprosy before the mid-19th century, nor was it present in New Guinea (Miles, 1997, Montgomerie, 1988). Thus, for Melanesia, it remains a question as to whether there was any leprosy in pre-contact times; we need skeletal evidence.

In Polynesia, our first accounts come from missionaries in Samoa who wrote of "a species of leprosy" and said it had "already greatly abated". They describe a chronic disease associated with loss of toes, fingers, noses and ears, and in the 1840s it is described as ulcerous sores until the fingers and toes "drop off" (Miles, 1997, Montgomerie, 1988). What is disturbing in these accounts, as well as in those from other islands, is that the description is clearly that of dry gangrene; the digit turns black, and the dead tissue separates from the living tissue and drops off - clearly not a description of leprosy.

In Tonga, we have a record of the first introduction of leprosy on the island of Niue, some time after 1850 (Dempster, 1949). In the Cook Islands it can be traced back to probably about the 1820s and there are no local legends, oral history or local names for it. In eastern Polynesia, early accounts from the Society Islands and the Marquesas make no mention of leprosy. In Hawai'i, we know it first appeared around 1840, and abundant pre-contact skeletal material confirms its absence. In New Zealand, there are written reports by physicians in the 1830s and 1840s describing health conditions, but they make no mention of leprosy. There are some native terms for it, but these same terms were used for something that does not fit the description of leprosy (Montgomerie, 1988). Furthermore, leprosy has not been found in any pre-contact skeletal material (P. Houghton, pers. comm.). Summing up, it seems there was no leprosy in pre-contact Polynesia.

For Micronesia, Spanish records go back to the 16th century but are not very helpful in determining whether leprosy was present (Carano and Sanchez, 1964, Rogers, 1995). In 1685, an Englishman travelling on a buccaneer ship noted many of the natives of Guam to be "troubled with a kind of leprosy", but his description does not fit leprosy (Roberts, 1928). Examination of pre-Spanish skeletal remains, however, has shown that leprosy was indeed present. It was the privilege of this author to examine bones exhumed in the course of salvage archaeology on the islands of Guam and Saipan, in the Marianas. During 1991-1993, remains of approximately 700 - 800 individuals (many very incomplete) were examined. Of these, seven individuals exhibited pathology which was determined to be leprosy (confirmed by K. Manchester, pers. comm.), for example, see Figures 2 and 3. Radiocarbon dating carried out on the bones of four of these individuals resulted in an early date of AD 830 ± 170; midpoints of the others lay between AD 1140 and 1370 (Trembly, 1995).

Micronesians were excellent navigators. They might have made many trips back and forth to the Philippines, or even Indonesia. There is no archaeological or linguistic evidence to suggest that they ever sailed to China or Japan, but it cannot be ruled out. Leprosy is believed to have existed in the Philippines since before the Spanish arrived, as the Franciscan friars in the 16th century built a shelter for people with "this and other diseases", and in 1632 the emperor of Japan shipped 134 "lepers" to Manilla for the friars to care for (Chapman, 1982). The incontrovertible evidence of pre-Spanish leprosy in the Marianas lends credence to the belief that it existed in the Philippines. According to one author, Filipinos had been having 'commerce with countries in Asia since before history began' (Chapman 1982:168). Considering the proximity to China, this would not be surprising. A China-Philippines-Marianas connection is very plausible.

3. The Historic Period

In the Philippines in 1830 three leprosy settlements were established, housing 400 patients, the purpose being simply to give the worst cases a place where they could be cared for; there was no attempt to isolate or control the disease. When the Philippines became an American possession in 1899, it was estimated that there were between 3,500 and 4,000 "lepers". As there was no cure, isolation seemed the best policy. The beautiful island of Culion was selected, because it was fertile, had good agricultural land, good fishing, fresh water, and an excellent harbour, and yet was sparsely populated. Construction began in 1905, and the first patients arrived the following year. The early years are described as chaotic, with inadequate facilities, lack of organization, and severely ill patients. However, by 1916, the 40 acre colony resembled a clean, well-organized village, with 400 homes, a theatre, town hall, school, electricity, a sewage disposal plant etc., and many patients who were not severely ill. They organized their own municipal government. Meanwhile, a campaign of education regarding leprosy was carried out throughout the islands. Initially, locating and moving people with leprosy to the hospital was not only difficult but perilous, and health officers were shot and stabbed. However, as word of the good life at Culion got around, patients came forth voluntarily. Diagnosis was confirmed bacteriologically before a patient was transferred to Culion (Chapman, 1982).

Patients were treated with chaulmoogra oil until 1921 (when it became too difficult to obtain the funds to buy it) and, whether as a result of this medicine or not, many became bacteriologically negative. Some of these returned to their homes, while others stayed on Culion, founding a new village named "Negative". The anticipated result, that with isolation the disease would be controlled and disappear, did not happen. There were constantly new cases. This, plus the fact that the number one health problem was tuberculosis, made it difficult to obtain funds for Culion in the 1920s. Although the population kept increasing (7,000 in 1935) and funds decreased, a visitor in 1928 described the village in glowing terms, stating there was "excellent morale, a spirit of enterprise, efficient organization and high professional standards". In the 1930s, regional centres were established and, thereafter, only the most serious cases went to Culion. The war years were devastating, as Culion was cut off from supplies, and 2,000 died from lack of food and medicine. With the advent of sulfones and the regional clinics, the Culion population continued diminishing, and in 1978 stood at 739. According to information from the Philippine consulate in Honolulu, the colony still exists.

Meanwhile, in Guam, yaws gangosa was prevalent and easy to confuse with leprosy. A leprosy colony (with either 40 or 110 patients, depending on the reference) was established in 1890, but many of the patients may have had yaws gangosa. It was destroyed by a typhoon in 1892, moved to another location, and destroyed again by a typhoon in 1900 (Carano and Sanchez, 1964, Ibanez and Resano, 1890, Roberts, 1928). In 1899, when Guam became a U.S. possession, the following recommendations were made:

1. To send all the Spanish priests back to Spain "because they are the moral "lepers" of the place"

2. To send four government doctors to examine every native and send all those with leprosy to the leprosy colony in Hawai'i

3. To establish a colony for all the syphilitics.

However, those with leprosy were not sent to Hawai'i, as a leprosy colony was established on Guam in 1902. By 1908, its population stood at 88. A separate colony was established for yaws patients. In 1912, the government of Guam ordered that the leprosy colony be closed and all patients sent to Culion, in the interest of economy. This was a very sad day for the people of Guam, as everyone turned out to watch and weep as their friends and relatives walked in procession from the colony to the ship. However, by 1941, leprosy was virtually eliminated from Guam.

On the Melanesian island of New Caledonia, the introduction of leprosy is attributed to a Chinese immigrant in 1863. In New Zealand, physicians writing in the 1850s stressed the inconsistencies between what was being called leprosy, and what they knew of as true leprosy, and the earliest reliable description is from 1890. Investigations in 1902 and 1904 of some 70 or 80 alleged cases confirmed only five, of which one was Chinese (Miles, 1997, Montgomerie, 1988). In both the Cook Islands and Niue (Tonga) the presumed first case has been traced through genealogies. In Niue, this was a man who had been living in Hawai'i and returned to Niue in the 1860s or 1870s. (Dempster, 1949). In the Cook Islands, although there are no confirmed reports before the 1880s, genealogy suggests an introduction in the 1820s. In Samoa, leprosy was documented by a physician in 1892 who believed it to be of Chinese introduction. In Nauru (slightly west of the Gilberts), there is an account of leprosy being introduced in 1912 by an immigrant from the Gilberts, and that by 1924 24% of the population of 1200 was infected. However, by 1927, the majority were going into spontaneous remission. Similarly, in New Guinea, the prevalence was reported to be 15% within 15 years of what was believed to be the first case. There, too, leprosy went into spontaneous remission. In both situations, the cases were virtually all of the tuberculoid type (Austin, 1949, Noordeen, 1985).

In Fiji, in 1899, a leprosy colony was established on the island of Bega, and moved to the island of Makogai in 1911. Over the next 40 years, 3,000 patients were admitted, 600 of whom came from outside of Fiji, particularly from Samoa, the Cook Islands, Tonga, the Solomons, and the Gilberts. These places helped financially to support the colony, in proportion to the number of patients they sent. The Indian population of Fiji (which by now outnumbered ethnic Fijians) was afflicted as much as the Fijians, but as they had all been examined at the time of immigration, it was not thought that they were the source of the disease (Austin, 1949). The rate of remission in tuberculoid patients, particularly those admitted while the disease was in an early stage, was high. Discharged cases spread the word that life was good at Makogai, and that isolation was not necessarily lifelong, thus allaying people's fears about revealing themselves to the authorities. As time went on, more and more cases were being admitted in the early stages, and it was felt that, at least within Fiji, few if any cases were being missed (Austin, 1949).

Leprosy was first reported in Hawai'i in the 1830s or 1840s, and has always been said to have come with Chinese immigrant labour. In 1865, a law was passed requiring mandatory reporting and a facility was established in Honolulu for diagnostic examination, detention, and treatment of light cases. For an isolation colony, 800 acres together with 20 houses was acquired on the Kalawao side of the Kalaupapa peninsula on the island of Moloka'i. This flat, low-lying peninsula juts out onto the north coast of Moloka'i and is separated from the rest of the island by an almost vertical cliff rising 600 metres. Triangular in shape, it is about 5 km across at the base where it backs up to the cliffs, and about 5 km from the base to the apex. In contrast to the rest of the world, Hawaiians never perceived leprosy as being contagious. They considered the tearing apart of families, and the shipping of the afflicted off to isolation, as cruel punishment for something they could not help.

The following history of the Kalaupapa colony is drawn from Bird (1890), Green (1985), Kuykendahl and Day (1948), Jarrett (1993), Somers (1985) and the author's own experience working there in 1960. The early years were very bad. The first patients to arrive tended to be the sickest and least able to care for themselves, although they were allowed to bring along a helper. The 20 houses were not sufficient, so many people were without shelter, other than what they might be able to construct with materials nature provided. Food shipments were insufficient, and there was no calm cove in which to bring landing boats ashore. When the sea was rough, supplies had to be floated and pulled ashore, and patients, likewise, might need to swim or be carried. However, to the government's credit, within two years they had built a great deal more housing, as well as a hospital and school and other buildings, and had increased food supplies. Some residents organized a church and atranged a visiting pastor.

Nevertheless, morale was still very low. There was a lot of drunkeness, gambling, theft, fighting, and general anarchy. Some of the able-bodied tried to get agriculture going, as had been the plan, but many refused to work because they got free food, anyway. Besides the fact that the earliest patients tended to be the sickest, the low morale probably contributed to the fact that in the first nine years, of 1145 patients sent, 442 (or 40%) had died. Things began to change in 1873 when the now famous Father Damien arrived. He had no financial resources, but what he had was a very deep caring and loving for these people, and an absolutely tireless energy and drive, and a charismatic personality. He was not there only to provide a religious perspective and comfort to the dying, he gave loving nursing care and, above all, was a morale builder. His tools included the saw and the hammer, the shovel and the hoe, as well as the Bible, and he worked from sunrise until long after sunset. He was one of these rare people who could take men who had no pride and spent their time drinking and gambling, and in a short time have them building, painting, cleaning, planting, weeding, volunteering, and taking pride in themselves. He created choirs, bands, clubs, and fraternities. However, he took no precautions, sharing patients' dishes and utensils, passing his pipe around, sleeping on mats they had slept on, carrying them, and so forth. This undoubtedly was part of what made him so effective, but it cost him his health. He died with leprosy in 1889. In all the history of Kalaupapa, he is the only worker known to have become infected.

In 1873, the same year that Damien arrived, the government acquired land on the west side of the peninsula, that is Kalaupapa proper, and many patients began moving to that side. The Kalawao side had been chosen partly because it was "cool and breezy" but to the patients it was "cold and windy"; the Kalaupapa side was sunnier. In 1879, the first semi-resident doctor arrived, in 1888 the sisters of St. Francis, a nursing order, and in 1886, Brother Joseph Dutton, who did not leave until 1930 at the age of 87. By 1888 there were 374 buildings including five churches, 12 hospital buildings, two dispensaries, a store, and a jail.

In 1900 Hawai'i was annexed to the United States. Almost immediately, the United States set out to build the finest, most modern, best equipped, state-of-the-art leprosy research facility in the world, equipped with bacteriological laboratory equipment, an animal laboratory, elaborate fencing and gates, and protocols to ensure staff would not become infected. This was built on the Kalawao side because it was "cool and breezy". It was a total failure because the patients would not come. Not only was it on the wrong side of the peninsula, but they were completely alienated by the "modern and scientific", the fencing and gates, the strict techniques and protocols, and the doctors in white coats. It was a culture clash. In a few years the place was abandoned. Equipment was moved to Honolulu, and the buildings and fences were left to the elements.

Thereafter, the Kalawao side was all but abandoned, while the Kalaupapa side was constantly upgraded with roads, piped water, an ice plant, dairying, a hog ranch, a slaughter house, cinema, bandstand, dock, airstrip, petrol stations, and telephone lines. During the war, the Honolulu facility was closed due to its vulnerable position. After the war, a new facility was opened in Honolulu. By this time, there was sulfone. Babies were separated from mothers at birth, new admissions to Kalaupapa ceased, and the name of the disease was changed to "Hansen's Disease". By 1960, 200 patients remained at Kalaupapa. Many could have moved out had they wished, but they had a good life there, it was home to them, and many were too self-conscious about their appearance. Many were also blind or infirm and, not being allowed to have children there, they kept a lot of pets. There were a lot of activities and comeraderie.

In 1980, Kalaupapa was made a National Historic Park but

residents were not forced to leave and, as of 1990, 80 remained. Visitors to the park arrive by plane, or come by mule down a narrow trail from the top of the cliffs. As of 1988, there were 639 Hansen's Disease patients on the state registry, mostly out-patients, although there is a facility in Honolulu (not exclusively for Hansen's Disease) to which they may be admitted if necessary. During the 1980s there were 30 to 50 new cases per year, two thirds from the Philippines, 15% from Samoa, 8% were born in Hawai'i, and 12% from elsewhere in Asia and the Pacific (Frankel, 1988, Dept. of Health, pers. comm.).

References

Austin, C.J. 1949 Leprosy in Fiji and the South Seas. *International Journal of Leprosy* Vol. 17,4: 399-409

Beaglehole, J.C. 1966 *The exploration of the Pacific*. Stanford, California: Stanford University Press.

Bellwood, P. 1979 *Man's conquest of the Pacific*. Oxford: University Press.

Bird, I. 1890 *Six months in the Sandwich Islands*. Rutland, Vermont: Charles Tuttle Co.

Browne, S.G. 1985 The history of leprosy. In R.C. Hastings(ed.): *Leprosy*. London: Churchill Livingstone, pp. 1-31.

Carano, P. and Sanchez, P.C. 1964 *A complete history of Guam*. Rutland, Vermont: Charles Tuttle Co.

Chapman, R.F. 1982 *Leonard Wood and leprosy in the Philippines*. University Press of America.

Daws, G. 1968 *Shoal of time*. Honolulu: University Press of Hawai'i.

Dempster, G.O.L. 1949 Leprosy in Niue Island. *International Journal of Leprosy* Vol.17(4): 411-414.

Frankel, R.J 1988 Hansen's Disease in Hawai'i: current status. *Hawai'i Medical Journal* Vol.47 (2)

Green, L.W. 1985 *Exile in paradise*. National Park Service, Denver: U.S. Dept. of Interior.

Howe, K.R. 1984 *Where the waves fall*. Honolulu: University of Hawai'i.

Ibanez, A. and Resano, F. 1890 *Chronicle of the Mariana Islands*. Translation by Driver. Micronesian Area Research Center Publication 5 (1976).

Jarrett, R.M. 1993 *Gifts from the shore: A Kalaupapa diary*. Beaverton, Oregon: Pacific Editions Co..

Jennings, J.D. (ed.) 1979 *The prehistory of Polynesia*. Cambridge, Massachusetts: Harvard University Press.

Kirch, P.V. and Hunt, T.L. 1988 The spatial and temporal boundaries of Lapita. In Kirch and Hunt (eds.): *Archaeology of the Lapita Cultural Complex*. Seattle: Burke Memorial Museum. Research Report 5, pp: 9-27

Kuykendahl, R.S. and Day, A.G. 1948 *Hawai'i: A History*. Englewood Cliffs, New Jersey: Prentice Hall Inc.

Miles, J. 1997 *Infectious diseases colonizing the Pacific*. Otago: University of Otago Press.

Montgomerie, J.Z. 1988 Leprosy in New Zealand. *Journal of the Polynesian Society* Vol. 97(1): 115-152.

Noordeen, S.K. 1985 The epidemiology of leprosy. In R.C. Hastings (ed): *Leprosy*. New York: Churchill Livingstone, pp. 15-30.

Roberts, L.J. 1928 Some historical notes on leprosy, with specific reference to the occurrence of the disease on Guam. *Guam Recorder*, March 1928: 48-49

Rogers, R.F. 1995 *Landfall: a history of Guam*. Honolulu: University of Hawai'i Press.

Somers, G. 1985 *Kalaupapa, more than a leprosy settlement*. National Park Service, Western Archaeology and Conservation Center Publication in Anthropology 30.

Trembly, D.L. 1995 On the antiquity of leprosy in Western Micronesia. *International Journal of Osteoarchaeology* 5:377-384.

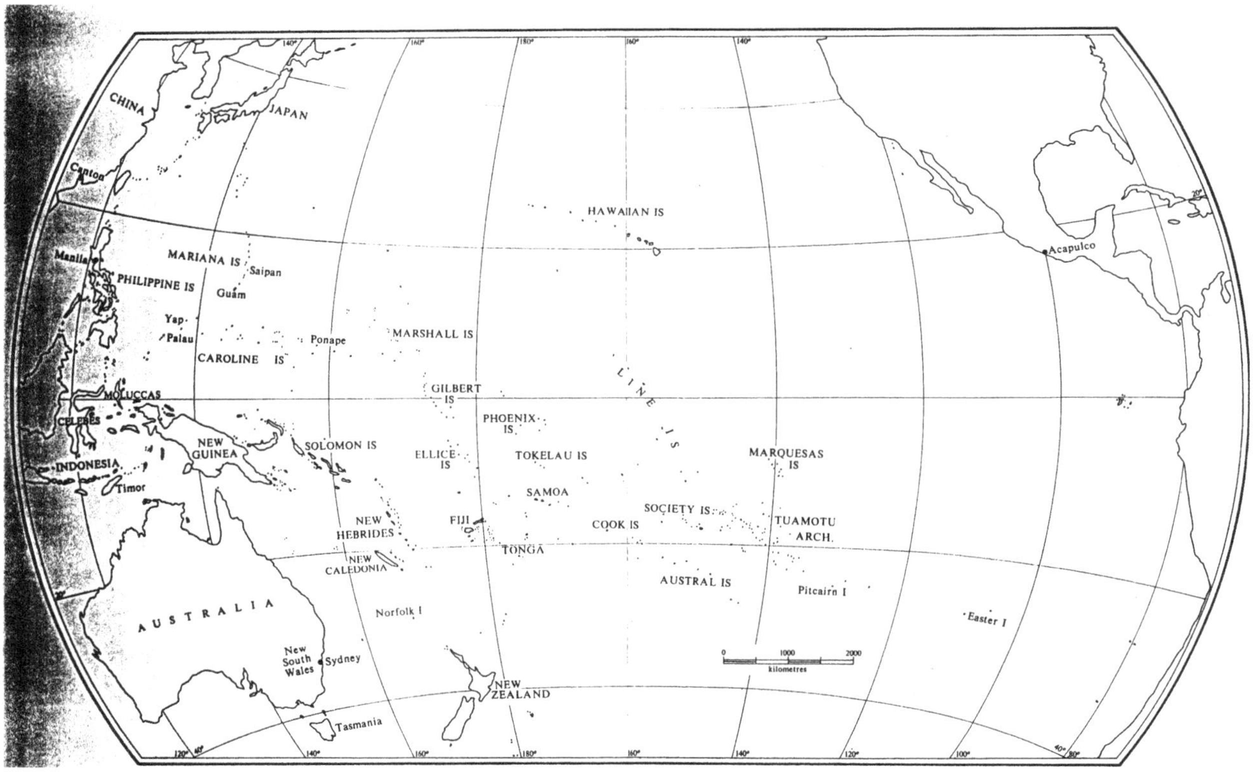

Figure 1: (from Howe, 1984). Note: the map is not referenced in the text as it is something to be referred to at any time throughout the text

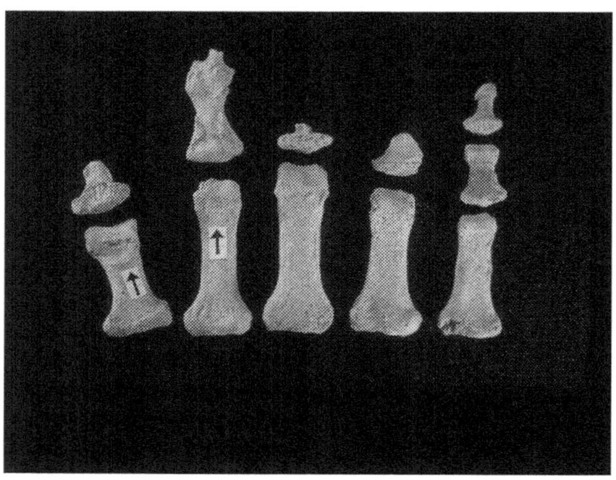

Figure 3: Hand phalanges from prehistoric burial on Guam. Note palmar grooves resulting from a claw-hand deformity and acro-osteolysis .

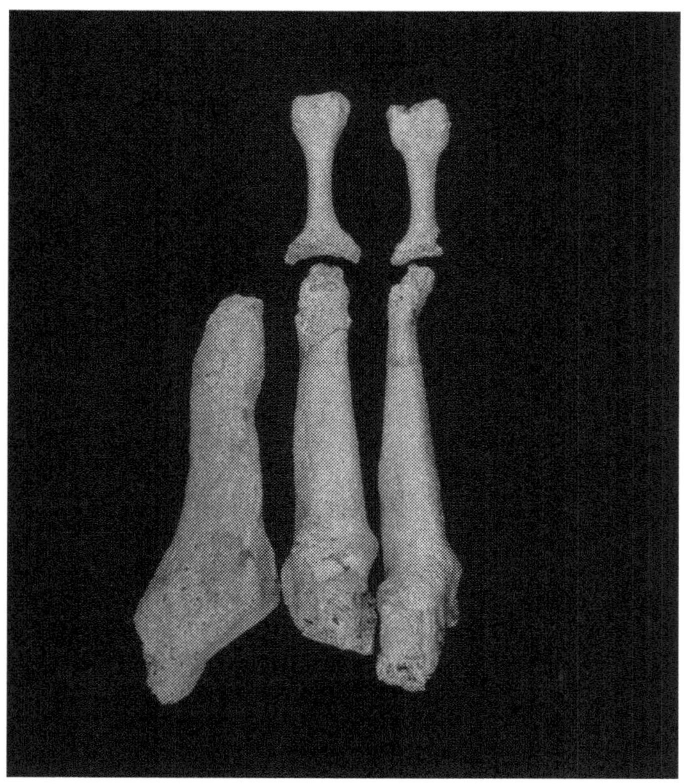

Figure 2: Foot bones from prehistoric burial on Guam showing “pencil and cup” deformity.

History of leprosy in Finland

H.S. Vuorinen
History Department
University of Tampere and
Department of Public Health
University of Helsinki
P.O.Box 41
00014 University of Helsinki
Finland

E-mail: heikki.vuorinen@helsinki.fi

Abstract

The earliest evidence for leprosy in Finland comes from 1355, when a leprosy hospital is mentioned in a will made in the town of Turku in the south-west of the country. There is no skeletal evidence for leprosy in Finland, but the literary references are more frequent from the 17th century onwards. Leprosy sufferers were poor, and their conditions, and those of the hospitals, were sometimes truly appalling. - Reliable identification and registration of leprosy sufferers started in the late 19th century. Leprosy was a disease of western and southwestern Finland; it was practically non-existent in eastern Finland. At the turn of the 20th century leprosy was already disappearing from some areas of formerly high prevalence. The last isolation hospital for leprosy sufferers operated in Finland in 1904-1953. The last diagnosis of "endemic" leprosy in Finland took place in the late 1950s.

Keywords

Leprosy, prevalence, Finland, history, hospitals

1. Introduction

Swedish kings began the conquest of Finland in the middle of the 12th century, and it remained part of Sweden until 1809. The earliest evidence of leprosy in Finland is from AD 1355, when a leprosy hospital dedicated to St. George is mentioned in a will made in the town of Turku (Åbo in Swedish), the Swedish capital of Finland in the south-west of the country (REA: 70). A hospital dedicated to St. Mary Magdalene in another coastal town of Viipuri (Viborg in Swedish) in southeast Finland was founded in 1475 (Fagerlund, 1901: 3-6).

At present there is no skeletal evidence for leprosy in Finland, but the literary references are more frequent from the 17th century onwards. There is a special problem with archival evidence in Finland as the gentry (including the few doctors) was Swedish speaking, while most of the common people spoke only Finnish. Lay people took care of the sick because there were few academically trained physicians in the country until the 20th century. In the 19th century, even for a medically trained person, leprosy was not clearly defined. The plethora of names and aetiological theories attributed to leprosy is well illustrated by Tapénius (1840) who studied at the Imperial Alexander University of Finland. Among the Finnish speaking common people leprosy tended to be confused with syphilis and other chronic diseases displaying skin lesions (Vuorinen, 1999). Leprosy was known as *spitelsko, spetälska, Elfkarlebysjukan, Älvkarlebysjukan* and *elephantiasis* in Swedish, and as *spitali, pitalitauti, pitaali, elewantin-kapi* and *oikia pitaalitauti* in Finnish during the 18th and early 19th century.

2. Hospitals

In addition to the two isolation hospitals for leprosy sufferers in two coastal towns, Turku and Viipuri (Figure 1) in the Middle Ages, another hospital was established in Helsinki in the mid-16th century. Although we lack detailed information about these early institutions, the common opinion is that they were small (Sandholm, 1973: 47-48). The earliest information about patient numbers is from 1551, when there were seven men and nine women with leprosy in the Viipuri hospital (Fagerlund, 1901: 40). The hospital dedicated to St. George in Turku (largest of the early institutions) cared for 39 people with leprosy in 1568 (the earliest year from which we have data), at least five of whom were admitted during that year (Fagerlund 1886: 120, 123). Patient numbers seem to have risen during the late 16th century, with 54 leprosy sufferers in the Turku hospital in 1600 (at least five admissions in that year) (Fagerlund, 1886: 120, 123-124). An urgent need for more leprosy isolation hospitals developed during the early 17th century, and several new ones were founded: on the islands of Seili (Sjählö in Swedish) and Gloskär in the southwestern archipelago, and in Kruunupyy (Kronoby), Korsholm and Carleborg in the province of Pohjanmaa (Österbotten). Only the Paltamo hospital was inland (Figure 1). An annexe of the Seili hospital operated in Turku in the late 17th century. Several of these institutions were small and existed for only a short time (i.e. those in Gloskär, Korsholm, Carleborg and Paltamo).

Typically, the 17th and 18th century hospitals consisted of several small wooden buildings: dwelling house(s), sauna(s), a church, barns for livestock etc. Buildings were small because they were easier and cheaper to keep warm in the cold winters. They were typically built directly on the ground or laid on cornerstones, and could have lasted for centuries if adequately cared for. However, there is evidence of neglect: for instance, we know from inspection reports in 1661 and 1664 that the wooden walls, floors and ceilings of Gloskär hospital deteriorated rapidly (Murenius, 1908: 503-508, 547-550), and the living conditions of the hospital inmates were generally poor.

In general, it was customary for medieval leprosy hospitals, such as those in Turku and Viipuri, to be situated outside the towns. The Helsinki hospital was an exception to this rule, being established within the urban area (Sandholm, 1974). This anomaly was corrected when the town was shifted to another location in 1640, and the hospital remained on its original site. The hospitals in Seili and Gloskär were established on isolated islands far from other communities. The hospitals admitted not only leprosy sufferers but also the elderly infirm, epileptics and the poor suffering from other chronic diseases; later an increasing numbers of patients with mental illness were admitted to the hospital. Normally, those with leprosy were separated from the "uncontaminated" inmates.

During the 18th and 19th centuries all the surviving leprosy hospitals were converted to lunatic asylums. The last person with leprosy in the Viipuri hospital died in 1705 (Fagerlund, 1901: 49), in the Helsinki hospital in 1749 (Sandholm, 1974: 53) and on Seili in 1766 (Fagerlund, 1886: 180, 202). The hospital in Kruunupyy also ceased having leprosy patients during the 18th century, although for a short period early in the next century leprosy sufferers were again admitted (Sandholm, 1973: 68). A small isolation hospital for leprosy operated transiently in the town of Oulu at the turn of the 19th century (Figure 1). Temporary isolation cottages in northern Finland (in Pudasjärvi and Kuusamo, Figure 1.) also operated during the early 19th century (Fagerlund, 1909, Förhandlingar vid Finska, 1900). Normally, however, patients suffering from leprosy were cared for in the ordinary hospitals (Westerlund, 1884, Fagerlund, 1891).

The need for an isolation hospital for leprosy was discussed at the General Assembly of the "Finska Läkaresällskapet" (Finnish Medical Society) in 1899. Whether the incidence of leprosy in Finland was actually increasing or decreasing was debated (Förhandlingar vid Finska, 1900). A temporary isolation hospital for leprosy sufferers was founded in Sörnäinen in Helsinki in July 1900 and functioned until December 1904 (af Hällström, 1910). The last dedicated hospital for leprosy patients operated in Orivesi (Figure 1) in central Finland in 1904-1953.

3. Life of a leprosy sufferer in the 17th century

The 17th century was the heyday of the Swedish Empire, but the burden of numerous (and mostly "victorious") wars fell heavily on the Finns. There were several fairly localized famines in Finland during the century, but in 1696 most of the crops failed and over a quarter of the population perished (Muroma, 1991). Academic medicine was established in 1640 when the university in Turku was founded (with one chair in the Faculty of Medicine). However, in practice the sick were all cared for by lay people during this century.

Murenius' (1908) reports of parish inspections give us a special insight into the life of leprosy sufferers in the parishes of the Åland (Ahvenanmaa) archipelago between 1637 and 1666 (Figure 1). Boëtius Murenius (Muur) was the rector of the parish of Saltvik and Dean of Åland at the time. The fate of a leprosy sufferer was hard in the 17th century. Those suspected of having contracted the disease were examined by all the parishioners present at the inspection meeting, and a formal declaration of leprosy was then in the hands of the parish meeting. A period of observation (a year or longer) was used if the diagnosis was not absolutely clear. During this time the suspect was isolated at home. If a formal declaration of leprosy was passed in the parish meeting, the church authorities were informed and they, in turn, informed the civic authorities. The person was then sent to a hospital: on Åland either to Gloskär (operating in 1653-1672) or to Seili. Leprosy sufferers were generally poor, and their conditions and those of the hospitals were sometimes truly appalling. In spite of this, some people survived for quite long period with their leprosy.

The available evidence indicates that more or less formalized ways were used to achieve a diagnosis, as illustrated in reports from the parishes of Sund, Lemland, and Föglö (Murenius 1908: 170, 181 and 184, respectively) in 1648. The diagnosis of leprosy was based on alterations to the 1) skin of the face, 2) skin in other areas, and 3) voice. Although it is impossible to evaluate how well-founded individual decisions concerning leprosy were, some confidence in the ability of lay people to reach a correct diagnosis is justified (see also Richards, 1977).

In theory, there was a well-formulated approach to a person suspected of having leprosy. From the parish inspection reports (Murenius, 1908) we can conclude that the system sometimes operated efficiently: a diagnosis was made, isolation at home arranged, and the person with leprosy was sent to the hospital without delay. On this basis the system can be justified as running quite smoothly, as Richards (1977) describes in his book. However, in many documented cases the approach did not work. People were often under observation for long periods without any formal decision, for example two suspected sufferers in Ekerö were isolated in their homes for four years (Murenius, 1908: 535). It was also hard for the authorities to persuade people to comply with isolation measures in daily life, for example a servant girl with leprosy in a farmhouse who normally took part in all the ordinary household activities (parish inspection in Jomala in 1640 - Murenius, 1908: 63-64). Some were admitted to Gloskär hospital several years after the formal diagnosis of leprosy, for example a daughter of Maths Bengtson in Främmanby, who was already considered to have leprosy in 1658, but not admitted to the hospital until 1665 (Murenius, 1908: 366, Fagerlund, 1886: 302). Some of those with leprosy who were sentenced to the hospitals were probably never admitted to them (at least not according to the surviving documents), for example Cnut in Lemböte (parish inspection in Lemland in February 1640 and 1642 - Murenius, 1908: 37, 96-97).

Dean Murenius (1908: 597) was deeply frustrated after 30 years of various parish inspections, as he admitted in his visit to Föglö in 1666; he complained about the repair of churches, noting that the parishioners promised help but nothing happened. His frustration over the fate of leprosy sufferers was equally great. He suggested the opening of a new hospital in his inspection report for the parish of Sund in 1648 (Murenius, 1908: 172-173), and consequently Gloskär was opened in 1653.

Thanks to Boëtius Murenius we have some idea of the prevalence of leprosy on the Åland islands in the mid- 17th century. Because identifying individuals is often difficult, the total of around 100 leprosy sufferers on the Åland islands mentioned in the 1637-1666 parish inspection reports is probably an overestimation. In many cases the short comment in an inspection report is inadequate as a reliable identification of a person in other reports. A typical example is "Spettelska: Lasse i Torp och en pijga i Ödningeby ..." (parish of Jomala in 1642 - Murenius, 1908: 93), which can be translated as: "Leprosy: Lasse in Torp and a servant girl in Ödningeby" (Torp and Ödningeby were villages in the parish of Jomala on the Åland islands).

We can identify 14 individuals with leprosy in 1649, and it is possible to recognize 10 or 11 of these same people in the 1650 records. Two or three of those with leprosy in 1650 had no entry in 1649. In all, there were probably 16 or 17 leprosy sufferers on the Åland islands in 1649 and 1650. Ten years later 18 can be identified, 16 of whom were in the Gloskär hospital in 1660-1661, while two remained outside the hospital in both years. The third report we have dates from August 1672, when a total of 23 leprosy sufferers was taken to Seili hospital from the Åland islands. These included the last nine inmates of Gloskär hospital (closed in 1672) and all those suspected of suffering from leprosy on the islands (Fagerlund, 1886: 302-303).

A new isolation hospital for leprosy was founded on Seili island in the southwestern archipelago in 1619, and the first sufferers were admitted in 1623/4 (Fagerlund, 1886). Around the same time the hospital dedicated to St. George in Turku was closed down, the buildings burned, and the inmates moved to Seili. The total number of identified leprosy cases admitted to the Seili hospital during the 17th century was 568, and during the 18th century 96; 359 were women and 305 men. The age of sufferers is rarely known, but at least twelve were children (some even infants). People hospitalized with leprosy usually stayed for life, although two women escaped in 1669 and one in 1686. Brita Rassmusdotter from Turku (Åbo) had the longest known period of isolation on Seili; she entered the hospital in 1709 and died there in 1748. Most of the inmates whose native parish is known (476 in the 17th century and 78 in the 18th century) came from the southwestern archipelago and the coastal area around Turku. Only a few were from further afield: two were from Helsinki and six from Ruovesi in the 17th century (Figure 1).

The personal data for Seili is most complete for the period 1675-1688 (Fagerlund, 1886: 182-202), so that fewer estimations of the length of isolation have to be made. One hundred and eight leprosy cases were isolated on Seili in 1675-1684 (Table 1); 53 were women and 55 men. In only seven cases is the length of isolation estimated and the minimum duration is used here. In one case the length of isolation could not be estimated. One man died accidentally, and one woman escaped, so their isolation did not end because of leprosy. Two of the men stayed in the hospital for 15 years and two for at least 17 years (in both latter men the maximum estimate of isolation was 29 years).

Throughout the 17th century information is scarce about the number of leprosy cases in the hospitals at any given time. However, estimates of inmate numbers for the middle of the century are presented in Table 2. Information from the hospitals in Paltamo, Korsholm and Carleborg is very sparse, but it can be concluded that they rarely admitted leprosy patients (Sandholm, 1974: 66-67).

The available evidence indicates that leprosy may have peaked in Finland in the 17th century. However, we can make only rough estimates of the numbers of leprosy sufferers. Based on the evidence from the Åland islands, and information from hospitals, we may safely argue that well over a hundred people had leprosy in Finland in the middle of the century. We have no reliable estimates of the total population in Finland at this point. The first more reliable estimate of the total population, based mainly on the surviving parish records, is a figure of about 440,000 persons in 1695 (Muroma, 1991: 179). This would imply a leprosy prevalence exceeding two sufferers per 10000 inhabitants in the whole country. However, the prevalence of leprosy was not uniform across Finland. We may conclude on the basis of 1) no positive evidence, and 2) knowledge of the distribution of leprosy in the late 19th century (see below), that leprosy was non-existent in vast areas of eastern and central Finland in the 17th century. We may speculate that this was due to the wide cultural differences between eastern and western Finland. For example, there was slash and burn agriculture in the east, and field agriculture in the west. In some areas of western and southwestern Finland the prevalence of leprosy might have been very high. For instance, on the Åland islands with a population of about 7600 (in 1695) there were about twenty leprosy cases in the mid- 17th century, giving a prevalence well in excess of two per 1000 inhabitants.

4. Leprosy in the late 19th and early 20th centuries

Finland was an autonomous Grand Duchy of Russia from 1809 until independence finally came in 1917. The 19th century was a period of peace, rapid population growth and accelerating economic development. The most serious setback was the severe famine of 1866-68. Academic medicine began to establish itself throughout the country and the number of physicians increased slowly but steadily.

Back at the beginning of the 19th century there were reports of leprosy in Oulu, Pudasjärvi and Kuusamo, which prompted temporary isolation facilities to be built (Figure 1). However, reliable identification and registration of leprosy sufferers started in the late 19th century. In 1891, 1897 and 1904 the Central Medical Board of Finland mailed all physicians, requesting them to inform the Medical Board of every leprosy case which came to their attention (Fagerlund, 1893, 1897 and 1904). Physicians responded to the first circular during 1891 and 1892. The Medical Board passed their letters to L.W. Fagerlund, who then collected further information by correspondence with the physicians and parish priests. Most cases of leprosy could be followed up to 1904, and occasionally up to 1910 (Table 3). Af Hällström (1909 and 1910) continued the work of Fagerlund. Only seldom could leprosy sufferers not be identified in the later surveys.

Only in the minority of leprosy cases do we have the actual cause-of-death data. One of those with leprosy in 1893 was considered healthy in 1904 (Fagerlund, 1904: 499, 580). Most of the known cases were men. The average age at death was, however, higher in women (Table 4). Farmers and their wives with leprosy tended to outlive other people with it. Leprosy was a disease of the poor: paupers, workers, "independent persons" (those who worked for housing) and crofters.

Leprosy was a disease of western and southwestern Finland (Figure 2 and Table 5); it was practically non-existent in eastern Finland. The distribution of the places of birth of leprosy sufferers (N=192) in Finland in 1892-1902 and the places of residence of those having clinical leprosy in 1892-1904 (N=232) is almost identical. At the turn of the 20th century, leprosy was already disappearing from some areas of formerly high prevalence (Kuusamo, Åland (Ahvenanmaa), the Turku region). However, in new urban areas such as Helsinki and Tampere (Figure 2.) new cases were emerging. If we consider that leprosy does not constitute a public health problem when there is less than 1 case per 10,000 of the population then in some areas of Finland leprosy definitely still persisted as a public health problem at the turn of the century (Table 5).

Leprosy was apparently not highly communicable in the late 19th century. Most spouses, children and parents of leprosy sufferers never developed clinical leprosy. This was in spite of the very poor living conditions, including small overcrowded cottages, sharing the same bed and eating from the same utensils. Leprosy can, however, have a very long incubation period, and thus people may have been infected but died from other reasons (accidents or illnesses) before having developed clinical leprosy. It seems that it was typical in many areas for only sporadic cases of leprosy to emerge for long periods (e.g. in Kuusamo, Figures 1 and 2).

5. Epilogue

Leprosy cases numbered 52 in 1893 and 95 in 1904, after which they rapidly declined (Table 6). After the closure of Orivesi hospital in 1953 the last three leprosy sufferers were moved to Helsinki. The last diagnosis of "endemic" leprosy in Finland took place in the late 1950s. This man made his final visit to the dermatological polyclinic of Helsinki University hospital in 1970 (Archives of Museum of Medical History in University of Helsinki), after which Finland ceased to be an endemic area for leprosy.

References

Fagerlund, L. W. 1886 Finlands leprosorier I. 1. S:t Jörans hospital, 2. hospitalet på Sjählö, 3. hospitalet på Gloskär. Bidrag till kännedom af Finlands Natur och Folk. *Utgifna af Finska Vetenskaps-Societeten* 43:107-311.

Fagerlund, L. W. 1891 Om spetelskan i Kuusamo. *Finska Läkaresällskapets Handlingar* (FLH) 33: 767-813.

Fagerlund, L. W. 1893 Spetälskan förekomst i Finland. *FLH* 35:495-555.

Fagerlund, L. W. 1897 De spetälska i Finland. *FLH* 39:1287-1345.

Fagerlund, L. W. 1901 *Finlands leprosorier II. 1. Maria Magdalenae, Wiborgs hospital*. Helsingfors: Finska Litteratursällskapets Tryckeri.

Fagerlund, L. W. 1904 Spetälskan i Finland. *FLH* 46:495-585.

Fagerlund, L. W. 1909 Die Lepra in Finnland. *Lepra* 8: Suppl.: 402-415 (II. Internationale Wissenschaftliche Lepra-Konferenz, Bergen 1909, Mitteilungen und Verhandlungen, Bd I, Lepzig 1909: 140-151)

Förhandlingar vid Finska 1900 *Läkaresällskapets sjuttonde allmänna möte i Helsingfors den 21, 22 och 23 september 1899*. Helsingfors: Helsingfors Centraltryckeri 1900: 90-144.

af Hällström, E. 1909 Tilastollisia lisiä leprataudista Suomessa. *Duodecim* 25:450-468.

af Hällström, E. 1910 Orihveden leprasairaala. *Duodecim* 26:480-493.

Murenius, B. 1908 *Boëtius Murenius´Acta Visitatoria 1637-1666 utgifna av Kaarlo Österbladh.* Borgå: Suomen Kirkkohistoriallisen Seuran Toimituksia VI.

Muroma, S. 1991 *Suurten kuolonvuosien (1696-1697) väestönmenetys Suomessa.* Helsinki: Suomen Historiallinen Seura, Historiallisia Tutkimuksia

161.
REA. Registrum ecclesiae Aboensis
Richards, P. 1977 *The medieval leper and his northern heirs*. Cambridge: D.S. Brewer.
Sandholm, Å. 1973 *Kyrkan och hospitalshjonen. En undersökning rörande omsorgen om de sjuka och fattiga i välfärdsanstalterna i Finland*. Helsingfors: Finska Kyrkohistoriska Samfundets Handlingar 88.
Sandholm, Å. 1974 *Hospitalet i Helsingfors 1550-1840. Huvudstadens äldsta social- och sjukvårdsanstalt*. Helsingfors i Forna Tider III. Helsingfors - Samfundet, Åbo.
Tapénius, C. N. A.1840 *Historiskt-Nosographiska Anmärkningar om Elephantiasis Nodosa*. Helsingfors: J. C. Frenckell and Son.
Westerlund, L. W. 1884 Om spetelskans förekomst inom Tyrvis distrikt. *FLH* 26: 137-152.
Vuorinen, H. S. 1999 Suomalainen tautinimistö ennen bakteriologista vallankumousta. In Finnish. Abstract in English. Finnish nomenclature for diseases before the bacteriological revolution. *Hippokrates, Suomen Lääketieteen Historian Seuran vuosikirja* 16: 33-61.

Table 1. Leprosy cases in Seili hospital in 1675-1684; length of isolation in years.

	Length of isolation									
	0-1		2-4		5-9		10+		Total	
	N	%	N	%	N	%	N	%	N	%
Women	22	41.5	15	28.3	16	30.2	0	0.0	53	100
Men	26	48.1	15	27.8	8	14.8	5	9.3	54	100
Total	48	44.9	30	28.0	24	22.4	5	4.7	107	100

Figures based on Fagerlund, 1886: 192-195.

Table 2. Number of inmates in the hospitals around the middle of the 17th century.

Hospital	Number	Year	Reference
Seili	37	1635	Fagerlund, 1886: 169
	57	1664	Fagerlund, 1886: 169
Helsinki	20	1651	Sandholm, 1973: 62 and 1974: 14
Viipuri	13	1639	Fagerlund, 1901: 45-46
	6	1658	Fagerlund, 1901: 45-46
Kruunupyy	25	1633	Sandholm, 1974: 67
Gloskär	9-13	1656-1661	Fagerlund, 1886: 300

Table 3. Collated material on 232 leprosy sufferers in 1892-1904.

Year	No.	Status	Source	Comments
1893	49	Living	Fagerlund, 1893	First survey
1892-93	15	Dead	Fagerlund, 1893	Died during data collection for the first survey
1892-97	16	Dead	Fagerlund, 1897	Died before the second survey, not included in the first survey.
1897	46	Living	Fagerlund, 1897	Second survey, not included in the first survey.
1897-1904	20	Dead	Fagerlund, 1904	Died before the third survey, not included in the first or second survey.
1904	65	Living	Fagerlund, 1904	Third survey, not included in the first or second survey.
< 1905	20	Living	af Hällström, 1909	Leprosy cases known to af Hällström, with symptoms before 1905 but not included in Fagerlund's surveys.
< 1905	1	Living	af Hällström, 1910	One case in Orivesi hospital, with symptoms before 1905 but not included in the surveys.

Table 4. Age at falling ill, age at death, duration of illness, and follow-up period (in years) by social class and sex in 1892-1904 (number of cases in brackets).

Social class	**Age at falling ill**			**Age at death**		
	♀	♂	**Total**	♀	♂	**Total**
Pauper	38.3 (4)	42.0(4)	45.1 (8)	52.5 (2)	53.0 (3)	52.8 (5)
Independent person	39.0(23)	36.3 (33)	37.4 (56)	51.6(18)	46.4(18)	49.0 (36)
Crofter, cottager	35.6 (8)	36.1 (20)	36.0 (28)	52.4 (5)	49.4(16)	50.1 (21)
Farmer	51.0 (1)	46.3 (14)	46.6 (15)	59.3 (3)	55.5(11)	56.4(14)
Worker	40.6(28)	38.0 (45)	39.0 (73)	56.5(10)	50.3(26)	52.1 (36)
Craftsman	39.0 (8)	42.9 (10)	41.2 (18)	51.0 (4)	51.4 (5)	51.2 (9)
Total	40.2(80)	39.4(128)	39.7(208)	53.0(49)	49.9(80)	51.1(129)

	Duration of illness			**Follow-up period**		
	♀	♂	**Total**	♀	♂	**Total**
Pauper	7.5 (2)	6.0 (3)	6.6 (5)	13.0 (4)	5.0 (4)	9.0 (8)
Independent person	8.4(16)	8.3(18)	8.4 (34)	8.1(22)	9.2 (31)	8.7 (53)
Crofter, cottager	17.0 (5)	9.3(12)	11.6 (17)	13.1 (7)	11.0 (19)	11.6 (26)
Farmer	- (0)	13.7 (9)	13.7 (9)	3.0 (1)	10.2 (14)	9.7 (15)
Worker	8.6 (8)	9.0(26)	8.9 (34)	10.3(28)	9.1 (44)	9.5 (72)
Craftsman	12.3 (4)	14.3 (4)	13.3 (8)	14.3 (8)	11.5 (10)	12.7 (18)
Total	9.6(41)	9.7(71)	9.7(112)	10.0(75)	9.7(124)	9.8(199)

Table 5. Areas of high prevalence of leprosy, leprosy sufferers by place of residence in 1892-1904.

Place of residence[1]	**Leprosy sufferers**	**Population in 1900**	**Prevalence per 1000 pop.**
Helsinki (town)	28	93576	0.30
Kärkölä (rural commune)	13	4500	2.89
Merikarvia (rural commune)	12	7049	1.70
Tampere (town)	10	36344	0.28
Huittinen (rural commune)	8	9758	0.82
Pori (town and rural commune)	7	20901	0.33
Tammela (rural commune)	6	14641	0.41
Kiikka (rural commune)	5	3739	1.34
Isojoki (rural commune)	5	5224	0.96
Rural areas with high frequency of leprosy[2]**:**			
I Merikarvia + Isojoki + Siipyy	18	15328	1.17
II Tammela + Renko + Urjala	11	27648	0.40
III Kärkölä + Mäntsälä + Hollola + Lammi	18	30821	0.58
IV Kiikka + Huittinen + Kiikoinen + Kauvatsa+ Tyrvää + Punkalaidun + Vampula + Alastaro + Köyliö	28	43558	0.64
Entire country			
Women	91	1310100[3]	0.07
Men	141	1270000[3]	0.11
Total	232	2580100[3]	0.09

[1] See Figure 2 for location in Finland.
[2] Same Roman numerals are used in Figure 2.
[3] Mean population in 1892-1904.

Table 6. Numbers of leprosy sufferers in Finland, 1893-1956.

Year	**Number**
1893	52
1897	67
1904	95
1924	57
1950	7
1956	4

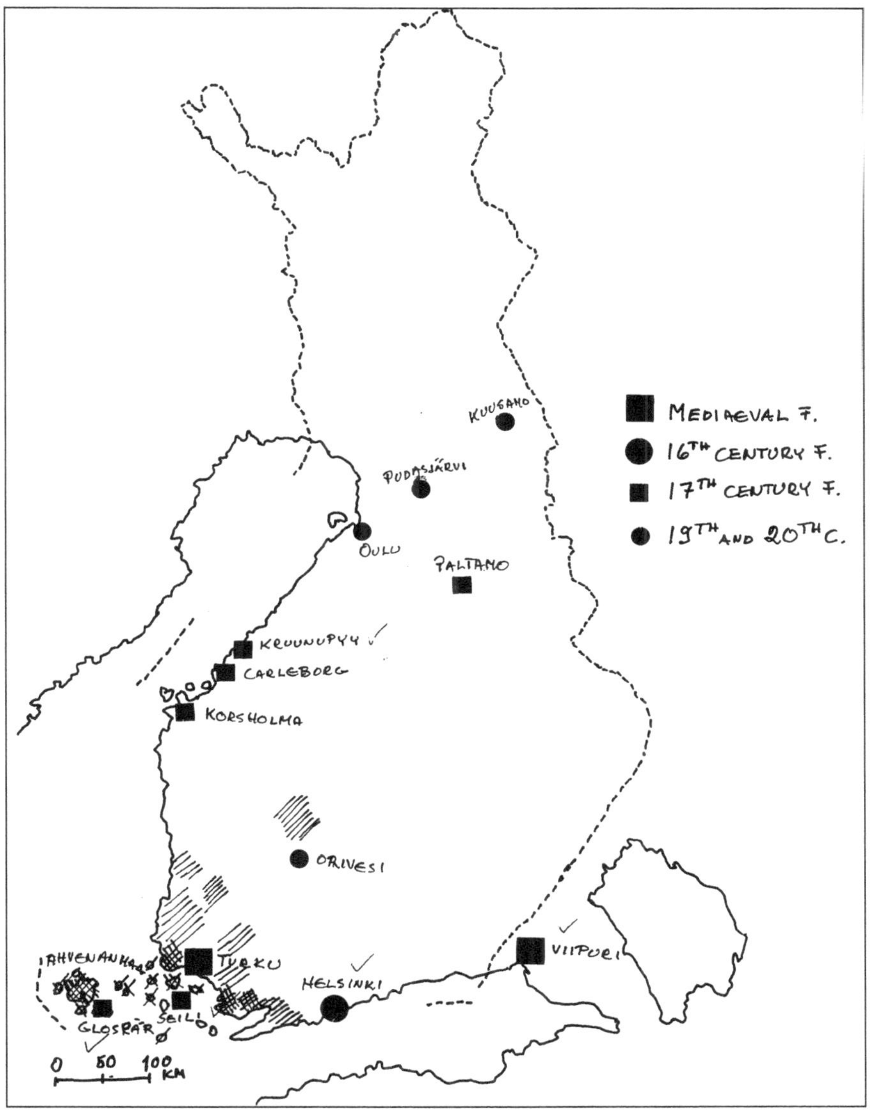

Figure 1: Isolation hospitals for leprosy in Finland and place of residence of sufferers (N= 554) isolated in Seili hospital during the 17th and 18th centuries (shaded areas).

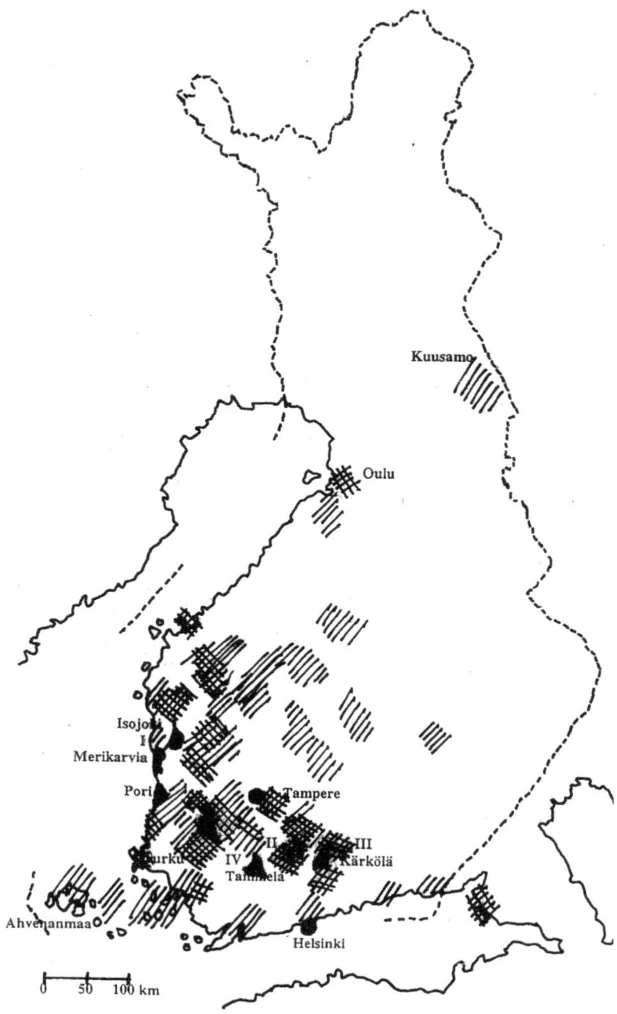

Figure 2. Leprosy sufferers (N=232) by place of residence in Finland in 1892-1904. Shading indicates the prevalence of leprosy: the darker the shading the more leprosy; areas of high prevalence of leprosy are colored black (for prevalence figures and Roman numerals see Table 5).

Mycobacterial disease in North America: an epidemiological test of Chaussinand's cross-immunity hypothesis

Alicia K. Wilbur, Jane E. Buikstra, Christopher Stojanowski
Department of Anthropology
University of New Mexico
Albuquerque
NM 87131-1086

Email: akwilbur@unm.edu
Buikstra@unm.edu
Cstoj@unm.edu

Abstract

Mycobacterial cross-immunity has been suggested as a mechanism to explain the simultaneous rise of tuberculosis and decline of leprosy in Medieval Europe. Most evidence for mycobacterial cross-immunization arises from the ability of the *M. bovis* BCG vaccine to effectively confer protection from tuberculosis and leprosy in some human groups. This issue can also be examined from an epidemiological perspective through analysis of incidence figures for mycobacterial disease.

In this study we first briefly consider the antiquity of leprosy and tuberculosis in North America, with special reference to the Gulf Coast of the United States. We then analyze retrospective incidence data from Texas where leprosy and tuberculosis have been endemic for nearly two and a half centuries. Using a simple correlation analysis, we examine the incidence data for evidence of cross-immunization. We calculated the correlations between leprosy and tuberculosis incidence, and between leprosy and tuberculosis incidence by year. Contrary to the expectations of the cross-immunization model, we found few significant negative trends in the data. In fact, the few significant correlation coefficients suggest the opposite, that the number of cases of both diseases increases or decreases in tandem.

Keywords

Leprosy, tuberculosis, cross-immunity, history, epidemiology, North America

1. Introduction

Historical and archaeological evidence indicates that leprosy spread throughout Europe during the Medieval period. Frequency rates of this dreaded disease apparently increased until the 13th century, when it began a precipitous decline (Andersen, 1969, Hunter and Thomas, 1984, Manchester, 1984, 1991, 1992, Manchester and Roberts, 1989, Møller-Christensen, 1953, 1969, Roberts, 1986, Roberts and Manchester, 1995). Northern Scandinavia, where the disease persisted into the 19th century, is a well-documented exception to this trend (Roberts and Manchester, 1995). While many explanations, ranging from social factors to climate have been proposed, one of the more intriguing suggestions developed in a series of articles published by Chaussinand (1944, 1948, 1953, 1964).

Chaussinand, developing his inferences from studies of humans and other animals, emphasized that while infection with *M. tuberculosis* offered immunity against leprosy, the converse was not universally true. Based upon clinical observations and experimental results, he argued that while prior infection with tuberculosis offers protection against leprosy, only those with tuberculoid leprosy would have conferred resistance to *M. tuberculosis* infection (Chaussinand 1944, 1948, 1953, 1964). He concluded that the antagonistic nature of the two diseases would eventuate in leprosy being driven out of any population affected by both leprosy and tuberculosis. Epidemiologically, an implication of his theory was that in regions where the spread of leprosy was recent or its prevalence high, the tuberculin-positive rate should be very low. Conversely, in regions where leprosy prevalence has declined, the tuberculin positive rate should be high. Chaussinand (1964) cited studies from Suriname and New Guinea in support of his model. Mycobacterial cross-immunity could also have serious epidemiological consequences for historical contexts, he proposed, including the disappearance of leprosy from western Europe, initially in urban centers (Chaussinand 1944, 1948, 1953, 1964).

Chaussinand's concept has been introduced into the anthropological literature by Manchester (1984, 1986, 1991, 1992, Manchester and Roberts, 1989, Roberts and Manchester, 1995), who has elaborated the archaeological, historical and cultural contexts for these mycobacterial maladies. Manchester argues that as urban populations of Medieval northern Europe became infected with both tuberculosis and leprosy, they became increasingly tuberculin-positive. Tuberculin-positive status served to confer a degree of immunity to leprosy, and those who later became infected with leprosy tended to develop the tuberculoid rather than the more highly contagious lepromatous form. Decreased infectivity led to extinction of the disease (Manchester, 1991). Recognizing the strength of this immunological argument, Roberts and Manchester (1995) also emphasize that the ultimate explanation for the decline and disappearance of leprosy no doubt involves a complex interaction of bacteriological, environmental, and social factors. Multiple possible agents, including cross-reactivity, are also described by Hunter and Thomas (1984).

Cross-reactivity between the two mycobacterial infections had been considered prior to Chaussinand's research, although the earlier studies were typically more concerned with vaccination than epidemiological arguments, e.g., Fernandez (1939). An exception was the work of Rogers (1924) who suggested that the protection against leprosy afforded by mild tuberculosis could have been a factor in the extinction of leprosy in Europe. An evolutionary argument similar to that proposed by Chaussinand, and by Manchester for western Europe, was advanced by Fernandez (1955, 1957) to explain the disappearance of leprosy in coastal Peru, where it had been introduced by the Spanish during the Conquest.

Also influenced by Chaussinand, Leitman *et al.* (1997) have mathematically formalized a competitive exclusion model in order to examine the evolutionary impact of cross-immunity. To assess the degree of cross-immunity necessary for *M. tuberculosis* to competitively exclude *M. leprae*, they formulate alternate models of mycobacterial transmission, tested through simulations. Their results emphasize the prolonged time necessary for eradication of leprosy. 'With reasonable parameter estimates in which the criteria for the competitive exclusion of leprosy are met, dynamic simulations of the...model indicated that centuries are required for tuberculosis to effectively eradicate leprosy' (Leitman *et al.* 1997:1926). A second conclusion involves the key role played by the basic reproductive rates for the two diseases, defined as 'the number of secondary infectious cases produced by a single infectious case in a totally susceptible population' (Leitman *et al.*, 1997: 1924). While the basic reproductive rate (BRR) cannot presently be estimated from prevalence data, the authors speculate that it was the relatively low BRR for leprosy during the great tuberculosis epidemic of the 17th and 18th centuries that 'played a major role in the disappearance of leprosy from western Europe' (Leitman *et al.,* 1997: 1926). As the authors emphasize, there are multiple factors that may confound their conclusions, not the least of which are the lack of prevalence data upon which to base an estimate of leprosy BRR and the immunological impact of other (environmental) mycobacteria (Leitman *et al.,* 1997). Another confounding variable is the fact that the authors do not distinguish between the lepromatous and tuberculoid forms of the disease.

Influenced by the work of Chaussinand, Manchester, and Leitman *et al.*, we too seek evidence for cross-immunity between leprosy and tuberculosis. The present study examines retrospective incidence figures from a single geographic region, the Gulf Coast of North America, where the presence of both leprosy and tuberculosis has been documented for nearly two and a half centuries. To contextualize our study, we begin with a brief history of leprosy and tuberculosis in North America, with emphasis upon the continental United States. We then consider the potential impact of environmental mycobacteria on cross-immunity models and review recent literature on cross-immunity. Our analytical methods and results follow.

2. Geographic and bacteriological contexts: Leprosy in the United States: a brief history

At the beginning of the 20th century, several regions within the continental United States were recognized as sites where immigrants had introduced leprosy. From Asia and the Pacific, the disease had spread to California. Scandinavians had brought leprosy to Wisconsin, Iowa, and Minnesota, while both African slaves and Canadians (Acadians) were implicated in Gulf Coast regions, including Florida, Louisiana, and Texas. New York's caseload reflected multiple points of origin (Faget, 1946: 3-4, Hudson and Genesse, 1982, Klingmuller, 1930, 65-67, Rogers, 1924).

Within the United States, Louisiana is distinguished by its long history of endemic leprosy infection (Feldman and Sturdivant, 1975, Hudson and Genesse, 1982). Although the disease was present prior to 1755, probably introduced by the Spanish and their slaves, the arrival of French Canadians (Acadians) from eastern Canada increased the presence of leprosy considerably (Faget, 1946, Feldman and Sturdivant, 1975, Hudson and Genesse, 1982, Klingmuller, 1930:66). Texas shares with Louisiana a high level of endemicity, with most Texas examples occurring in individuals of Spanish descent (Hudson and Genesse, 1982). The third Gulf Coast state with significant leprosy prevalence is Florida, where concentrations in Miami and Key West reflect a Caribbean-West Indies source area (Hudson and Genesse, 1982).

Recognizing leprosy as an indigenous disease, the State of Louisiana in 1891 selected a former smallpox hospital as the site of its first institution dedicated to the treatment of "lepers" (Gussow, 1979: 601). Conditions at this site proved unhealthy, however, and three years later, through the efforts of Dr. Isadore Dyer, the first president of the Louisiana "State Control Board for the Leper Home", a "temporary" institutional home was established at Indian Camp Plantation, Iberville Parish, approximately 85 miles from New Orleans (Gussow, 1979: 601, Moschella, 1997, Stein, 1973). Located "at a dead end of the River Road in Louisiana, between New Orleans and Baton Rouge", the site was regularly inundated by the Mississippi River prior to levee construction (White, 1996: 15). The first eight inmates were transferred from New Orleans on a coal barge during the night of the 30th November, 1894. 'The appalling details of the trip were depicted in the daily press' (Denney, 1926: 924). During 1905, the Indian Camp institution became the permanent Louisiana State Home for "Lepers" and, finally, in 1921, the United States Marine Hospital No. 66, known as Carville (Faget, 1946, Gussow, 1979, White, 1996).

In response to the perceived threat of widespread infections, the U.S. Federal Government imposed regulations that severely limited the freedom of those suffering from Hansen's disease. Leprosy was, in fact, the first and apparently only disease for which specific regulations were made about transport and inter-state immigration, enacted in 1912 (The Interstate Quarantine Regulations). People with leprosy had to be transported in strict isolation and only after obtaining a permit from the Surgeon General of the Untied States. People with leprosy who crossed state lines were either forcibly returned to their home states or transported to 'such Federal station as the Secretary of the Treasury may designate' (Denney, 1926: 5). By an ensuing act of Congress that Federal station became the leprosarium at Carville. In 1926, O. E. Denney, the first Marine Medical Officer in Charge of Carville, assured the American public that 'leprosy is not a crime...Once a leper is in detention, however, it is a crime against society for him to abscond and subject his fellow human beings to the risk of contracting a malady that is practically incurable. To restrain such an individual is for the public good. This the law does with justice'. Denney goes on to say that:

'With few exceptions, the lepers at Carville are contented with their lot. In comfortable quarters located on a beautiful 358-acre tract of land, with good food, excellent medical and surgical attention and nursing, and a diversity of amusements, these unfortunates, the wards of the Government, are living out their lives without worry and in full realization of the fact that they are no longer a menace to the health and contentment of their fellow beings'. (Denney 1926: 5)

Autobiographical statements, however, suggest that many of Carville's inmates led lives that belied Denney's comments. From the moment of arrival, those suffering from Hansen's disease experienced a wrenching isolation and depersonalization. Stanley Stein, whose journalistic efforts as editor of the Carville Sixty-six Star (1931-34) and Star (1941+) changed public opinion concerning leprosy, remarked upon entering Carville:

'Then I noticed the high metal cyclone fence with the three strands of barbed wire running along the top. I first saw the uniformed guards at the gate--and I realized that the dreadful moment that I had been fighting against for nearly ten years had come at last. I had arrived at the U.S. Marine Hospital no. 66, Carville, Louisiana...At ten o'clock on that Sunday morning, March 1, 1931, I became an exile in my own country'.
(Stein,1973: 4-5)

'Henceforth I would always be No. 746, a number identified not only with my case history but with everything I would do at Carville for the next third of a century, from seeking laxative tablets to making a laundry list. I was sinking into the quagmire of anonymity which society reserves for the victims of leprosy, mental illness, or crime. We were no more entitled to individuality than a convict in a penitentiary.

I was not completely submerged, however, until a doctor stuck his head into the Record Room and administered the coup de grace...'New patient, I see. What name is he taking?

Another name? What was wrong with my own name? Did I have to hide under an alias like a hunted criminal? Could I keep nothing of my old life to clothe my naked ego?

'Have you decided on your new name, young man?' Sister Laura asked sweetly.

I shook my head. I was too crushed to speak'.
(Stein, 1973:11)

Betty Martin, whose two autobiographies reached international audiences, entered Carville in 1928, aged 19. She later commented that she then began 'an elaborate network of duplicity under which I was to cower for the rest of my life' (Martin, 1953). She never forgot the 'horror of having that New Orleans doctor shouting the news of my illness to my poor, terrified mother: Get her out of here--before she infects the entire city' (Martin, 1959: 26). Even after leaving Carville and living in California for several years, Martin commented that she and her husband could not publicly disclose their birth names,

'We belong with the secret people. There are thousands like us, who for one reason or another must walk carefully, that no one may know we walk in a secret world'.
(Martin, 1959: 23)

Even though the built walls that had physically separated the sexes during the pre-1921 era had been dismantled, Carville inmates were not allowed to cohabit with spouses within Carville's grounds (Martin, 1953, Stein, 1973). This rule, like so many others, was finally reversed during the post-World War Two era, following the development of increasingly effective drug therapies and in the face of activism by Carville's inmates. Rather than having to "go through the hole in the fence" as Betty Martin and Harry Parker did in order to live as man and wife during the 1930s, couples were allowed to live together on the property. Influential in this regard was the widely publicized Hornbostel case, where a Colonel in the U.S. Army sought to follow his wife into Carville during 1946 (Martin, 1953, Stein, 1973).

During the post-World War Two era, Carville finally became a voluntary hospital. Outgoing mail stopped being sterilized (Martin, 1959, Stein, 1973). Disease treatment had progressed from the distasteful concoction of chaulmooga oil, hyperthermia, and vaccines of the 1920s and 1930s to the effective sulfones of the 1940s. Ensuing multiple drug therapies and the experimental use of laboratory animals such as the mouse and the nine-banded armadillo are much to Carville's credit (Moschella, 1997, Trautman *et al.*, 1981). Even so, the history of Carville and its inhabitants reflects the oppressive social history accorded people with leprosy.

A comparison of U.S. public policies directed toward leprosy and the closely related mycobacterial disease, tuberculosis, reveals striking differences. Carville as a national leprosarium had opened its doors in 1921 to 80 patients, with the total number of inmates averaging no more than 400-500 in a given year. During 1907, as plans for Carville were being promulgated by Congress and the Interstate Quarantine Legislation was being composed, an estimated 400 US citizens died of tuberculosis every day (Gallos, 1985). Theodore Roosevelt, President of the United States, accepted election to the presidency of the International Congress on Tuberculosis that convened on September 21st, 1908. His letter of acceptance forcefully defined a campaign against tuberculosis, 'The importance of the crusade against tuberculosis, in the interest of which this Congress convenes, cannot be overestimated when it is realized that tuberculosis costs our country two hundred thousand lives a year, and the entire world a million lives a year...The modern crusade against tuberculosis brings hope and bright prospects of recovery to hundreds and thousands of victims of the disease, who, under old teachings were abandoned to despair' (Moore *et al.*, 1908: 16). Thus, a "modern crusade" was launched against one mycobacterial disease while victims of another suffered incarceration reminiscent of Medieval leprosaria.

3. Tuberculosis in North America

Tuberculosis has a history either longer or much longer than leprosy in the New World (Allison *et al.*, 1973, Buikstra, 1981a, 1999, Buikstra and Williams, 1991, Salo *et al.*, 1994). Despite archaeological and genetic evidence for the disease in the Americas before European contact, sceptics cite a lack of animal reservoirs and the high prevalence of tuberculosis among Native Americans following the European entrada (Stead *et al.*, 1995). While the latter could reflect the response of populations to a novel infectious organism, the pattern is more easily attributable to the enforced lifestyle changes that accompanied colonization (Buikstra, 1981b). Whether a European introduction or earlier development, however, tuberculosis was clearly a widespread malady in North America at the time leprosy was introduced.

Thus, in contrast to the European situation of widespread leprosy upon introduction of tuberculosis, North American populations were suffering from endemic tuberculosis by the time leprosy was introduced. Following an hypothesis of cross-immunization against leprosy by tuberculosis, we would expect that there would be an initial high incidence of leprosy in the immigrant populations that introduced it, with very limited spread to surrounding groups. Further, over time the incidence of leprosy should decrease as the incidence of tuberculosis increased and *vice versa.*

4. Environmental Mycobacteria

As Leitman *et al.* (1997) emphasized, the role of environmental mycobacteria in epidemiological tests of the tuberculosis-leprosy cross-immunity model is potentially confounding. A high proportion of persons with low-grade tuberculin sensitivity has been seen in several studies, which has been construed to indicate infection with mycobacteria other than those of the *M. tuberculosis* complex (see, for example, Comstock *et al.,* 1974). Palmer and Edwards (1968) suggested that nonpathogenic environmental mycobacteria act as natural vaccines against tuberculosis by inducing low levels of tuberculin sensitivity. Fine *et al.* (1994) observed that, among test subjects in northern Malawi, lowest tuberculosis rates were found among people with low or intermediate levels of tuberculin sensitivity. Because BCG vaccination provides little or no protection against tuberculosis in this area, Fine *et al.*'s observation is consistent with Palmer and Edward's hypothesis of natural vaccination via environmental mycobacteria. Unfortunately, systematically collected comparative incidence data for the environmental mycobacteria for the period in question are not available.

4.1 Experimental evidence for mycobacterial cross-immunity

Evidence for mycobacterial cross-immunization arises from the ability of the *M. bovis* BCG and the *M. microti* vole bacillus vaccines to effectively protect against tuberculosis, and even leprosy, in both humans and other animals from certain parts of the world. Palmer and Long (1966), for example, found that infection with non-tuberculous mycobacteria has a weak vaccination effect in guinea pigs. In a 15 year controlled trial in England, both BCG and vole bacillus vaccines conferred substantial protection against all forms of tuberculosis (Hart *et al.*, 1967, Medical Research Council, 1972). Retrospective analysis of vaccination studies in various populations indicates that BCG effected protection to newborn and infant children. BCG does not protect against tuberculosis in all populations, however (Ten Dam and Hitze, 1980). Palmer and Edwards (1968) suggest that this may be due to sensitization of cross-reactive environmental mycobacteria, which Lowrie *et al.* (1997) hypothesize to prevent development of a protective response upon subsequent BCG vaccination. Comstock and colleagues (Comstock and Edwards, 1972, Comstock *et al.*, 1974, 1976) noted no or very little significant protective efficacy of BCG against tuberculosis in Puerto Rico or in the southern United States. Pönninghaus *et al.* (1992) found, for example, that BCG did not protect against tuberculosis in northern Malawi, although it did provide at least 50% protection against leprosy in the same populations.

Vaccination against leprosy began in the late 1930s when Fernandez (1939) noted that BCG injected into lepromin-negative children caused lepromin conversion. Several studies in the following years indicated that BCG could protect against leprosy in at least some populations (see, for example, Convit, 1956; Fernandez, 1955; Yanagisawa, 1958). Convit *et al.* (1992) found that in Venezuela, BCG vaccination is associated with 56% protective efficacy against leprosy, with additional protection conferred by repeated vaccinations. No significant differences were observed when a mixture of *autoclaved M. leprae* from nine-banded armadillos and BCG were used for the vaccinations. In a WHO-controlled trial in Burma, BCG vaccination offered overall protection against leprosy of around 20%, but this varied with age, sex, contact status of the child, and vaccine batch (Lwin *et al.*, 1985). Vaccination did not appear to have an effect on the form of leprosy that some children later contracted. In a case-control study in Myanmar, Bertolli *et al.* (1997) also observed increasing vaccine effectiveness with increasing dosage of BCG, as well as higher efficacy when first vaccination occurred during infancy. Much higher efficacy, 80%, was found in a long-term study of children exposed to leprosy in Uganda (Brown *et al.*, 1968; Stanley *et al.*, 1981), and age at vaccination had no effect on protection. In India, however, BCG has not been seen to protect against leprosy (Muliyil *et al.*, 1991), and may even induce mild forms of the disease in individuals with subclinical leprosy (Stoner *et al.*, 1981). In a long-term study of BCG, human-derived *M. leprae*, and mixed BCG-*M. leprae* vaccines, Chaudhury *et al.* (1994) found no difference between the three vaccines, which reduced overall incidence of leprosy from 13.7 to 7.8% among individuals with household contact with leprosy. The incidence of lepromatous leprosy was not affected, and the authors postulated that vaccination affected individuals at risk of developing borderline leprosy by shifting them toward the tuberculoid end of the spectrum; research by other workers in New Guinea and in India confirm this observation. Chaudhury *et al.* (1994) also find that spontaneous (natural) lepromin reactivity is lifelong, while BCG vaccination induces a weaker and transient reactivity. Further, the grades of lepromin reactivity were found to be more intense in naturally lepromin-positive individuals than in vaccine-induced lepromin-positive individuals. In a vaccination study which used BCG, *M. leprae*, and several non-tuberculous mycobacteria, Kwapinski *et al.* (1975) found that, while BCG increased late lepromin reactivity (and presumably resistance to leprosy), the other antigens (including one which contained *M. leprae*) depressed lepromin reactivity. In sum, experimental evidence indicates support for cross-immunization in some human groups in certain world areas.

5. Materials

To evaluate the cross-immunity model for leprosy and tuberculosis, we have chosen to focus upon incidence data for two data sets from 20th century Texas. As noted above, Texas and Louisiana present the highest leprosy endemicity of the continental United States (Hudson and Genesse, 1982). The Centers for Disease Control and Prevention (CDC 2000) reported that, in 1998, the incidence of tuberculosis in Texas was 9.2 cases per 100,000 population, the fourth highest rate in the United States. Figures 1 and 2 present the overall incidence of tuberculosis and leprosy for the entire state of Texas over the period 1940 to 1998 by decade. Although the statewide incidence in 1998 is fairly low, it is clear that the rates were much higher in 1940 and 1950 (45.33 and 55.44 cases per 100,000, respectively). Further, these are statewide averages. As discussed below, these statewide data are analyzed by county in order to examine changes in incidence in smaller regional populations, which might be expected to approach more closely population conditions in pre-industrial Europe. Incidence in a number of these counties was well in excess of 100 cases per 100,000 population, with some populations reaching rates as high as 300 or more cases per 100,000.

While the incidence of leprosy in the state overall was not high relative to the incidence of tuberculosis, in some counties it reached rates as high as 25 cases per 100,000 for some years. It should thus be the case that sufficient numbers of the population were exposed to either tuberculosis or leprosy for us to evaluate Chaussinand's hypothesis.

As a final note, we have chosen incidence rather than prevalence data because we are interested in the number of new cases over time rather than the total number of individuals affected with the disease. Prevalence data would be confounded by such factors as medical intervention or duration of illness. While we would have preferentially chosen to analyze data from Louisiana, the Louisiana State Department of Health reported to us that they have lost the relevant information concerning tuberculosis incidence (C. DeGraw, pers. comm. to AKW, March 1999). The difficulties we experienced in obtaining relevant data parallel those experienced by Chaussinand (1948: 435), who lamented the absence of data to support (*appuyer*) his theory.

5.1 Test 1: Texas Years 1938-1980

The first data set that we analyzed was obtained from the Texas Department of Health and contains tuberculosis and leprosy incidence data for the years 1938-1980. The lower cutoff of 1938 represents the earliest data available through the Texas Department of Health. The upper cutoff of 1980 was chosen to avoid the potentially confounding effects of HIV infection which has been shown to increase susceptibility to tuberculosis (Fertel and Pitchenik, 1989, Fitzgerald *et al.*, 1991). The data consist of the number of new cases of leprosy and tuberculosis per year for each Texas county. We calculated incidence by dividing the number of new cases by the recorded population size for that county and year. We therefore have three primary variables: weighted tuberculosis incidence (IncTB), weighted leprosy incidence (IncLEP), and year (Yr).

The county data allowed us to further define our study populations by examining changes in incidence in smaller, regional populations. In addition, each county was classified into one of two contrasting categories to investigate the potential effects of environmental mycobacteria. The comparison focused on differences in urban versus rural counties, inland versus coastal counties, and east versus west counties. Decisions on which group a county belonged to were based on statistics compiled by the U.S. Department of Commerce, Economics and Statistics Administration, Bureau of the Census. For our purposes, an urban environment was defined as any county designated as a Metropolitan Statistical Area with a major city with a population of 100,000+ individuals. Coastal counties were defined as those counties adjacent to the Gulf of Mexico from Cameron County in the south to Jefferson County in the north. The east/west division was based upon longitude with counties west of 100 degrees designated "west" and counties east of 99 degrees designated "east". Counties located between these longitudes were not coded for this category. This was done to avoid counties in transitional climate zones.

The 1938-1980 data set provides us with the ability to evaluate changes in tuberculosis and leprosy incidence, while avoiding the potentially confounding effects of HIV infection. This data set also provides considerable time depth to our analyses (42 years) which allows us to evaluate relatively longterm trends in the incidence data.

5.2 Test 2: Texas Years 1991-1998

To explore the possible confounding effects of population-specific disease response and immigration across the U.S.-Mexican border, we obtained a second data set from the Texas Department of Health. This data set contains tuberculosis and leprosy incidence data for the years 1991-1998. The lower cutoff date of 1991 represents the first year in which population affiliation was recorded by the Texas Department of Health. The upper limit of 1998 represents the extent of the data available at the inception of our research.

The 1991-1998 data set contains information on tuberculosis and leprosy incidence by year. Again, we calculated the weighted incidence data by dividing the numbers of cases by the county-specific population estimate for each year. As with the 1938-1980 data set, these data were further partitioned into urban versus rural, inland versus coastal, and east versus west based on the same criteria as above. In addition to being partitioned by county, this data set also contains data on ethnic/population group (African-American, Asian/Islander, Caucasian, Hispanic, Native American). This allows us to control for immigration of individuals from areas with high endemicity of tuberculosis and leprosy as well as potential population differences in host immunity. In exchange for this benefit we sacrifice time depth in our analyses as well as allowing for potential HIV effects.

6. Analytical methods

Our statistical procedure to test for cross-immunity is a simple correlation analysis on each of the two Texas data sets. Our analysis consists of calculating the Pearson correlation coefficient and testing whether it differs significantly from zero. We utilized the UNIX-based SAS statistical package to perform all analyses.

6.1 Tests and Expectations: Texas Years 1938-1980

This data set has three primary variables (IncTB, IncLep, Yr) and one grouping variable (county). Using the entire data set pooled across all counties, we calculated three correlations to examine overall Texas trends. The first two correlations examined how the incidence of each disease changed through time (IncTB*Yr; IncLep*Yr). We calculated these tests to determine if disease incidence exhibited opposing trends through time. Based upon the cross-immunization hypothesis, we expect the diseases to be correlated in an opposite manner with year. For example, if IncTB*Yr is positive (indicating an increasing incidence) the cross-immunization model predicts that IncLep*Yr will be negative.

The third correlation we calculated for the pooled Texas sample is that between the incidence data themselves (IncTB*IncLep). For the comparative incidence analyses (IncTB*IncLep), the cross-immunization hypothesis predicts that, all other factors being equal, all correlations will be negative. A negative correlation indicates that the incidence of one disease is inversely related to that of the other disease. Because these data are grouped by year, a negative correlation between leprosy and tuberculosis would indicate an inverse temporal trend for the diseases as predicted by the cross-immunization theory.

Although all three correlations are related to the temporal variable (year), each test measures different things. The direct correlations with year (IncTB*Yr, IncLep*Yr) measure the change in disease incidence over time for each disease independently. The interpretation of these correlations is post-hoc and we expect the two correlations to have opposite signs under the cross-immunity model. The third correlation directly compares the incidence figures (IncTB*IncLep). This test differs from the former two in that we are not examining trends in disease incidence independently, rather we are examining how disease incidence changed through time concordantly. This test is a formal hypothesis test of the cross-immunity model because the p-value associated with the correlation is directly related to the hypothesis.

We repeat the correlation analysis on the incidence data (IncTB*IncLep) using the county as the unit of analysis rather than the aggregated Texas sample. These analyses were performed to better define the study population as well as to help control for the potential effects of differences in environmental mycobacteria. As with the analysis of the aggregate Texas data set, we expect the correlations to be negative. For this series of analyses, we recorded the number of counties with positive and negative correlations and then considered only those counties with correlations that differ significantly from zero. Based upon the cross-immunity model we expect all correlations to be negative. These data were then examined for patterns in terms of the three environmental factors defined above. Overall, we calculated 84 correlations for the 1938-1980 data set.

6.2 Tests and Expectations: Texas Years 1991-1998

This data set has three primary variables (IncTB, IncLep, Yr) and two grouping variables (county and ethnic group). We repeated all the analyses discussed above for the 1991-1998 data set. The one difference, however, was the ability to differentiate the study populations by ethnic group. This allowed us to evaluate and partially control the effects of differential host response and immigration across the U.S.-Mexico border, something not possible with the 1938-1980 data set.

We first consider the correlations with year (IncTB*Yr, IncLep*Yr) where we again expect the disease incidence correlations to be of opposite signs. We calculated an aggregate (all of Texas) correlation in addition to the population-specific correlations (Caucasian, African-American, Hispanic, Asian/Islander, Native American). Any differences in significance could then be interpreted in terms of host response and immigration.

We also calculated the correlations directly between disease incidence figures (IncTB*IncLep). Again, we calculated an aggregate group in addition to the population-specific correlations. The expectations for these analyses were that all correlations would be negative, suggesting the increasing incidence of one disease at the expense of the other.

As with the 1938-1980 data set, we recalculated the incidence correlations (IncTB*IncLep) using each county as the unit of analysis. We repeated the correlations for an aggregate group in addition to the population-specific data. For each county, therefore, we recorded an IncTB*IncLep correlation for the entire county, Caucasian residents, African-American residents, Hispanic residents and, where the sample sizes allowed, for the Asian/Islander and Native American residents. For each grouping we recorded the number of positive and negative correlations in addition to emphasizing only those that differed significantly from zero. This allowed us to compare population-specific differences in disease incidence which could then be related to historical or genetic factors of interest. Overall, we calculated 160 correlations for the 1991-1998 data set.

7. Results

7.1 Results: Texas 1938-1980

7.1.1 Temporal analyses

If we consider the data for the entire state of Texas for the period 1938-1980, we find limited temporal evidence for the cross-immunization hypothesis. The correlation for IncLep*Yr is small though significant ($R=.038$, $p=.0001$). Similarly, the correlation for IncTB*Yr is also significant ($r=-.181$, $p=.0001$) though, again, the magnitude of the correlation is quite small. When considered together, the correlations suggest that, as the number of new tuberculosis cases decreased through time, new leprosy cases increased. The trend is slight but significant, and therefore offers modest support for the cross-immunization hypothesis. Given the theoretical perspectives developed by Chaussinand and Manchester, this is the less likely alternative. It should be noted, however, that the significance of the correlation coefficients is a factor of our large sample ($n \sim 10{,}900$) which ensures high statistical power. The value of the actual correlations is quite low.

7.1.2 Disease incidence

Another method to evaluate the cross-immunization hypothesis is to calculate the correlation between the disease incidence (IncTB*IncLep). If we consider all of Texas, the results indicate a small, positive, significant correlation for IncTB*IncLep ($r=.021$, $p=.027$). These results are contrary to the temporal analyses (IncLep, IncTB*Yr) and suggest that as the incidence of one disease increases so does the other, a finding contrary to the expectations of the cross-immunization hypothesis.

7.1.3 County-specific analyses

The county-specific correlations for IncTB*IncLep do not support the expectations of the cross-immunization hypothesis. A total of 21 counties exhibited a positive correlation while 60 exhibited a negative correlation. At first, these data seem to support the expectations of the cross-immunization model, but only three correlation coefficients reach statistical significance. The three significant correlations are all positive and large in magnitude. These counties are Bexar ($r=.407$, $p=.007$), Crockett ($r=.414$, $p=.006$), and Gregg ($r=.783$, $p=.0001$). If we consider the environmental context for these counties we find that all were inland counties, and two of the three were counties located in Eastern Texas (Bexar and Gregg). Besides a preference for inland counties (a result likely resulting from small sample sizes) the environmental factors are unremarkable. Therefore, the county-specific incidence data lend little support to the cross-immunization hypothesis.

7.2 Results: Texas 1991-1998

7.2.1 Temporal analyses

The results of the temporal analyses are presented in Table 1, sorted by population group. Despite the fact that the Caucasian and aggregate samples were correlated with year as predicted by the cross-immunization hypothesis, none of these values differ significantly from zero. The temporal correlations for the African-American, Hispanic, and Asian/Islander populations are all negative, suggesting a declining incidence as we approach the present. None of these correlations, however, differ significantly from zero. The Native American sample lacked sufficient leprosy incidence data to evaluate any trends. Interestingly, the Native American group did exhibit the largest positive correlation for IncTB*Yr for any group (r=.211), suggesting an increasing incidence of tuberculosis in this population.

7.2.2 Disease incidence

The correlations for IncTB*IncLep are presented in Table 1, sorted by population group. As predicted by the cross-immunization model, four of the five groups for which we have data exhibit a negative correlation suggesting an inverse relationship between tuberculosis and leprosy incidence. None of these results, however, are statistically significant and all correlation coefficients are very small. The Caucasian population had an insignificant positive correlation that was likewise very small. These data provide no support for the cross-immunization hypothesis.

7.2.3 County-specific results

As an additional measure to help control for environmental effects, we recalculated the IncTB*IncLep correlations for all individual Texas counties, again sorted by population group. The county-specific analyses helped define the study population and allowed us to examine the incidence data on a smaller, regional level. For example, our study sample, previously defined as all of Texas, could now be specified as Hispanic residents of Bexar County, for example. We were also able to again evaluate our three environmental factors (urban/rural, coast/inland, east/west) by considering the pattern of significant county-specific correlations. The results for the IncTB*IncLep correlations are presented in Table 2. As a whole, these data provide little support for the cross-immunity hypothesis. Where significant correlations were recorded, 16 of 16 were positively correlated. This suggests that an increase in tuberculosis incidence was often accompanied by an increase in leprosy incidence. Population-specific data by county is discussed further below. The Asian/Islander and Native American population groups were omitted from this discussion due to small sample sizes.

7.2.4 Aggregate population

The aggregate population sample exhibits seven counties with significant positive correlations (Atacosa, Dallas, Denton, El Paso, Fort Bend, Nueces, and Tarrant) and zero counties with significant negative correlations. The average significant correlation was .423 (sd=.091) indicating the correlations were relatively large with little variation. Of the significant IncTB*IncLep correlations, seven of seven were coastal, two of seven were urban, and six of seven were eastern.

7.2.5 Caucasian population

The Caucasian population exhibited seven significant positive correlations by county (Atacosa, Dallas, Denton, El Paso, Fort Bend, Harris, and Tarrant) and zero significant negative correlations for the incidence data. The average significant correlation for the incidence data was .412 (sd=.107). Of the significant IncTB*IncLep comparisons, seven of seven were coastal, zero of seven were urban, and six of seven were eastern.

7.2.6 African-American Population

The African-American population exhibited no significant positive or negative correlations for the IncTB*IncLep comparisons.

7.2.7 Hispanic population

The Hispanic population exhibited two significant positive correlations (Atacosa, Fort Bend) and zero significant negative correlations. The average significant correlation was .693, which is very large. Both counties are coastal and eastern, however, and Atacosa County is urban and Fort Bend rural.

8. Summary of results

We have presented the results of multiple tests designed to evaluate the cross-immunization hypothesis. We expected to find negative correlations in incidence figures for leprosy and tuberculosis over time. For example, the correlations for IncTB*Yr should be negative while the correlations for IncLep*Yr are positive. This would indicate that while we approach the present, the incidence of one disease increases and the other decreases. Our results, however, do not support these expectations. Rather we find few significant differences in disease incidence through time, a result not being driven by lack of sample size and low statistical power. We therefore find this negative evidence compelling and contrary to the cross-immunization hypothesis.

In addition to the time-sensitive portion of our analysis, we also directly examined the relationship between disease incidence figures directly. In these tests, the cross-immunization theory predicts that all correlations are negative, as one disease incidence increases and the other decreases. Our results do not support this expectation. Rather, we found that only positive correlations were statistically different from zero, a result that suggests the increased incidence of one disease was paralleled by a corresponding increased incidence of the other. Although we recognize the potential effects of confounding variables in our analysis, these results actually contradict the cross-immunity hypothesis. In addition, our results remain robust for the pre-HIV era (1938-1980), controlling for immigration (1991-1998 data set), and controlling for the effects of environmental mycobacteria by partitioning our data set by county (1938-1980 and 1991-1998 data sets).

9. Conclusions

Although theoretical work by numerous researchers indicates that mycobacterial cross-immunity could have caused the decline of leprosy in Medieval Europe following the spread of tuberculosis, this hypothesis is difficult to test. Vaccination studies have provided suggestive but conflicting results, which are most likely due to the presence of several confounding factors such as population-specific host susceptibility, the environmental mycobacterial species present in affected areas, and types and perhaps even virulence of the mycobacterial strains involved in disease outbreaks. In this epidemiological study, we attempted to control for at least these first two potential confounding variables by stratifying the analyses by ethnicity/population and by rural/urban, coast/non-coast, and east/west.

Our own preliminary analysis of incidence data for leprosy and tuberculosis across Texas offers possible though not conclusive evidence against a model of cross-immunity. In general, we see that an increase in tuberculosis incidence was often accompanied by an increase in leprosy incidence. The one time-transgressive negative incidence correlation that achieves statistical significance runs counter to expectations. Our results may, of course, be confounded by a number of factors, including a lack of diagnostic accuracy, medical intervention, underreporting and other forms of reporting biases. Greater control of these variables and an ability to focus upon endemic regions of Louisiana could provide a more rigorous test of Chaussinand's and Manchester's cross-immunity model.

Acknowledgements

Data from Texas counties was provided by Virginia Enriquez, Peggy Perkins, and Gail Shevik of the Texas Department of Health. This work was supported in part by a National Science Foundation Graduate Research Fellowship (AKW).

References

Allison, M.J., Mendoza, D. and Pezzia, A. 1973 Documentation of a case of tuberculosis in pre-Columbian America. *Am. Rev. Resp. Dis.* 107: 985-991.

Andersen, J.G. 1969 *Studies in the Mediaeval diagnosis of leprosy in Denmark.* Copenhagen: Costers Bogtrykkeri.

Bertolli, J., Pangi, C., Frerichs, R. and Halloran, M.E. 1997 A case-control study of the effectiveness of BCG vaccine for preventing leprosy in Yangon, Myanmar. *International Journal of Epidemiology* 26: 888-896.

Brown, Y.A.K., Stone, M.M. and Sutherland, I. 1968 BCG vaccination of children against leprosy: results at end of second follow-up. *British Medical Journal* 1: 24-27.

Buikstra, J.E. (ed) 1981a *Prehistoric tuberculosis in the Americas.* Chicago: Northwestern University Archeological Program. Scientific Papers, Number 5.

Buikstra, J.E. 1981b An historical introduction. In J.E. Buikstra (ed): *Prehistoric Tuberculosis in the Americas.* Chicago: Northwestern University Archeological Program. Scientific Papers, Number 5, pp. 1-23.

Buikstra, J.E. 1999 Paleoepidemiology of tuberculosis in the Americas. In G. Pálfi, O. Dutour, J. Deák, I. Hutás (eds): *Tuberculosis Past and Present.* Szeged, Hungary: Tuberculosis Foundation and Budapest, Hungary: Golden Book Publisher Ltd., pp. 479-494.

Buikstra, J.E. and Williams, S. 1991 Tuberculosis in the Americas: current perspectives. In D.J. Ortner and A.C. Aufderheide (eds): *Human paleopathology: current syntheses and future options.* Washington D.C.: Smithsonian Institution Press, pp. 161-172.

Chaudhury, S., Hazra, S.K., Saha, B., Mazumder, B., Biswas, P.C., Chattopadhya, D. and Saha, K. 1994 An eight-year field trial on antileprosy vaccines among high-risk household contacts in Calcutta metropolis. *International Journal of Leprosy* 62: 389-394.

Chaussinand, R. 1944 Tuberculose et leprè, maladies antagoniques. Eviction de la lepré par la tuberculose. *R. Med. Fr. d'Extreme-Orient* 22: 677-685.

Chaussinand, R. 1948 Tuberculose et leprè maladies antagoniques: eviction de la leprè par la tuberculose. *International Journal of Leprosy* 16: 431-438.

Chaussinand, R. 1953. Tuberculosis and leprosy: mutually antagonistic diseases. *Leprosy Rev.* 24: 90-94.

Chaussinand, R. 1964 Quelques remarques concernant la theorie de l'antagonism entre tuberculose e leprè. *Acta Trop. Basel* 21: 82-87.

Comstock, G.W. and Edwards, P.Q. 1972 An American view of BCG vaccination, illustrated by results of a controlled trial in Puerto Rico. *Scandinavian Journal of Respiratory Disease* 53: 207-217.

Comstock, G.W., Livesay, V.T. and Woolpert, S.F. 1974 Evaluation of BCG vaccination among Puerto Rican children. *American Journal of Public Health* 64: 283-291.

Comstock, G.W., Woolpert, S.F. and Livesay, V.T. 1976 Tuberculosis studies in Muscogee County, Georgia. *Public Health Reports* 91: 276-280.

Convit, J. 1956 Studies of leprosy in the German ethnic group of Colonia Tovar, Venezuela V. The morbidity rates in BCG-vaccinated and unvaccinated groups. *International Journal of Leprosy* 24: 269-274.

Convit, J., Sampson, C., Zuniga, M., Smith, P.G., Plata, J., Silva, J., Molina, J., Pinardi, M.J., Bloom, B.R. and Salgado, A. 1992 Immunoprophylactic trial with combined Mycobacterium leprae/BCG vaccine against leprosy: preliminary results. *Lancet* 339: 446-450.

Denney, O.E. 1926 The leprosy problem in the United States. *Reprint No. 1080 Public Health Reports* May 14th: 923-929.

Faget, G.H. 1946. The story of the national leprosarium. The United States Marine Hospital Carville, La. Reprint No. 2764. *Public Health Reports* 61: 1871-1883.

Feldman, R.A. and Sturdivant, M. 1975 Leprosy in Louisiana, 1855-1970: an epidemiologic study of long-term trends. *Amer. J. Epidemiol.* 102: 303-310.

Fernandez, J.M.M. 1939 [Comparative study of the Mitsude reaction with the tuberculin reaction]. *Rev. Argent. Dermatosifilis* 23: 425-453 (in Spanish).

Fernandez, J.M.M. 1955 Influence of the tuberculosis factor on the clinical and immunological evolution of child contacts with leprosy patients. *International Journal of Leprosy* 23: 243-258.

Fernandez, J.M.M. 1957 Leprosy and tuberculosis: antagonistic diseases. *A.M.A. Arch. Dermat.* 75: 101-106.

Fertel, D. and Pitchenik, A.E. 1989 Tuberculosis in acquired immune deficiency syndrome. *Semin. Repir. Infect.* 4: 198-205.

Fine, P.E.M., Sterne, J.A.C., Ponnighaus, J.M. and Rees, R.J.W. 1994 Delayed-type hypersensitivity, mycobacterial vaccines, and protective immunity. *Lancet* 344: 1245-1249.

Fitzgerald, J.M., Gryzbowski, S. and Allen, E.A. 1991 The impact of human immunodeficiency virus on tuberculosis and its control. *Chest* 148: 191-200.

Gallos, P.L. 1985 *Cure cottages of Saranac Lake: architecture and history of a pioneer health resort.* Saranac Lake, New York: Historic Saranac Lake.

Gussow, Z. 1979 Notes on the history of leprosy in Louisiana. *Soc. Med. J.* 72: 600-604.

Hart, PD., Sutherland, I. and Thomas, J. 1967 The immunity conferred by effective BCG and vole bacillus vaccines, in relation to individual variations in induced tuberculin sensitivity and to technical variations in the vaccines. *Tubercle, Lond.* 48: 201.

Hudson, T. and Genesse, J. 1982 Hansen's disease in the United States. *Soc. Sci. Med.* 16: 997-1004.

Hunter, J.M. and Thomas, M.O. 1984 Hypothesis of leprosy, tuberculosis, and urbanization in Africa. *Soc. Sci. Med.* 19: 27-57.

Klingmuller, V. 1930 *Die Lepra.* Handb. d. Haut- und Geschiechtskt, Bd. X. Berlin.

Kwapinski, J.B.G., Bechelli, L.M., Haddad, N. and Simao, E.T. 1975 Impairment of reactivity to lepromin by mycobacterial antigens related to, or identical with, *Mycobacterium leprae. Canadian Journal of Microbiology* 21: 896-901.

Leitman, T., Porco, T. and Blower, S. 1997 Leprosy and tuberculosis: the epidemiological consequences of cross-immunity. *American Journal of Public Health* 87: 1923-1927.

Lowrie, D.B., Silva, C.L., Colston, M.J., Ragno, S. and Tascon, R.E. 1997 Protection against tuberculosis by a plasmid DNA vaccine. *Vaccine* 15: 834-838.

Lwin, K., Sundaresan, T., Gyi, M.M., Bechelli, L.M., Tamondong, C., Gallego Garbojosa, P., Sansaricq, H. and Noordeen, S.K. 1985 BCG vaccination of children against leprosy: fourteen-year findings of the trial in Burma. *Bulletin of the World Health Organization* 63: 1069-1078.

Manchester, K. 1984. Tuberculosis and leprosy in antiquity: an interpretation. *Medical History* 28: 162-173.

Manchester, K. 1986 *Tuberculosis: an evolutionary model.* Proceedings of the Paleopathology Association 6th European Meeting, Madrid, pp. 189-198.

Manchester, K. 1991 Tuberculosis and leprosy: evidence for interaction of disease. In D.J. Ortner and A.C. Aufderheide (eds): *Human paleopathology: current syntheses and future options.* Washington D.C.: Smithsonian Institution Press, pp. 23-35.

Manchester, K. 1992 Leprosy: the origin and development of the disease in antiquity. In D.F. Gourevitch (ed.): *Maladie et maladies: historie et conceptualisation.* Geneve: Libraire Droz S.A., pp. 31-49.

Manchester, K. and Roberts, C. 1989 The palaeopathology of leprosy in Britain: a review. *World Archaeology* 21: 265-272.

Martin, B. 1953 *Miracle at Carville.* Garden City, NY: Doubleday and Company.

Martin, B. 1959. *No one must ever know.* Garden City, N.Y: Doubleday and Company.

Medical Research Council (Vaccines Clinical Trials Committee) 1972 BCG and vole bacillus vaccines in the prevention of tuberculosis in adolescence and early adult life. *Bulletin of the World Health Organization* 46: 371-385.

Møller-Christensen V. 1953 Location and excavation of the mediaeval leper church yard at Næstved in Denmark. *International Journal of Leprosy* 21: 616-617.

Møller-Christensen, V. 1969 The history of syphilis and leprosy - an osteological approach. *Abbotempo* 1: 20-25.

Moore, H., Strouse, I.R., Babcock, J.N., Holman, W.S. and Bennett, B.F. 1908 *State of Indiana: Report of the State Tuberculosis Hospital Commission 1907-1908.* Indianapolis: Wm. B. Burford.

Moschella, S.L. 1997 Carville: 100 years of excellence. *J. Amer. Acad. Dermatol.* 36: 126-130.

Muliyil, J., Nelson, K.E. and Diamond, E.L. 1991 Effect of

BCG on the risk of leprosy in an endemic area: a case control study. *International Journal of Leprosy* 59: 229-236.

Palmer, C.G. and Edwards, L.B. 1968 Identifying the tuberculous infection. The dual-test technique. *Journal of the American Medical Association* 205: 117-119.

Palmer, C.E. and Long, M.W. 1966 Effects of infection with atypical mycobacteria on BCG vaccination and tuberculosis. *American Review of Respiratory Disease* 94: 553-568.

Pönnighaus, J.M., Fine, P.E.M., Sterne, J.A.C., Wilson, R.J., Msosa, E., Gruer, P.J.K., Jenkins, P.A., Lucas, S.B., Liomba, N.G. and Bliss L. 1992 Efficacy of BCG vaccine against leprosy and tuberculosis in northern Malawi. *Lancet* 339: 636-639.

Roberts C. 1986 Leprosy and leprosaria in Medieval Britain. *M.A.S.C.A. Journal* 4: 15-21.

Roberts, C. and Manchester, K. 1995 *The archaeology of disease.* Second edition. Ithaca, New York: Cornell University Press.

Rogers, L. 1924 The epidemiology of leprosy. *Ann. Trop. Med. Parasitol.* 18: 267-322.

Salo, W.L., Aufderheide, A., Buikstra, J. and Holcomb, T.A. 1994 Identification of *Mycobacterium tuberculosis* DNA in a pre-Columbian mummy. *Proc. Natl. Acad. Sci. U.S.A.* 91: 2091-2094.

Stanley, S.J., Howland, C., Stone, M.M. and Sutherland, I. 1981 BCG vaccination of children against leprosy in Uganda: final results. *Journal of Hygiene, Cambridge* 87: 233-248.

Stead, W.W., Eisenach, K.D., Cave, M.D., Beggs, M.L., Templeton, G.L., Thoen, C.O. and Bates, J.H. 1995 When did *Mycobacterium tuberculosis* infection first occur in the New World? *American Journal of Respiratory and Critical Care Medicine* 151: 1267-1268.

Stein, S. 1973 *Alone no longer.* Carville, Louisiana: The Star.

Stoner, G.L., Belehu, A., Nsibambi, J. and Warndorff, J. 1981 Borderline tuberculoid leprosy following BCG vaccination; a case report. *International Journal of Leprosy* 49: 16-20.

Ten Dam, H.G. and Hitze, K.L. 1980 Does BCG vaccination protect the newborn and young infants? *Bulletin of the World Health Organization* 58: 37-41.

Trautman, J.R., Kirchheimer, W.F., Prabhakaran, K., Hastings, R.C., Shannon, E.J., Jacobson, R.R. and Brand, P.W. 1981 An overview of Carville research. *Acta Leprol.* 84: 1-29.

White, C. 1996 Physical and metaphorical confinement: the experience of Hansen's disease (leprosy) in the United States. *Human Mosaic* 30: 12-28.

Yanagisawa, K. 1958 The effect of BCG vaccination upon the occurrence of leprosy in nursery children. *International Journal of Leprosy* 26: 325-327.

Table 1: Pearson Correlation Coefficients by Population Group

Group	Correlation	Pearson r	p-Value
All	IncLep*Yr	-.011	.5438
	IncTB*Yr	.004	.8390
	IncTB*IncLep	-.008	.6548
Caucasian	IncLep*Yr	.052	.0991
	IncTB*Yr	-.041	.1837
	IncTB*IncLep	.034	.2698
African-American	IncLep*Yr	-.043	.2915
	IncTB*Yr	-.001	.9796
	IncTB*IncLep	-.008	.8434
Hispanic	IncLep*Yr	-.051	.0962
	IncTB*Yr	-.012	.6898
	IncTB*IncLep	-.043	.1607
Asian/Islander	IncLep*Yr	-.061	.2995
	IncTB*Yr	-.058	.3215
	IncTB*IncLep	-.047	.4218
Native American	IncLep*Yr	------	------
	IncTB*Yr	.211	.1697
	IncTB*IncLep	------	------

Table 2: Number of Counties with Positive and Negative Incidence Correlations

Group	IncTB*IncLep		
	+	-	# significant
Caucasian	19	30	7 positive
African American	6	2	0
Hispanic	18	9	2 positive
Asian	8	3	0
Aggregate	19	30	7 positive

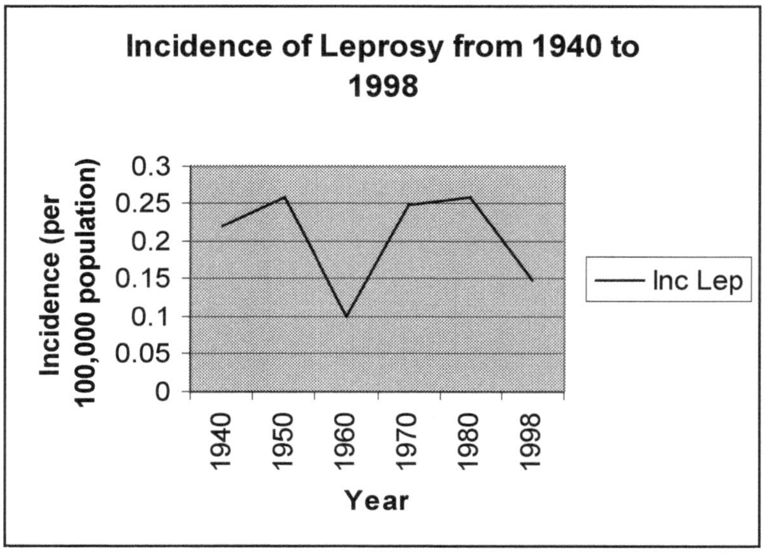

Figure 1: Incidence of leprosy from 1940-1998

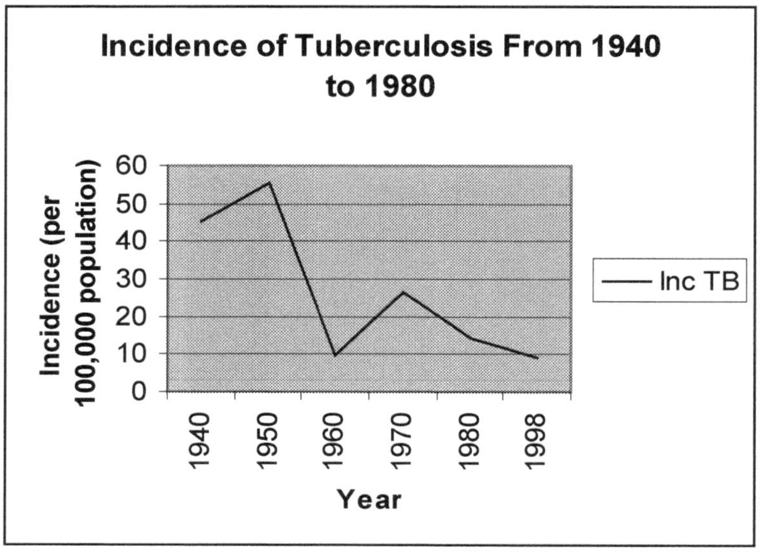

Figure 2: Incidence of tuberculosis from 1940-1980

New evidence for the history of leprosy in the Ancient Near East: an overview

Joe Zias
Science and Antiquity Group
The Hebrew University of Jerusalem
PO Box 37209
Jerusalem
Israel

Email: zias@inter.net.il

Keywords

Near East, leprosy, skeletal evidence

1. Introduction

"Ask not what disease the person has, but what person the disease has"

The above phrase, attributed to Hippocrates, the father of medicine, perhaps best describes the disease leprosy, or what today is medically referred to as Hansen's Disease. There are few infectious diseases in the modern world which correlate so closely with poverty as leprosy does. As much of what we witness today mimics the social conditions of antiquity, the disease which today affects an estimated 800,000 is a sad reminder of our inability to cope over the millennia with conditions of extreme poverty, and the degrading conditions which contribute to and foster this disease.

2. Leprosy: the disease

Leprosy is a disease caused by chronic infection caused by *Mycobacterium leprae*. The non-heritable disease involves principally the superficial peripheral nerves and skin, while the clinical features vary, and are determined by the "vigor" of the host's defensive reaction to the infection. Though highly infectious, particularly for young children, only a minority of those infected will actually acquire the disease clinically. For adults, the chances of being infected clinically are even less; by 1952 over 100 attempts to infect volunteers in prison with live bacteria were attempted, all having failed (Lendrum 1952). In 1975 Stanley Browne, one of the worlds leading experts on the disease wrote that 'if existing knowledge about leprosy were conscientiously and persistently applied, the disease could be controlled in our generation and eradicated in the next'. Despite these sentiments, expressed one generation ago, the disease still rages in many parts of the world, with an estimated 790,000 new cases discovered in 1997 alone (WHO). While the WHO estimated that 10-12 million persons suffered from the disease in the 1980s, Multi-Drug Therapy (MDT) instituted in the 1980s now renders nearly all the *M. leprae* bacilli non-viable within 48 hours and cures patients in 6-24 months time. Thus, the patient of the past, who was a leprosy sufferer forever, is now curable. Of those 10-12 million who were suffering from the disease, leprosy control clinics cured over 10 million cases between 1982 and 1998. As to whether or not Browne's 1975 prediction will eventually come true, only time will tell.

3. History of the disease in the Ancient Near East

Preuss (1911) wrote that there were as many articles on the subject of leprosy as there were letters in the Bible. Today, while there is still an enormous amount of written material appearing on the subject by palaeopathologists, physicians and medical historians, what may be one of the earliest cases ever recorded has remained largely unnoticed by colleagues. Angel (1974:256), the noted physical anthropologist from the Smithsonian Institution in Washington DC, published a find from a site in Anatolia (Karatas) in which he reported finding one adult male infected with bilateral spindling atrophy of the distal metatarsal shafts, suggestive of leprosy. Angel and other scientists apparently overlooked the importance of this discovery, as it is not reported in any of the literature dealing with the history of the disease by later researchers. The early date of this specimen 2,700-2,300 BC temporally coincides, though somewhat earlier, with an article by Wilson (1982) who argues from ancient texts that leprosy existed and was described in Babylonia in the Old Babylonian period. (2,000-1600BC). Rastogi and Rastogi (1984) believed that the disease can be found in the ancient Vedas manuscripts of India as early as 1,500 BC, while Manchester (1992) wrote that convincing literary evidence for the disease certainly existed in India as early as 600 BC. While there is no East Asian skeletal evidence due in large part to cremating the dead, these 6th century BC descriptions are probably medically accurate inasmuch as they prescribed chaulmooga oil for relief which, medically effective to a degree, was widely used until 1943 when the treatment was replaced by sulfones (see Oommen, this volume for a discussion on the use of chaulmoogra oil).

Preuss (1978), in his research on Biblical and Talmudic medicine devotes several passages to the subject of medical amputation of diseased limbs in antiquity, during the era of the Jewish Temple spanning seven centuries (1,000 BC-70AD). Preuss specifically pointed out that these medical procedures were carried out by physicians *(ibid. 12)* not only on the leprous but others which suggests that the disease was not only present in the Ancient Near East (ANE) but well known to the Hebraic medical profession (Kerithoth 3:7). Unfortunately, for our purposes, whereas early medical texts of the Ancient Egyptians (2,000 BC) and Greeks (500 BC) exist, there are no ancient Hebrew medical texts earlier than the third century AD in which to accurately judge the extent and meaning of these pathological conditions leading to eventual amputation.

Both the authors of Chronicles II 26:23 and Josephus (Ant. IX, 10,4) report that the Old Testament Judean King Uzziah (808-739 BC), believed to have been suffering from leprosy, was buried in a separate cemetery away from other graves. According to Jewish tradition it was forbidden to displace the graves of the House of David. Therefore, this suggests that Huzziah, because of his dermatological condition, was not buried along with his fellow Jews, but in a special field of burial which belonged to the ancient Kings of Israel (Enc. Judaica. vol. 44). This specific mention of a King being buried separately from his people is unique in terms of Jewish burial practices. Therefore, despite the extensive archaeological excavation in modern day Israel, the absence of human skeletal remains presenting clinical signs of Hansen's disease from the archaeological record, prior to the Byzantine period (4th-7th centuries AD) is not surprising. Thus, disease epidemiology based upon evidence from human skeletal remains in the archaeological record of the ANE, particularly in Syria and Palestine, will always be plagued by this problem. Therefore, unlike many European countries, literary texts may invariably take precedence over the osteoarchaeological record, even though the latter is the primary evidence.

By the fifth century BC true leprosy, as opposed to a bewildering constellation of dermatological conditions specifically unrelated to Hansen's disease was, according to Andersen (1969), eventually recognized by Greek physicians under the term *elephas* or *elephantiasis Graecorum.* However, it was not until 1300 years later that the Syria-Palestinian monk, John of Damascus (777-857), used the word "lepra" for the disease which we now know as Hansen's disease (Andersen, 1969:48).

4. Biblical evidence for leprosy

Of all the Biblical diseases described, none is so widely known or misunderstood as the dermatological conditions which are listed in Leviticus:13-14 under the term "tsara'ath". The confusion surrounding the exact meaning of the Hebrew term and its Greek translation "lepra" from a historical, religious and medical perspective has generated a vast literature over the millennia. Despite this enormous literary effort, the exact meaning of the term is today still shrouded in confusion with numerous attempts to clarify the exact meaning of this skin condition. "Tsara'ath", the Hebraic term for Biblical leprosy, like all other diseases in the Old Testament, was considered a divinely ordained punishment for sin, and the bible makes it clear what sins are involved: slander, tale-bearing, infringing on the rights of ones neighbour and fraud, all of which are crimes committed by one against one's neighbour (Feldman, 1966:988).

The authors of Leviticus devoted an enormous amount of space to this condition, which was primarily intended to provide the priestly class with the diagnostic skills in order to recognize the dermatological conditions which fell under the generic umbrella of "tsara'ath". The ambiguity surrounding this vague term has prompted some scholars to suggest that it's vagueness was intentional, in that it gave wide latitude to the priestly class who bore the enormous burden of interpreting these skin conditions (Gramberg *et al.,* 1961:130). Diagnosing dermatological conditions, with their bewildering variety of causes, is even today difficult. Therefore, this assumption by Gramberg, as to the reasons for the vagueness would appear correct. In fact, one of the most remarkable things about leprosy is the enormously wide variation in the way the disease expresses itself in different people (Jopling and McDougall, 1988:22). Secondly, one must be cognizant of the fact that the Old Testament was never intended to be a medical text, but rather a collection of myths, history, theology and law devoted to a specific ethnic group, the Israelites.

Whatever the reason for the vagueness, there exists today a consensus of opinion among medical historians of all three monotheistic faiths that the Biblical term "tsara'ath" or it's Greek translation in the Septuagint "lepra" have no relation to Hansen's disease, the term used today for modern day leprosy (Cochrane, 1963, Andersen, 1969, Hulse, 1975). Rather, these terms are generic and non-scientific in the sense that they pertain to a variety of fungal and bacterial skin diseases such as psoriasis, eczema, favors and pityriasis which were prevalent in the Eastern Mediterranean during the first millennium BC.

The complete absence of any mention of sensory loss, particularly in the main extremities, which is one of the main features of the disease is central to the argument against linking true leprosy with Biblical "tsara'ath". Moreover, the Levitical disease as described in chapters 13 and 14 of Leviticus was a temporary condition which placed the individual in a state of cultic impurity devoid of any medical significance. Thus, in a Biblical sense, the flaking of skin was recognized as a bodily discharge similar to a woman in menstruation, who was to be avoided at all costs because of spiritual contamination. Other diagnostic criteria associated with true leprosy which would have been clearly visible to the priest such as hoarseness of the voice, swelling of the face and paralysis of the hands and feet were never equated with the biblical disease "tsara'ath". (Gramberg *et al.*, 1961:129) Those diagnostic criteria associated with the disease in Leviticus 13-14 make it clear that modern day leprosy can definitely be excluded from the list of possible disease subsumed under the generic term "tsara'ath".

Centuries later, during the post-Biblical period, Rabbinical literature also mirrored this seemingly obsessive concern with the disease, with the Talmud alone devoting 14 chapters to the disease, whereas in the Mishna one finds nine chapters devoted to the laws of "tsara'ath". Much has been written on the subject from the view of Levicitus, particularly chapters 13 and 14, which medically a very problematical. There are other chapters which suggest, via inference, that the disease of true leprosy is discussed, though indirectly. If, for example, one takes the following passage from Leviticus (Chapter 21:18) which is ignored by most scholars:

'for whatsoever man he be hath a blemish, he shall not approach: a blind man, or a lame, or he that hath a flat nose...'

one can see that the crippling effects of leprosy, the blindness which will affect many of its sufferers, and the flattened nose which is a result of *facies leprosa,* clinically describes those individuals suffering from Hansen's disease. Therefore, although the clinical syndrome which presents these features is not named, the wording in the narrative certainly suggests true leprosy. Many later scholars have glossed over this passage preferring to concentrate on "tsara'ath" which, affecting the skin as well as walls and clothing, clearly does not refer to Hansen's disease whereas, in the opinion of the author and that of others, the subject in Leviticus (Chapter 21:18), which the authors of the bible are referring to is probably leprosy (Brim, 1936, Preuss, 1911). Interestingly enough, the Danish leprologist and physician Johs Andersen, who has compiled the most extensive and thorough scholarly work on the subject to date, mentions Brim's (1936) work. However, for some reason he omits from his analysis the passage from Leviticus (Chapter 21) in his treatise, as do a host of other researchers.

While there is general unanimity among all scholars that the Levitical descriptions of "tsara'ath/lepra" do not pertain to modern day leprosy, there is a lack of consensus on whether or not the leprosy mentioned in the New Testament (seven times) is actually related to modern day leprosy. Hulse (1975), for example, argues that the authors of the New Testament continue to use the term "lepra" in its Old Testament meaning, that is as a generic non-scientific term, although the disease existed in the New Testament period, as true leprosy was mentioned by Pliny the Elder (AD 23-79) Celsus (25 BC-37AD), Galen (130-201), and Aretaios (200 AD). Andersen (1969:15) disagrees, believing that, as the Gospel of Luke was reputed to have been written by a physician, Luke would have been aware of the distinction between "tsara'ath /lepra" and "elephantiasis". Had Luke used the term "elephas" or "elephantiasis" when describing those individuals cleansed, one could argue that he believed that the New Testament was in fact referring to today's true leprosy. Other authors (Browne, 1979), however, suggest that since the disease was known in the Alexandrian and Greek medical schools, albeit under the term "elephas", true leprosy probably existed in Syria and Palestine and was referred to in the New Testament under the generic term "lepra"; however, physicians, being more exact in their terminology, used the word "elephas" or "elephantiasias" when referring to the disease. Thus, the layperson, being unconcerned with the medical aspects of the condition, used the common, centuries old, generic terms "tsara'ath /lepra". This assumption by Browne (1979), in light of the literary evidence, certainly appears plausible, if not correct.

The publication of the Essene Temple Scroll from Qumran by Yadin (1983) provided fresh new evidence on the controversy surrounding the meaning of the term "tsara'ath" and its relationship to true leprosy. This document, which Yadin regarded as 'the basic Torah of the Dead Sea sect known as the Essenes' (1983:229), was probably compiled during the reign of John Hyranacous in the second century, or according to some authorities as early as the middle of the third century BCE. While the document reflects Essene theology, it also reflects to a certain degree beliefs which were applicable to Judaism as a whole. In Judaism, for example, it is generally accepted that the Old Testament Pentateuchal laws were intended specifically for the priestly class, or the Cohenim. There was, however, from the earliest times another tradition which stated that these laws applied to all people, thereby extending priestly holiness to the entire people of Israel (Yadin, 1983:277). In the Temple Scroll, the section dealing with the laws of uncleanness and purity expressively state that those suffering from "tsara'ath" must not enter the Temple City of Jerusalem as well as all the cities (Yadin 1983:294). The authors of the Temple Scroll further state that "lepers" must be east of the city, which corroborates the Markian tradition (Mk 14) that Bethany, which lay east of Jerusalem, was the village inhabited by those suffering from various skin disorders. The Hebraic concept "east of the city" stems from the halachic belief that wind blowing over an individual suffering from leprosy is, in and of itself, contaminating. For example, the Talmud states that one can walk four cubits east of a "leper" if the wind is not blowing; however, if the wind is blowing from the West, which is normal in the eastern Mediterranean, one must walk 100 cubits to the east of the "leper". What is more remarkable in the Temple Scroll is a portion of the text found in the section designated Messianic Rule (1Qsa ii:3-9) which Yadin (1983) translates as follows:

'And no men smitten with any human uncleanness shall enter the assembly of God; no man smitten with any of them shall be confirmed in his office in the congregation. No man smitten in his flesh, or paralyzed in his feet or hands or lame, or blind, or deaf, or dumb, or smitten in his flesh with a visible blemish, or an old man so crippled that he cannot hold himself up in the congregation of the men renown for the Angels of Holiness are with their congregation.'

This prohibition, in which certain categories of individuals are banned from the Holy City of Jerusalem, describe those diagnostic features which today one associates clinically with those unfortunate victims of the disease who go untreated. Interestingly enough, while these diagnostic criteria do not appear in Leviticus (Chapters 13 and 14), they do appear later in Chapter 21 (Verse 18). What is significant here, however, is the addition of the term paralysis of the hands, feet and blindness which is highly suggestive that, by the third century BC, if not earlier, the disease has already become part and parcel of those categories of disease which were now subsumed under the generic term "tsara'ath". Jopling and McDougall in their handbook on leprosy correctly state that a combination of skin and neurological disorders is strongly indicative of leprosy (1988:22).

Wilkenson (1978), writing from a theological perspective, believed that certain passages in Leviticus 13 may refer to the early signs of the disease which were important in terms of the priestly class excluding the individual "from the camp of Israel", while Leviticus 21:18 refers to later stages of the disease. Secondly, Wilkenson (*ibid.*) stated that anaesthesia or sensory impairment, which is absent from the Biblical account, is perhaps understandable in light of Jewish law as it would have forced the priestly class to touch the individual which, culturally and ritually speaking, would make the priest unclean. While one can argue that the more scientific term "elephas" does not appear in Jewish documents, one must remember that neither the Bible nor later sectarian writings of the Second Temple period were meant to be medical treaties. The editors of these works were actually concerned with the cultic aspects of the disease rather than the medical terminology, since what is important here is ritual contagion irrespective of its medical origin. Moreover, the priestly class was obligated to diagnose the condition not as a physician would, but as one charged with interpreting the Biblical Laws (De Vaux, 1965:462). Unfortunately, this vague and often confusing Old Testament nomenclature makes it exceedingly difficult to match modern medical terminology with the Hebraic text as "tsara'ath" can mean many things, both medically and in a religious framework.

5. The Hellenistic Period (4th-1st cent. BC)

This period in which the Temple Scroll was compiled also coincides temporally with the anthropological evidence for the earliest discovery in the eastern Mediterranean of the disease. Archaeologically, Hansen's disease makes its first appearance in Egypt in ca. 200 BC with the discovery, in 1980, of four Caucasoid males which were interred in a Nubian cemetery located in the Dakhleh oasis (Dzierzkyray-Rogalski, 1980). What is perhaps most significant regarding these Nubian burials is that those Caucasians afflicted were most probably ostracized and banished because of their disease and were forced to live and be buried amongst strangers. While it is now clear that the disease was in the region of the Eastern Mediterranean, the lack of any corroborating evidence in Syria and Palestine is not surprising in view of the Biblical evidence in Chronicles II 26:23 that people with leprosy, as well as Kings, were buried in separate cemeteries, away from other graves.

6. The Roman Period (37BC-324 AD)

Despite the lack of firm osteoarchaeological evidence during the Roman period for the presence of Hansen's disease in Syria and Palestine, the literary evidence would appear to confirm that true leprosy was certainly known in the population of Judea during the first century BC to the 4th century AD. Not only do we possess New Testament references which corroborate the Temple Scroll written two to three centuries earlier, that people with leprosy were living east of the city in Bethany, but other references attest to the fact that they were forced to dwell outside of cities, as ordered in Leviticus. The fact that we have direct osteoarchaeological evidence from the Dakhleh Oasis some two centuries earlier, presupposes that the disease was well known to the populations of the Eastern Mediterranean of the Roman Period.

Dauphine (1997) pointed out that both Lucretius (*De Rerum Natura* VI, 1114-15) and Flavius Josephus (*Contra Apionem* I, xxvi, 229 and I, xxx14, 304-11) wrote that leprosy was endemic in Roman Egypt and widespread in Palestine during the first centuries of our era. Therefore, despite the semantic difficulties in nomenclature and lack of osteoarchaeological evidence, the possibility of its existence appears very convincing. Similarly, the writings of Josephus speak of societal expulsion: 'persons afflicted with gonorrhea or leprosy were excluded from the entire city' (War 5:227). The distinction between gonorrhea and leprosy which Josephus makes, is also later elaborated on in the Midrashic literature (XVI:3) which states categorically that people with leprosy must live alone, that is that they were not allowed to mingle with other unclean persons. This forced segregation, which Josephus and the Midrash authors later refer to, implies that there was a further separation of those regarded as suffering from leprosy, in that not only did they suffer the stigma of the disease but were forced to live alone outside of towns and villages. This spatial segregation suggests that religious and medical authorities had recognized both leprosy and gonorrhea as categories of disease that were contagious, both clinically and spiritually, which they attempted to contain, lest they accidentally infected other unfortunate individuals, such as those suffering from psoriasis. This separation, which originally was interpreted as a temporary state of affairs until the dermatological condition disappeared, may also have been intended to permit a more definite diagnosis insomuch as dermatologists may fail to recognize leprosy in its early stages, being unfamiliar with the bewildering variety of its clinical manifestations (Browne, 1975:280). Two of the four clinical forms of true leprosy recognized today, tuberculoid and indeterminate will, in fact, occasionally arrest spontaneously without any medical treatment (Krupp and Chatton, 1984, Macaulay and Knell, 1987). This further confuses the issue for those responsible in antiquity for diagnosing the disease. Fortunately, for those living outside the "camp of Israel" east of Jerusalem, and other Biblical towns, cities and villages, one always had the possibility of returning to society if, spontaneous or self limiting forms of Hansen's disease occurred. For those with other self-limiting dermatological conditions, which fell under the generic term "tsara'ath", likewise this would be true.

Yadin (1983) speculated that Josephus may have been influenced by the writings of the Essenes, with whom he lived, when he wrote his text on leprosy. While it may be possible to see Essene influence in this passage, the following third century AD Midrashic passage sums up similar societal sentiments regarding the separation of the leprous from society:

'if such an unfortunate individual ever dared to enter the city, they even educated and pious people would throw stones at him and call out to him: go to your place and do not defile the people' (Leviticus Rabbah 16:3).

Similar thoughts were also expressed by other writers such as Galenos (2nd century AD) who mentions an account of euthanasia, in which the sufferer was poisoned by friends who regarded his death as preferable to living with the incurable disease (Andersen, 1969:43). Aretaios (2nd century AD), in a similar fashion, mentions the revulsion and social ostracism associated with the disease in which even 'beloved relatives were exposed to the wilderness' (Andersen, 1969:37).

While the literary evidence for the existence of the disease in the Holy Land (8th century BC-1st century AD) is compelling, there are experts such as Andersen (1969:15) who would disagree. Luke, he maintains, was reputed to have been a physician and, therefore, would have been well aware of the differences between "elephas" (true leprosy) and "lepra" the latter term being the Greek translation of "tsara'ath". However, when Luke refers to the disease (4:27, 17:12-19) he always uses the Greek word "lepra" in its Old Testament sense. Had true leprosy existed, he maintains, Luke would have referred to those being cleansed and suffering from elephas or *elephantiasis graecoreum*. However, Luke, like the rest of the New Testament writers, was writing for the masses and not for a medical audience. In fact, those outside of the medical profession would probably not be aware of, or even concerned with, the distinction between various dermatological conditions, as what was important was ritualistic and not medical contagion. Furthermore, the assertion that Luke was a physician, or that the Gospel of Luke was actually written by Luke, are unsubstantiated by modern Biblical scholarship. Medically, wherever lies the truth as to its origins, true leprosy had by now crossed the Mediterranean into the continent of Europe, as distant as England, where a fifth century find has been reported (Reader, 1974).

7. The Byzantine-Islamic Periods (324-1516 AD)

During the Byzantine period (4th -7th centuries AD) it becomes clear to medical historians that true leprosy had reached endemic proportions in the Roman Empire, and strict legislative measures were now enacted to deal with the problem. For example, we know that not only were people with leprosy forbidden to enter the Roman public forum in the 4th century, but also to use the public baths (Avi-Yona, 1963). The Early Church also was becoming aware of the acute problem and from the middle of the same century numerous leprosaria were being constructed. The fact that special leprosy hospitals were now being established early in the 4th century by Byzantine authorities attests to the severity of the problem which clearly had reached epidemic proportions. (Browne, 1979:486, Avi-Yona, 1963).

According to Gramberg *et al.* (1961:136) the early Christian community took a special interest in those suffering from the disease in the belief that those sufferers were chosen by God to suffer in this world for the world to come. It appears, however, that this theological perspective was not accepted by all, since the Council of Ancyrus in 314 AD decreed that people with leprosy must be allowed to worship among the public in the open air i.e. physically outside the church. That this decree should now be enacted appears to suggest that early in the Church history those suffering from the disease were excluded from public mass and the sacraments.

Here in the Holy Land literary evidence reports that the Judean Desert monastery of St. Theodosis had separate hospital facilities for the mentally ill and sick (Usener, 1890:35) Although no mention is made of what specific illnesses were treated there, the separate facilities, along with the mention that St. Theodosis himself attended to the needs of those with leprosy, provides evidence that a leprosy hospital was established at the site east of Bethlehem. Gregory of Tours, who travelled extensively in the Holy Land in the 4th century, also mentions that by this time it was customary for those suffering from the disease to bathe in the Jordan River, particularly at the site of the baptism of Jesus in the hope of a miraculous cure. While no mention of the Old Testament tradition of Naaman is made, the commander of the Syrian army (2 Kings 5:10-14) being cured in the Jordan River appears in the narrative; there are strong theological precepts /underpinnings for this belief. The Judaeo-Christian attitude regarding illness which appears throughout the Bible, is that all sickness is a divinely ordained punishment for sin, (Deuteronomy 28:22, Corinthians 11:29-30) is perhaps best expressed in the often quoted New Testament phrase: 'The wages of sin is death' (Romans 6:23). If this is the accepted theological belief, then the act of baptizing in which the Christian washed away one's sin should, therefore, cure not only the soul but one's afflictions as well.

According to 14th century writings of the Church historian Nicephorus Callistus the Empress Eudokia (d.460) established a leprosarium in the Phordisia. The exact location of the place is lost today, however, since most leprosaria were located in areas far from the populace. However, it most probably lay somewhere in the Judean Desert. While there as been controversy regarding the exact location of the Phordisia with Milik (1961) suggesting that the site lay west of Jerusalem, Zias (1986) argues that Herodium, East of Bethlehem was probably the location for the leprosarium. Basing his conclusions on archaeological and literary evidence, and the linguistic similarity between the name Furadis, the Arabic term for Herodium and Eudocia's Phordisia, both of which are borrowed from the Persian word meaning "garden" and later corrupted into meaning "paradise", the designation of Herodium as the location for the leprosarium appears plausible. The four small churches located at the foot of the palace fortress, and the lack of any apparent associated Byzantine dwellings, suggests that the site may indeed be the Phordisia established by the Empress Eudocia for 400 individuals suffering from the "holy sickness". From the Byzantine period onward, Christianity portrayed those suffering from the disease as being chosen by God to suffer in this world for the world to come, that is a heavenly paradise. Furthermore, to the nearby Monastery of St. Theodosius may be added corroboration of this possible identification, as hospital facilities were established there, and skeletal evidence for the existence of the disease is present (Zias, 1985). Furthermore, as Herodium does not appear anywhere in the Byzantine literature, nor was it resettled in the Islamic period despite its proximity to Bethlehem, suggests that the social stigma of the site, once having been a leprosarium, may have been the one factor for its permanent abandonment following the Byzantine period.

As we now have direct paleopathological evidence for the existence of the disease in three Judean Desert monasteries (Mar Theodosis, Dier Hagla, and St. John the Baptist) the founding of a leprosarium in the Phordisia may have been intended to relieve the pressure in these monastic communities, which seemed to have functioned as quasi monasteries/hospitals. In fact, already in the Byzantine period, it became customary to build hospitals in association with monasteries and churches at those places frequented by pilgrims. The monasteries at Mt. Sinai, St. Theodosius, as well as the Nea Church in Jerusalem, are examples of this Byzantine tradition of combining ecclestical institutions with hospitals. These monasteries were ideal places not only for those suffering from the disease but also society as a whole. Not only were the spiritual and physical needs of those afflicted administered to, but their isolation in somewhat isolated desert communities minimized contact with the population at large, and thus reduced the spread of the disease. As the monasteries were inhabited mainly by adults, and the possibility of contagion or cross infection from those adults infected with the disease was extremely low, the disease would effectively be controlled, if not entirely eliminated within a generation. We know today that in an isolated adult community, where no children are present, the disease will eventually die out; therefore, these places of refuge, with laws of enforced celibacy, also served the needs of society at large. In the Judean Desert Monastery of Saint Saba, Saint Saba himself was reported to have drawn two pictures, one of Jerusalem and one of Bethlehem, alongside the monastery gate leading to Bethlehem and nearby Jerusalem. It was here that he instructed all inhabitants of the monastery that one need not make the trek to either holy city, as praying in front of the picture alone would suffice in place of pilgrimage. Ideas such as this, although not intended to prevent crowd infection, would certainly minimize the transmission of contagious diseases between those in the monasteries and town dwellers, and *vice versa*. Unfortunately, the centuries old tradition of bathing in the Jordan River to cure one's illness attracted pilgrims far and wide with contagious disease, thus diminishing and/or minimizing any public health measures which these monastic communities in Palestine would have afforded.

We know from the writings of Procopius, that the Byzantine Emperor Justinian also constructed xenodochia (houses for the poor and diseased) in Jerusalem as well as Jericho in AD 535 (Stewart, 1986). Interestingly, by the 8th century, public laws were now enacted (Cochrane, 1963:14) which paralleled the conditions of the monastic orders in that marriage was expressly forbidden among people with leprosy, and their enforced social isolation now mimicked those conditions established centuries earlier in Jerusalem and the monasteries of the Judean Desert. As the disease is known today to be one of low infectivity for adults and its dissemination within the adult population slow, its appearance in three of five monasteries surveyed, along with the literary evidence, suggests that many of those suffering from the disease came from great distances. In fact, a mass grave at the traditional Jordan River site (Monastery of St. John the Baptist) for the washing of leprous individuals, and radiocarbon dated to the beginning of the 10th century, showed clear evidence that those individuals attracted to the site travelled enormous distances in the hope of "washing away" their illness. Seeds from the Egyptian tree *Balantines aegyptiaca*, enclosed in the hands of the deceased (a centuries old Egyptian burial custom), along with additional anthropological evidence, indicated that these individuals in the mass grave were probably Egyptian in origin. In addition to the palaeobotanical evidence was the finding of a walnut shell placed on the thorax of one of the deceased which, theologically, symbolized paradise, along with its oil, which was prescribed for various human illnesses, including leprosy (Pope, 1977:579). Whatever the reason for these long and difficult migrations, over vast distances and under harsh conditions, it seems clear that the Judean Desert communities were accepting those suffering from the disease on a much wider scale than what was previously known or believed.

This charitable attitude towards the unfortunate sufferers of the disease, which began in the early Christian period, continued for several centuries in the Near East with additional hospital facilities specifically for leprosy patients established in the Crusader Period (AD 1099) in cities such as Jerusalem and Bethlehem (Mitchell, 1993:288). The 12th century Latin Kingdom of Jerusalem was now ruled by Baldwin the IV (1160-1185) who himself suffered from leprosy, while there was even an order of knights (the Knights of St. Lazarus), comprised of many who were suffering from leprosy. While conditions in the region of Jerusalem appeared to have improved, European attitudes actually worsened during the middle ages with leprosy sufferers being burned at the stake or excluded from church by a funeral mass along with a symbolic burial, their property was confiscated and they were banished from towns (Lendrum, 1952:1004). In England, Edward the 1st (14th century) buried people with leprosy alive, albeit with a Christian funeral (Browne, 1979). To gauge the severity of the disease in Europe alone, there were 19,000 leprosaria in the 11th century (Lendrum 1952: 1004). In the Holy Land it is difficult to assess the situation following the period of the Crusades. With the fall of Jerusalem to the Mamluke Arabs in AD 1187, many Crusader institutions ceased to function and the leprosy hospital of Bethlehem was transferred to Acre where the Crusaders ruled until their defeat by the Mamlukes in 1291.

In marked contrast to Europe was the more tolerant and accepting attitude prevalent in Islamic society towards those suffering from leprosy which prevented the widespread abuse and social ostracism meted out towards the sufferers of leprosy in the lands west of the Mediterranean. While most of the literature on leprosy in the Ancient Near East deals with the Biblical periods, the medical historian Dols (1979:319) has written on the subject during the Islamic and medieval periods which presents an interesting and somewhat different perspective. He, like others, disagrees with Andersen (1969) in the belief that NewTestament leprosy did not exist, and argues that true leprosy may have been an early manifestation of "elephantiasis". Interestingly, one of the issues pointed out by many writers is the loss of sensation, one of the main diagnostic criteria for Hansen's disease which, aside from one mention in India about 700 BC (Manchester, 1992:38), does not appear in the writings of the ancients, even in the work of Aretaios whom Andersen (1969:33-39) believed presents the first reliable clinical description of the disease, albeit under the term "elephas/elephantiasis". This silence is regarded as one of the major arguments for the absence of the disease in antiquity, whereas none of the later writers prior to the Islamic period mention sensory deprivation when describing the disease, even when there is direct osteoarchaeological evidence that leprosy did exist in the ANE. Therefore, this silence appears to be a moot and irrelevant point in clinical descriptions of the disease in antiquity.

Dols (1979:333) was of the opinion that, although leprosy was common among the Arabs of the Middle East, those suffering from the disease were not banished from society because the disease, as unlike the earlier Judeo-Christian belief, was not considered a divine punishment. In fact, nothing in the Qur'an is comparable to the Levitical law, which was so decisive for the Western Christian view of leprosy (Dols 1983:913). Furthermore, the Koran is specific on this fundamental theological break from earlier Judeo-Christian thought in the belief, 'that there is no fault in the blind, and there is no fault in the lame and there is no fault in the sick' (Qur'an 24:60). Thus, Arab doctors did not advise flight from a person with leprosy, or their isolation from the community, as was normal in the Judeo-Christian world; they may have been, on occasion, separated but not stigmatized to the extent seen in Medieval Europe. (*ibid.*: 914). Moreover, Islamic medical knowledge, much of it based on earlier Greek writings, was more exact than the classical writers of antiquity and eventually formed the basis for Western knowledge of the disease up until the 17th century.

The long established tradition of bathing in the sulphurous waters of the Holy Land for pilgrims afflicted with a wide range of physical ailments, often travelling vast distances, continued during the Islamic period. Mid -11th century documents found in the Jewish Geniza in Cairo attest to Jews coming from far as away as Egypt to bathe in the hot springs of the Tiberias (Mitchell, 1993:287), which was an impressive feat when one realizes that the debilitating nature of the disease would have made travel by land or sea difficult.

8. Decline of leprosy

With the rise of tuberculosis in Europe during the 12th and 13th centuries and its possible cross-immunity to leprosy (but see Wilbur *et al.*, this volume), those individuals surviving initial infection with tuberculosis may have been increasingly resistant to leprosy (Manchester, 1991). Later in the 14th century the Black Death (1349) ravaged Europe, thus effectively eliminating much of the disease from the European continent. As a result of these two highly infectious diseases and their cross immunity, leprosy now began a rapid decline in Europe.

9. Osteoarchaeological evidence

Despite the enormous amount of archaeology activity in the Ancient Near East this century, there has only recently been any reported osteoarchaeological evidence for leprosy. This is not surprising in light of the negative sentiments and ostracism of those afflicted with various dermatological conditions which led not only to their banishment from the community in life, but also in death. As most burial tombs accidentally discovered are those for the extended family the possibility of finding someone afflicted with the disease is remote. Our initial discovery of the disease in the Byzantine Monastery of Martyrius, 10 kilometres east of Jerusalem (Zias, 1985), later proved in fact to be psoriatic arthritis (Zias and Mitchell, 1996). Three individuals interred here in a mass grave of nine males and one female presented clinical features limited to a certain extent with those osteological changes found in Hansen's disease. Bilateral erosive changes in the hands and feet, along with erosive lesions in the axial skeleton, were initially suggestive of Hansen's disease. As they were found in a Judean Desert monastery, the skeletal evidence at first seemed persuasive until colleagues from Denmark and England later confirmed that what we had perceived as Hansen's disease was, in fact, a medical misdiagnosis (Zias and Mitchell, 1996). It was not until several years later that the possibility of psoriatic arthritis, a seronegative erosive arthropathy, came to our attention and we were able to make a more definite and correct diagnosis. Our difficulty, and that of colleagues, in making a correct medical diagnosis on dried skeletal material suggests that these individuals may also have been misdiagnosed by society as suffering from leprosy, as the term was for centuries a generic term for a wide variety of dermatological conditions. The psoriatic lesions, coupled with the debilitating and mutilating arthritic features of chronic arthritis (*arthritis mutilans)*, certainly would have been confusing to lay society at large. We know today that dermatologists today may fail to recognize the disease because they are unfamiliar with the bewildering variety of its cutaneous manifestations. Thus, the number of skin and nerve diseases that have been mistakenly diagnosed as leprosy is legion (Browne, 1964:282-284). An alternative possibility would be the fact that many individuals suffering from diseases, other than true leprosy, were known to have taken advantage of the charitable attitude of the Christian and Islamic communities and feigned the disease for personal gain. Both Dols (1983:110), and Andersen (1969), point out this fact that individuals were on occasion prone to disfigure themselves or feign the disease in order to receive public charity. According to Andersen (1969), this abuse of charitable public trust eventually led St. John of Damascus (777-857 AD) to develop better diagnostic criteria in order to determine who, in fact, was truly suffering from the disease, lest the monasteries be over run by "malingerers" and "medical impostors" looking for a free meal. Evidently, this attempt at sorting out those truly suffering from the disease was only partially successful. The accidental discovery of a 9th century mass grave along the Jordan River near the site of the "washing of the leper" in 1985 had evidence both of Hansen's disease and tuberculosis. It also included two individuals with severe facial disfigurement, neither of whom were suffering from true leprosy. This site, the Monastery of Saint John the Baptist, produced a mass grave of an estimated 300 individuals, 90% of which was destroyed by road construction. This mass burial is unique in that it probably represents a hospital population of outcasts, including people with tuberculosis, leprosy, and facial/nasal disfigurement. The latter condition, whether or not it was the result of trauma, or true infectious disease, resulted in the individual suffering severe social restrictions in Judaic society, particularly during the time of the monastery (Rosen, 1971). Unfortunately, the skeletal remains with pathological changes indicative of Hansen's disease and tuberculosis were found commingled and disassociated. Therefore, it was impossible to determine the prevalence of the disease in the population, but confirmatory evidence of our initial diagnosis was later provided through a PCR (polymerase chain reaction) analysis of the material which showed positive evidence of mycobacterial DNA from the ancient bone (Spigelman and Lemma, 1993, Rafi *et al.,* 1994).

Following our discovery of the disease, a survey of the Byzantine desert monasteries provided further evidence that the main reason for our inability to locate Hansen's disease in the archaeological record was our failure to study the vast amounts of skeletal material housed in these ancient monasteries. Subsequent surveys in the nearby Greek Orthodox monasteries of Deir Hagla and St. Theodosis provided direct palaeopathological evidence for the disease. Researching this material is difficult due to the fact that the remains have been moved numerous times over the centuries, and have been disarticulated, and the cranial remains are spatially not stored with the post-cranial material. Oral tradition, suggested that these skeletal remains can be dated to the Byzantine period (4th to 7th centuries AD) when the monasteries were first founded. These findings attest to the written record that the "lame, blind and the halt" found refuge in these monasteries on a much wider scale than originally believed. Not only was tuberculosis and leprosy found, but additional paleopathological conditions. These findings from the Judean Desert monasteries with leprous bone changes constitute the only human skeletal remains yet recovered, aside from one controversial find described by Hershkovitz *et al.* (1992) as Madura foot which some palaeopathologists have regarded as Hansen's disease (Hershkovitz *et al.,* 1993). Recent ancient DNA studies by colleagues in England confirmed this latter diagnosis as Hansen's Disease (Spigelman and Donoghue, this volume).

10. Conclusions

Without direct osteoarchaeological evidence, diagnosing disease in antiquity, in which one can posit a definite diagnosis, is difficult, if not at times impossible (Neiburger, 1991). For the medical historian, there is no magical diagnostic technique. One can often only infer on the basis of circumstantial evidence, often literary, that the probability that the condition existed is high. As in archaeology, the evidence is often inferred and not directly observed. Aside from the 6th century BC Indian Vedic description, we have few if any undisputed early clinical descriptions, until the 5th century BC Hellenistic writers, which may imply that we have to rely entirely on inferential information, as problematical as that is. Using these early sources one sees Babylonian, Vedic, Old

Testament, and Chinese literary descriptions which suggest that, despite the semantic problems and nomenclature involved, the picture that emerges is that Hansen's Disease, which has been referred to by many names, has been around for a considerable time.

By the end of the 3rd century BC, there is both literary as well as archaeological evidence that the disease existed in the Eastern Mediterranean region. Straton, a disciple of Erisastratos (300-250) and an Alexandrian physician, provides the first clinical description which is regarded by Andersen (1969) as medically accurate. As both individuals are physicians in the modern sense of the word, their clinical descriptions can thus be viewed as reliable whereas earlier descriptions by lay people are more problematic and subject to much debate (Manchester, 1992).

The strong negative sentiments found in the ancient narratives towards those suffering from leprosy, infers that societal expulsion must have been in effect long before the finding of four Caucasoid males in the Egyptian Dakhleh Oasis. However, one wishes to define the dermatological conditions described in ancient texts, beginning with the Babylonian inscription of about 2,000 BC published by Wilson (1996), that those afflicted with this dermatological condition were "rejected by mankind and his god", and the reburial of the Old Testament Judean King Uzziah (808-739) (Chronicles II 26:26) inevitably suggests that Hansen's Disease is a disease of long duration. The finding by Angel in Karatas (Turkey) in 1974 would appear to confirm this view that the disease in the Ancient World precedes the find in the Dakhleh Oasis by as much as two millennia. These cities of the ANE, situated astride ancient trade routes had by 2,000 BC, reached population figures running into the tens of thousands, thus easily permitting the spread of infectious disease by caravaners travelling these ancient routes. Ideas put forth by leprologists (Andersen, 1969:45) that the disease was probably brought by the returning armies of Alexander the Great in the late 4th century, though reasonable, have neglected to take cognizance of the fact that trade routes from the Far East to the Ancient Near East existed long before the rise of Alexander the Great in the latter half of the 4th century BC.

One must, however, be aware of the fact that while there is no European osteoarchaeological evidence for the disease before the 5th century AD, it is clear from the historical record that the disease had already reached alarming proportions several centuries before, with laws passed segregating these individuals away from public places, not to mention hospitals being constructed for their care. This disconformity between the historical and the archaeological record can perhaps be best explained by the marginal or superficial role physical anthropology has held in Europe and the ANE this century. The vast number of human skeletal remains recovered in archaeological excavations which have not been examined by the critical eye of a trained physical anthropologist has led to a situation where much scientific and medical information regarding ancient disease has been irrevocably lost. This loss to science is immeasurable as ancient human skeletal remains, like archaeological sites, are not "born again". Each and every human skeleton, tomb and grave uncovered from the past contributes in a direct way to a better understanding of disease and disease processes. Thus, the reburial or wanton destruction of these remains, through looting or pillaging ancient tombs, constitutes damage to a valuable resource which is irreplacable.

References

Andersen, J.G. 1969 *Studies in the_Mediaeval diagnosis of leprosy in Denmark.* Danish Medical Bulletin 16 (Supplement 9).

Angel, J.L. 1974 Human skeletal remains at Karatas (Turkey). *American Journal of Archaeology* 74: 253-259.

Avi-Yonah, M. 1963 The bath of the lepers at Scythopolis. *Israel Exploration Journal* 13: 325-326

Brim, C.J. 1936 *Medicine in the Bible.* New York: Froben Press.

Browne, S.G. 1975 Some aspects of the history of leprosy: the leprosy of yesterday. *Proc. Royal Soc. Med.* 68: 485-493

Browne, S.G. 1979 *Leprosy in the Bible.* London: Christian Medical Fellowship Publication.

Brim, C. J. 1936. *Medicine in the Bible.* New York: Faroben Press.

Cochrane, R.C. 1963 *Leprosy in Theory and Practice.* Bristol: Wright.

Dauphine, C 1997 Health and Hygine in Byzantine Palestine. *Bulletin of the Anglo-Israel Archaeological Society*. London.

De Vaux, R. 1965 *Ancient Israel.* New York – Totonto: McGraw-Hill Book Co.

Dols, M.W. 1979 Leprosy in Medieval Arabic medicine. *J. of Medical History:* 34 313-333.

Dols, M.W.1983 The leper in Medieval Islamic society. *Speculum* 58 (4): 891-916.

Dzierzykray-Rogalski, T. 1980 Paleopathology of the Ptolemaic inhabitants of Dakhleh Oasis (Egypt) *J. of Human Evolution,* 9:71-71.

Feldman, Y. 1966 Dermatology. *Hebrew Bible* 2: 984-988.

Gramberg, K.P., Swellenbrebel, L., Nida, E. and Wallington, D.H. 1961 *Leprosy and the Bible.* Amsterdam: MedWave.

Hershkovitz, I., Speirs, M., Katznelson, A. and Arensburg, B. 1992 Unusual pathological condition in the lower extremities of a skeleton from ancient Israel *Amer. J. Phys. Anthropol.* 88:23-26.

Hershkovitz, I, Arensburg, B. and Speirs, M. 1993 Leprosy or Madura Foot? The ambiguous nature of infectious disease in paleopathology: Reply to Dr. Manchester. *Amer. J. Phys. Anthropol.* 91: 251-253.

Hulse, E.V. 1975 The nature of Biblical "leprosy" and the use of alternative medical terms in modern translations of the Bible. *Palestine Exploration Quarterly 107:* 87-105.

Jopling, W.H. and McDougall, A.C. 1988 *Handbook of Leprosy.* 4th edition. Oxford: Heinemann.

Krupp, M. and Chatton, M. 1984 *Current medical diagnosis and treatment.* Lange Medical Publication.

Lendrum, F. C. 1952 The name leprosy. *The American Journal of Tropical Medicine and Hygiene* 1: 999-1,008.

Macaulay D.E. and Knell, A. J. 1987 *Leprosy.* London: The Wellcome Tropical Institute.

Manchester, K. 1991 Tuberculosis and leprosy: evidence for the interaction of the disease. In D.J. Ortner and A.C. Aufderheide (eds): *Human paleopathology. Current syntheses and future options.* Washington, D.C.: Smithsonian Institution Press, pp. 23-35.

Manchester, K. 1992 Leprosy: The origin and development of the disease in antiquity. In D. Gourevitch (ed): *Maladie et maladies: histoire et conceptalisation.* Geneva: Librairie Droz SA., pp. 31-49.

Milik, J. T. 1961 La topographia de Jerusalem ver la fin de l'epoque Byzantine. *Melanges de l'Univeste St. Joseph* 37:127-89.

Mitchell, P.D. 1993 Leprosy and the case of King Baldwin IV of Jerusalem: mycobacterial disease in the Crusader States of the 12th and 13th Centuries. *International J. of Leprosy* 61 (2): 283-291.

Neiburger, E.J. 1991 *The impossibility of diagnosing old diseases.* Abstract of a paper presented to the Paleopathology Association Annual Meeting, Milwaukee April 2nd to 3rd.

Pope, M.H. 1977 *Song of songs.* Garden City New York, Double Day and Co.

Preuss, J. 1978 *Biblical and Talmudic Medicine.* Translated by F. Rosnar. New York – London: Sanedrin Press.

Rafi, A., Spigelman, M., Stanford, J., Lemma, E., Donoghue, H. and Zias, J. 1994 DNA of *Mycobacterium leprae* detected by PCR in ancient bone. *International Journal of Osteoarchaeology* 4:287-290.

Rastogi R.C. and Rastogi, N. 1984 Leprosy in ancient India. *International Journal of Leprosy* 32: 541-543.

Reader, R. 1974. New evidence for the antiquity of leprosy in early Britain. *J. of Archaeological Science* 1: 205-207.

Rosen, Z. 1971. Nasal deformities in Biblical times. *Arch Otolaryng* 93:211-212.

Spigelman, M. and Lemma, E. 1993 The use of the polymerase chain reaction to detect *Mycobacterium tuberculosis* in ancient skeletons. *International Journal of Osteoarchaeology* 3:137-143.

Stewart, A. 1986 *Procopius, On the Buildings of Justinian.* London:Translation by Palestine Pilgrims Text Society.

Usener, H. 1890 *Vita S. Theodosii a Cyrillo Scythopolitano Scripta.* George University, Bonn: Typis Typoger.

Wilkenson, J. 1978 Leprosy and Leviticus. *Scottish J. of Theology* 31:156-166.

Wilson, J.V. 1982 Medicine in the land and times of the Old Testament. In I. Tomoo (ed): *Studies in the Period of David and Solomon and other Essays.* Tokyo: Eisenbrauns, pp. 337-365.

Yadin, Y. 1983 *The Temple Scroll.* Three volumes. London: Weidenfeld and Nicolson.

Zias, J. 1985 Ancient leprosy in the monasteries of the Judean Desert. *Korith* 9: 242-48.

Zias, J. 1986 Was Byzantine Herodian a leprosarium? *Biblical Archaeology* 49: 182-186. (Article reprinted in *Teva VeAretz*, pp. 17-18).

Zias, J. and Mitchell, P. 1996 Psoriatic arthritis in a fifth century Judean Desert monastery. *American Journal of Physical Anthropology* 101(4):491-503.

Molecular Diagnosis of Leprosy in Skeletal Material

Mycobacterium leprae in archaeological samples

Helen D. Donoghue[1], Judyta Gladykowska-Rzeczycka[2], Antónia Marscik[3], John Holton[1] and Mark Spigelman[1]
[1]Department of Medical Microbiology
University College London
46 Cleveland Street
London W1T 4JF
England
E-mail: h.donoghue@ucl.ac.uk, j.holton@ucl.ac.uk, spigelman@btinternet.com

[2]Ul. Marusarzewny 3/8
80-288 Gdansk
Poland
E-mail: jgr@kawiarenka.com.pl

[3]Department of Anthropology
Natural Sciences Faculty of the University of Szeged
P.O. Box 660
H-6701 Szeged
Hungary
Email: marcsik@bio.u-szeged.hu

Abstract

There is a greater likelihood of detecting ancient DNA with smaller target DNA sequences, so two *Mycobacterium leprae* -specific nested PCRs were devised. The first was based on a repetitive sequence (RLEP) which gives a 129bp outer product and 99bp nested product. Nested primers were also devised for the single-copy 18kD Ag gene which target an outer sequence of 136bp and an inner sequence of 110bp. Several bones, which appeared from their morphology to be from individuals afflicted by leprosy (Hansen's disease), were examined by PCR for the presence of species-specific *M. leprae* DNA with two newly-designed nested primer pairs and a pair in current clinical use. This proved successful in samples dating from the 10th to 11th and 14th to15th centuries AD. *Carum nasale* samples were strongly positive, with one of the largest pieces of aDNA ever recovered from ancient bone, isolated and provisionally sequenced.

Keywords

DNA, *Mycobacterium leprae*, PCR, palaeomicrobiology.

1. Introduction

Recent advances in the field of palaeomicrobiology enable palaeopathologists to confirm scientifically many previously provisional diagnoses of diseases made on skeletal remains from the past. The use of the PCR reaction has enabled species-specific segments of DNA from known human pathogens to be amplified and identified. This is particularly so in lesions which leave characteristic morphological changes on the skeleton. A number of papers on this subject have now been published. *Mycobacterium tuberculosis* is the major pathogen whose DNA has been extracted from ancient specimens (Spigelman and Lemma, 1993, Dixon *et al.,* 1994, Salo *et al.,* 1994, Taylor, *et al.*, 1996, Crubézy *et al.,* 1997, Faerman *et al.,* 1997, Donoghue *et al.,* 1998). Ancient DNA from other pathogens has been reported by Rafi *et al.* (1994) who described *M. leprae*, Fricker *et al.* (1997) who detected *Escherichia coli* DNA in a bog body and Taylor *et al.* (1997) who reported finding *Plasmodium falciparum* DNA in a 60 year old specimen. Sequences of influenza virus RNA have been obtained from archival fixed sections and from a specimen taken during the 1918 influenza pandemic (Taubenberger *et al.*, 2000). This is the only group to have published details of viral genetic material from ancient specimens, although others have sought to do so (Davis *et al.*, 2000).

Before this conference we could find no published reports, apart from the early work of Rafi *et al.* (1994), on the finding of *M. leprae* DNA from antiquity in skeletal or soft tissue remains, although some of the conference presentations have since been published elsewhere (Donoghue *et al.*, 2001; Haas *et al.,* 2000; Spigelman and Donoghue, 2001). There are distinctive palaeopathological lesions associated with leprosy, although differential diagnoses can still pose problems. The detection of *M. leprae* DNA in archaeological specimens may, therefore, provide the palaeopathologist with a useful confirmation of diagnosis (Spigelman and Donoghue, 2001). In addition, where the pathology is less distinctive, clinical questions may be answered as to the nature and type of infection responsible for a bony lesion. Archaeological specimens are likely to contain fragmented *M. leprae* DNA and around 200 base pairs has been regarded as the maximum size of amplifiable mammalian DNA (Hagelberg and Sykes, 1989).

There are several pairs of primers with different target sequences that are used clinically to detect *Mycobacterium leprae* DNA. However, these are not necessarily of use for the examination of archaeological samples as the target sequences are quite large. For example, the 36kD antigen (Ag) gene primers (Hartskeerl *et al.*,1989) yield a 531bp product. Williams *et al.* (1990) described primers for the *M. leprae* 18kD Ag gene which give a 360bp product and could detect as few as 100 *M. leprae* in seeded uninfected human skin biopsy preparations. Woods and Cole (1989) based their PCR on an *M. leprae*-specific repetitive element (RLEP). This was developed by Jamil *et al.* (1994), who devised a two-stage nested PCR based on the RLEP repetitive sequence with an outer product of 455bp and inner product of 320bp.

Therefore, the initial aim of the present study was to increase the sensitivity of PCR detection methods for *M. leprae*-specific DNA by designing suitable nested primers based on the RLEP and 18kD Ag gene target sequences. Thereafter, our aim was to determine the feasibility of finding *M. leprae* DNA in morphologically obvious specimens from various times in the past, bearing in mind that, whilst the skeletal changes around the facial skeleton are inevitably caused by the organism itself, the peripheral changes typical of leprosy are often the result of secondary infection (Spigelman and Donoghue, and others, this volume).

2. Materials

The skeletal material utilized for the study derived from a variety of sources:

1. A skeleton from 624 AD, obtained from the monastery of John the Baptist from Israel. One bone had already been shown to contain *M. leprae* DNA using primers for the 36kD antigen gene (Rafi *et al.*, 1994). The sample consisted of another metatarsal bone from the same skeleton.

2. A skeleton with leprosy from a mediaeval burial ground in Suraz, Poland. Samples of this specimen were examined from two areas – the metatarsals and around the nasal region (Gladykowska-Rzeczycka, 1976). This was the skeleton of a 40-50 year old male with characteristic changes in the nose on a radiograph (Figure 1) and in the leg and foot bones (Figures 2a and 2b).

3. A metatarsal ("pencilled") and a fibula (the tibia showed periostitis) were examined from a skeleton in grave 923 in a 14th to 15th century site from Ópusztaszer-Monostor, Hungary (Figure 3).

4. Two skeletons from graves 222 and 503 in a 10th to 11th century site at Püspökladány, Hungary, were examined. Both had pitting and new bone formation on the nasal surface of the palate. Skeleton 503 was an adult female, whose right cavum nasale was sampled (Figure 4). Identical changes were noted on the left side (Figure 5). The left side of the *cavum nasale* was sampled from skeleton 222 (Figure 6). Similarly, identical changes were noted on the right side (Figure 7).

5. Sample HZ was a foot bone from the skeleton described by Hershkovitz *et al.* (1992).

3. Methods

3.1 Precautions against cross-contamination

Stringent precautions were taken to avoid cross-contamination, based on the departmental procedures used routinely to avoid cross-infection with pathogenic bacteria. Prior to this study no work with *M. leprae* had been done in the premises. Clean protective clothing was worn. Gloves were changed at the end of every procedure and whenever there were any interruptions. Pipettors and surfaces were cleaned with neat household liquid detergent, rinsed with ultrapure water and dried with ethanol before use. Sterile tubes and plugged tips were used. A separate room was used for handling PCR product, with a different set of pipettors and protective clothing. In addition, a third room was used to set up the second stage of nested PCR, with a further set of pipettors and protective clothing.

3.2 DNA extraction

A small quantity of each sample was crushed in a sterile pestle and mortar in a clean room and stored in sterile screw-capped 1.5ml Eppendorf tubes. An extended time was allowed for cell lysis, as preliminary work showed that this was necessary for reliable DNA extraction (Donoghue *et al.*, 1998). Negative extraction controls were always included. For each sample 100 µl of demineralization solution, consisting of 0.5 mol^{-1} EDTA at pH 8.0 plus proteinase K (1 mg ml^{-1}), was dispensed into a sterile 1.5 ml tube containing glass beads. Approximately 25 mg of sample was added, each tube was mixed on a mini bead beater and incubated at 56 °C overnight. Lysis buffer L6 (250 µl)) based on guanidium thiocyanate/EDTA/Triton X-100 (Boom *et al.*, 1990) was added, tubes were mixed briefly and incubated at 37 °C for 2 h. After re-mixing, tubes were centrifuged for 1 min at 13000g and the supernates collected into sterile tubes. Silica suspension (12% SiO_2 w/v, pH 2.0) was added (25 µl/tube), mixed and gently shaken for 2 h to allow cell lysis and adsorption of DNA to the silica. The silica was washed once with washing buffer L2 (Boom *et al.*, 1990), twice with cold (-20 °C) 70% ethanol and once with acetone, then dried and the DNA eluted into water by incubation at 65 °C for 1h. The silica supernates were also processed. These were chilled on ice and 200µl Puregene® Protein Precipitation

Solution (Flowgen Instruments Ltd, Lichfield, U.K.) was added to each tube. Tubes were vortex mixed for at least 20 s and spun for three mins. To precipitate DNA, the supernates were added to 600μl isopropanol (-20 °C) per sample in clean, sterile tubes, mixed by inversion 50x and spun 3 mins. After discarding the supernates, tubes were drained on clean, absorbent paper, and washed once with 600μl 70% ethanol (-20 °C). Tubes were drained once again and dried with loosened lids in a 56 °C heating block for 2h. DNA was rehydrated in 100μl water by placing tubes into a 65 °C water bath for 1h. All DNA preparations were mixed and spun briefly, aliquoted and stored at -20 °C.

3.3 Oligonucleotides used as PCR primers

Two nested primer pairs were devised, which comprise outer and inner nested pairs based on the *M. leprae*-specific RLEP and 18kD Ag gene sequences (Table 1). Outer primers LP1 and LP2 amplify a 129bp sequence of the RLEP element. The inner primers LP3 and LP4 bind internally with some overlap to LP1 and LP2, and amplify a 99 bp product. Outer primers LP7 and LP8 amplify a 136bp sequence of the 18kD Ag gene and inner primers LP5 and LP6 amplify a 110bp length of the 136bp product. The performance of these primers was compared with the 36kDAg gene primers (Hartskeerl *et al.,* 1989) S13 (5'-CTCCACCTGGACCGGCGAT-3') and S62 (5'-GACTAGCCTGCCAAGTCG - 3') which give an amplicon of 531bp.

3.4 Bacterial cell suspension

An autoclaved suspension of *M. leprae* bacilli extracted from armadillo tissues was used as the source of *M. leprae* DNA. This is the reagent known as lepromin, used for skin testing in epidemiological studies. This had an estimated bacterial load of 6×10^9 organisms/ml and was kindly supplied by Dr M.J. Colston (National Institute for Medical Research, London, U.K.). The suspension was aliquoted and stored at – 20°C. Serial 10-fold dilutions were prepared with sterile water (Sigma-Aldrich Co. Ltd, Poole, Dorset, UK) immediately before use.

3.5 DNA amplification by PCR

Pre-aliquoted double-strength PCR mix was purchased (Advanced Biotechnologies Ltd, Epsom, UK). The final composition of the PCR mixture (50 μl) was 75 $mmol^{-1}$ Tris-HCl (pH 8.8); 20 $mmol^{-1}$ $(NH_4)_2SO_4$; 1.5 $mmol^{-1}$ MgC_{12}; 0.01% (v/v) Tween 20; 200 μmol^{-1} (each) dATP, dCTP, dGTP and dTTP; and 1.25 units *Taq* DNA polymerase. Bovine serum albumin (BSA) was normally added to give a final concentration of 2mM. The primer pair and DNA preparation (5.0 μl) were added to each pre-aliquoted tube plus sufficient water (Sigma-Aldrich, molecular biology grade) to bring the volume to 50μl. Tubes with water in place of template were always included as negative controls. Primers were used at a final concentration as follows: LP1, LP2, LP3 and LP4 (60nM); LP5 and LP6 (100nM); LP7 and LP8 (200nM); S13 and S62 (1.0μM initially, then 100nM).

3.6 Single-stage PCR to detect M. leprae DNA

Amplification consisted of one cycle of denaturation at 95 °C for 5 min, primer annealing at 58 °C (S13/S62 and RLEP primers) or 55 °C for 2 min, and DNA extension at 72 °C for 2 min. This was followed by 35-45 cycles of: 30 s at 94 °C; 30 s at 58 °C or 55 °C; and 1 min at 72 °C; followed by 10 min at 72 °C.

3.7 Nested PCR to detect M. leprae DNA

A two-tube procedure was used. In the first stage the outer primers (LP1/LP2 or LP7/LP8) were added to the PCR mix. Amplification was carried out as follows: after strand separation at 94 °C for 4 min there were 25 amplification cycles of 94 °C for 40 s; annealing at 55 °C or 58 °C for 1 min; then strand extension at 72 °C for 20s + 1 s increment per cycle. Finally there was 1 min at 72 °C for further strand extension. In the second stage of nested PCR, new tubes of PCR mix were used. The inner primers (LP3/LP4 or LP5/LP6) were transferred into each reaction tube, together with 0.5 μl of product which was used as the DNA template. The total volume was made up to 50μl with molecular biology grade sterile water. This second stage PCR cycle was identical to the first.

3.8 Gel electrophoresis

PCR product (8 μl) was added to 2.5 μl loading buffer (Sigma) and electrophoresed in a 3.0% (w/v) NuSieve 3:1® agarose gel (FMC Bioproducts, Flowgen) in TBE buffer (0.09 mol^{-1} Tris-borate and 0.002 mol^{-1} EDTA) at 8.8 volts cm^{-1} for 80 min. Amplified DNA was visualised by ethidium bromide staining plus ultraviolet light and was recorded with a Polaroid camera.

3.9 DNA sequencing

PCR product (8-15 μl) previously shown to contain amplified DNA was mixed with 4 μl loading buffer and separated on a gel, as above, by electrophoresis with TAE buffer (0.04 mol^{-1} Tris-acetate and 0.001 mol^{-1} EDTA). The selected band (150-250 mg) was removed from the gel with a sterile scalpel blade into a sterile 0.5 ml tube. DNA was extracted from the gel slices using a MERmaid™ Spin Kit (10-200 nucleotides) (Anachem Ltd, Luton, U.K.). The purified DNA was sequenced by MWG- Biotech AG, Ebersberg, Germany or by an in-house service provided by the UCL Department of Biology.

4. Results

4.1 PCR and primer sensitivity

There was no evidence of any cross-contamination as negative extraction and PCR controls were consistently negative. Initially, single-stage PCR was used with dilutions of the lepromin reagent, in order to compare the sensitivity of the different primers with each other and the Hartskeerl primers S13 and S62. The outer RLEP primers

LP1/LP2 gave a strong product with a 1 x 10^{-7} dilution whilst the inner primers LP3/LP4 were less sensitive when used alone, giving a very weak positive with a 10^{-6} dilution of the *M. leprae* suspension (Figure 8). Figure 8 also shows the *cavum nasale* sample 222 was strongly positive with the outer primers, yet gave a scarcely visible band with LP3/LP4. The level of BSA was important and maximal sensitivity of the RLEP outer primers was obtained with 2mM BSA in the PCR mix (Figure 9). Under these conditions, a positive result was obtained with a 1 x 10^{-8} dilution of the *M. leprae* suspension, and the Püspökladány *cavum nasale* samples 222 and 503 each gave a strongly positive result.

The highest dilution of *M. leprae* suspension that gave a visible product with the 18kD Ag gene outer primers LP7/LP8 was 1 x 10^{-6} (Figure 10). Figure 10 also shows clear positive results from the Suraz *cavum nasale* sample (Lep II), 222 and 503, plus an extremely weak positive result with the HZ sample (Spigelman and Donoghue, this volume).

The highest dilution of the *M. leprae* suspension that gave a visible product after 45 cycles with S13/S62 was 1 x 10^{-5} (Figure 11), but only when 2mM BSA was included in the PCR mix. The two samples from Püspökladány were also positive by single-stage PCR for the large 531bp site in the 36kD Ag gene. Specimen 222 gave a very strong reaction without added BSA, but that from sample 503 was weak and showed evidence of DNA fragmentation. The inclusion of BSA in the PCR mix with these primers resulted in a less clear positive result for 222 and non-specific bands were seen, which earlier work indicated were primer-dimers and chimaeric primer products.

Nested PCR increased the sensitivity a further 10-fold (Fig. 12) and under these conditions 1 x 10^{-1} dilutions of 222 and 503 gave a visible product, as did the silica supernate DNA extracts. No other specimens from Hungary, Poland or the metatarsal from the Monastery of St John the Baptist yielded any positive results. Comparison of the nested primer pairs confirmed that those based on the 18kD Ag gene were less sensitive than the RLEP primers.

4.2 DNA sequence data

With the RLEP primers LP1 and LP2, the diluted lepromin reagent yielded a 102-base consensus sequence when results from both strands were combined. There was one mismatch of C for T in part of the sequence obtained from LP1 only. In contrast, the complete sequence of 129 bases was obtained from sample 503 (Figures 13 and 14). Similarly, a consensus sequence of 127 bases was obtained from sample 222. Samples 503 and 222 gave no discrepancies from the *M. leprae* sequence held in GenBank. Using the National Center for Biotechnology Information BLAST search engine, there were only four 18-base sequences from other organisms that showed any homology (Figures 13 and 14). Using the inner RLEP primers a provisional chimaeric 120-base sequence was obtained from sample 222, which included part of the outer primer LP1. Where a consensus sequence could be obtained from both primers it was identical to that of *M. leprae*. However, where the sequence was obtained from only one primer there were occasional indeterminate (N), missing (-) or substituted bases. Excluding those that occurred in the primer sequences, there was one missing value and one substitution of G for C.

A 101-base sequence was obtained from the diluted lepromin reagent using the outer primer LP7 for the 18 kD antigen gene (Figure 15). Sample 222 yielded a consensus 109-base sequence with LP7, and the inner primers LP5 and LP6 (Figures 15 and 16). Both lepromin and sample 222 sequences were identical to *M. leprae*. The Suraz *casum navale* sample LepII gave a 108-base sequence with the reverse outer primer LP6, which was also identical to *M. leprae* except for one undetermined base (Figure 16). There were three short sequences (18-22 bases) from other organisms that showed homology with the LP7/LP8 PCR product and some others which were homologous with the LP7 product from sample 222. However, as there was no homology with the inner product from LP5 and LP6 it is assumed that there were sequencing artefacts from the LP7 amplicon.

The 531bp amplicons of the 36kD antigen gene were also sequenced (Figures 17 and 18). A 518-base sequence was obtained from the lepromin reagent. There were two indeterminate results and one missing value, but comparison of sequences obtained from both S13 and S62 primers resulted in a consensus sequence that was identical to *M. leprae*. The sequences from sample 222 were of much lower quality. Using S13, sample 222 yielded two sequences of 223 bases and 90 bases. A 78-base section of the target sequence was missing. However, the reverse primer S62 gave a continuous sequence of 496 bases and the total length of the amplicon with both primers was 517 bases.

Where both strands were sequenced, there were three sites with indeterminate results: at 5592 and 5713 indeterminate results were obtained in place of GC, and at 5721 GN was obtained in place of CG. Otherwise the consensus sequence was identical to that of *M. leprae*. Similarly, the regions 5924-5951, sequenced by S13 only, and 5785-5862, sequenced by S62 only, agreed with that of *M. leprae*. A 128-base sequence obtained from the reverse primer S62 alone, contained one missing, five indeterminate and three mismatched bases. These results are of work in progress and require further confirmation. However, they are included because of the interest in obtaining an amplicon of this size from archaeological material.

5. Discussion

Four new pairs of primers, comprising two nested pairs, were used to successfully amplify *Mycobacterium leprae* DNA. The primers were based on *M. leprae*-specific DNA sequences and other PCRs which include the same target sequence have been shown to be specific for this organism. The primers and target sequences were found to be specific for *M. leprae* in BLAST nucleotide searches of GenBank, and DNA sequencing of the amplified target sequence showed that these were of *M. leprae* DNA. Nested PCR increased the sensitivity of PCR as it enables a total of 50 amplification cycles to be used, but avoids over-amplification artefacts. The inclusion of BSA in the PCR mix has been shown to improve the yield from clinical material (Abu Al-Soud and Rådström, 2000), and mycobacterial DNA (Forbes and Hicks, 1996) and has long been used by those studying ancient mammalian DNA (Pääbo *et al.*, 1988).

Both the new outer primer pairs were of greater sensitivity than the commonly used primers described by Hartskeerl *et al.* (1989). This is probably due to their smaller target size, as this is more likely to result in successful PCR from damaged and fragmented DNA. The most sensitive PCR was with primers LP1 and LP2 which target the repetitive sequence (RLEP) of which there are 29 copies in the *M. leprae* chromosome (Eiglmeier *et al.*, 1993). The PCR with primers LP7 and LP8, based on the single-copy 18kD Ag gene, was approximately 10 times less sensitive. The PCR with the Hartskeerl primers, which result in a 531bp product, was 1000-fold less sensitive than the new RLEP PCR.

The lepromin reagent contains approximately 6×10^9 organisms per ml, but it is not known if all the DNA is available for amplification. Therefore, as a positive result was obtained from a PCR with 5µl of a 1×10^{-8} dilution, it is calculated that primers LP1/LP2, in a single-stage PCR with 45 amplification cycles, detected at least 0.3 of an organism, and it is very likely that the sensitivity is even greater.

As these newly-described primers lower the detection level of *M. leprae*-specific DNA it is unsurprising that they successfully detected *M. leprae* DNA in several archaeological specimens with the typical morphology associated with leprosy. The DNA sequences gave good agreement with those of *M. leprae* held in GenBank. Comparison of sequences from the highly diluted lepromin reagent and the archaeological samples indicates that occasional indeterminate results or mismatches occurred, primarily when results were available from one strand only. Considering the great age of the archaeological specimens, we believe that the discrepancies from the *M. leprae* DNA sequence result from the surviving DNA being damaged by the passage of time. In addition, there may well have been technical artefacts, such as those caused by the difficulty of sequencing GC-rich DNA.

All three *cavum nasale* samples, from graves 222 and 503, at Püspökladány, Hungary, and from Suraz, Poland were positive for *M. leprae* DNA using the primers for both the multicopy RLEP site and the single-site 18kD Ag gene. *M. leprae* is known to localize to Schwann cells and nasal epithelial cells (Andersen and Manchester, 1992), and this can result in an overwhelming rhinitis associated with very high numbers of the organism. The three *cavum nasale* specimens demonstrated the typical palaeopathological changes associated with leprosy, so the presence of *M. leprae* DNA in these samples was anticipated. However, the demonstration of *M. leprae* DNA from three different sites on the chromosome, including a large amplicon of 531bp, suggests extremely good DNA preservation.

It is unlikely that this was due to the characteristics of the burial site as the samples were from two different locations. Similarly, age appears unimportant as the two samples from Püspökladány were 500 years older than those from Suraz. Mycobacteria have a resistant cell wall with a high lipid content (Draper, 1982), which is believed to be responsible for the persistence and survival of tubercle bacilli throughout the lifetime of a mammalian host and the initial decay process after death. A preliminary report of human DNA in mummified Hungarian remains (Matheson *et al.*, 2000) confirmed that it is much more difficult to detect host nuclear DNA than mycobacterial DNA in the same samples of mummified tissue, indicating that mycobacterial DNA is better preserved. This, plus the extremely high numbers and localised distribution of the organism in individuals with lepromatous leprosy, and the association of bacterial remains with calcified tissue (Donoghue *et al.*, 1998; Lassen *et al.*, 1994) may be some of the reasons why it was possible to amplify an *M. leprae* DNA fragment of >500bp.

The specimen from the Monastery of John the Baptist, although previously being found positive in a different bone in the foot by Rafi *et al.* (1994), failed to yield any *M. leprae* DNA in the present study, despite several attempts. The demonstration of *M. leprae* DNA in peripheral bones of a similar age was shown to be difficult (Spigelman and Donoghue, this volume) and near the limits of detection of the method. The failure to find *M. leprae* DNA in a different bone, although disappointing, is not surprising. This shows that the old adage of "absence of proof is not proof of absence" to be very real in palaeomicrobiology. Similarly, the absence of detectable *M. leprae* DNA in the mediaeval metatarsal sample from Suraz, the 14th to 15th century metatarsal specimens, and the fibula sample from Ópusztaszer-Monostor, Hungary does not necessarily imply that the remains were from individuals who were not suffering from the disease. Although many skeletons have been found with changes associated with lepromatous leprosy, the peripheral bones are likely to contain relatively little *M. leprae* DNA compared with the nasal region. In addition, most observers consider the changes in the limbs to be due to secondary infection, which is the result of loss of sensation due to nerve damage by the organism (Møller-Christensen, 1961:42-43) so some affected limbs may contain no *M. leprae* DNA at all.

Møller-Christensen (1969) found that around 70% of skeletons from Naestved Denmark showed evidence of the lepromatous state, which seems to be a very high figure.

We suggest that use of the methods we have described may help prove or disprove the veracity of these figures. However, the presence of palaeopathological changes due to secondary infections ensures that there will always be some uncertainty about whether or not *M. leprae* infection was the primary cause of the pathological changes observed, if no DNA can be identified. We believe it is important that in any specimen suspected of harbouring *M. leprae* that microbiological sampling must include the nasal area and not just sites of the more obvious peripheral changes.

Acknowledgements

Thanks are due to Jo Colston, National Institute for Medical Research, London, UK, who supplied the lepromin skin test reagent as a source of *M. leprae*. Archaeological speciments were kindly provided as indicated: Joe Zias, Hebrew University of Jerusalem Science and Antiquity Group, Israel : the specimen from the Monastery of John the Baptist; Israel Hershkovitz, Department of Anatomy and Anthropology, Tel Aviv University, Tel Aviv, Israel: specimen HZ. Figures 1, 2a and 2b, and specimens Lep I and Lep II were provided by Judyta Gladykowska-Ryyeczycka. Figures 3 to 7, specimens 222, 503 and two samples from 923 were provided by Antónia Marcsik.

M.S. and J.H. acknowledge a grant from The Wellcome Trust for the archaeological studies.

References

Abu Al-Soud, W. and Rådström, P. 2000 Effects of amplification facilitators on diagnostic PCR in the presence of blood, feces and meat. *J. Clin. Microbiol.* 38: 4463-4470.

Andersen, J.G. and Manchester, K. 1992 The rhino-maxillary syndrome in leprosy: a clinical radiological and paleopathological study. *Int. J. Osteoarchaeology* 2:121-129.

Boom, R., Sol, A., Salimans, M., Jansen, C., Wertheim-van Dillen, P. and van der Noordaa, J. 1990 Rapid and simple method for purification of nucleic acid. *J. Clin. Microbiol.* 28: 495-503.

Crubézy, E., Proveda, J.D., Montagnon, D. and Ludes, B., 1997 *Identification of Mycobacterium tuberculosis or bovis DNA in an Egyptian Pott's disease of 5400 years old.* Proceedings of the 4th Ancient DNA Conference, Göttingen, Germany, 5-7 June 1997.

Davis, J.L., Heginbottom, J.A., Annan, A.P., Daniels, R.S., Berdal, B.P., Bergan, T., Duncan, K.E., Lewin, P.K., Oxford, J.S., Roberts, N., Sjehel, J.J. and Smith, C.R. 2000 Ground penetrating radar surveys to locate 1918 Spanish flu victims in permafrost. *J. Forensic Sci.* 45: 68-76.

Dixon, R., Ensor, S., Lewis, E. and Roberts, C.A. 1994 *The detection of Mycobacterium tuberculosis by PCR in ancient human bone.* Proceedings of the 10th European Meeting of the Paleopathology Association, Gottingen, Germany, 1994.

Donoghue, H.D., Holton, J. and Spigelman, M. 2001 PCR primers that can detect low levels of *Mycobacterium leprae* DNA. *J. Med. Microbiol.* 50: 177-182

Donoghue, H.D., Spigelman, M., Zias, J., Gernaey-Child, A.M. and Minnikin, D.E. 1998 *Mycobacterium tuberculosis* complex DNA in calcified pleura from remains 1400 years old. *Letters in Applied Microbiology* 27: 265-269.

Draper, P. 1982 The anatomy of mycobacteria. In C. Ratledge and J. Stanford (eds): *The biology of the mycobacteria*, Volume 1. London: Academic Press, pp. 9-52.

Eiglmeier, K., Honore, N., Woods, S.A., Caudron, B. and Cole, S.T. 1993 Use of an ordered cosmid library to deduce the genomic organisation of *Mycobacterium leprae. Mol. Microbiol.* 7: 197-206.

Faerman. M., Jankauskas, R.J., Gorski, A., Bercovier, H. and Greenblatt, C.L.1997 Prevalence of human tuberculosis in a Mediaeval population of Lithuania studied by ancient DNA analysis. *Ancient Biomolecules* 1: 205-214.

Forbes, B.A. and Hicks, K.E. 1996 Substances interfering with direct detection of *Mycobacterium tuberculosis* in clinical specimens by PCR: effects of bovine serum albumin. *J. Clin. Microbiol.* 34: 2125-2128.

Fricker, E.J., Spigelman, M. and Fricker, C.R. 1997 The detection of *Escherichia coli* DNA in the ancient remains of Lindow Man using the polymerase chain reaction. *Letters in Applied Microbiology* 24: 351-354.

Gladykowska-Rzeczycka, J. 1976 A case of leprosy from a Medieval burial ground. *Fol. Morphol. Warsz.* 35:253-264.

Haas, C.J., Zink, A., Palfi, G., Szeimies, U. and Nerlich, A.G. 2000 Detection of leprosy in ancient human skeletal remains by molecular identification of *Mycobacterium leprae. Amer. J. Clin. Pathol.* 114: 428-436.

Hagelberg, E. and Sykes, B. 1989 Ancient bone DNA amplified. *Nature (Lond.)* 342: 485.

Hartskeerl, R.A., De Wit, M.Y.L. and Klatser, P.R. 1989 Polymerase chain reaction for the detection of *Mycobacterium leprae. J. Gen. Microbiol.* 135:2355-2364.

Hershkovitz, I., Spiers, M., Katznelson, A. and Arensberg, B. 1992 Unusual pathological conditions in the lower extremities of a skeleton from ancient Israel. *Amer. J. Phys. Anthropol.* 88:23-26.

Jamil, S., Wilson, S.M., Hacket, M., Hussein, R. and Stoker, N.G. 1994 A colorimetric PCR method for the detection of *M. leprae* in skin biopsies from leprosy patients. *Int. J. Leprosy* 62: 512-520.

Lassen, C., Hummel, S. and Herrmann B. 1994 Comparison of DNA extraction and amplification from ancient human bone and mummified soft tissue. *Int. J. Leg. Med.* 107: 152-155.

Matheson, C., Donoghue, H.D., Fletcher, H., Holton, J., Thomas, M., Pap, I. and Spigelman, M. 2000 *Tuberculosis in ancient populations: a linkage study*. Paper presented at the 5th International Ancient DNA Conference, Manchester, England, 12th-14th July 2000 (Abstract 33).

Møller-Christensen, V. 1961 *Bone changes in leprosy*. Copenhagen: Munksgaard.

Møller-Christensen, V 1969 Provisional results of the examination of the whole Naevsted leprosy hospital churchyard-ab. 1250-1550 A. D. *Nordisk Medicin-Historisk Arsbok*. Part 4: 29-36.

Pääbo, S., Gifford, J.A. and Wilson, A.C. 1988 Mitochondrial DNA sequences from a 7000-year old brain. *Nucl. Acids Res.* 16: 9775-9787.

Rafi, A., Spigelman, M., Stanford, J., Lemma, E., Donoghue, H. and Zias, J. 1994 DNA of *Mycobacterium leprae* detected by PCR in ancient bone. *Int. J. Osteoarchaeol.* 4: 287-290.

Salo, W.L., Aufderheide, A.C., Buikstra, J. and Holcomb, T.A. 1994 Identification of *Mycobacterium tuberculosis* DNA in a pre-Columbian Peruvian mummy. *Proc. Natl. Acad. Sci. USA* 91: 2091-2094.

Spigelman, M. and Donoghue, H.D. 2001 Brief communication: unusual pathological condition in the lower extremities of a skeleton from ancient Israel. *Amer. J. Phy. Anthropol.* 114:92-93.

Spigelman, M. and Lemma, E. 1993 The use of the polymerase chain reaction to detect *Mycobacterium tuberculosis* in ancient skeletons. *Int. J. Osteoarchaeol.* 3: 137-143.

Taubenberger, J. K., Reid, A H. and Fanning, T. G. 2000 The 1918 influenza virus: a killer comes into view. *Virology* 274: 241-245.

Taylor, G.M., Crossey, M., Saldanha, J. and Waldron,T. 1996 DNA from *Mycobacterium tuberculosis* identified in mediaeval human skeletal remains using polymerase chain reaction. *J. Archaeological. Sci.* 23:789-798.

Taylor, G.M. Rutland, P. and Molleson, T. 1997 A sensitive polymerase chain reaction method for the detection of *Plasmodium* species DNA in ancient human remains. *Ancient Biomolecules* 1: 193-203

Williams, D.L., Gillis. T.P., Booth, R.J., Looker, D.L. and Watson, J.D. 1990 The use of a specific DNA probe and polymerase chain reaction for the detection of *Mycobacterium leprae*. *J. Infect. Dis.* 162: 193-200.

Woods, S.A. and Cole, S.T. 1989 A rapid method for the detection of potentially viable Mycobacterium leprae in human biopsies: a novel application of PCR. *FEMS Microbiology* 53:305-309.

Table 1 Sequences of oligonucleotide primers for *M. leprae*

Primer	Residue[1,2]	Sequence (5' to 3')
LP1	490-509	TGCATGTCATGGCCTTGAGG
LP2	618-599	CACCGATACCAGCGGCAGAA
LP3	505-522	TGAGGTGTCGGCGTGGTC
LP4	603-586	CAGAAATGGTGCAAGGGA
LP5	1050-1070	ATCGACTGTTGTTTGCGCAAC
LP6	1159-1139	CCAGCAACCGAAATGTTCGGA
LP7	1037-1057	TCATAGATGCCTAATCGACTG
LP8	1172-1154	GGCACATCTGCGGCCAGCA

[1] For primers LP1-LP4 the primer sequence is found in the *M. leprae* RELP3 sequence X17153.
[2] For primers LP5-LP8 the primer sequence is found in the *M. leprae* 18 kD Ag gene sequence MSGANT18K.

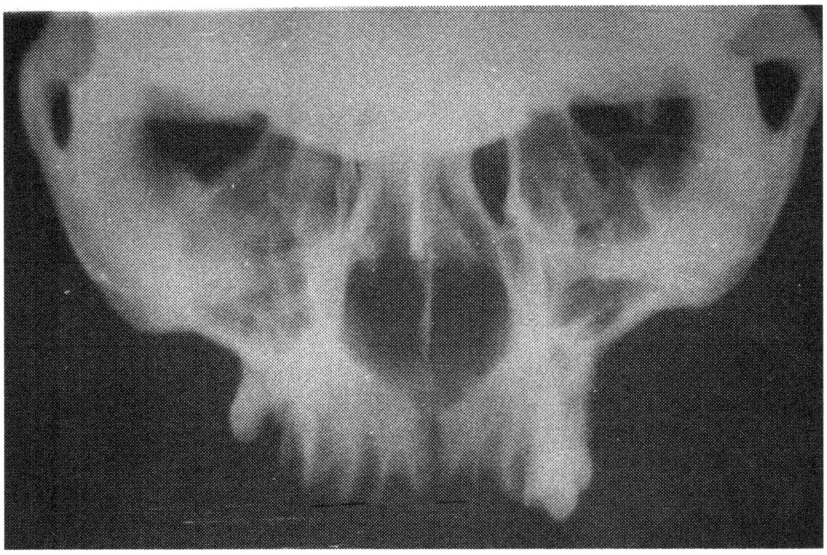

Figure 1: Skull from mediaeval burial ground, Suraz, Poland. Nasal changes on anterior radiograph

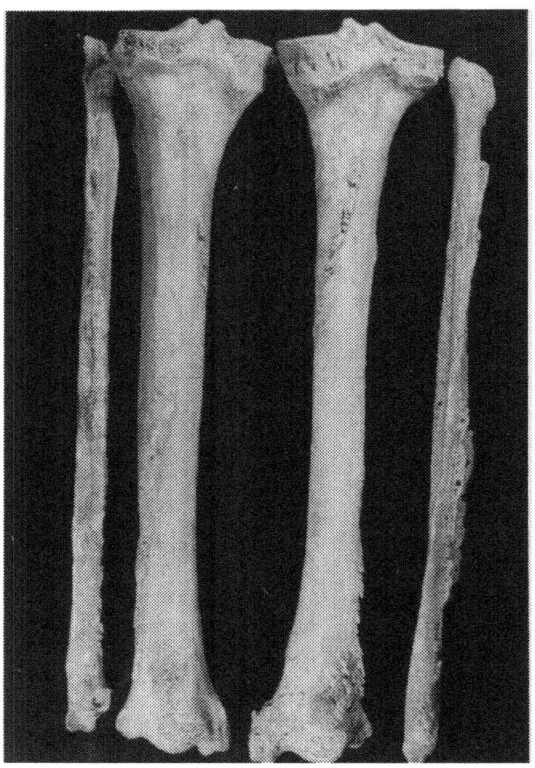

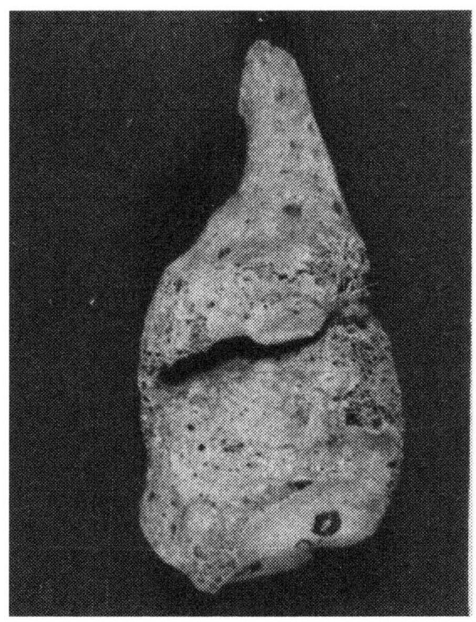

Figure 2: a: New bone formation on the tibae and fibulae in the Suraz sample
b: "Pencilling" of a phalanx in the Suraz sample

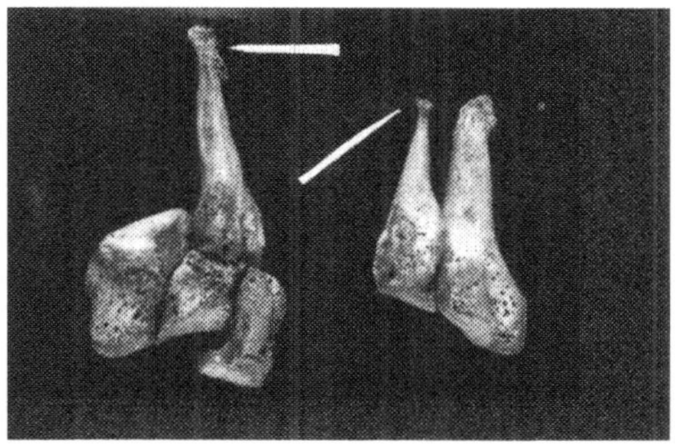

Figure 3: "Pencilling" of metatarsals in a skeleton from grave 923, from Ópusztaszer-Monostor, Hungary.

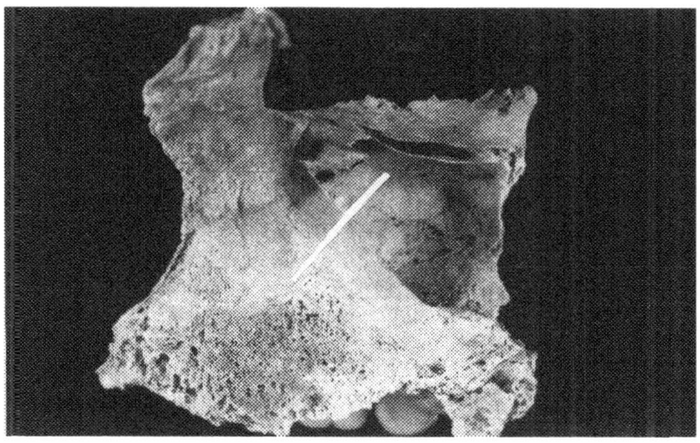

Figure 4: Right maxilla showing periostitis on the nasal surface of skeleton 503, Püspökladány, Hungary

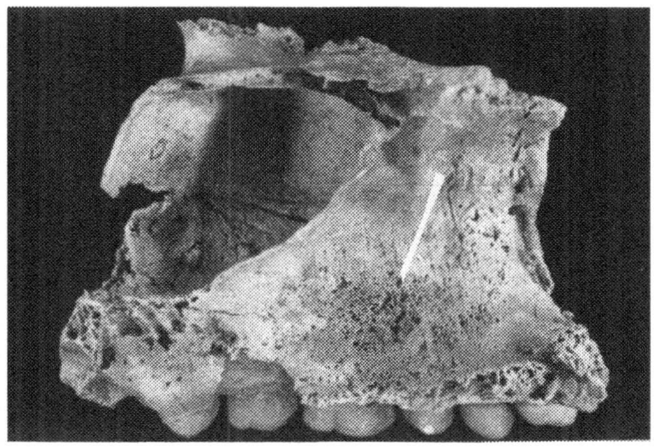

Figure 5: Skeleton 503: Identical changes on the left side.

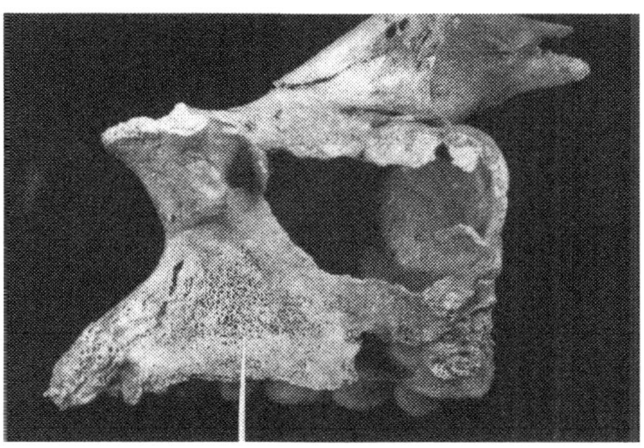

Figure 6: Left side of the maxilla showing periostitis on the nasal surface from skeleton 222, Püspökladány, Hungary.

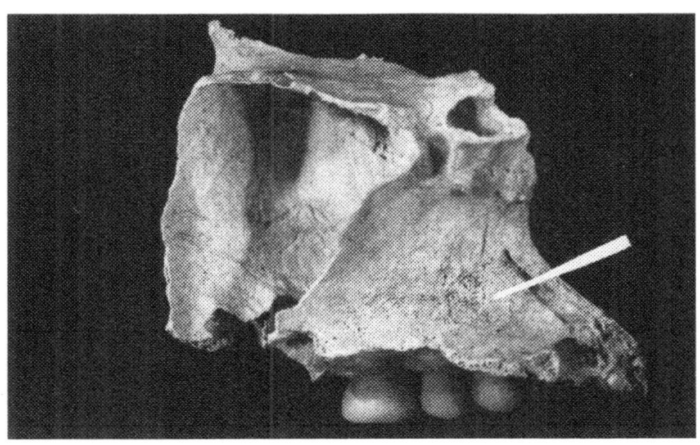

Figure 7: Skeleton 222: Similar changes on the right side.

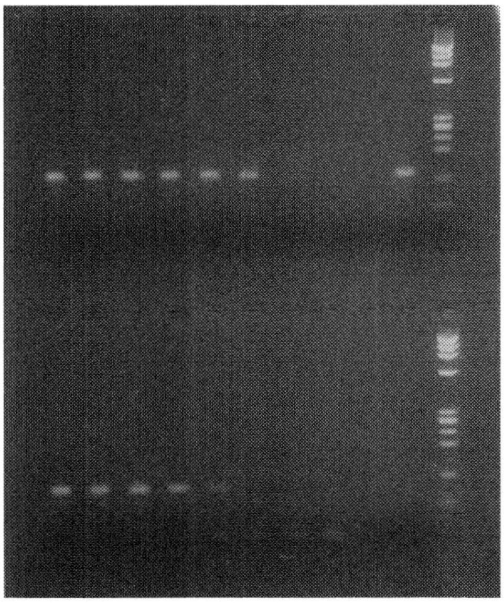

Figure 8: *M. leprae*-specific PCR after 45 cycles of amplification with (top row) outer RLEP primers (129bp product) and (bottom row) inner RLEP primers (99bp product). Left to right: lanes 1-9: *M. leprae* 10^{-2} - 10^{-10} dilutions; lane 10: sample 222 10^{-1} dilution; lane 11: ØX174*Hae*III digest molecular markers.

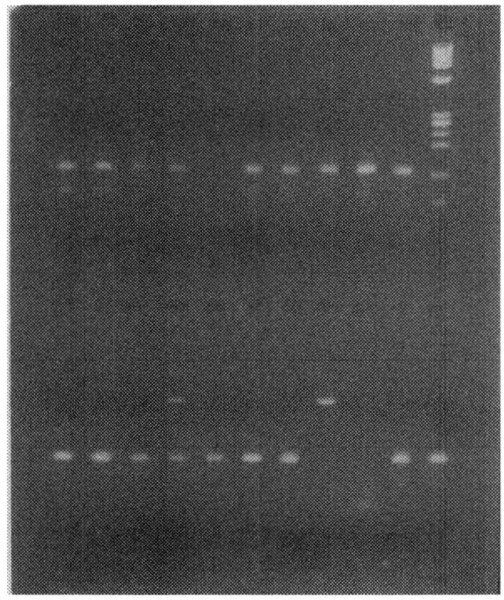

Figure 9: *M. leprae*-specific PCR after 45 cycles of amplification with outer RLEP primers (129bp product) and different concentrations of BSA. Top row (left to right): lanes 1-7: no BSA; lanes 1-5: *M. leprae* 10^{-4} to 10^{-8} dilutions; lane 6: sample 222 (neat); lane 7: sample 503 (neat); lanes 8-10: 4mM BSA, *M. leprae* 10^{-4} to 1 x 10^{-6} dilutions; lane 11: ØX174*Hae*III digest molecular markers.
Bottom row (left to right): lanes 1-7: 2mM BSA; lanes 1-5: *M. leprae* 10^{-4} to 10^{-8} dilutions; lane 6: sample 222 (neat); lane 7: sample 503 (neat); lanes 8-11: 4mM BSA; lanes 8 to 9: *M. leprae* 10^{-7}, 10^{-8} dilutions; lane 10: sample 222 (neat); lane 11: sample 503 (neat).

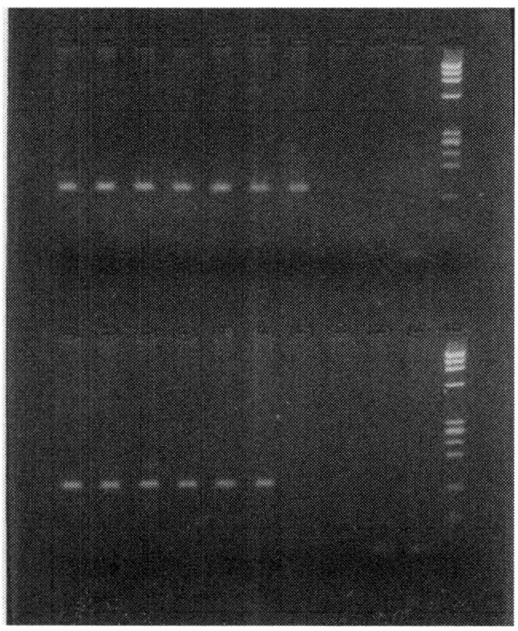

Figure10: *M. leprae*-specific PCR after 45 cycles of amplification with outer primers for the 18kD Ag gene (136bp product). Top row (left to right): lanes 1-6: *M. leprae* 10^{-1} to 10^{-6} dilutions; lanes 7-10: sample 222 (extract 1) 10^{-1} to 10^{-4} dilutions; lane 11: ØX174*Hae*III digest molecular markers.
Bottom row (left to right): lanes 1-2:sample 222 (extract 2) neat and 10^{-1} dilution; lanes 3-5: sample 503 neat, 10^{-1} and 10^{-2} dilutions; lanes 6-7: sample LepII neat and 10^{-1} dilution; lanes 8-9: sample HZ neat and 10^{-1} dilution; lane 10: PCR negative control; lane 11: ØX174*Hae*III digest molecular markers.

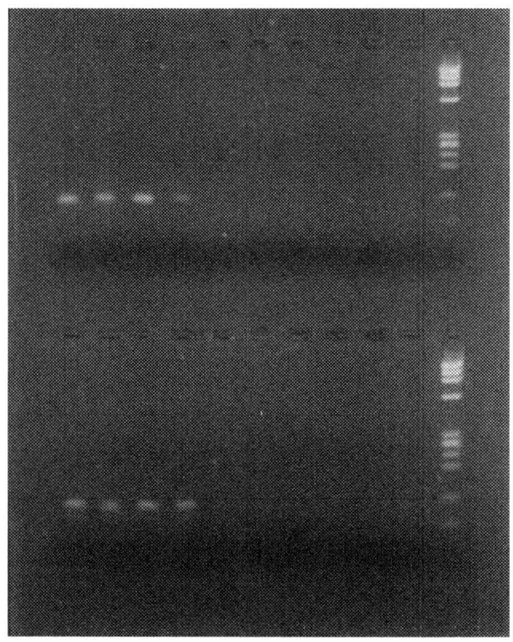

Figure 11: *M. leprae*-specific nested PCR after 2 x 25 cycles of amplification with (Top row) 18kD Ag gene outer then inner primers (110bp product) and (Bottom row) RLEP outer then inner primer pairs (99bp product). Left to right: lane 1: *M. leprae* 10^{-5} dilution; lane 2: sample LepII; lanes 3-4: sample 222 neat and 10^{-1} dilution; lane 5: sample 222 (supernate); lanes 6-9: Monastery of St John the Baptist and other negative samples; lane 10: PCR negative control; lane 11: ØX174*Hae*III digest molecular markers.

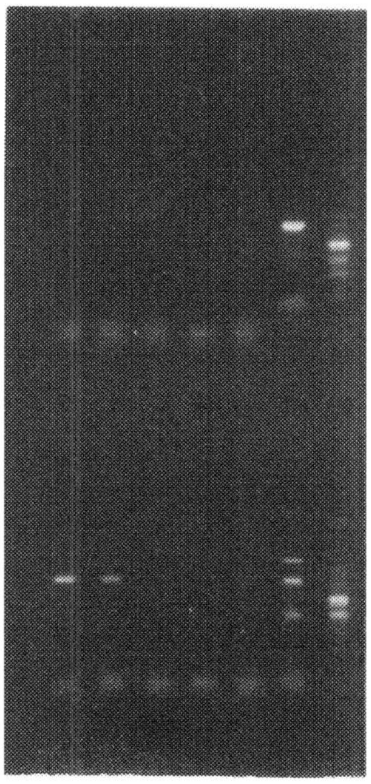

Figure 12: *M. leprae*-specific PCR after 45 cycles of amplification with primers for the 36kD antigen gene (531bp product) and different concentrations of BSA. Top row (left to right):
lanes 1-7: no BSA; lanes 1-5: *M. leprae* 10^{-4} to 10^{-8} dilutions; lane 6: sample 222; lane 7: sample 503. Bottom row (left to right): lanes 1-7: 2mM BSA; lanes 1-5: *M. leprae* 10^{-4} to 10^{-8} dilutions; lane 6: sample 222; lane 7: sample 503.

```
M.leprae:       517   GTGGTCAATGTGGCCGCACCTGAACAGGCACGTCCCCGTGCACGGTATAA    566
Lepromin:       1     .****A****C***************************************     49
503:            1     **************************************************     50
222(LP1):       1     ......********************************************     44
222(LP3):       1     .................****************N******NC********     33
H.sapiens1:     59383 .......................******************.........  59400


M.leprae:       567   CTATTCGCACCTGATGTTATCCCTTGCACCATTTCTGCCGCTGGTATCGG    616
Lepromin:       50    **************************************************     99
503:            51    **************************************************    100
222 LP1:        45    **************************************************     94
222 LP3:        34    *******_*****************************N******......     76
X.sulcatipes2:  127   ....................................**************    140


M.leprae:       617   TG................................................    618
Lepromin:       100   **................................................    101
503:            101   **AA..............................................    104
X.sulcatipes:   141   **AA..............................................    144
```

Underlined sequence corresponds to outer primer, italics indicates inner primer (Table 1)
****** sequence homology
[1]HS322B1 Human DNA sequence from clone 322B1 on chromosome 22q11-12
[2]AY005287.1 *Xylocopa sulcatipes* elongation factor-1 alpha (EF-1a) gene

Figure 13: Alignment of genome sequences (5' - 3') from X17153.1|MLREP3 *M. leprae* repetitive element, RLEP3; PCR product from diluted lepromin reagent, samples 503 and 222 obtained with the outer RLEP primer LP1; PCR product from sample 222 with the inner RLEP primer LP3; and other sequences with some homology.

```
M.leprae:      589    GGGATAACATCAGGTGCGAATAGTTATACCGTGCACGGGGACGTGCCT-G      540
Lepromin:      1      *****************N*************N**********N*T***-*       49
503:           1      *******N****************************************-*       49
222(LP2):      1      .***********************************************-*       48
222(LP4):      1      ................**************N***-*************T*       33
H.sapiens1:    59400  ................................****************-*    59384
H.sapiens2:    54114  ...........................................*****T*    54120
```

```
M.leprae:      540    TTCAGG--TGCGGCCACATTGACCACGCCGACACCTCAAGGCCATGACATGCA   490
Lepromin:      50     **...................................................    51
503:           50     ******--******************N**************************   100
222(LP2):      49     ******--*********************************************    99
222(LP4):      34     ****A*GT*****G**************************NG***********    86
H.sapiens1:    59383  *....................................................  59383
H.sapiens2:    54121  ****A*GT***..........................................  54131
S.fredii3:     4360   ............*G****************.......................   4377
```

Underlined sequence corresponds to outer primer, italics indicates inner primer (Table 1)

****** sequence homology

[1]HS322B1 Human DNA sequence from clone 322B1 on chromosome 22q11-12

[2]HS568C11 Human DNA sequence from clone RP4-568C11 on chromosome 20p11.21-11.23

[3]AF039956 *Sinorhizobium fredii* cytochrome c oxidase CcoO subunit (fixO) gene

Figure 14: Alignment of genome sequences (3' - 5') from X17153.1|MLREP3 *M. leprae* repetitive element, RLEP3; PCR products from diluted lepromin reagent, samples 503 and 222 obtained with the outer RLEP reverse primer LP2; PCR product from sample 222 with the inner RLEP reverse primer LP4; and other sequences with some homology.

```
M.leprae:          1065    CGCAACAAGCTAAGCACGAAGCTAAGCACGCCGC---TGCGGTCAAAAGC    1111
Lepromin:          1       **********************************---*************      47
222(LP7):          1       .........********************CG***GTC************T      41
222(LP5):          1       ...........................*******---***N*A*******      20
A.orientalis1:     9430    ..........................***CG***GTC******.......    9414
D.melanogaster2:   241607  .......................******CG***GTC***.......... 241591
```

```
M.leprae:          1112    CCGTCTTAGCCATGGATCTTTAAAGTATCCGAACATTTCGGTTGCTGGCCG   1162
Lepromin:          48      ***************************************************     98
222(LP7):          42      ***A*****..........................................     50
222(LP5):          21      ************************************************...     68
H.sapiens3:        103504  ..*******************..............................103486

M.leprae:          1163    CAGATGTGCC.........................................   1172
Lepromin:          99      ***................................................    101
```

Underlined sequence corresponds to outer primer, italics indicates inner primer (Table 1)

****** sequence homology

[1]AOPCZA361 *Amycolatopsis orientalis* cosmid PCZA361

[2]AE003499 *Drosophila melanogaster* genomic scaffold 142000013386053 section 16 of 30

[3]AC017028 *Homo sapiens* clone RP11-275G7

Figure 15: Alignment of genome sequences (5' - 3') from MSGANT18K *M. leprae* 18 kD antigen gene; PCR products from diluted lepromin reagent and sample 222 obtained with the outer RLEP primer LP7; PCR product from sample 222 obtained with the inner primer LP5; and examples of sequences showing some homology.

```
M.leprae:            1146  TGTTCGGATACTTTAAAGATCCATGGCTAAGACGGGCTTTTGACCGCAGC    1097
LepII:               1     *****N********************************************      50
222(LP6)(a):         1     .............................****N**N**********N**      21
222(LP6)(b):         1     .....................................**********N**      13
H.sapiens[1]:        103486 ..............******************* ................  103504
D.radiodurans[2]:    8560  .........................................*********    8552

M.leprae:            1096  GGCGTGCTTAGCTTCGTGCTTAGCTTGTTGCGCAAACAACAGTCGATTAG    1047
LepII:               51    **************************************************     100
222(LP6)(a):         22    *-**************N****N**************************...      67
222(LP6)(b):         14    *****NN****************************************...      60
M.musculus[3]:       110762 ....***********C**********........................  110741
D.radiodurans:       8551  *********.........................................    8543

M.leprae:            1046  GCATCTAT..........................................    1039
LepII:               101   ********..........................................     108
```

Underlined sequence corresponds to outer primer, italics indicates inner primer (Table 1)

****** sequence homology

[1]AC017028 *Homo sapiens* clone RP11-275G7

[2]AE001931 *Deinococcus radiodurans* R1 section 68 of 229 of the complete chromosome 1

[3]AC020969 *Mus musculus* chromosome 18 clone RP23-19K11

Figure 16: Alignment of genome sequences (3' - 5') from MSGANT18K *M. leprae* 18 kD antigen gene; PCR products from sample LepII obtained with the outer reverse primer LP8; PCR product from sample 222 obtained with the inner reverse primer LP6; and sequences showing some homology.

M.leprae:	5465	CCTACACATTCTGGGTCACCCGGGTGCTGGCTTATGTCATCGACAACATC	5514
Lepromin:	1	****************_****N****************************	49
M.leprae:	5515	CCAGCCACGGTCCTGCTCGGCATTGGCATGTTGATTCAGACGCTCACGAA	5564
Lepromin:	50	**	99
222:	1	....**	2
M.leprae:	5565	GCAAGAGGCGTGCGTCACTGATATCA-CGCAGTACAATGTTAATCAGTAC	5613
Lepromin:	100	**************************_***********************	148
222:	3	******A********N********N*C*N***************N*****	52
M.leprae:	5614	TGTGCTACTCAGCCTACCGGCATCGGCATGTTGGCGTTCTGGTTCGCATG	5663
Lepromin:	149	**	198
222:	53	**N******N*************N*************NN***********	102
M.leprae:	5664	GTTGATGGCGACGGCCTACCTGGTCTGGAACTACGGCTATCGCCAGGGCG	5713
Lepromin:	199	**	248
222:	103	*********N*************GN**N********N***********NN	152
M.tuberculosis[1]:	31585	................*********************************A	31618
M.leprae:	5714	CCACCGGCTCCAGCATTGGCAAGACGGTAATGAAGTTCAAGGTGATCAGC	5763
Lepromin:	249	**	298
222:	153	*****N*G*T*********N***********N**********N*******	202
M.tuberculosis:	31619	***T***G**G*****C******T****GC**************G*****	31668
H.sapiens[2]:	90517	.*******************..............................	90499
M.musculus[3]:	102289	.....*******************..........................	102271
D.melanogaster[4]:	23018	....................................**************	23031
M.leprae:	5764	GAGGCTACTGGGCAGCCAATCGGTTTCGGTATGTCGGTGGTGCGCCAGCT	5813
Lepromin:	299	**	348
222:	203	*N*******************.............................	222
M.tuberculosis:	31669	***A*C**C*****A********C*****G***********A********	31718
D.melanogaster[4]:	23032	*****...	23036
M.leprae:	5814	CGCGCACTTCGTCGATGCGGTCATCTGTTGCATCGGGTTCTTGTTCCCGC	5863
Lepromin:	349	**	398
222:	1	....*	1
M.tuberculosis:	31719	T**C*****TA****C***A*******C*T*G********C****T****	31768
M.leprae:	5864	TGTGGGACTCCAAACGGCAAACCCTGGCGGACAAGATCATGACGACGGTG	5913
Lepromin:	399	**	448
222:	2	*****N*********T*******N**************************	51
M.tuberculosis:	31769	********G*T***********GT**************************	31818
M.leprae:	5914	TGTCTGCCGATCTGACACAGTGCCACGACTTGGCAGGC............	5951
Lepromin:	449	************************************..............	484
222:	52	**NT*NT****N*******N*************N****............	89
M.tuberculosis:	31819	**CG***********...................................	31833

Underlined sequence corresponds to primer

****** sequence homology

[1] MTV017 *Mycobacterium tuberculosis* H37Rv complete genome; segment 48/162
[2] AC020916 *Homo sapiens* chromosome 19 clone CTD-3252C9, complete sequence
[3] AC026385 *Mus musculus* 11 BAC RP23-413K14 complete sequence
[4] AE003511 *Drosophila melanogaster* genomic scaffold 142000013386053 section 28 of 30

Figure 17: Alignment of genome sequences (5' - 3') from MLU15183 *M. leprae* cosmid B1740 (36 kD antigen gene); PCR products from diluted lepromin reagent and sample 222 obtained with primerS13; and sequences showing some homology.

```
M. leprae:              5928 TCAGATCGGCAGACACACCGTCGTCATGATCTTGTCCGCCAGGGTTTGCC   5879
Lepromin:               1    **********************N***************************     50
M.tuberculosis[1]:      31833 ***********CG****************************AC*******  31784

M. leprae:              5978 GTTTGGAGTCCCACAGCGGGAACAAGAACCCGATGCAACAGATGACCGCA   5829
Lepromin:               51   **************************************************    100
M.tuberculosis:         31783 ****A*C************A****G********C*A*G*******T***G  31734

M. leprae:              5928 TCGACGAAGTGCGCGAGCTGGCGCACCACCGACATACCGAAACCGATTGG   5779
Lepromin:               101  **************************************************    150
222:                    1    .....******N**************************************     44
M.tuberculosis:         31733 ****TA*****G**A********T***********C*****G********  31684

M. leprae:              5778 CTGCCCAGTAGCCTCGCTGATCACCTTGAACTTCATTACCGTCTTGCCAA   5729
Lepromin:               151  **************************************************    200
222:                    45   ***G****N*N***T***********************************    194
M.tuberculosis:         31683 T*****G**G*T********C**************GC****A******G*  31634
H.sapiens[2]:           90499 .............................................*****  90503
M.musculus[3]:          102271 .........................................********* 102279
D.melanogaster[4]:      23036 ..........******************* ....................  23018

M. leprae:              5728 TGCTGGAGCCGGTGGCGCCCTGGCGATAGCCGTAGTTCCAGACCAGGTAG   5679
Lepromin:               201  **************************************************    250
222:                    195  *******N*******N*******NA**N*N********************    244
M.tuberculosis:         31633 ****C**C***A***T*********************************.  31585
H.sapiens:              90504 **************....................................  90517
M.musculus:             102280 **********........................................ 102289

M. leprae:              5678 GCCGT-CGCCATCAACCATGCGAACCAGAACGCCAACATGCCGATGCCGGT  5629
Lepromin:               251  *****-*********************************************   300
222:                    245  *****G********************************************N   295

M. leprae:              5628 AGGCTGAGTAGCACAGTACTGATTAACATTGTACTGCGTGATATCAG-TGA  5579
Lepromin:               301  ***********************************************-***   350
222:                    296  N******N**NN***T************N**N****N*********NAN**   346

M. leprae:              5578 CGCACGCCTCTTGCTTCGTGAGCGTCTGAATCAACATGCCAATGCCGAGC   5529
Lepromin:               351  **************************************************    400
222:                    347  *T***N**********N*****************************A***    396
A.thaliana[5]:          29881 ................................******************  29898

M. leprae:              5528 AGGACCGTGGCTGGGATGTTGTCGATGACATAAGCCAGCACCCGGGTGAC   5479
Lepromin:               401  **************************************************    450
222:                    397  **********************NN*************************N    446
A.thaliana:             29899 *.................................................  29899

M. leprae:              5478 CCAGAATGTGTAGGCCTCCTTAGGCAGTGAGCGGATCGCCGGTCCA....   5433
Lepromin:               451  **********************************************....    496
222:                    447  ***T***********A**-*************N*****N*****......    489
```

Underlined sequence corresponds to primer

****** sequence homology

[1]MTV017 *Mycobacterium tuberculosis* H37Rv complete genome; segment 48/162
[2]AC020916 *Homo sapiens* chromosome 19 clone CTD-3252C9, complete sequence
[3]AC026385 *Mus musculus* 11 BAC RP23-413K14 complete sequence
[4]AE003511 *Drosophila melanogaste*r genomic scaffold 142000013386053 section 28 of 30
[5]ATF18D22 *Araidopsis thaliana* DNA chromosome 5, BAC clone F18D22

Figure 18: Alignment of genome sequences (5' - 3') from MLU15183 *M. leprae* cosmid B1740 (36 kD antigen gene); PCR products from diluted lepromin reagent and sample 222 obtained with primerS62; and sequences showing some homology.

Molecular evidence of *Mycobacterium leprae* in historic bone samples from South Germany

C.J. Haas[1], A. Zink[2], U. Szeimies[3] and A. G. Nerlich[4]

[1]Department of Pathology
Zentralklinikum Augsburg
PF 101920
D-86009 Augsburg
Germany
e-mail: molekular.pathologie@klinikum-augsburg.de

[2]Department of Pathology
Ludwig-Maximilians-Universität München
Thalkirchner Str. 36
D-80337 München
Germany
e-mail: Albert.Zink@lrz.uni-muenchen.de

[3]Department of Diagnostic Radiology
Ludwig-Maximilians-Universität München
Ziemssenstr. 1
D-80336 München
Germany
e-mail: szeimies@medinn.med.uni-muenchen.de

[4]Ludwig-Maximilians-Universität München
Thalkirchner Str. 36
D-80337 München
Germany
e-mail: Andreas.Nerlich@lrz.uni-muenchen.de

Abstract

A method was established to identify mycobacterial DNA in ancient skeletal remains. The investigated specimens were derived from an ossuary in Rain/Lech, Bavaria, dating from approximately AD 1400-1800, and comprise two skulls with macro-morphological alterations suggestive of leprosy and one mandible with pathological changes suggestive of facial tuberculosis. These specimens were analyzed for the presence of DNA from *M. leprae* and *M. tuberculosis* by molecular techniques. Using the PCR (Polymerase chain reaction) segments of the *M. leprae* specific sequences RLEP 1 and RLEP 3 were amplified in the skull samples and their specificity was confirmed by direct sequencing. In the case of the mandible, amplification of RLEP1 and RLEP3 was negative, but instead the *M. tuberculosis* specific sequence IS6110 was amplified. Thus, the unambiguous identification of *M. leprae* and *M. tuberculosis* was demonstrated in historical samples, allowing for the distinction between different mycobacterial infections and more profound research on the palaeoepidemiology of these infectious diseases.

Keywords

Ancient DNA, *Mycobacterium leprae,* PCR, palaeopathology, historic bone samples

NB: another version of this paper has since been published in *Am. J. Clin. Pathol.* 114:428-436 (2000)

1. Introduction

Infectious diseases such as tuberculosis and leprosy often result in characteristic morphological alterations in the skeleton which can be relatively easily identified in ancient human remains. However, in cases with less significant bone changes it is more difficult to come to a clear diagnosis of the underlying disease. It is particularly in these cases that the analysis of genetic material in ancient tissue can help to clarify the uncertainty of the morphological analysis. With the development of modern methods of molecular biological analysis, such as PCR (polymerase chain reaction) and sequencing, this has become possible during recent years. These techniques allow for the identification of ancient bacterial DNA in ancient human remains, and thus provide direct evidence for the occurrence and frequency of infectious diseases in historic populations. Furthermore, they may yield information about the evolution of microorganisms and their associated diseases.

In this respect the molecular analysis of skeletons with

possible infection with *M. leprae* is of particular interest, as several aspects of the palaeopathology and palaeoepidemiology of leprosy are still under discussion (Aufderheide and Rodríguez-Martin, 1998). However, only one previous report deals with the extraction of *M. leprae* DNA from ancient bone samples (Rafi *et al.*, 1994), although others are reported in this volume. We therefore established a method for the unambiguous identification of *M. leprae* in ancient bone samples by means of molecular biology. For our study we chose historic bone samples from an ossuary in South Germany, which were between 200 and 400 years old.

2. Materials and methods

2.1 Materials

The molecular analysis was conducted on historic bone samples with some obvious pathological abnormalities. This material derived from an ossuary located in the town Rain/Lech in Southern Germany which had been used between 1400 and 1800 AD to house the skeletal remains of an in-town cemetery. Despite historical evidence for a leprosy hospital in Rain/Lech between 1481 and 1632 AD, it is not known whether the leprosy patients were buried in this urban cemetery. The analysed specimens comprised two skulls (Numbers R1788 and R2208), one mandible (R180), and one non-pathological humerus (R424). The following describes the samples:

R1788: The skull of an adult female individual with typical pathological lesions *of facies leprosa*, i.e. atrophy of the alveolar bone with loss of the upper incisors, thinning of the hard palate with an irregular bone surface, osseous remodelling of the margins of the pyriform aperture and absence of the anterior nasal spine (Figure 1). The sample for the DNA extraction was taken from the hard palate.

R2208: This was also the skull of an adult female (Figure 2A). The morphological alterations suggesting leprosy were less extensive: only minor osteitis of the hard palate and slight osseous remodelling of the margins of the pyriform aperture were noted (Figure 2B). The anterior nasal spine was reduced. In this case the alveolar bone was not preserved. Part of the hard palate was used for the DNA-extraction.

R180: The mandible of an adult male individual (Figure 3), characterized by extended osteolysis in the region of the anterior teeth with reactive new bone formation. These morphological changes were probably due to an infectious inflammatory process other than leprosy. This individual may have suffered from facial tuberculosis (*lupus vulgaris*). The bone sample for the DNA extraction was derived from the osteolytic region.

R424: As a control, the right humerus of an adult individual was used. It showed no evidence of pathological alterations.

2.2 Methods

For the analysis, samples of bone were taken from the hard palate of the skulls (R1788, R2208), from the osteolytic region of the mandible (R180), and from the shaft of the humerus (R424). After elimination of possible contamination by treatment with sodium hypochlorite solution and abrasion of the outer surface of the bone, the samples were pulverized and an extraction buffer (0.5 mol/L EDTA, 0.25 g/L proteinase K) was added (Evison *et al.*, 1997). Following an incubation for 2 days in a rotatory mixer the samples were centrifuged for 15 min at 3000 g and the supernatants were transferred into new vials. To the supernatants two volumes of freshly prepared guanidine isothiocyanate solution (4 M guanidine isothiocyanate, 0.1 M Tris-HCl (pH 6.4), 25 mM EDTA (pH 8.0), 20 mM Triton X-100) containing diatomaceous earth (Boom *et al.*, 1990) were added. After incubation for 2 hours the diatomaceous earth was pelleted by centrifugation and washed twice with 70 % ethanol and once with acetone. The DNA was then eluted with sterile water and concentrated with microconcentrators (Microcon-30, Millipore, Bedford, MA). The final DNA-solution was diluted to 20 µL with sterile water and stored at 4 °C until used in the PCR.

For the specific identification of *M. leprae* DNA segments of the two repetitive elements RLEP1 and RLEP3 were amplified by PCR (Yoon *et al.*, 1993, Jamil *et al.*, 1994). The expected size of the PCR products were 372 bp and 320 bp for RLEP1 and RLEP3, respectively. For the identification of *M. tuberculosis* the insertional sequence IS6110 was amplified resulting in a 123 bp fragment. According to the original publication, this PCR product is specific for *M. tuberculosis*, *M. bovis* and *M. simiae* (Eisenach *et al.*, 1990).

The amplifications of RLEP1 and RLEP3 were performed in reaction volumes of 20 µL containing PCR Buffer II (PE Biosystems, Foster City, CA), 2.0 mmol/L $MgCl_2$, 200 µmol/L of each deoxynucleotide trisphosphate (Amersham Pharmacia, Uppsala, Sweden), 1 µmol/L of each primer, 0.025 U/µL AmpliTaq Gold (PE Biosystems) and 0.5 µL DNA solution. The PCR-conditions were as follows: after activation of the Taq-polymerase for 10 min at 95 °C the amplification was performed for 45 cycles, each consisting of denaturation at 94 °C for 1 min, annealing at 61 °C (RLEP1) or 52 °C (RLEP3) for 1 min, and extension at 72 °C for 3 min. The final cycle was 8 min at 72 °C. The PCR conditions for the amplification of the IS6110 specific fragment have been described elsewhere (Haas *et al.*, 1999).

As a control for successful DNA extraction and exclusion of PCR inhibition a 202 bp segment of the human β-actin gene was amplified (Ghossein *et al.*, 1994). The PCR mixture was as described for the amplification of the *M. leprae* specific sequences and the cycling conditions were as follows: 10 min at 95 °C, 45 cycles of 94 °C for 1 min,

60 °C for 1 min and 72 °C for 3 min, and final extension at 72 °C for 8 min. After amplification, the PCR products were electrophoresed on a 4 % agarose gel and visualized on a UV-screen after staining with SYBR Green (Molecular Probes, Eugene, OR). For direct sequencing of the PCR products the respective bands were eluted with a purification kit (Freeze'n Squeeze, Bio-Rad, Hercules, CA). Following this, a cycle sequencing reaction was performed with dye terminators according to the manufacturer's instructions (PE Biosystems) followed by automatic sequencing on an ABI PRISM 310 Genetic Analyzer (PE Biosystems).

To confirm the specificity of the of IS6110 amplification, the 123 bp PCR product was digested with the restriction enzyme Hae III (Bohle *et al.*, 1994). For the digestion, 8 μL of the PCR product were incubated with 10 U Hae III (Roche Diagnostics, Mannheim, Germany) for 2 h at 37 °C. This resulted in a 94 bp and a 29 bp fragment, which were analyzed by agarose gel electrophoresis.

3. Results

For the unambiguous identification of *M. leprae* infection of ancient skeletal remains a molecular analysis with several bone samples with changes suggesting leprosy was performed. From these cases DNA was extracted and, subsequently, PCRs with primers specific for various mycobacterial genes were run: with primers directed against the repetitive element 1 of the *M. leprae* genome (RLEP1), the expected amplification product of 372 bp (Yoon *et al.*, 1993) was obtained for the cases R1788 and R2208 (Figure 4A). In these samples, the PCR for the amplification of the 320 bp fragment (Jamil *et al.*, 1994) of the repetitive element 3 of *M. leprae* (RLEP3) was also positive (Figure 4B). No specific PCR products with these primer pairs were obtained with the sample of the male mandible (R 180) nor the control sample of the macroscopically unremarkable humerus (R424). The blanks were negative for all amplifications. To exclude the possibility that the negative results were due to the presence of inhibitors of PCR or the absence of any amplifiable DNA, a 202 bp fragment of the human β-actin gene (Ghossein *et al.*, 1994) was amplified from all samples (data not shown). The specificity of the amplified RLPEP1 and RLEP3 segments was confirmed by direct sequencing of these PCR products (Figure 5), which were identical to the respective published sequences of recent *M. leprae*.

Additionally to the specific analysis of *M. leprae* DNA, the possible infection with *M. tuberculosis* was invesigated with primers targeting the insertional sequence IS6110 (Eisenach *et al.*, 1990). The expected segment of 123 bp was amplified in case R180, whereas no PCR product of that size was obtained with the other cases (Figure 6A). For confirmation of the specificity of the 123 bp amplification product of case R180 a digestion was performed with the restriction enzyme Hae III (Bohle *et al.*, 1994), which resulted in the expected fragments of 94 bp and 29 bp (Figure 6B). With the amplification products of cases R1788 and R2208 no Hae III digestion was seen (data not shown).

4. Discussion

Infectious diseases such as tuberculosis or leprosy are sometimes manifested by characteristic pathological alterations in ancient skeletal remains of humans. However, more often the pathological changes are less significant, and therefore a definite diagnosis of the underlying disease is not possible by macromorphological criteria alone. In these cases modern methods of molelcular biology offer new opportunities to clarify the uncertainty of morphological analysis.

The development of PCR, and its application to ancient material, has made the unambiguous identification of ancient bacterial DNA possible (Pääbo, 1989; Spigelman and Lemma, 1993). This technique has been repeatedly applied in palaeopathology for the identification of *M. tuberculosis* (*e.g.* Salo *et al.*, 1994; Nerlich *et al.*, 1997; Haas *et al.*, 1999). However, previous to this volume, only one report had been published in which DNA of *M. leprae* could be identified in bone samples from ancient sources (Rafi *et al.*, 1994). Nevertheless, the broader application of PCR technology to the identification of *M. leprae* seems to be a promising tool for the investigation of diverse aspects of the palaeopathology and palaeoepidemiology of *M. leprae.* Apart from the definite diagnosis of leprosy in individual cases of historic relevance, questions concerning the historic spread of *M. leprae*, i.e. time and place of early occurrence of leprosy, as well as its disappearance, may be addressed. In addition, the frequency of infection of historic populations with *M. leprae* may be estimated more precisely. The possible co-infection of leprosy with other infectious diseases, especially *M. tuberculosis*, could also be investigated. Finally, in connection with sequencing of the amplified DNA from *M. leprae,* information about the evolution of this infectious bacterium may also be obtained.

Therefore, a protocol for the successful amplification of DNA from *M. leprae* from ancient bone samples has been developed, and by using this method several historic bone samples were analysed for infection with *M. leprae*. Additionally these samples were tested for infection with *M. tuberculosis*. The material derived from an ossuary in Southern Germany, dating from approximately AD 1400-1800. The two skulls from the ossuary (R1788 and R2208) showed morphological alterations suggestive of infection with *M. leprae*. The pathological alterations of the third case (R180), a male mandible, were not characteristic for leprosy but may have been due to facial tuberculosis. The sample taken from a humerus (R424) from this archaeological site was used as negative control for the amplification of DNA from *M. leprae* and *M. tuberculosis*. With the samples taken from the cases of suspected leprosy

(R1788 and R2208) DNA of *M. leprae* was amplified and the identity of the PCR products was confirmed by sequencing. Thus, infection with *M. leprae* was specifically proved for these cases. As expected, in the samples where leprosy was not suspected, no *M. leprae* specific DNA could be amplified. A positive result of IS6110 specific PCR supported the morphological diagnosis of possible facial tuberculosis for R180 (extensive osteolysis and reactive bone formation). In contrast, the typical leprosy cases did not show additional infection with *M. tuberculosis*. In summary, the results of this study demonstrate that, with the protocol described here, it is possible to isolate DNA of high quality from ancient bone samples containing *M. leprae*, as can be concluded from the relatively long segments we were able to amplify, and from the high quality of the electropherograms of the sequencing reactions. This is in agreement with other reports which have demonstrated that PCR products of more than 300 bp can be obtained from ancient tissue, especially bones (Haack, 1998, Lassen *et al.*, 1994). It is assumed that the very good preservation, especially of *M. leprae* DNA, may be due to the extraordinary resistance of the cell walls of these bacteria to degradation (Rafi *et al.*, 1994). Furthermore, in connection with our experimental approach of DNA isolation, the application of specific PCR protocols allows for the discrimination between *M. leprae* and *M. tuberculosis* infections in ancient bones. This makes possible the unambiguous identification of leprous infection in historic samples where the morphological diagnosis is unclear. As a consequence, various aspects of the palaeopathology and palaeoepidemiology of leprosy may be addressed more easily. Finally, the high quality of the sequencing results suggests that further investigations of ancient mycobacterial DNA may considerably contribute to the unravelling of the phylogenetic tree of mycobacteria.

References

Aufderheide, A.C. and Rodríguez-Martin, C. 1998 *The Cambridge encyclopedia of human paleopathology*. Cambridge: University Press.

Bohle, R.M., Heidemann, A., Borkhardt, A., Velcovsky, H-G. and Altmannsberger, H-M. 1994 Detection of mycobacterial DNA from paraffin-embedded tissue with epitheloid granulomas of lung and lymph nodes. *Verh. Dtsch. Ges. Path.* 78:189-194.

Boom, R., Sol, C.J.A., Salimans, M.M.M., Jansen, C.L., Wertheim-van Dillen, P.M.E. and van der Noordaa, J. 1990 Rapid and simple method for purification of nucleic acids. *J. Clin. Microbiol.* 28:495-503.

Eisenach, K.D., Cave, M.D., Bates, J.H. and Crawford, J.T. 1990 Polymerase chain reaction amplification of a repetitive DNA sequence specific for *Mycobacterium tuberculosis*. *J. Infect. Dis.* 161:977-981.

Evison, M.P., Smillie, D.M. and Chamberlain, A.T. 1997 Extraction of single-copy nuclear DNA from forensic specimens with a variety of postmortem histories. *J. Forensic Sci.* 42:1032-1038.

Ghossein, R.A., Ross, D.G., Salomon, R.N. and Rabson, A.R. 1994 A search for mycobacterial DNA in sarcoidosis using the polymerase chain reaction. *Am. J. Clin. Pathol.* 101:733-737.

Haack, K. 1998 Screening auf aDNA-Sequenzen von über 300 bp Länge. *HOMO* 49/S:40.

Haas, C.J., Zink, A., Molnár, E., Marcsik, A., Dutour, O., Nerlich, A.G. and Pálfi, G. 1999 Molecular evidence for tuberculosis in Hungarian tissue samples. In G. Pálfi, O. Dutour, J. Deák and I. Hutás (eds): *Tuberculosis: Past and Present*. Szeged, Hungary: Tuberculosis Foundation and Budapest, Hungary: Golden Book Publisher Ltd., pp. 385-391.

Jamil, S., Wilson, S.M., Hacket, M., Hussain, R. and Stoker, N.G. 1994 A colorimetric PCR method for the detection of *M. leprae* in skin biopsies from leprosy patients. *Int. J. Lepr.* 62:512-520.

Lassen, C., Hummel, S. and Herrmann, B. 1994 Comparison of DNA extraction and amplification from ancient human bone and mummified soft tissue. *Int. J. Leg. Med.* 107:152-155.

Nerlich, A.G., Haas, C.J., Zink, A., Szeimies, U. and Hagedorn, H.G. 1997 Molecular evidence for tuberculosis in an ancient Egyptian mummy. *Lancet* 350:1404.

Pääbo, S. 1989 Ancient DNA: extraction, characterization, molecular cloning, and enzymatic amplification. *Proc. Natl. Acad. Sci. USA* 86:1939-1943.

Rafi, A., Spigelman, M., Stanford, J., Lemma, E., Donoghue, H. and Zias, J. 1994 DNA of *Mycobacterium leprae* detected in ancient bone. *Int. J. Osteoarchaeol.* 4:287-290.

Salo, W.L., Aufderheide, A.C., Buikstra, J.E. and Holcomb, T.A. 1994 Identification of *Mycobacterium tuberculosis* DNA in a pre-Columbian Peruvian mummy. *Proc. Natl. Acad. Sci. USA* 91:2091-2094.

Spigelman, M. and Lemma, E. 1993 The use of polymerase chain reaction to detect *Mycobacterium tuberculosis* in ancient skeletons. *Int. J. Osteoarchaeol.* 3:137-143.

Yoon, K.-H., Cho, S-N., Lee, M-K., Abalos, R.M., Cellona, R.V., Fajardo, T.T., Guido, L.S., Dela Cruz, E.C., Walsh, G.P. and Kim, J-D. 1993 Evaluation of polymerase chain reaction amplification of *Mycobacterium leprae*-specific repetitive sequence in biopsy specimens from leprosy patients. *J. Clin. Microbiol.* 31:895-899.

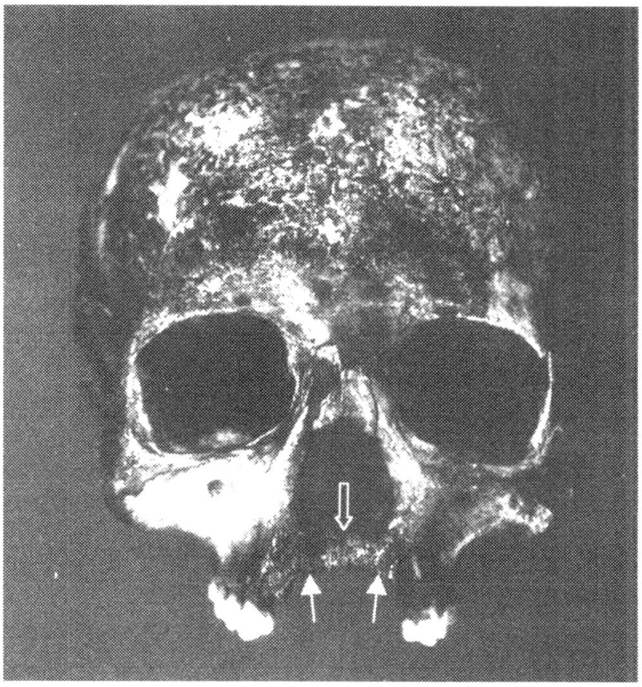

Figure 1: Skull of case R1788 with typical pathological changes of *facies leprosa*. Note the atrophy of the alveolar bone with loss of the upper incisors (white arrows), the osseous remodelling of the margins of the pyriform aperture and absence of the anterior nasal spine (open arrow).

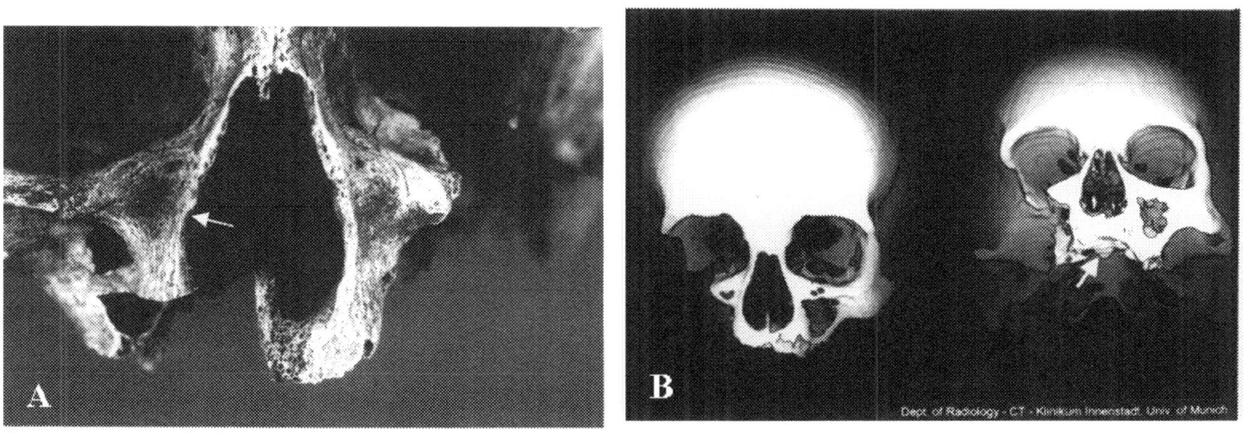

Figure 2: A) Detailed view of skull R2208 showing minor pathological alterations suggestive of leprosy (arrow). B) Three-dimensional reconstructions based on high-resolution CT images rendered with SSD (surface shaded display) revealing irregular surface and slight thinning of the hard palate (arrow) of case R2208.

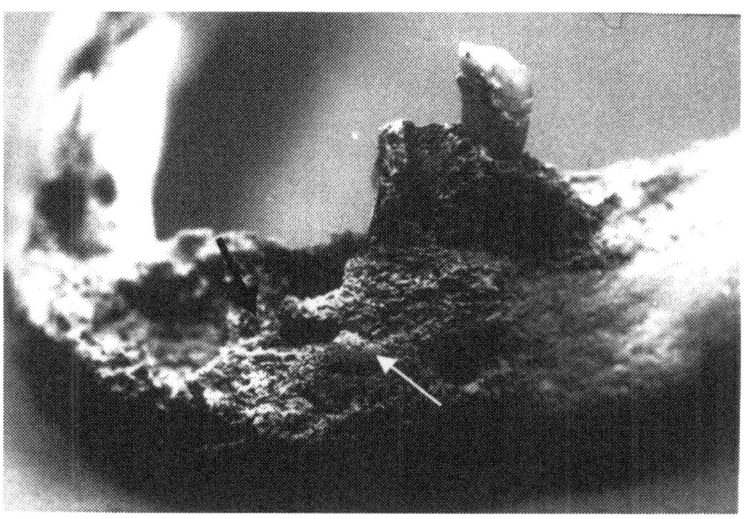

Figure 3: Mandible of case R180 showing extended osteolysis (black arrow) and reactive new bone formation (white arrow).

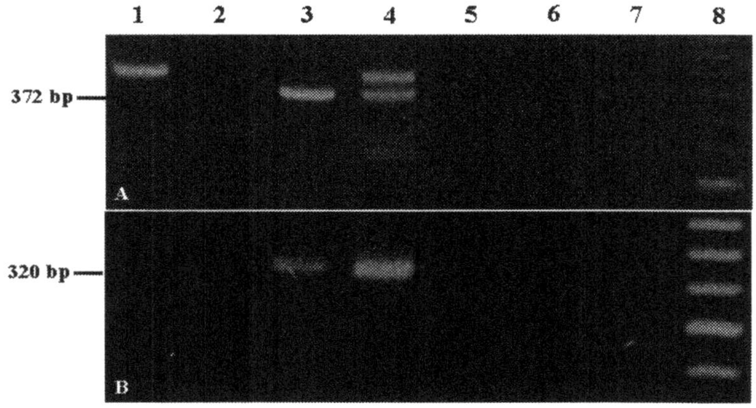

Figure 4: Amplification of DNA specific for *M. leprae*. A) Amplification of a 372 bp segment of the repetitive element RLEP1. B) Amplification of a 320 bp segment of the repetitive element RLEP3. Lane 1: case R180; lane 2: case R424; lane 3: case R1788; lane 4: case R2208; lanes 5-7: blank controls; lane 8: 50 bp ladder.

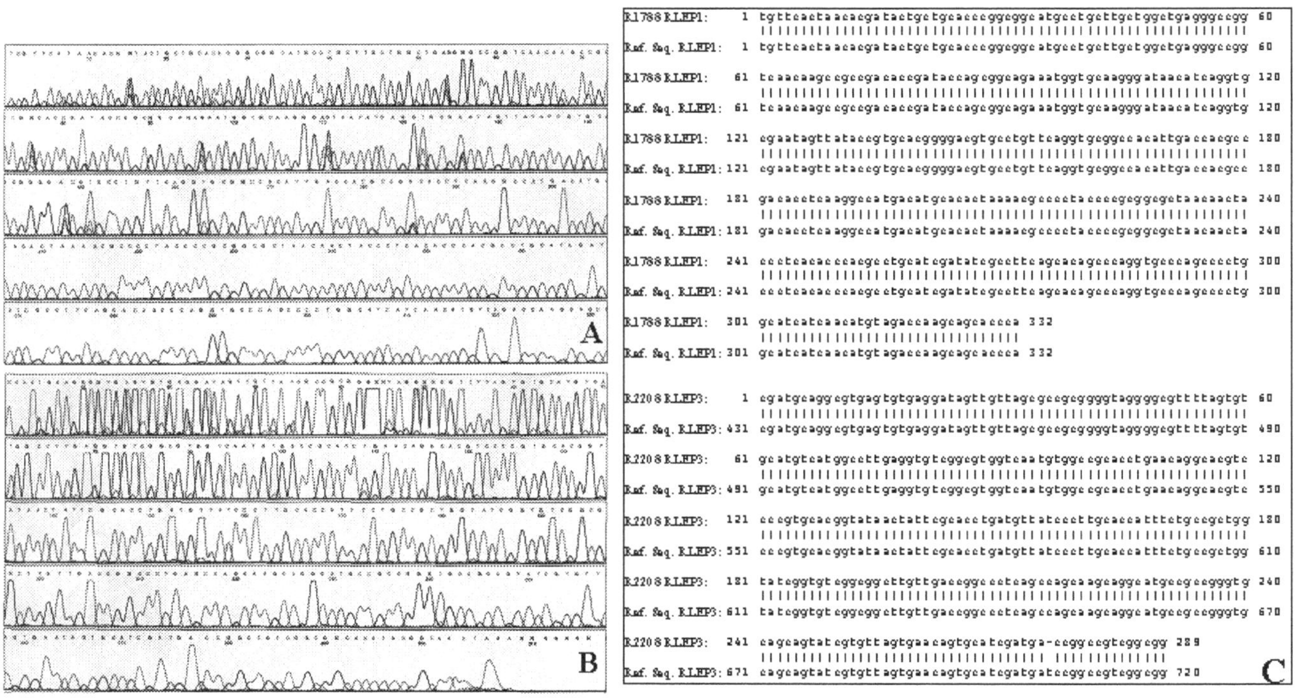

Figure 5. Electropherograms of the *M. leprae* specific amplification products RLEP1 of case R1788 (A) and RLEP3 of case R2208 (B): They represent the intensities of each fluorescent band which have been determined during automated sequencing and from which the DNA sequences are derived. C) Alignment of the determined sequences with the respective reference sequences.

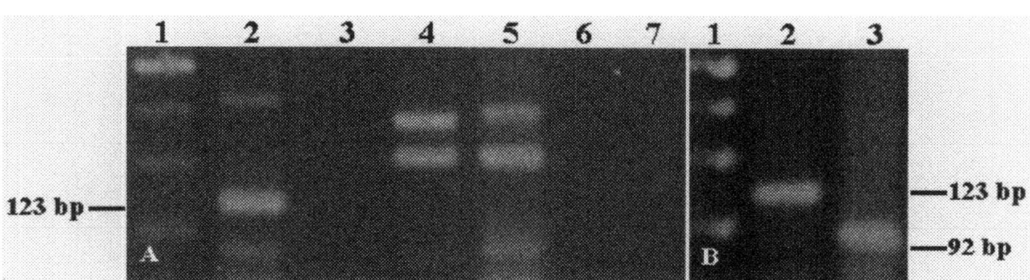

Figure 6. A) Amplification of the 123 bp segment of the insertional sequence IS6110, specific for the *M. tuberculosis* complex. Lane 1: 50 bp ladder; lane 2: case R180; lane 3: case R424; lane 4: case R1788; lane 5: case R2208; lanes 6-7: blank controls. B) Hae III-digest of the amplification product of IS6110 of case R180. Lane 1: 50 bp ladder; lane 2:

The study of ancient DNA answers a palaeopathological question

Mark Spigelman and Helen D. Donoghue
Department of Medical Microbiology
University College London
46 Cleveland Street
London W1T 4JF
England

E-mail: Spigelman@btinternet.com
h.donoghue@ucl.ac.uk

Abstract

Unusual pathology in a 1500-year-old skeleton was reported previously and presented diagnostic difficulties. Both Madura foot and leprosy were suggested as the likely cause. A sample from this skeleton was examined for the presence of ancient microbial DNA in an attempt to determine the responsible microorganisms. *Mycobacterium leprae* DNA was detected which indicates that leprosy was the underlying cause of the pathology, but the possibility of a mixed or secondary infection could not be excluded.

Keywords

Ancient DNA, leprosy, Madura foot, *Mycobacterium leprae*, PCR

NB: A version of this paper has been published since the congress in *American J. Physical Anthropology* 114:92-93 (2001)

1. Introduction

Hershkovitz *et al.* (1992) reported on a skeleton from Bet Guvrin, Israel, dating from the Byzantine period (300-600 AD). There were pathological changes in the lower extremities, which presented significant diagnostic problems. Their original diagnosis was of a Madura foot, based on the absence of typical medullary changes associated with osteomyelitis, and on the presence of severely ankylosed (left talo-calcaneal joint) and deformed bilateral foot bones, with erosion of the *sustentaculum tali* and head of left talus. The right ankle contained a normal talo-calcaneal joint, but the bones showed numerous osteophytes, with the calcaneum showing prominent osseous "loops" (Figure 1). The metatarsal bones demonstrated subperiosteal plaques and multi-surface deformities, with ankylosis of the left 2nd, 3rd and 4th metatarsals proximally on the left side (Figure 2). The right 5th metatarsal was hypertrophic (Figure 3). Also noted were changes in both tibia and fibular diaphyses with spicules and plaques, particularly at the diaphyseal edges, and significant subperiosteal bone deposition (Figures 4 and 5).

However, in correspondence, Manchester (1993) noted that Madura foot is usually a predominantly unilateral condition and that many of the surface changes were consistent with secondary pyogenic infections following plantar ulceration associated with leprous peripheral neuropathy. There was evidence that at least one joint was a typical Charcot's joint, suggestive of the presence of some form of neuropathy. In addition, he noted that ankylosis was more likely from secondary pyogenic sepsis, as it is not a feature of Madurosis *per se*, neither did he feel was proliferative new bone formation. Manchester noted that this skeleton was from the same period as one reported by Zias (1985). Indeed, this latter skeleton was the first ancient skeleton from which *M. Leprae* DNA was extracted and amplified (Rafi *et al.*, 1994).

In reply, Hershkovitz *et al.* (1993) admitted that the pathological features could fit both Madura foot and leprosy but still considered Madura foot more likely. They cited the fragmentary nature of bones other than those from the foot, and particularly the facial bones, as the reason for preventing a full palaeopathological assessment. Those parts of the naso-maxillary bones that had survived did not exhibit any of the pathognomonic signs of lepromatous leprosy. The authors noted also that there were no *sequestra* (diagnostic of osteomyelitis) found in any of the affected bones and that all changes appeared superficial without any medullary involvement.

Here the controversy rested and has remained unresolved. Since then, with the development of palaeomicrobiological techniques it was felt that some resolution of the problem could be achieved by examining the bones using modern microbiological techniques. One of the bones from the foot was provided for further examination by Professor Israel Hershkovitz of the Department of Anatomy and Anthropology, Tel Aviv University, Israel. The aim was to examine the specimen to determine whether *M. Leprae*-specific DNA was present, as mycobacterial DNA has been shown to persist in samples of a similar age (Rafi *et al.*,1994, Donoghue *et al.*,1998).

Although most of the peripheral changes in leprosy are not the direct result of the *M. Leprae* organism itself, this is not the case in lepromatous leprosy. Although the numbers of bacilli in peripheral sites are not believed to be high, at times lepromatous leprosy skin lesions may be found full of bacilli (Carmichael, 1993). Therefore, the specimen was examined using a specific nested polymerase chain reaction (PCR), to determine whether *M. Leprae*-specific DNA could be detected.

2. Methods

Stringent precautions were taken to avoid cross-contamination. The detailed method is described by Donoghue *et al.* in this volume. A small quantity of the bone sample was crushed in a sterile pestle and mortar in a clean room. About 25mg of crushed sample was demineralised in Proteinase K and EDTA, incubated in lysis buffer L6 (Boom *et al.,* 1990) which contained guanidium thiocyanate, EDTA and Triton X-100, and the DNA captured onto silica. This was washed, then dried and the DNA eluted into a small volume of water. Negative extraction controls were processed in parallel with sample tubes.

Two sets of *M. Leprae* specific nested primers were designed. Those based on the 18kD antigen (Ag) gave a product of 136bp (outer) and 110bp (inner). The RLEP primers amplify a 129bp sequence of the RLEP element, which is reported to be present at 28 copies/cell, and yield a 129bp product (outer) and 99bp-nested product. The sequences and locations of these primers are listed in Donoghue *et al.* (this volume, Table 1).

Details of the PCR mix and experimental conditions are described elsewhere (Donoghue *et al.*, this volume). A two-tube nested PCR was carried out. In the first stage the outer primers were added to the PCR mix. Amplification was carried out with 25 cycles. An annealing temperature of 58°C was used for the RLEP primers and of 55 °C for the 18 kD Ag gene primers. In the second stage of nested PCR, new tubes of PCR mix were used. The inner primers were transferred into each reaction tube, together with 0.5 µl, 1.0µl or 5.0µl of first stage product which was used as the DNA template. The total volume was made up to 50µl with sterile water. This second stage PCR cycle was identical to the first. PCR products were electrophoresed in an agarose gel. Amplified DNA was visualized by ethidium bromide staining plus ultraviolet light and was recorded with a Polaroid camera. Amplified DNA PCR product was sequenced directly by an in-house sequencing service in the U.C.L. Department of Biology.

3. Results

A weak positive result was obtained using the primers for both the multi-copy RLEP and single-copy18kD Ag gene. This was much clearer when 5.0µl of the first stage product was used as the template in the nested reaction (Figure 6). Because of the low quantity of amplicon, only a partial sequence from the RLEP primers was obtained. However, those obtained from other archaeological samples show a sequence identical with that of *M. Leprae*. A preliminary sequence of 123bp was obtained from the reverse inner primer which includes part of the outer forward primer (underlined), the forward and reverse inner primers (*italics*) and all the intervening DNA:

GCATGTCATG[C]NNT*TGAGGTGTCGGCGTGGTC*AA
TGT-GCCGCACC[N] TGAACAGGCAC[G]GTCCCCG-
GCACGGTATA[G]CTATTCGCACCTGATGTTA*TCCC*
TTGCACCATTTCTG

There are three missing or indeterminate values and two doubtful results in the sequence between the inner primers, plus two indeterminate and one doubtful result in the forward primer region. Otherwise, the sequence was identical to 30 sequences of *M. leprae* held in the gene data bank. With the exception of an internal 18bp sequence, a BLAST search of the gene database held by the National Center for Biotechnology Information showed that this product is unique to *M. Leprae.*

4. Discussion

The primers used have been described elsewhere (Donoghue *et al.*, this volume) and have been shown to be both specific and extremely sensitive, capable of detecting less than one leprosy bacillus. The presence of *Mycobacterium leprae* DNA in the specimen, together with the pathological changes, suggests that the individual was suffering from leprosy (Hansen's disease). The DNA sequence of the amplified target sequence is of *M. Leprae*, although there was too little target DNA in the original extraction to obtain sufficient nested PCR product for a definitive sequencing result. The need to use 5µl of the outer primer product in the nested reaction led to the amplification of a chimaeric sequence which is intermediate in size between the outer and inner product.

A connection between leprosy and mycetomas has been speculated upon in the literature at various times (Pavithran, 1988, Jayakumar *et al.*, 1993, Grossman *et al.*, 1994, Suttner *et al.*, 1990). It is highly probable that more cases with this connection occur but are not necessarily reported. We are aware from discussions with workers in modern day Hansen's disease hospitals that the diagnosis is often made on the grounds of sighting the classical granules in pus (white or black). Based purely on economic grounds no culture is usually undertaken in these cases (Jal Mehta and Tom Oommen, pers.comm.). In such cases, treatment is instituted with standard anti-fungal drugs and, with healing, the cases are never fully diagnosed or reported. Both doctors expressed the opinion that they were aware of mycetoma/leprosy cases in their hospitals.

Mycetoma, also known as Madura foot, is a local, chronic, slowly progressive disease with the classic presentation involving tumefaction, multiple draining sinuses, and grain-filled pus. Either a bacterium (actinomycetoma) or a fungal (eumycetoma) organism is the primary cause. Many different organisms have been reported as causal agents in the literature. The majority are soil environmental saprophytes (McGinnis, 1996).

The original authors (Hershkovitz *et al.*,1992) considered that as only nine cases had been reported at that time in the literature in Israel, and because the condition was more common in tropical or subtropical climates, it was likely that this person was an immigrant rather than a local resident. A review of the current literature has found, however, that cases have been noted in Iran, Great Britain, Bulgaria, the U.S.A. as well as the Middle East (Griffiths *et al.*,1975, Warintarawej, 1975, Grigorov, 1965, Hay and Mackenzie, 1983). If this is the case, it is possible that a local fungus or bacterium could have caused a coincidental or secondary case of Madura foot to a native of the area.

We believe that it is possible that the specimen came from an individual with a mixed or secondary infection that is ubiquitous in such peripheral *M. Leprae* infections. We can conclude that the underlying disease in this case was leprosy. Having said that, we cannot, and we believe it is impossible, to exclude a secondary infection, which has given some of the features of Madura foot. The basic facts are that if an individual walks barefoot and has foot ulcers, it is not uncommon for them to contract secondary infections such as Madura foot from the normal microbiological components of the soil.

Molecular analysis can help find the DNA of any uncommon bacteria, particularly human pathogens, in samples, but is currently unable to help the archaeologist with distinguishing infection from common soil organisms. Thus, palaeopathological disagreements can at times be helped by the application of more recently developed techniques but they cannot fully answer all question posed. This as always rests with the skills and knowledge of the palaeopathologist.

Acknowledgements

Professor Israel Hershkovitz, Department of Anatomy and Anthropology, Tel Aviv University, Israel kindly provided the bone sample and Figures 1-6.

References

Boom, R., Sol, A., Salimans, M., Jansen, C., Wertheim-van Dillen, P., van der Noordaa, J. 1990 Rapid and simple method for purification of nucleic acid. *J. Clin. Microbiol.* 28: 495-503.

Carmichael, A.J. 1993. Leprosy. In K.F. Kiple (ed): *The Cambridge world history of human disease.* Cambridge: University Press, pp. 834-839.

Donoghue, H.D., Spigelman, M., Zias, J., Gernaey-Child, A.M., Minnikin, D.E. 1998 *Mycobacterium tuberculosis* complex DNA in calcified pleura from remains 1400 years old. *Letters in Applied Microbiol.* 27: 265-269.

Griffiths, W.D., Kohout, E., Vessal, K. 1975 Mycetoma in Iran. *Int. J. Dermatology* 14:209-213

Grigorov, G. 1965 Five cases of mycetoma. *Khirurgia* 18: 602-605.

Grossman, D., Rapini, R.P., Osborne, B. and Duvic, M. 1994 Emergence of leprosy in a patient with mycosis fungoides. *J. Am. Acad. Dermatol.* 30:313-315.

Hay R.J. and Mackenzie D.W. 1983 Mycetoma (Madura foot) in the United Kingdom - a survey of forty-four cases. *Clin. Exp.Dermatol.* 8:553-562.

Hershkovitz, I., Spiers, M., Katznelson, A. and Arensberg, B. 1992 Unusual pathological conditions in the lower extremities of a skeleton from ancient Israel. *Am. J. Phys. Anthropol.* 88:23-26.

Hershkovitz, I., Spiers, M. and Arensberg, B. 1993 Leprosy or Madura foot? The ambiguous nature of infectious disease in paleopathology: reply to Dr Manchester. *Am. J. Phys. Anthropol.* 91:251-253.

Jayakumar, J., Aschhoff, M. and Renuka, G. 1993 Mycetomas in leprosy. *Indian J. Lepr.* 65:229-233.

Manchester, K . 1993 Unusual pathological condition in the lower extremities of a skeleton from ancient Israel. *Am. J. Phys. Anthropol.* 91:249-250.

McGinnis, M.R. 1996 Mycetoma. *Dermatol. Clin.* 1:97-104.

Pavithran, K. 1988 Chromoblastomycosis in a residual patch of leprosy. *Indian J. Leprosy* 60:444-447.

Rafi, A., Spigelman, M., Stanford, J., Lemma, E., Donoghue, H., Zias, J. 1994 DNA of *Mycobacterium leprae* detected by PCR in ancient bone. *Int. J. Osteoarchaeol.* 4: 287-290.

Suttner, J.F., Wirth, C.J., Wulker, N. and Seeliger, H. 1990 Madura foot. A report of two cases. *Int. Orthop.* 14:217-219.

Warintarawej, A., Winter, W.G. Jr, Goodman,N.L. 1975 Maduromycosis (Madura foot) in Kentucky. *Southern Med J.* 68:1570-1575.

Zias, J. 1985 Paleopathological evidence of leprosy in Palestine during the Talmudic period. *Koroth* 9:242-248.

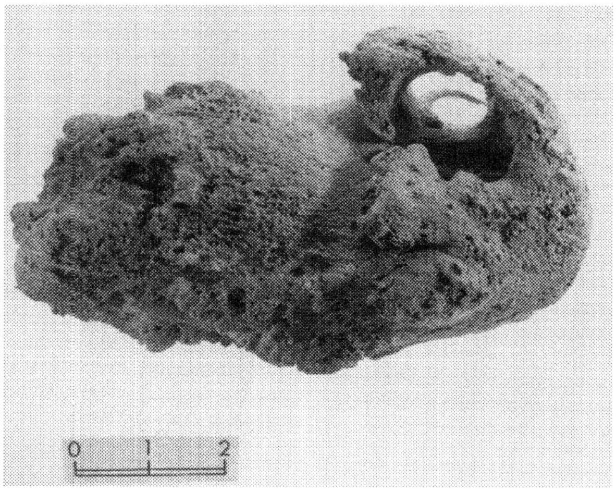

Figure 1: Right calcaneum, plantar view. Sinus-tunnels due to destruction of bone and subsequent new bone formation.

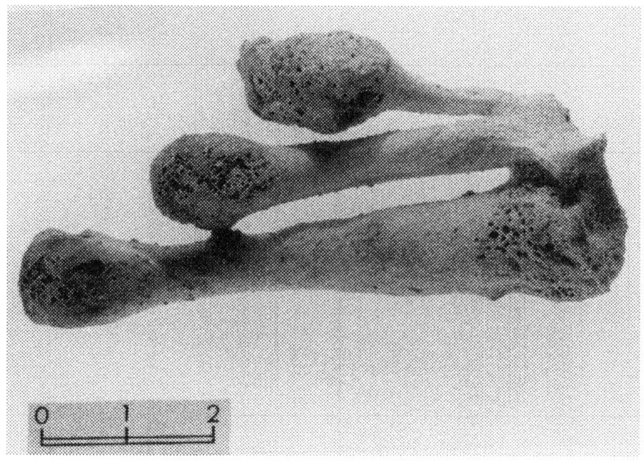

Figure 2: Fusion of the second, third and fourth metatarsals of the left foot, plantar view.

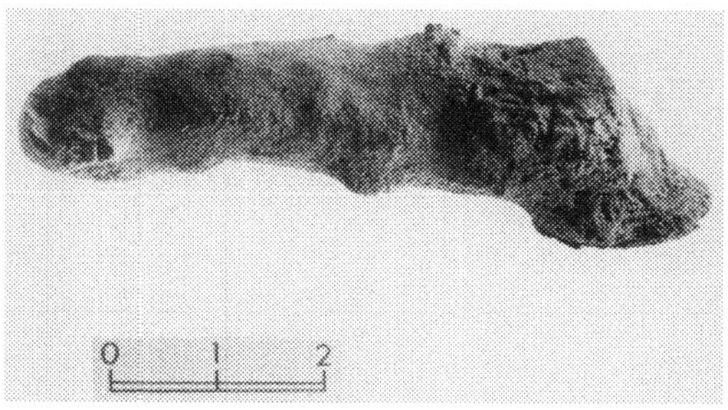

Figure 3: Right fifth metatarsal, dorsal view.

Figure 4: Left tibia, anterior view, and right tibia, lateral view.

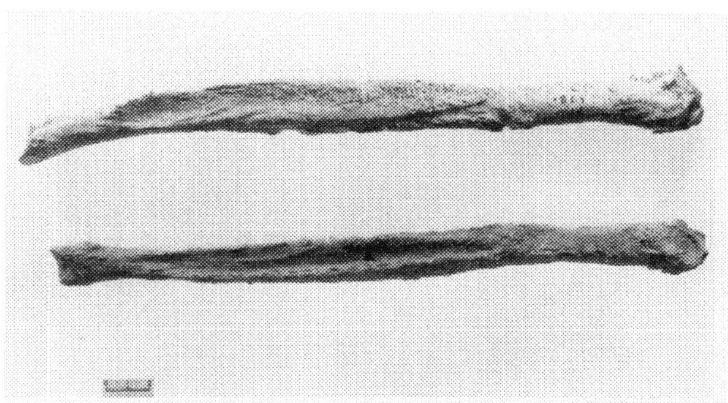

Figure 5:Left and right fibulae, medial view.

Figure 6: *M. leprae*-specific nested PCR of Hershkovitz's bone sample after 2 x 25 cycles of amplification with RLEP outer then inner primer pairs (99bp product). Left to right: lane 1: 0.5μl stage 1 product used as nested DNA template; lane 4: 1.0μl stage 1 product used as nested DNA template; lane 7: 5.0μl stage 1 product used as nested DNA template; lane 10: negative extraction control; lane 11: ØX174*Hae*III digest molecular markers.

Index

www.ingramcontent.com/pod-product-compliance
Ingram Content Group UK Ltd.
Pitfield, Milton Keynes, MK11 3LW, UK
UKHW061845041026
15591UKWH00015B/916

9 781841 714349